# PLASMA PHYSICS

## Nonlinear Theory and Experiments

## NOBEL SYMPOSIUM COMMITTEE (1976)

| | |
|---|---|
| STIG RAMEL, *Chairman* | Executive Director, Nobel Foundation |
| ARNE FREDGA | Chairman, Nobel Committee for Chemistry |
| TIM GREVE | Director, Norwegian Nobel Institute (Peace) |
| BENGT GUSTAFSSON | Secretary, Nobel Committee for Medicine |
| LARS GYLLENSTEN | Member, Swedish Academy (Literature) |
| LAMEK HULTHÉN | Chairman, Nobel Committee for Physics |
| ERIK LUNDBERG | Chairman, Prize Committee for Economic Sciences |
| NILS-ERIC SVENSSON | Executive Director, Bank of Sweden Tercentenary Foundation |

# PLASMA PHYSICS

## Nonlinear Theory and Experiments

Edited by

Hans Wilhelmsson

*Chalmers University of Technology*
*Göteborg, Sweden*

SPRINGER SCIENCE+BUSINESS MEDIA, LLC

Library of Congress Cataloging in Publication Data

Nobel Symposium, 36th, Lerum, Sweden, 1976.
Plasma physics.

Includes index.
1. Plasma (Ionized gases)—Congresses. I. Wilhelmsson, Hans. II. Title.
QC717.6.N6 1976 530.4'4 76-46955

ISBN 978-1-4757-1573-6 ISBN 978-1-4757-1571-2 (eBook)
DOI 10.1007/978-1-4757-1571-2

Proceedings of the thirty-sixth Nobel Symposium on Nonlinear Effects in Plasmas held at Aspenäsgården, Lerum, Sweden, June 11-17, 1976


Originally published by Plenum Press, New York in 1977

# PUBLISHED NOBEL SYMPOSIA

1 • **Muscular Afferents and Motor Control**—*Edited by Ragnar Granit*
2 • **Prostaglandins**—*Edited by Sune Bergström and Bengt Samuelsson*
3 • **Gamma globulins**—*Edited by Johan Killander*
4 • **Current Problems of Lower Vertebrate Phylogeny**—*Edited by Tor Ørvig*
5 • **Fast Reactions and Primary Processes in Chemical Kinetics**—*Edited by Stig Claesson*
6 • **Problems of International Literary Understanding**—*Edited by Karl Ragnar Gierow*
7 • **International Protection of Human Rights**—*Edited by Asbjörn Eide and August Schou*
8 • **Elementary Particle Theory**—*Edited by Nils Svartholm*
9 • **Mass Motions in Solar Flares and Related Phenomena**—*Edited by Yngve Öhman*
10 • **Disorders of the Skull Base Region**—*Edited by Carl-Axel Hamberger and Jan Wersäll*
11 • **Symmetry and Function of Biological Systems at the Macromolecular Level**—*Edited by Arne Engström and Bror Strandberg*
12 • **Radiocarbon Variations and Absolute Chronology**—*Edited by Ingrid U. Olsson*
13 • **Pathogenesis of Diabetes Mellitus**—*Edited by Erol Cerasi and Rolf Luft*
14 • **The Place of Value in a World of Facts**—*Edited by Arne Tiselius and Sam Nilsson*
15 • **Control of Human Fertility**—*Edited by Egon Diczfalusy and Ulf Borell*
16 • **Frontiers in Gastrointestinal Hormone Research**—*Edited by Sven Andersson*
17 • **Small States in International Relations**—*Edited by August Schou and Arne Olav Brundtland*
20 • **The Changing Chemistry of the Oceans**—*Edited by David Dyrssen and Daniel Jagner*
21 • **From Plasma to Planet**—*Edited by Aina Elvius*
22 • **ESR Applications to Polymer Research**—*Edited by Per-Olof Kinell and Bengt Rånby*
23 • **Chromosome Identification Technique and Applications in Biology and Medicine**—*Edited by Torbjörn Caspersson and Lore Zech*
24 • **Collective Properties of Physical Systems**—*Edited by Bengt Lundqvist and Stig Lundqvist*
25 • **Chemistry in Botanical Classification**—*Edited by Gerd Bendz and Johan Santesson*
26 • **Coordination in the Field of Science and Technology**—*Edited by August Schou and Finn Sollie*
27 • **Super-Heavy Elements**—*Edited by Sven Gösta Nilsson and Nils Robert Nilsson*
28 • **Somatomedins and Some Other Growth Factors**—*Edited by Rolf Luft and Kerstin Hall*
29 • **Man, Environment, and Resources**—*Edited by Torgny Segerstedt and Sam Nilsson*
30 • **Physics of the Hot Plasma in the Magnetosphere**—*Edited by Bengt Hultqvist and Lennart Stenflo*
31 • **The Impact of Space Science on Mankind**—*Edited by Tim Greve, Finn Lied, and Erik Tandberg*
33 • **Molecular and Biological Aspects of the Acute Allergic Reaction**—*Edited by S. G. O. Johansson, Kjell Strandberg, and Börje Uvnäs*
34 • **Structure of Biological Membranes** – *Edited by Sixten Abrahamsson and Irmin Pascher*
36 • **Plasma Physics: Nonlinear Theory and Experiments**—*Edited by Hans Wilhelmsson*

Symposia 1-17 and 20-22 were published by Almqvist & Wiksell, Stockholm and John Wiley & Sons, New York; Symposia 23-25 by Nobel Foundation, Stockholm and Academic Press, New York; Symposium 26 by the Norwegian Nobel Institute, Universitetsforlaget, Oslo; Symposium 27 by Nobel Foundation, Stockholm and Almqvist & Wiksell International, Stockholm; Symposium 28 by Academic Press, New York; Symposium 29 by Nobel Foundation, Stockholm and Trycksaksservice AB, Stockholm; and Symposia 30, 31, 33, 34, and 36 by Plenum Press, New York.

## Preface

A Nobel Foundation Symposium on the subject: "Nonlinear Effects in Plasmas", was held at Aspenäsgården, Lerum, in the Göteborg area of Sweden from June 11-17, 1976. The Symposium was the 36th in the series of Nobel Foundation Symposia, which have been held mainly within the areas of physics, chemistry, medicine, literature and peace prizes. Some 30 leading experts from the United States, Soviet Union, Japan and Western Europe attended the Symposium.

The purpose of the Symposium was to discuss various topics in the field of modern plasma physics. We had to select from this vast area of active research a suitable common theme with a great number of new and interesting contributions. We decided to devote our Symposium in particular to nonlinear effects in plasmas and to emphasize some areas where important developments seemed to be taking place.

In recent years basic theory and experiments in nonlinear plasma physics have been stimulated largely by the need for an energy source based on fusion of light nuclei. In many laboratories all over the world attempts are being made to come closer to the final goal by studying magnetically confined plasmas and systems of inertial confinement. Heating of plasmas to fusion temperatures remains a key problem. There are good reasons to believe that the nonlinear effects in plasmas will play an important rôle for fusion, a long-range program which is still largely in its basic research phase. Astrophysics also offers exceedingly interesting applications of the nonlinear plasma. Since from recent experiments and observations new results accumulate which need interpretation, plasma theory has to be developed to describe the plasma in new and nonlinear situations.

The new developments in nonlinear basic plasma theory, which are of considerable interest from the point of view of pure theory, may well be the natural tools for interpreting many new experiments. However, analytic approaches to describe the plasma are not always possible. Modern computers offer facilities for simulating plasma situations which are highly nonlinear and where analytic approaches fail or are lacking. Considerable progress has, in fact, been made in recent years concerning the comparison between experimental results (or computer simulations) and nonlinear theory. Various aspects of these questions entered the discussions during the Symposium which was mainly devoted to the following areas:

- Laser-Plasma Interaction and Laser-Driven Fusion
- Strong Langmuir Turbulence, Solitons and Cavitons
- Parametric Interactions
- Nonlinear Effects in Magnetically Confined Plasmas
- Relativistic Electron Beam-Plasma Interactions
- Plasma Astrophysics

During the Symposium it turned out that the discussions became very lively even during the presentation of the contributions. This often led to clarifying the respective points of view and certainly helped and stimulated the audience.

The various areas listed above have, in fact, many features in common from a nonlinear point of view. Often the different contributions could be refered to several of the areas. In the following presentation of the material we find it natural therefore to let the contributions follow the sequence in which they were given at the Symposium. It is our hope that this will transmit the contents in a reasonably systematic way and to some extent mirror the authentic atmosphere during the Symposium, as induced already by the introductory talk by Professor D. ter Haar, and as stimulated successively by the many fascinating contributions.

The Organizing Committee for the Symposium consisted of the following Swedish scientists: L. Hulthén, S. Lundqvist, L. Stenflo and H. Wilhelmsson.

The Institute for Electromagnetic Field Theory and Plasma Physics, Chalmers University of Technology, Göteborg, was responsible for the local arrangements.

The Symposium was financed by the Nobel Foundation through grants from the Tercentenary Foundation of the Bank of Sweden, by the Swedish Atomic Research Council, and by the Royal Swedish Academy of Sciences, which is gratefully acknowledged. The City Council of Göteborg showed us graceful hospitality.

An excellent collaboration of the contributors and the publisher enabled rapid publication of the material.

A collective effort of the members of the Institute for Electromagnetic Field Theory and Plasma Physics, at Chalmers University of

Technology, Göteborg, enabled us to carry through a most successful meeting. The qualified help of Dr Dan Anderson, Mrs Agneta Mannheimer and Dr Jan Weiland is particularly acknowledged.

Hans Wilhelmsson

# Contents

## Parametric Interaction, Three-Wave Coupling, and Ponderomotive Forces



## Strong Plasma Turbulence, Solitons and Cavitons

## Plasma Astrophysics



## Magnetically Confined Plasmas



## Basic Theory and Experiments on Nonlinear Effects in Plasmas

# NON-LINEAR EFFECTS IN PLASMAS

D. ter Haar

Magdalen College

Oxford

Plasma physics is nowadays a fast-developing field. This is to a large extent due to its importance in the search for terrestrial thermonuclear energy-sources. Of course, it is also a subject which is of interest both in its own merit and because of its applications to such more academic fields as astrophysics. All these aspects are well represented at this symposium. The programme before us promises a rich feast. The first day we shall hear about experimental results, a useful reminder to all theorists present that physics is an empirical science and that we should be sure never to stray too far from our experimental base. The remainder of the programme looks like having mainly a theoretical bias. No doubt some of the theoretical talks will be dealing with possible explanations of experiments, but I suspect that many of us will indulge in unbridled theorizing. As a theorist with a great interest in astrophysics I have also noted a few talks about astrophysical applications.

Professor Wilhelmsson has asked me to give this introductory talk. As it is clearly impossible, even if I were qualified to do so, to cover the whole field of non-linear plasma theory in such a talk, I shall restrict myself to reminding you of some of the effects which seem to be important and which are likely to be covered in this conference. I shall dwell a little longer on two aspects which will perhaps not be covered so thoroughly and in which I have some personal interest: renormalization theories and non-linear density waves in galaxies. The last subject is not strictly speaking plasma physics, but the Coulomb forces and gravitational forces derive from the same form of potential energy and as a result some of the effects which are important in plasma physics also turn up in systems of gravitating particles.

It was natural that for a long time plasma theorists worked with linear theories. After all, linear equations allow the superposition of solutions, so that if we have found the various eigenmodes in the linear approximation, we can consider a general state of the plasma to be a superposition of a number of those eigenmodes. The next obvious step is the quasi-linear stage in which these elementary excitations of the linear theory are allowed to interact while still retaining their individual character to a very large degree. Most plasmas in which physicists or astrophysicists are interested are far from in a thermodynamic equilibrium state. Often the external circumstances are such that instabilities occur, giving rise to a high level of field fluctuations or - to put it into the language of elementary excitations - to the presence of elementary excitations with very large amplitudes, such large amplitudes that the linear approximation is certainly no longer applicable; the quasi-linear approach may also become questionable, even though notably Tsytovich has applied this approach to such situations with a fair modicum of success. The question then is what the next stage in the theory should be. If one wants to retain as far as possible the concepts introduced in the linear theory one tries to "renormalize" the elementary excitations. Professor Balescu and Ichimaru of those present at the present symposium have contributed to this approach to strong plasma turbulence. On the other hand, one might argue that in a strongly turbulent plasma this approach still smacks too much of a linear theory and one can look for specific solutions of the non-linear equation themselves. This leads to what may be called the soliton approach to plasma turbulence. Several participants of the present symposium have contributed to this approach and the programme before us contains a number of papers on this subject. I shall briefly describe these two alternative approaches.

Let us remind ourselves how one obtains the elementary excitations of the linear theory. We shall assume that the particle (electron) distribution function for the plasma satisfies the Vlasov equation (for the sake of simplicity we consider a plasma with a uniform positive background and neglect magnetic fields).

$$f_t + (\underset{\sim}{v}.f_{\underset{\sim}{x}}) + \frac{q}{m}(\underset{\sim}{E}.f_{\underset{\sim}{v}}) = 0, \qquad (1)$$

where subscripts indicate partial differentiation everywhere in this paper, while the electric field $\underset{\sim}{E}$ satisfies the Poisson equation

$$(\nabla.\underset{\sim}{E}) = 4\pi q\,[\int f\, d^3\underset{\sim}{v} - n_0], \qquad (2)$$

where $f(\underset{\sim}{x},\underset{\sim}{v},t)$ is the electron distribution function, normalized to $n_o$, and q and m are the electron charge and mass; $\underset{\sim}{x}$ and $\underset{\sim}{v}$ are here vectors, as is $\underset{\sim}{k}$ to be introduced presently. Once we have solved equations (1) and (2) and found f, we can evaluate the dielectric constant $\varepsilon(\underset{\sim}{k},\omega)$ and from the equation

$$\varepsilon(\underset{\sim}{k}, \omega) = 0. \tag{3}$$

we can find the dispersion relations for the elementary excitations, $\omega = \omega(\underset{\sim}{k})$.

To solve (1) we follow Kono and Ichikawa (1973); I realize that the work by Misguich and Balescu goes beyond this work, but because of the limited amount of time available for my exposition I have preferred the approach by Kono and Ischikawa to the more detailed work of Misguich and Balescu (1974), Dupree (1966), or Weinstock (1969). Unfortunately, I am also unable to discuss Ichimaru's work, about which we may hear something next week, I hope. We introduce the retarded Green function $G(\underset{\sim}{x},\underset{\sim}{v},t;\underset{\sim}{x}',\underset{\sim}{v}',t')$:

$$G_t + (\underset{\sim}{v}.G_{\underset{\sim}{x}}) + \frac{q}{m}(\underset{\sim}{E}.G_{\underset{\sim}{v}}) = \delta^3(\underset{\sim}{x}-\underset{\sim}{x}')\delta^3(\underset{\sim}{v}-\underset{\sim}{v}')\,\delta(t-t'), \tag{4}$$

$$G = o, \text{ if } t < t'. \tag{5}$$

If $f_o(\underset{\sim}{x},\underset{\sim}{v})$ is the value of $f(\underset{\sim}{x},\underset{\sim}{v},t)$ at t = o, we have

$$f(\underset{\sim}{x},\underset{\sim}{v},t) = \int G(\underset{\sim}{x},\underset{\sim}{v},t;\, \underset{\sim}{x}',\underset{\sim}{v}',\, 0)\, f_o(\underset{\sim}{x}',\underset{\sim}{v}')\, d^3\,\underset{\sim}{x}'d^3\,\underset{\sim}{v}'. \tag{6}$$

If there is no field, the unperturbed Green function $G_o$ satisfies the equation

$$G_{ot} + (\underset{\sim}{v}.G_{o\underset{\sim}{x}}) = \delta^3(\underset{\sim}{x}-\underset{\sim}{x}')\delta^3(\underset{\sim}{v}-\underset{\sim}{v}')\,\delta(t-t'), \tag{7}$$

with the solution

$$G_o = \theta(t-t')\delta^3(\underset{\sim}{v}-\underset{\sim}{v}')\delta^3(\underset{\sim}{x}-\underset{\sim}{x}' - \underset{\sim}{v}\,[t-t']), \tag{8}$$

where $\theta$ (t) is the Heaviside step-function. The second delta-function in (8) shows that one is dealing here with particles moving along unperturbed orbits. Using (4) and 7) we get the following integral equation for G:

$$\begin{aligned} G(\underset{\sim}{x},\underset{\sim}{v},t;\underset{\sim}{x}',\underset{\sim}{v}'t') &= G_o(\underset{\sim}{x},\underset{\sim}{v},t;\underset{\sim}{x}',\underset{\sim}{v}',t') \\ &- \frac{q}{m}\int_o^t dt'' \int d^3\,\underset{\sim}{x}''\, d^3\underset{\sim}{v}''\, G_o(\underset{\sim}{x},\underset{\sim}{v},t;\, \underset{\sim}{x}'',\underset{\sim}{v}''t'') \\ &\qquad (\underset{\sim}{E}(\underset{\sim}{x}'',t'').G_{\underset{\sim}{v}''}(\underset{\sim}{x}'',\underset{\sim}{v}'',t'';\, \underset{\sim}{x}',\underset{\sim}{v}',t')) \end{aligned} \tag{9}$$

In order to have a self-consistent solution $\underset{\sim}{E}$ must satisfy the equation

$$(\nabla.\mathbf{E}(\mathbf{x},t)) = 4\pi q[\int d^3\mathbf{v}\int d^3\mathbf{x}'d^3\mathbf{v}'G(\mathbf{x},\mathbf{v},t;\mathbf{x}',\mathbf{v}',0)f_0(\mathbf{x}',\mathbf{v}')-n_0].\qquad(10)$$

We now Fourier transform and get from (8) and (9)

$$G(\mathbf{k},\mathbf{v},\omega;\ \mathbf{k}',\mathbf{v}',\omega') = G_0(\mathbf{k},\mathbf{v},\omega)\delta^3_{\mathbf{k}\mathbf{k}'}\delta^3(\mathbf{v}-\mathbf{v}')\delta(\omega-\omega')$$
$$-\frac{q}{m}\sum_{\mathbf{k}''}\int\frac{d\omega''}{2\pi}G_0(\mathbf{k},\mathbf{v},\omega)(\mathbf{E}(\mathbf{k}'',\omega'').\mathbf{G}_{\mathbf{v}}(\mathbf{k}-\mathbf{k}'',\mathbf{v},\omega-\omega'';\ \mathbf{k}',\mathbf{v}',\omega')),\qquad(11)$$

$$G_0(\mathbf{k},\mathbf{v},\omega) = \frac{i}{\omega-(\mathbf{k}.\mathbf{v})}\ .\qquad(12)$$

It is convenient to introduce a diagram representation, as follows: a thin line ——— represents $G_0$, a thick dot • represents the "coupling" $-(q/m)\,\partial/\partial\mathbf{v}$, and a wavy line ∼∼ the field $\mathbf{E}$, that is, the plasma excitations. The procedure is now the standard one from Feynman diagram theory. If we write

$$G(\mathbf{k},\mathbf{v},\omega) \equiv G(\mathbf{k},\mathbf{v},\omega;\mathbf{k},\mathbf{v},\omega),\qquad(13)$$

and use expression (8) for $G_0$ and through iteration solve (11) we get diagrammatically, up to terms in fourth order in q:

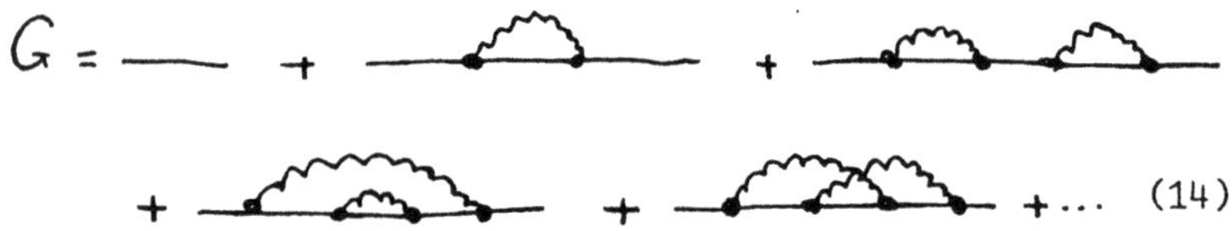

(14)

If we define G' as that part of G which is obtained by taking only the first of the fourth order terms and doing the same for the higher order terms, we find that G' (represented by a double line ═══ ) satisfies the Dyson equation:

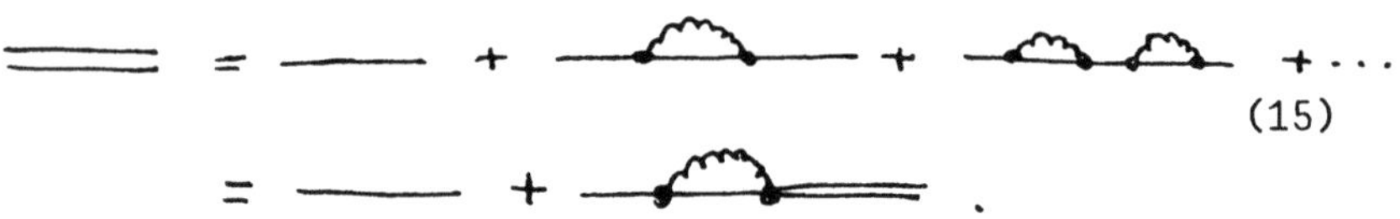

(15)

Solving this equation by iteration we get

$$G'(\mathbf{k},\mathbf{v},t) = \exp\{-it(\mathbf{k}.\mathbf{v})$$
$$-\frac{1}{3}t^3\frac{q}{m}\sum_{\mathbf{k}'}\int\frac{d\omega'}{2\pi}|(\mathbf{k}.\mathbf{E}(\mathbf{k}',\omega'))|^2G_0(\mathbf{k}-\mathbf{k}',\mathbf{v},(\mathbf{k}\cdot\mathbf{v})-\omega')\},\qquad(16)$$

which shows that the renormalization carried out so far - which corresponds to the removal of the most secular terms in t which occur due to the resonance denominators introduced by the $G_o$ – accounts for the distortion of the unperturbed particle orbit by the presence of the plasma excitations.

So far we have only looked at the renormalization of the particle propagator. It is also possible to consider the renormalized dielectric constant, and hence to renormalize the wave propagator. Once this is done one can in the usual way (see, e.g., Harris 1969) consider various wave-wave and wave-particle processes. The diagram for Compton scattering, for instance, will now be

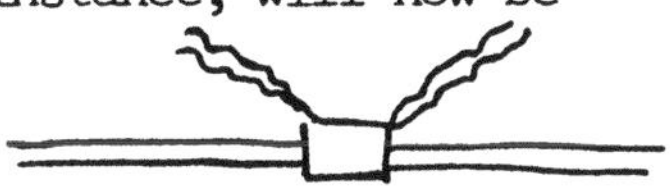

, where the double lines indicate renormalized propagators, while the square indicates the relevant coupling. Of course, we are assuming here tacitly that it still makes sense to use the concept of elementary excitations.

A different approach is the soliton approach to plasma turbulence. I hope later at this conference to show why this may not be as fruitful an approach as was hoped earlier, but I want all the same to consider it briefly. I just remind you that quasi-linear theory shows that in a Langmuir-turbulent plasma the energy tends to accumulate at very small wavenumbers; however, this practically uniform plasma state is modulationally unstable and Zakharov (1972) has formulated the equations describing the non-linear stage of this instability in which the high-frequency Langmuir oscillations are coupled to the ion motion by the Miller force. In the one-dimensional case the equations have the form

$$iE_t + E_{xx} = nE, \quad n_{tt} - n_{xx} = (|E|^2)_{xx}, \tag{17}$$

where n is the deviation of the ion-density from its equilibrium value; the variables in (17) are all suitably normalized. Equations (17) have the so-called soliton solutions

$$E = E_o \operatorname{sech}[\alpha^{\frac{1}{2}} E_o(x-Mt)] \exp i[\tfrac{1}{2}M x - \tfrac{1}{4} M^2 t - \alpha E_o^2 t] \;, \tag{18}$$
$$\alpha = \tfrac{1}{2} (1-M^2)^{-\frac{1}{2}}, \quad n = - |E|^2/(1-M^2).$$

We note that if we neglect the term $n_{tt}$ in the second equation, the first equation reduces to the non-linear Schrödinger equation

$$iE_t + E_{xx} + |E|^2 E = 0, \tag{19}$$

which also has soliton solutions. There are several talks planned about solitons so that I shall not say any more about them, but I want to finish my talk with a brief discussion of some recent, related work by Norman (1976). Norman's work deals with density waves in galaxies. To see the connection between this work and Zakharov's equations we remind ourselves that equations (17) are derived by averaging over the high-frequency (Langmuir) oscillations. Similarly one can consider a time-averaged Lagrangian (Dewar 1972) to describe the behaviour of stars in a stellar disc, where the averaging is over wave, epicycle, and rotational periods. Using the Lagrangian approach is particularly suitable for the linear approximation so that to a certain extent this approach is inconsistent if one uses it to derive an essentially non-linear equation. To see more clearly what is involved, I shall use a different approach, also due to Norman, which leads to the same equation.

In a flat disc we can describe the motion in terms of the planar polars r and $\theta$. We also use the hydrodynamic approximation so that, if we neglect pressure terms and magnetic effects our variables are the hydrodynamic velocity, which in the unperturbed motion will only have a component in the $\theta$ direction, equal to $r\Omega(r)$, and the surface density $\sigma$. The equations to be satisfied by these quantities are the equation of continuity (subscripts again denote partial differentiation)

$$\sigma_t + \frac{1}{r}\,[\sigma(v+r\Omega)]_\theta + \frac{1}{r}\,[\sigma r u]_r = 0, \tag{20}$$

where we have written u for the velocity component in the r-direction and $v + r\,\Omega$ for the component in the $\theta$ direction where $\Omega(=\Omega(r))$ is the unperturbed angular velocity, the equation of motion, which is a vector equation,

$$u_t + u\,u_r + \tfrac{1}{2}\,(v + r\,\Omega)\,u_\theta - (v + r\,\Omega)^2/r = -U_r, \tag{21}$$

$$[v + r\,\Omega]_t + u\,[v + r\Omega]_r + \tfrac{1}{2}\,[v+r\Omega]\,[v+r\Omega]_\theta + u(v+r\Omega)/r = -\frac{1}{r}\,U_\theta, \tag{22}$$

where U is the gravitational potential which satisfies the Poisson equation

$$U_{rr} + \frac{1}{r}\,U_r + \frac{1}{r^2}\,U_{\theta\theta} + U_{zz} = 4\pi\,G\,\sigma\delta(z), \tag{23}$$

where G is the gravitational constant.

Looking for a quasi-linear approach, we have in zeroth approximation $u_0 = v_0 = 0$, $\sigma = \sigma_0$, $(U_0)_r = r\,\Omega^2$. We now look for solutions behaving as $\exp\,(i\omega t - im\theta + ikr)$. We also look for solutions in the so-called tight-winding approximation, that is, we assume that $kr \gg 1$ (see Lin 1970 for a detailed discussion of these equations and the assumptions involved). Equations (20) to (22) then become in the linear approximation

$$i\kappa\nu\,\sigma_1 + i\,k\,\sigma_o\,u_1 = 0, \tag{24}$$

$$i\kappa\nu\,u_1 - 2\,\Omega v_1 - \frac{2\pi G i|k|}{k}\sigma_1 = 0, \tag{25}$$

$$i\kappa\nu\,v_1 + \frac{\kappa^2}{2\Omega}\,u_1 = 0, \tag{26}$$

where $\kappa$ is the epicycle frequency given by the equation

$$\kappa^2 = 4\Omega^2\,[1 + \frac{r}{2\Omega}\,\frac{d\Omega}{dr}], \tag{27}$$

and $\nu\kappa$ is the frequency at which a particle encounters the periodic gravitational field,

$$\nu\kappa = \omega - m\Omega. \tag{28}$$

From these equations we get the dispersion relation of the linear theory

$$1 - \nu^2 = \frac{|k|}{k_o}\,,\; k_o = \frac{\kappa^2}{2\pi G\sigma_o}\,. \tag{29}$$

To get the non-linear dispersion relation we write the equations in a slightly different form. The equation of continuity is written as

$$ik\nu\sigma + \sigma u_r = 0, \tag{30}$$

and from this we get

$$\frac{\sigma_1}{\sigma_0} = \frac{i\,u_r}{\kappa\nu}\,, \qquad \frac{\sigma_2}{\sigma_1} = \frac{i\,u_r}{\kappa\nu}\,, \tag{31}$$

or

$$\sigma_2 = -\,\sigma_0\,\frac{\overline{|u_r^2|}}{\kappa^2\nu^2}\;; \tag{32}$$

we note that $\sigma_1$ averages to zero. If we put $2\pi\,G\sigma_1 = E$, so that $E$ is a measure of the amplitude of the perturbation, and $2\pi G\sigma_0 = E_0$, and use the fact that in our approximation $|u_r^2| = k^2\,u_1^2$, we can replace equation (24) by

$$i\kappa\nu\,\sigma_1 + i\,k\,\sigma_0\,(1 - \frac{|E|^2}{E_0^2})\,u_1 = 0. \tag{33}$$

Proceeding as before we are then led to the non-linear dispersion relation

$$1 - \nu^2 = (1 - \frac{|E|^2}{E_0^2})\frac{|k|}{k_0} . \qquad (34)$$

If we introduce the variable $\psi = (\omega_t - m\theta)/\kappa$ and suitably renormalize our variables, equation (34) leads, when we go back to $\psi, r$ from the $\nu$, k representation, to the differential equation

$$E_{\psi\psi} + E + 1\ (1 - |E|^2)E_r = 0, \qquad (35)$$

which is a particular form of the non-linear Schrödinger equation. The soliton solutions of equation (35),

$$E = \Phi_0 \text{ sech } [\Phi_0(\psi - 2\nu r)]\text{expi}[\omega t - m\theta + kr], \qquad (36)$$

correspond to trailing spirals, and the stability of the solitons corresponds, one hopes, to the persistence of the spiral structures.

## References

J.H. Misguich and R. Balescu 1974 Phys. Lett. 48A, 426; Report EUR. CEA. FC. 738.

R.L. Dewar 1972 Ap. J. 174, 301.

T.H. Dupree 1966 Phys. Fluids 9, 1773.

E.G. Harris 1969 Adv. Plasma Phys. 3, 157.

M. Kono and Y.H. Ichikawa 1973 Progr. Theor. Phys. 49, 754.

C.C. Lin 1970 Galactic Astronomy (Eds. H.Y. Chiu and A. Muriel) Gordon and Breach, Vol. 2, p.1.

C.A. Norman 1976 Non-linear spiral density waves, unpublished.

J. Weinstock 1969 Phys. Fluids 12, 1045.

V.E. Zakharov 1972 Sov. Phys. JETP 35, 908.

# INTERPRETATION OF EXPERIMENTS IN LASER-DRIVEN FUSION

Keith A. Brueckner

Univcrsity of California, San Diego

La Jolla, California 92093 U.S.A.

## I. INTRODUCTION

The number and energy distribution of superthermal electrons produced in a laser-heated plasma can be quantitatively obtained directly from the experimental x-ray spectrum using only the assumption that the fast electrons lose energy by bremsstrahlung and electron-electron collisions. The result is independent of the spatial and temporal distribution of electron density and temperature. This analysis is given in Sec. II.

The effects of electrostrictive forces on a laser-heated plasma have been calculated for different laser wavelengths using a hydrodynamic code of standard form. Large density discontinuities and marked fluctuations in reflectivity are predicted. The analysis is given in Sec. III.

Hydrodynamic calculations of shell implosions show a marked dependence on laser wavelength for wavelengths less than one micron. The coupling to the dense imploded shell rises from 6% at one micron to 33% at 0.1 microns, increasing possible energy multiplication in a pellet from approximately 60 to 400. The details are given in Sec. IV.

## II. ENERGY DEPOSITION IN LASER-HEATED PLASMAS

The x-ray spectra from laser-irradiated plasmas typically show an athermal component which cannot be attributed to the radiation from a plasma with a well defined temperature and hence Maxwellian electron distribution. The hard radiation is believed to be due to energetic electrons produced by resonant absorption, or by laser-plasma instabilities. The correlation between the observed x-ray spectrum and the athermal electron population is usually made by the use of computer models which incorporate various models of the absorption process. In this section we show that a simple direct correlation exists which allows essentially model-independent determination of the fast electron distribution and also direct determination of the energy deposited in fast electrons. We base our analysis on the simple and reasonable assumption that electrons produced in the absorption process, with initial energies well above the temperature of the bulk population, lose energy by radiation and by collisional exchange with the much colder thermal electrons, that collisions among the superthermal electrons can be ignored, and that the fast electron density is much less than the thermal electron density.

The radiation emitted into all solid angles by a single electron of energy $\epsilon$ per unit path length per unit radiated energy [1] is

$$\frac{d^2\epsilon_{rad}}{dx\,dh\nu} = \frac{8}{3}\frac{e^2}{\hbar c}\frac{e^4}{mc^2}\frac{\langle Z^2\rangle N_i\, g_{rad}}{\epsilon}\,, \tag{2.1}$$

with $\langle Z^2\rangle$ the average value of $Z^2$ and

$$g_{rad} = \frac{x_1 x}{(1-\exp-x_1)(\exp x-1)}\ln\frac{x_1+x}{x_1-x}$$

$$x = \frac{2\pi z e^2}{\hbar}\left(\frac{m}{2\varepsilon}\right)^{1/2}, \quad x_1 = \frac{2\pi z e^2}{\hbar}\left(\frac{m}{2(\varepsilon-h\nu)}\right)^{1/2}. \tag{2.2}$$

The rate of energy loss by electron-electron collisions [2] is

$$\frac{d\epsilon}{dx} = -\frac{4\pi \langle Z \rangle N_i e^4}{\epsilon} \ln \Lambda_{coll} , \tag{2.3}$$

with

$$\Lambda_{coll} \cong \frac{2}{\hbar k_D} (2m\epsilon)^{1/2} . \tag{2.4}$$

The total energy radiated by the electron of initial energy $\epsilon_0$ in stopping may be obtained by combining Eqs. (2.1) and (2.3), with the result

$$\frac{d\epsilon_{rad}}{dh\nu} = \frac{2}{3\pi} \frac{e^2}{\hbar c} \frac{\langle Z^2 \rangle}{mc^2 \langle Z \rangle} \int_{h\nu}^{\epsilon_0} \frac{g_{rad}\, d\epsilon}{\ln \Lambda_{coll}} . \tag{2.5}$$

The ratio of the two coulomb logarithms is slowly varying with $\epsilon$ and $h\nu$; we approximate the ratio by taking the average value for $g_{rad}$ of 2 and by evaluating the collision logarithm at $\epsilon = 10$ kev, $\theta = 1$ kev, and $n_e = 10^{21}/cm^3$, where $\ln \Lambda_{coll} = 7.85$. The result is

$$\frac{d\epsilon_{rad}}{dh\nu} = \lambda_Z \frac{\epsilon_0 - h\nu}{mc^2}$$

$$\lambda_Z = 3.95 \times 10^{-4} \frac{\langle Z^2 \rangle}{\langle Z \rangle} \tag{2.6}$$

The energy radiated by a distribution of electrons produced with initial energy $\epsilon_0$ is

$$\frac{dE_{rad}}{dh\nu} = \int_{h\nu}^{\infty} n(\epsilon_0) \frac{d\epsilon_{rad}}{dh\nu} d\epsilon_0 \tag{2.7}$$

$$= \frac{\lambda_Z}{mc^2} \int_{h\nu}^{\infty} n(\epsilon_0)(\epsilon_0 - h\nu)\, d\epsilon_0 \quad . \tag{2.8}$$

From this result we find

$$\frac{d^2}{d(h\nu)^2} \frac{dE_{rad}}{dh\nu} = \frac{\lambda_Z}{mc^2} n(h\nu) \tag{2.9}$$

and

$$N_{fast} = \int_0^{\infty} n(\epsilon_0)\, d\epsilon_0$$

$$= \frac{mc^2}{\lambda_Z} \left( \frac{d}{dh\nu} \frac{dE_{rad}}{dh\nu} \right)_{h\nu = 0} \quad , \tag{2.10}$$

$$E_{fast} = \int_0^{\infty} n(\epsilon_0)\, \epsilon_0\, d\epsilon_0$$

$$= \frac{mc^2}{\lambda_Z} \left( \frac{dE_{rad}}{dh\nu} \right)_{h\nu = 0}$$

$$= 2.28 \times 10^{-10} \frac{\langle Z \rangle}{\langle Z^2 \rangle} \left( \frac{dE_{rad}}{dh\nu} \right)_{h\nu=0} \text{joules.} \tag{2.11}$$

The distribution of radiation emitted by the fast electrons at $h\nu \to 0$ may be readily estimated by extrapolation of the energetic spectrum [see Figs. 2.1 and 2.2].

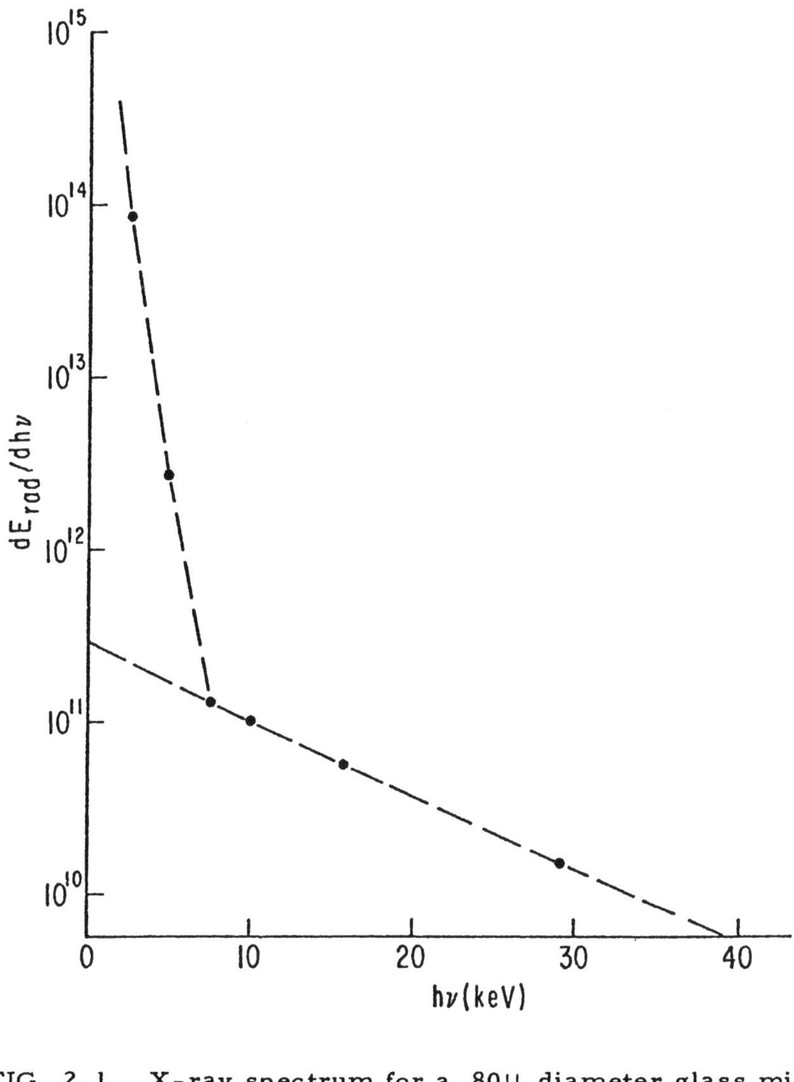


FIG. 2.1 X-ray spectrum for a 80μ diameter glass micro-balloon target, with 30 joules incident in a 70 picosecond pulse. For uniform illumination, the power is $2 \times 10^{15}$ w/cm$^2$. The dashed curves are best fits to the data and have been used in the estimates given in the text.

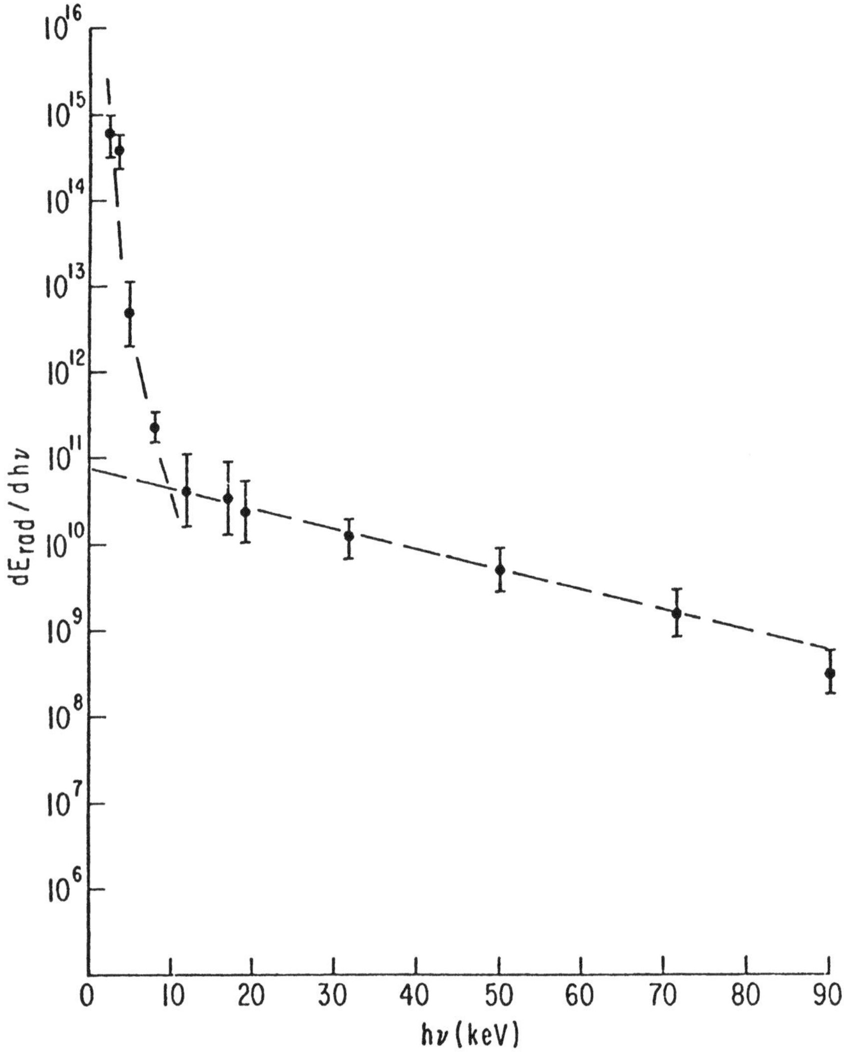


FIG. 2.2 X-ray spectrum from a parylene disk with 9.0 joules incident and a focal spot of 30μ diameter. The pulse length is approximately 130 picoseconds, giving a power of $10^{16}$ w/cm$^2$.

The preceding calculation is valid only if the fast electrons slow down in a time considerably less than the time for the hydrodynamic expansion of the corona. If the slow-down is too long, the electrons can lose energy in work done in accelerating the plasma. The slow-down time depends on the density which is determined by the motion of the electron resulting from scattering. At the critical density for one micron radiation, the slowing-down time as determined from Eq. (2.2) is $\tau = 1.26 \times 10^{-13}\ \epsilon_0(\text{kev})^{3/2}$ seconds or, for an energy of 20 kev, 12 picoseconds. The actual time is considerably less since the electron scattering causes diffusion into the dense plasma. This conclusion also holds in the presence of magnetic fields since scattering, drift resulting from the inhomogeneity of the fields, and motion along the fields will bring the electron into regions of high density where rapid energy loss occurs.

As examples, we consider two measurements reported by the Lawrence Livermore Laboratory [3,4] shown in Figs. 2.1 and 2.2. For the glass micro-balloon experiment, we find (for $\overline{Z} = 10$, $\overline{Z^2} = 108$) $E_{fast} = 6.3$ joules and $\langle\epsilon\rangle = 10$ kev. The measured absorbed energy is about 20% of 30 joules incident, or 6.0 joules. For the parylene ($C_2H_5$) disk experiment, we find (for $\overline{Z} = 2.43$, $\overline{Z^2} = 11$) $E_{fast} = 4.0$ joules and $\langle\epsilon\rangle =$ 30 kev. The measured absorption is 30 - 40% of 9.0 joules incident, or 2.7 - 3.6 joules, again in reasonable agreement with the spectral estimate. These results strongly imply that nearly all of the absorbed energy is in fast electrons, consistent with the previous interpretation [3,4]. If the electrons are produced by resonant absorption, the average energy is [5]

$$\langle\epsilon\rangle = \left(\frac{m_e \omega_0 \phi_{abs} L}{n_c}\right)^{1/2} , \qquad (2.12)$$

with $\phi_{abs}$ the absorbed laser flux per unit area and L the electron density scale height evaluated at the critical density surface. For the glass micro-balloon, the absorbed flux is about $4 \times 10^{14}$ watts/cm$^2$, giving $\langle\epsilon\rangle = 18$ kev $[L\,(\text{microns})]^{1/2}$. For the parylene disk, the absorbed flux is about $3 \times 10^{15}$ w/cm$^2$, giving $\langle\epsilon\rangle = 49$ kev $[L\,(\text{microns})]^{1/2}$. These estimates are in approximate agreement with the x-ray spectrum only if the scale height is of the order of one micron, a surprising and somewhat implausible result. The average energy may, however, be reduced from Eq. (2.12) by the varying angles of incidence and polarization of the laser beam.

This analysis shows that the number of fast electrons produced in laser-plasma interaction can be directly determined from the x-ray spectrum. The production rate is approximately determined by the laser pulse length. The spatial origin is given by the illuminated area and may be reasonably assumed to be at the critical density surface. This provides a highly useful nearly model independent input to computer codes and hence a semi-empirical procedure for determining other effects of the fast electron production such as shell and fuel pre-heat.

## III. RADIATION-PRODUCED FORCES IN LASER-HEATED PLASMAS

The radiation pressure and ponderomotive forces [6] on a laser-heated plasma can produce substantial changes in the plasma density and expansion under conditions easily produced in present experiments. The radiation field interacts with the electrons through the average value of the $j \times B$ force, which gives

$$F_{es} = (1 - \epsilon) \nabla \frac{E^2}{8\pi} \tag{3.1}$$

$$\epsilon = 1 - \frac{\omega_{p\ell}^2}{\omega_L^2}$$

$$= 1 - \rho/\rho_c \quad ,$$

with $\rho_c$ the critical density. Included in Eq. (3.1) is a force resulting from momentum deposition [7] by absorbed photons which results from the variation of the flux due to absorption. In the WKB approximation, neglecting absorption, the field is given by

$$\frac{E^2}{8\pi} = \frac{E_v^2}{4\pi} \frac{1}{\sqrt{\epsilon}} \quad , \tag{3.2}$$

including the factor of two arising from the average value of the incident and reflected flux.

The effect of the electrostrictive force on the plasma conditions is difficult to estimate analytically since the interaction is time-dependent and the plasma adjusts the density profile and the critical-density radius in response to the electrostrictive pressure. R. E. Kidder [8] estimated the density discontinuity and plasma density variation resulting from radiation pressure (omitting the electrostrictive forces) in a simplified model for one-dimensional isothermal expansion. His results are, however, only qualitatively similar to the numerical calculations of this paper. We have evaluated the effects of electrostriction by numerical solution of the hydrodynamic equations, using a hydrodynamic code [9] of standard form (Lagrangian with implicit heat conduction) simplified by the use of classical transport coefficients and by the assumption of a single temperature of electrons and ions.

The electric field was computed in under-dense zones according to Eq. (3.2), with $\epsilon \geq 0.2$. This limit roughly simulates the restriction [10] on the peak field given by solution of Maxwell's equations. The precise treatment of the variation of $\epsilon$ near $\rho = \rho_c$ is not important since the expanding plasma passes very quickly through the critical density. In the over-dense zones, the value of the WKB estimate of $E^2/8\pi$ averaged over the zone was used. The effect of the electrostrictive pressure on the plasma energetics was determined by evaluation of the overall plasma energy balance. The electrostrictive pressure was found to affect the energy balance by only a few percent. The absorbed laser energy was deposited in the first over-dense zone.

Typical results are given in Fig. 3.1 for one-micron radiation and in Fig. 3.2 for eight-micron radiation, for the conditions indicated in the figure captions. The detailed structure in the critical-density region could be obtained only if fine-zoning was used in the hydrodynamic calculations. In the one-micron case, the zoning in the initial glass shell in the region of the critical density surface shown in Fig. 3.1 was 35 Å at the initial

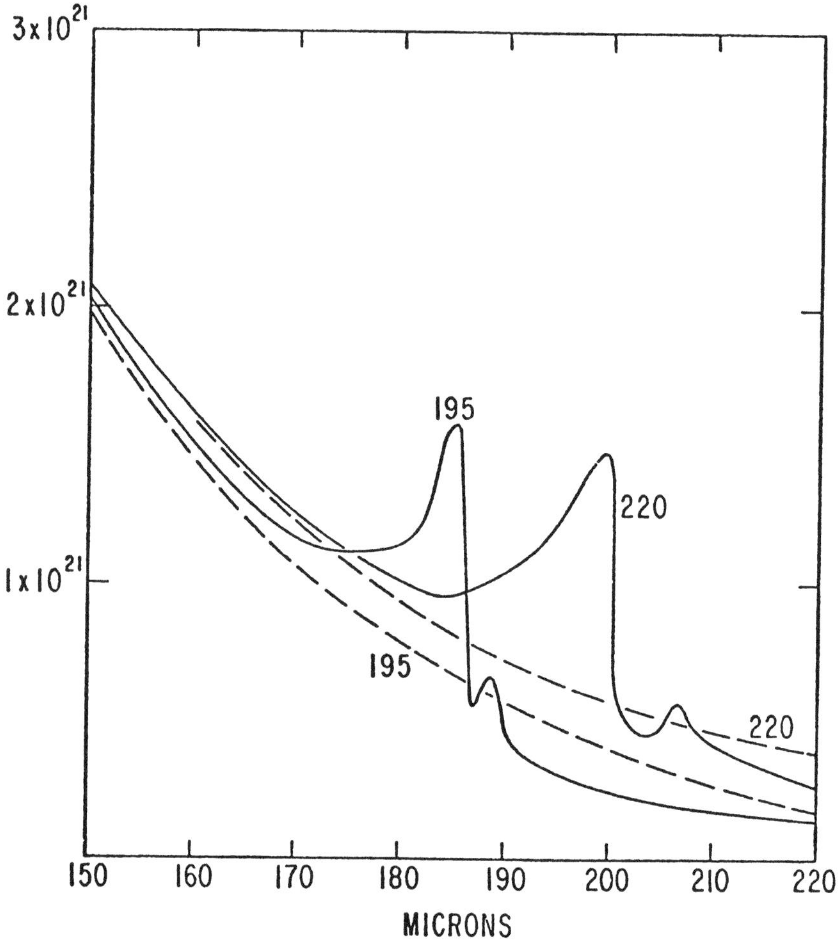


FIG. 3.1 Density (electrons/$cm^3$) profile in the vicinity of the critical density ($10^{21}/cm^3$) surface for one-micron radiation. The dashed curves give the density in the absence of the radiation pressure. The incident flux at the critical radius of 100 microns was $10^{16}$ watts/$cm^2$. Ten percent of the incident flux was assumed to be absorbed. The pellet configuration was a glass shell ($Z = 10$) with an initial wall thickness of 4 microns. Over the density range shown, the temperature was approximately constant at 3.3 kev. The material velocity was supersonic but subsonic with respect to the moving critical density interface. The curves are labelled with the time in picoseconds.

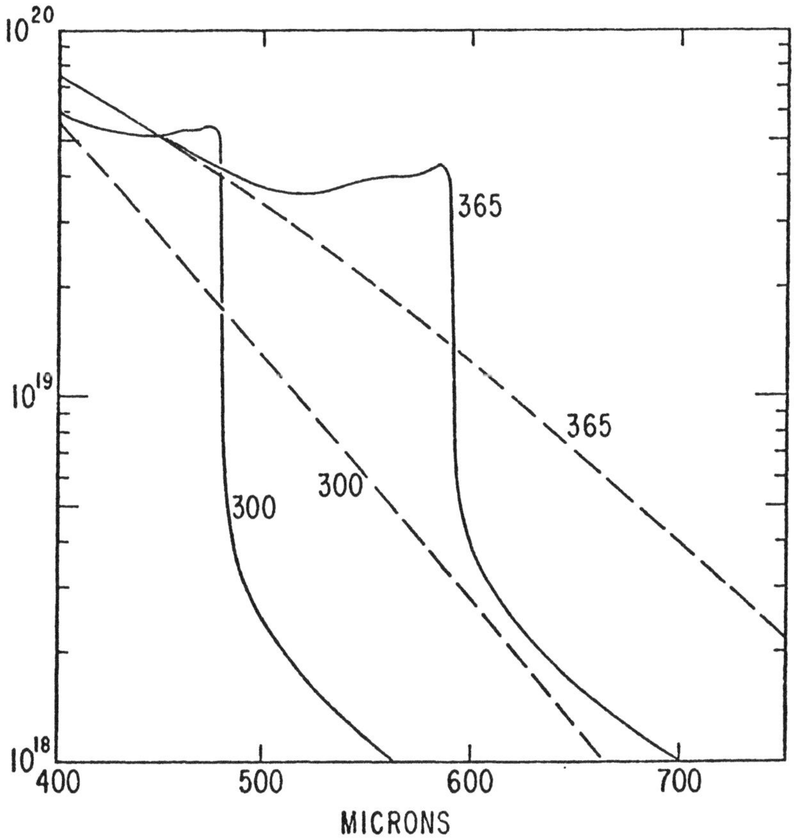


FIG. 3.2 The same as Fig. 3.1 for eight-micron radiation, but with the absorption increased to 20%. The temperature in the plasma was approximately constant at 5.3 kev. The material velocity was again supersonic, but subsonic with respect to the moving critical density ($1.56 \times 10^{19}/cm^3$) interface.

glass density. For both cases, a marked density discontinuity forms which is convected outward at a velocity close to the natural velocity. The possibility of rapid convection of the critical density surface appears to have been omitted in plasma calculations [11] using numerical simulation. For the results shown in Figs. 3.1 and 3.2, the flow is supersonic into the critical density surface. The flux at the critical density surface as a result drops rapidly with time, particularly for the eight-micron case where the critical density surface moves to radii much larger than the critical radius.

As examples, for the one-micron case at 195 picoseconds, the critical density radius has moved outward to 186 microns and the incident flux has dropped to $2.9 \times 10^{15}$ w/cm$^2$. The radiation pressure (assuming density much less than the critical density) is about 2.0 megabars while the pressure in the over-dense plasma is 8.3 megabars. For the eight-micron case at 365 picoseconds, the critical density radius has moved from 100 microns to 590 microns, reducing the laser flux to $2.87 \times 10^{14}$ w/cm$^2$. The radiation pressure is 0.19 megabars while the pressure in the over-dense plasma is 0.36 megabars.

The fine structure near the critical density surface for the one-micron case persists at later times and is independent of the zoning if fine zoning is used. The effect may, however, be an artifact resulting from the numerical-difference scheme used. A striking phenomenon appears at late time in the one-micron case, as the density behind the density maximum near the critical density surface falls below critical density. This is shown in Fig. 3.3. As the critical density surface is convected outward into regions of decreasing electrostrictive pressure, the density maximum drops below critical density and the deposition point abruptly shifts inward to a much smaller radius. This effect, shown in this calculation which is spherically symmetric, may not persist in a 2-D calculation, particularly with non-uniform irradiation. Fluctuations in the critical density radius may, however, occur; the phenomenon would be conspicuous experimentally as a source of fluctuations in reflectivity.

The striking phenomena shown in this relatively simple model of the laser-plasma interaction are of particular interest in their effect on the processes of energy deposition and possible growth of unstable plasma and/or ion waves in the sub-critical density plasma. The electrostrictive pressure, which markedly increases the plasma density gradient and depresses the density of the sub-critical density plasma, can be expected to have a profound effect on the laser-plasma interaction [12].

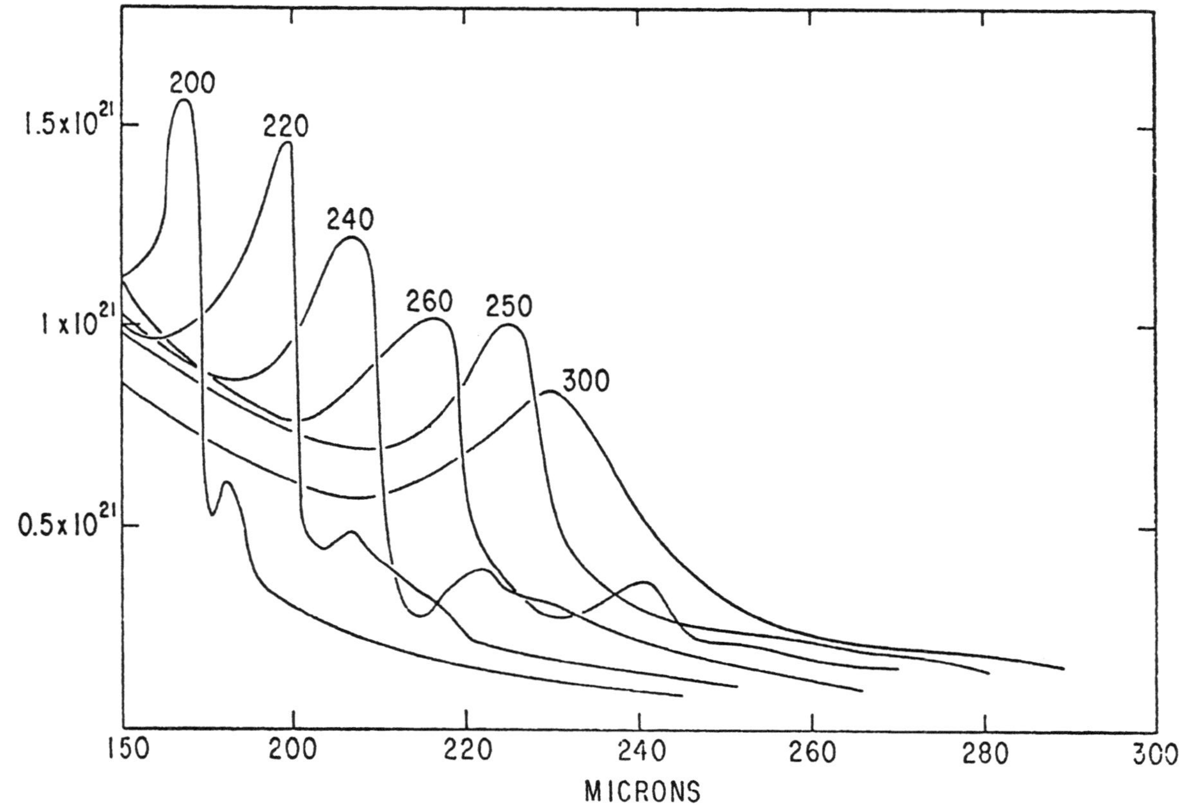


Fig. 3.3 Late-time variation of the density, for the same conditions as in Fig. 3.1. The laser deposition is in the outermost over-dense zone, allowing the density at smaller radii to become less than the critical density of $10^{21}/cm^3$. The deposition shifts abruptly inward at about 380 picoseconds to about 165 microns, as the radiation pressure at the larger radii becomes insufficient to keep the outer density maximum above critical density.

## IV. WAVELENGTH DEPENDENCE OF LASER COUPLING TO PELLET IMPLOSIONS

The coupling of laser energy into hydrodynamic motion in a laser-driven implosion is of critical importance in determining the possible fusion energy output per unit of absorbed laser energy. Analytic estimates [9] and numerical calculations [13] have indicated that the hydrodynamic coupling is in the range of 5% to 10% for laser wavelengths less than a few microns but may become considerably lower at larger wavelengths [14].

A recent suggestion by Dr. Raymond Grandey (private communication) that considerably stronger coupling can occur for shorter wavelengths has stimulated further study of the problem. A time-independent calculation by Kitomer et al. [15], ignoring acceleration and hence the dynamics of the implosion, suggested that the coupling is inversely proportional to the wavelength. These results are, however, inconclusive because of the limited model used, and are in disagreement with the numerical calculations described in this section.

We have calculated a series of implosions using a hydrodynamic code of standard form [9] (Lagrangian hydrodynamics with implicit heat condition) simplified by the assumption of a single electron and ion temperature and by the use of an ideal gas equation of state. Classical heat conduction by electrons was assumed. The calculations were for a 100 micron radius glass shell with wall thickness of 4 microns. The incident absorbed flux was 1.26 terawatts, which was deposited in the first overdense plasma zone. The effects of electrostrictive pressure were included but had negligible effect on the hydrodynamic coupling.

The pressure profiles for several wavelengths from 0.1 to 8 microns are given in Fig. 4.1 at times selected to give the peak pressure at approximately the same radius. A significant variation of peak pressure with wavelength for wavelengths less than one micron is shown, which produces greater acceleration of the shell and hence increased coupling at short wavelength. The ratio of kinetic energy of the imploding shell material (at greater density than the initial glass density) to the absorbed energy at the times corresponding to the pressure profiles in Fig. 4.1 is given in Fig. 4.2. The ratio increases only slightly between 8 microns and 1 micron but then rises rapidly from 6.4% at 1 micron to

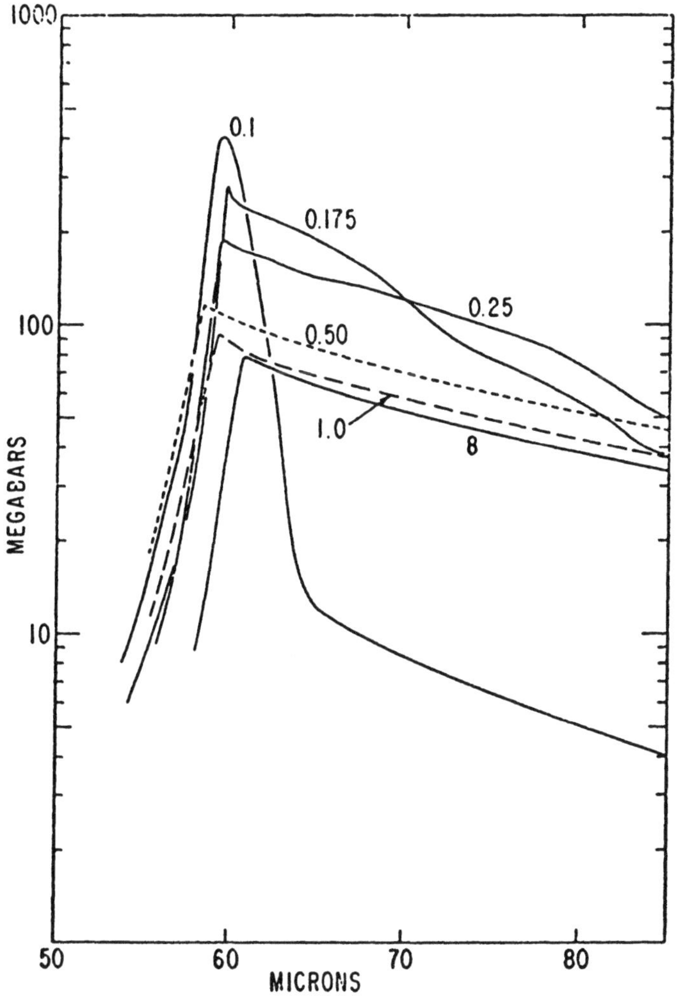


FIG. 4.1 Pressure as a function of radius. The curves are labelled by the wavelength in microns. The time in the implosion at which the peak pressure occurs at about 60-micron radius is 300, 260, 250, 210, 180, and 130 picoseconds at 8, 1, 0.5, 0.25, 0.175, and 0.1 micron wavelength.

28.6% at 0.1 micron. Further analysis of the short wavelength implosions shows peak transfer efficiency to the dense shell of 33.4% at 0.1 micron and 19.4% at 0.175 micron.

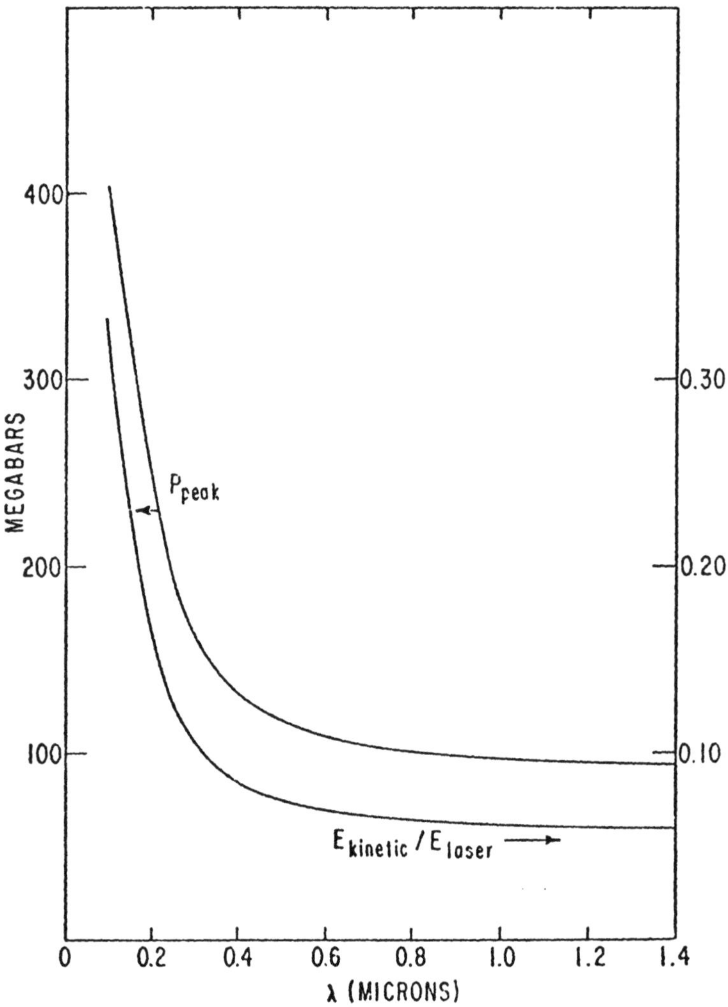


FIG. 4.2 Peak pressure from Fig. 4.1 and ratio of the kinetic energy of the dense shell (density greater than initial) to the absorbed laser energy. The variation from 1 micron to 8 microns is less than 20%.

The variation of coupling with wavelength arises from the deposition of the laser energy in increasingly dense material, steepening the temperature gradient in the ablation front and increasing the ablation rate. The temperature profile for the

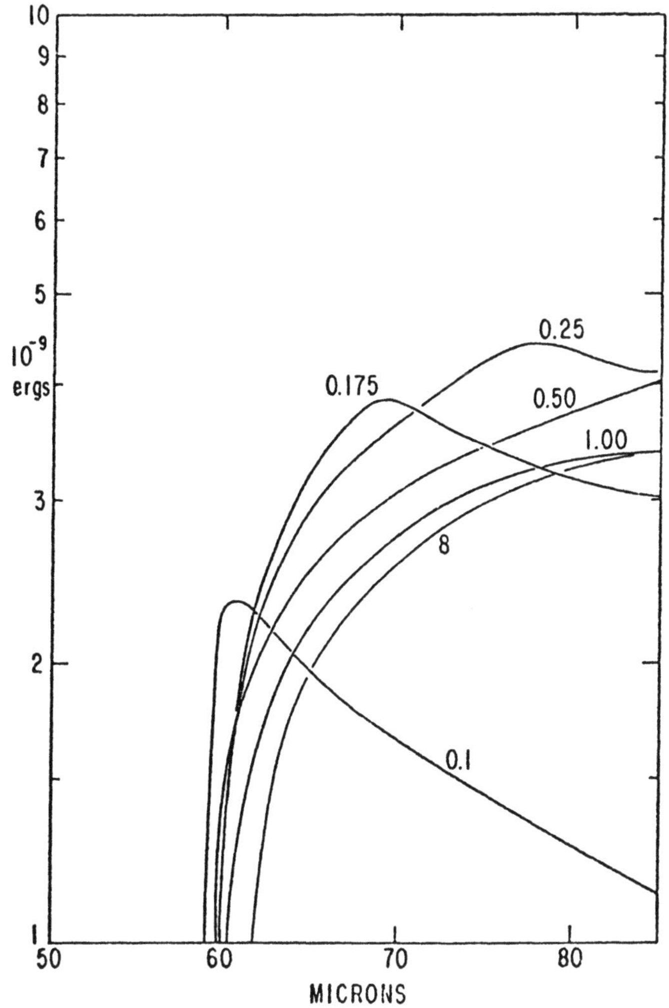


FIG. 4.3 Variation of temperature with radius for the evaluated wavelengths (in microns). The temperature is given in energy units ($1.60 \times 10^{-9}$ erg = 1 kev).

cases given in Fig. 4.1 is shown in Fig. 4.3. The temperatures in the ablation front are also increased by the reduced corona temperatures and the corresponding decrease in the kinetic energy loss into the corona expansion as the laser wavelength decreases.

The possible increase in the efficiency of hydrodynamic coupling of the absorbed laser energy at short wavelengths can have a large effect on laser energy requirement for high pellet gain. The coupling of the incident laser energy to the pellet may also improve at short wavelength; we do not consider that effect in this paper.

Detailed implosion calculations [9] have shown that a practical limit in pellet gain, set by details of the hydrodynamics of compression and fuel burn-up, is about 1200 times the energy delivered to the compressed fuel core. This limits pellet gain (ratio of fusion energy output to absorbed laser energy) to about 60 for a coupling efficiency of 5% but gives a gain of 400 for the coupling efficiency of 33.4% we have computed at 0.1 microns. The practical usefulness of this high possible gain is limited by present laser technology. Our results emphasize the importance of development of lasers with wavelength considerably less than one micron.

## REFERENCES

1. C. W. Allen, Astrophysical Quantities (University of London, The Athlone Press, 1955).

2. Lyman Spitzer, Jr., Physics of Fully Ionized Gases, Interscience Tracts on Physics and Astronomy No. 3 (Interscience Publishers, Inc., New York, 1956).

3. H. G. Ahlstrom, UCRL Report 77094 Rev 1, Oct. 8, 1975; Bull. Am. Phys. Soc., Oct. 1975 (papers 3C9, 3C10, and 3C 11).

4. W. C. Mead, W. L. Kruer, J. D. Lindl, and H. D. Shay, Bull. Am. Phys. Soc., Oct. 1975 (paper 2C5).

5. J. P. Freidberg, R. W. Mitchell, R. L. Morse, and L. I. Rudsinski, Phys. Rev. Letters 28, 795 (1972).

6. See, for example, Landau and Lifshitz, Electrodynamics of Continuous Media (Pergamon Press, New York, 1960).

7. John A. Stamper, "Field Generation by a Generalized Ponderomotive Force," NRL Memorandum Report 3186, Naval Research Lab., Washington, D.C. (Dec. 1975).

8. R. E. Kidder, "Interaction of Intense Photon Beams with Plasmas (II)," Proc. Japan-U.S. Seminar, Laser Interaction with Matter, Sept. 24-29, 1972.

9. K. A. Brueckner and S. Jorna, "Laser-Driven Fusion," Rev. Mod. Phys. 46, 325 (1974).

10. J. D. Lindl and P. K. Kaw, Phys. Fluids 14, 371 (1971).

11. K. Lee, D. Forslund, J. Kindel, and E. Lindman, Los Alamos Scientific Lab. Report LA-UR-75-2097 (submitted to J. Phys. Fluids).

12. Experiments at the Naval Research Lab., Washington, D.C. (Dr. Stephen E. Bodner, private communication) and at the University of Rochester (Professor Moshe Lubin, private communication) have suggested that marked density gradient anomalies exist with one-micron radiation in the $10^{15}$-$10^{16}$ watts/cm$^2$ range.

13. K. A. Brueckner, P. M. Campbell, and R. A. Grandey, "Laser-Driven Implosion of Spherical Shells," Nucl. Fusion 15, 471 (1975).

14. R. E. Kidder, "Laser Compression of Matter: Optical Power and Energy Requirements," Nucl. Fusion 14, 797 (1974).

15. S. J. Kitomer, R. L. Morse, and B. S. Newberger, "Structure and Scaling Laws of Laser-Driven Ablative Implosions," Los Alamos Scientific Lab. Report #LA6079MS, to be published in Physics of Fluids.

# EXPERIMENTS ON THE ABSORPTION OF HIGH INTENSITY LASER LIGHT AND SUBSEQUENT COMPRESSION OF SPHERICAL TARGETS

Roy R. Johnson

KMS Fusion, Inc.

Ann Arbor, Michigan 48106

## INTRODUCTION

Laser fusion experiments have been performed with the goal of understanding the physics of the laser target interaction so that the neutron scaling laws with laser power, laser wavelength and target material can be verified[1]. An important consideration of this interaction is the transfer of laser energy into the thermonuclear fuel. The experimental features which provide information on the amount of energy absorbed by the fuel are target compression and particle generation. The complementary measurements of laser light absorption by the target are used to obtain information on the efficiency of the hydrodynamic coupling of this energy to the fuel.

The laser required for laser-fusion energy production must be designed to maximize the efficiency of the coupling of the laser energy to the fuel. This includes optimization with respect to laser wavelength as well as considerations with respect to polarization effects. Pulse shaping of the laser energy must also be optimized. The target illumination system is important in order to make the target implode with the greatest degree of spherical symmetry possible.

Nearly all laser-fusion implosion experiments have used glass-shell targets with an effective $\bar{Z} = 10$. Numerical simulations indicate that the relatively high-Z corona may be radiatively preheating the inner wall of the shell and degrading the implosion[2]. Experiments at KMSF with lower-Z composites are performed in order to minimize this corona-radiation preheat as well as determining the laser-energy absorption in lower-Z targets.

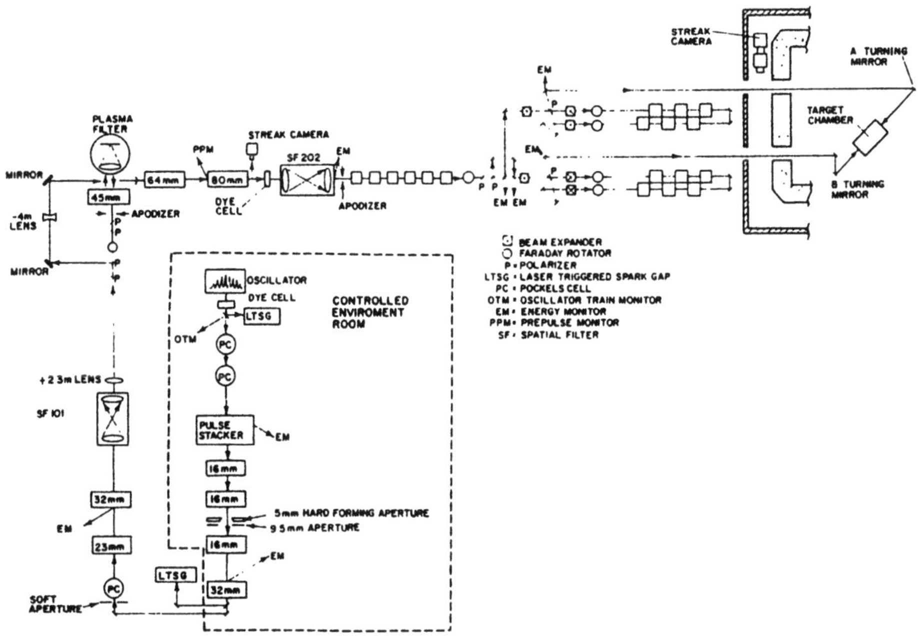


Figure 1 The KMSF Laser Layout in Circulating Double Pass. The Plasma Filter is Optional with a Totally Reflecting Normal Mirror.

## LASER

The basic KMSF laser system is shown in Figure 1. The laser pulse originates in a mode-locked Nd:YAG oscillator and is then shaped temporally by a pulse stacker[3]. In this configuration the laser beam is split to pass through the final six amplifiers in each channel in a circulating double-pass arrangement[4]. A variation of this system is to single-pass the final six amplifiers in each channel. The principal difference between these two variations affecting their performance in target experiments is the existence of cross-polarized laser beams in the arrangement shown in Figure 1; which can be important for certain laser plasma phenomena such as resonant absorption.

## PELLET

The pellets used for most neutron-producing experiments are spherical glass shells filled with a mixture of deuterium and tritium gas[5]. Pellets in the diameter range from 30 to 200 micrometers filled with 10 to 100 atmospheres of D-T gas are used in these experiments with wall thickness in the range of 0.4 to 2.0 micrometers.

The targets used in some experiments were coated glass shells and polymeric shells. Coatings of lithium, boron, polyvinyl alcohol, and lithium borohydride were used on glass shells. Plastic shells of both polymethyl-methacrylate (PMMA) and polyvinyl alcohol (PVA) were also used.

## ILLUMINATION

The spherical pellet is illuminated by a lens-mirror combination that provides orthogonal illumination over 70% of the target surface (see inset in Figure 2). Near-normal illumination is necessary in order to minimize refractive effects and increase the amount of absorbed laser light. The lens-mirror system provides excellent illumination using two laser beams[6].

## DIAGNOSTICS

The 50 mm gap between the ellipsoidal mirrors provides a direct-view diagnostic region (Figure 2). Plasma, optical, X-ray, and particle diagnostics are located in this region and are used to determine emission orthogonal to the laser axis[7]. The target X-ray emission and spectral distribution are measured using thermoluminescent dosimeters (TLD's), X-ray PIN diodes, X-ray pinhole cameras, and an X-ray spectrograph[8,9]. Charge collectors and an ion mass spectrometer[10] are used for measuring the total ion yield and the ion charge distribution. The charge collectors are biased to collect ions. Numerical integration of the charge-collector oscilloscope traces provides a measurement of the target-mass ablation and the amount and distribution of the ion kinetic energy. The charge-collector measurements also provide a determination of the degree of disassembly symmetry.

Total-yield measurements of the neutrons have been made using activation counters[11] and neutron scintillation counters. Time-of-flight measurements with the neutron scintillation counters identified the neutrons as 14.1 MeV neutrons from the D-T reaction.

## IMPLOSION MEASUREMENTS

The amount of compression that occurs in the target is inferred by means of photographs of the X-ray emission made with X-ray pinhole cameras[12]. The pinholes are covered with various thicknesses of beryllium in order to observe the spectral X-ray energy distri-

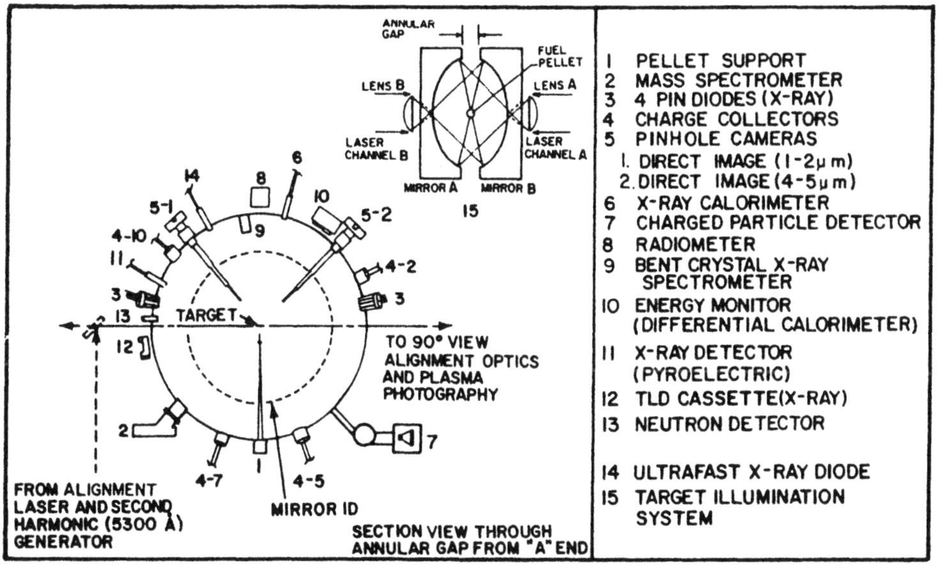


Figure 2 Section View Between the Ellipsoidal Mirrors of the Direct Viewing Diagnostic Region (DVDR).

bution with spatial resolution and determine the electron temperature of the tamper.

The X-ray photograph is a time-integrated record of the X-rays emitted by a target as a consequence of the laser irradiation, and provides a history of the implosion, showing where heating and compression were greatest[13]. Two regions are commonly observed for which X-ray emission is strong. The first is near the initial location of the glass shell where rapid absorption of energy from the laser results in a hot, dense layer approximately at the original diameter of the pellet. As the implosion progresses to the kinetic phase, a portion of the original shell moves rapidly inward acting as a pusher, entrapping the fuel, until it is stopped by the back pressure of the highly compressed and heated DT gas. The second region where X-ray emission is strong occurs when the pusher is stopped and its kinetic energy becomes thermal energy. During the kinetic phase, as a result of a lower temperature in the pusher and the relatively short dwell time at any given radius, emission is weaker, and there is a decrease in film density on the X-ray photographs. Two circular regions of higher film density are observed, one representing the original pellet diameter and the other the final diameter of the compressed pellet (or an upper limit for the final diameter, depending on the resolution of the X-ray camera). The measurement of the diameter of this still-intact pusher which acts as a tamper at peak compression compared to the initial shell diameter permits calculation of the volume reduction and hence determines the amount of compression in the D-T fuel.

The amount of X-ray emission observed in the central region of the pinhole photographs is strongly dependent on the fuel and tamper constituents. The radiative process for the DT fuel is bremsstrahlung, while for a higher-Z tamper the contribution of free-bound and bound-bound transitions becomes significant. For glass-shell targets where the average atomic number is 10 times that of the fuel, the tamper produces more X-ray emission than the fuel does.

The photographs taken by the X-ray pinhole camera show two regions of strong X-ray emission. From microdensitometer traces along a diameter of the X-ray pinhole photographs, the volume reduction of the laser-driven shell is directly determined.

## ABSORPTION MEASUREMENTS WITH DIFFERENT SHELL MATERIALS

A series of experiments was conducted whose purpose was to explore the relationship between the fraction of laser energy absorbed and the average atomic number ($\overline{Z}$) of the pellet shell. By means of low-Z coatings on the exterior of glass-shell pellets as well as plastic shell pellets containing thermonuclear fuel it was possible to study coupling of the laser energy to the fuel for a large number of low-Z materials.

The incident laser power in these measurements was between 0.2 and 0.5 terawatt with pulse lengths between 30 and 200 psec.

The absorbed laser energy is converted by means of electron-ion collisions to ion kinetic energy and soft X-ray energy. The ion kinetic energy was measured using charge collectors.

The charge collectors determine both the ion mass and energy flowing into detectors located at four circumferential positions around the annular gap between the ellipsoidal mirrors. The charge collector data reduction procedure incorporates a correction for secondary electron emission from the collector surface.

Ion Mass Spectrometer measurements of the ion flow have been made for glass and protons. Knowing the secondary emission corrections for hydrogen and glass, it is possible to estimate the correction for other materials. Only the coating, which is approximately 1 μm thick, is assumed to be involved in the laser energy absorption process.

The total X-ray yield measurements have been made using bare $CaF_2(D_y)$ thermoluminescent dosimeters (TLDs). The TLD sensitivity has been determined accurately down to photon energies of 0.8 keV. However, the bare TLDs are used, assuming equal sensitivity, to determine the total X-ray yield, the largest share of which lies below 0.8 keV.

Target-shot data were obtained for the low-Z-coated targets and plastic-shell targets in experiments with the laser both in the single-pass and the circulating double-pass configurations.

To make a larger statistical grouping, all the low-Z coated targets and plastic-shell targets were lumped together and an average Z determined for the group of shots under consideration. Table I lists the average absorbed energy fraction for the low-Z coated targets and the plastic shell targets in both laser configurations. As baseline comparisons, results for standard glass-shell targets are also listed, as well as the X-ray energy fractions (relative to energy on target), the fast ion energy fraction, the DT neutron yield, and a group-averaged Z.

## DISCUSSION OF RESULTS ON LOW-Z TARGETS

Light is absorbed classically in a fully ionized plasma by the inverse bremsstrahlung process. The ratio of the absorbed to incident flux is given by

$$\phi_{abs}/\phi_o = 1 - e^{-A} \qquad (1)$$

with A defined by

$$A = \int K_B d\ell \approx K_B L \qquad (2)$$

wherein L is the density gradient length and $K_B$ the absorption coefficient. For inverse bremsstrahlung,

$$K_B \sim Z/\lambda^2 \theta^{3/2} \qquad (3)$$

where Z is the ionic charge and θ the electron temperature. For A much less than unity, which corresponds to the experimental observations, ε, the absorbed energy fraction is given by

$$\varepsilon \equiv \phi_{abs}/\phi_o \approx A = K_B L \sim ZL/\lambda^2 \theta^{3/2} \qquad (4)$$

The energy flux, F, transported by the electrons cannot exceed

$$F = n_e V_e \theta \sim \theta^{3/2} \qquad (5)$$

TABLE I

| | Total Absorbed Energy Fraction | X-Ray Energy Fraction of Incident Energy | Fast Ion Energy Fraction of Kinetic Energy | Neutron Yield ($n \times 10^{-7}$) | Group Averaged $\overline{Z}$ | Power TW |
|---|---|---|---|---|---|---|
| 1. Standard XL - Single-Pass (12) | | | | | | |
| AV | 8.5% | 2.5% | 71% | 1.10 | 10 | 0.53 |
| SD | 2.3% | 0.5% | 6.8% | 0.43 | | 0.17 |
| 2. Standard XL - Double-Pass (12) | | | | | | |
| AV | 17.1% | 4.1% | 73% | 2.5 | 10 | 0.62 |
| SD | 3.1% | 1.3% | 6.9% | 1.4 | | 0.09 |
| 3. Coated XL and Low-Z Targets - Double-Pass (8) | | | | | | |
| AV | 11.5% | 3.7% | 71% | 1.46 | 3.54 | 0.40 |
| SD | 3.3% | 1.3% | 7.3% | 0.78 | | 0.08 |
| 4. Large Glass Targets - Single-Pass (3) | | | | | | |
| AV | 9.6% | 3.5% | 61% | 0.22 | 10 | 0.42 |
| SD | 0.3% | 0.5% | 4.0% | 0.17 | | 0.06 |
| 5. Large Glass Targets - Double-Pass (5) | | | | | | |
| AV | 18.4% | 4.3% | 73% | 0.53 | 10 | 0.44 |
| SD | 5.8% | 0.5% | 8.2% | 0.33 | | 0.09 |
| 6. Large Low-Z Targets - Double-Pass (4) | | | | | | |
| AV | 17.1% | 3.7% | 65% | --- | 3.6 | 0.40 |
| SD | 3.1% | 0.7% | 4.5% | --- | | 0.10 |

and furthermore energy conservation requires that

$$F = \phi_{abs} \quad . \tag{6}$$

Thus

$$\frac{ZL}{\lambda^2 \theta^{3/2}} \phi_o \sim \theta^{3/2}$$

or

$$\theta \sim (ZL\phi_0/\lambda^2)^{1/3} \quad . \tag{7}$$

Consequently,

$$\varepsilon = \phi_{abs}/\phi_o \sim (ZL/\lambda^2\phi_0)^{1/2}. \tag{8}$$

As is shown in Table I, the absorption by coated and low-Z targets decreased with respect to comparable glass targets in the double-pass configuration. The fractional mass pick-up by the charge collectors corresponds approximately to the mass of the coating; the charge collector energy analysis is based on this amount. Furthermore, the results on the large plastic shells (Item 6, Table I) show a decrease in absorption as compared with similar-size targets (Item 5, Table I).

## IMPLOSION SYMMETRY MEASUREMENTS

A series of experiments was performed using glass-shell pellets with diameters in the range 100 to 200 μm. These target shots were used to determine the influence that departures from spherical symmetry in an implosion have on neutron production. Control of symmetry is expected to become more crucial in future experiments as the pellet is scaled up with increased laser power on target.

Two types of experiments were performed in order to deliberately change the symmetry of the implosion. The first set of experiments involved changing the relative timing of the two laser beams that illuminate the target. The second set of experiments involved changing the location of the paraxial focal position of the illumination system relative to the center of the pellet.

There was a number of target shots in which the timing of the two laser beams was displaced from simultaneity at the pellet. There is approximately 10 psec uncertainty in the timing of the simultaneous arrival of the laser beams as determined by two-photon fluorescence. The timing was deliberately changed by inserting 35-mm-thick slabs (58 psec delay) of BK-7 glass.

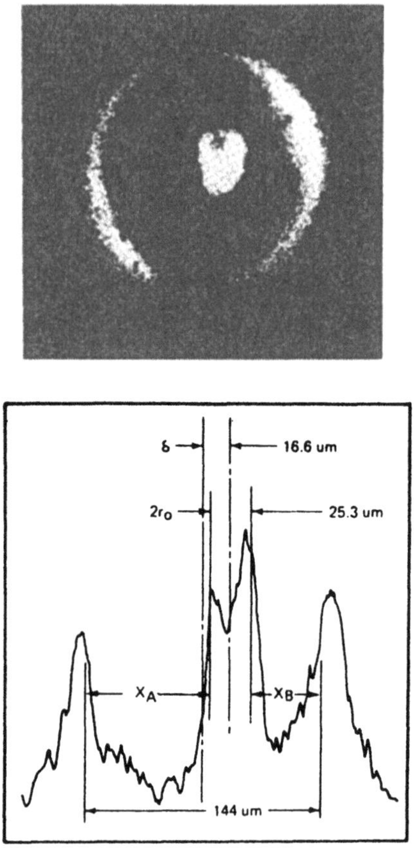


Target Shot 2369 144 um Diameter

Figure 3 Microdensitometer trace of the X-ray pinhole photograph for target shot No. 2369. The original target diameter was 144 μm and the energy on target was 81.8 joules in 240 psec. The 16.6 μm displacement of the center structure caused by a 58 psec delay in channel B corresponds to a pusher velocity of $5.7 \times 10^7$ cm/sec.

In one group of experiments, glass-shell pellets, nominally 145 μm in diameter and containing 7.5 atmospheres of DT, were imploded. The energy on target was approximately 70 joules and the pulse length 240 psec. The channel delays were 0, 58, 117, and 176 psec. In the X-ray pinhole photographs the central structure of the X-ray pattern was clearly displaced from the geometric center of the pellet. Figure 3 shows an X-ray photograph and the microdensitometer trace from one of the implosions with a 58 psec delay (Shot 2369).

In a second group of experiments, pellets having a nominal 100-μm-diameter and 7.5 atmospheres of DT gas were imploded. The energy on target was approximately 80 joules in 240 psec. The displacements in the X-ray pinhole photographs were similar to those obtained with the larger targets. The results from both groups of experiments are shown in Table II. In every case the neutron yield was below threshold except for the zero-delay experiments, where approximately $10^6$ neutrons were observed. Shot 2360 exhibits the measured compression of a DT filled pellet. The pellet had an initial diameter of 124 μm and the inner ring in the X-ray pinhole photograph was 11.3 μm, corresponding to a volume reduction of 1320; the yield was $2.2 \times 10^6$ neutrons.

These experiments are very useful for examining the behavior of the imploding pusher. In particular, they provide a direct geometric approach to measuring pusher velocity. The terminal implosion velocity of the glass shell can be calculated from pinhole pictures that show central spikes or rings offset along the laser-axis when a delay is introduced into one beam. It is assumed that the implosion velocities on the A and B sides are equal. Suppose the B beam is delayed by $T_D$ with respect to the A beam. Each portion of the wall requires an acceleration time $T_A$ to reach terminal velocity V, and the collapse time to the offset position is $T_C$. The deceleration time at peak compression is very small and can be neglected. The wall displacements are then given by

$$X_A = \int_0^{T_A} \frac{dX}{dt}\, dt + V(T_C - T_A) \tag{10}$$

$$X_B = \int_0^{T_A} \frac{dX}{dt}\, dt + V(T_C - T_A - T_D) \tag{11}$$

provided

$$T_C > T_A + T_D \quad . \tag{12}$$

Subtracting (11) from (10) yields

$$X_A - X_B = VT_D \tag{13}$$

so

$$V = \frac{X_A - X_B}{T_D} \quad . \tag{14}$$

Since $X_A$, $X_B$, and $T_D$ are measured, V is determined. Equation (14) can be rewritten in terms of the offset σ by noting that $\sigma = (X_A + X_B + 2r_o)/2 - X_B - r_o$ as shown in Figure 3. It follows that $X_A - X_B = 2\sigma$. Therefore,

$$V = \frac{2\sigma}{T_D} \quad . \tag{15}$$

TABLE II

| Shot | Target OD/Wall (μm) | $E_{LA}/E_{LB}$ (J) | Delay $T_D$ (ps) | $X_A$ (μm) | $X_B$ (μm) | $r_o$ (μm) | $V_{imp}$ (cm/sec) |
|---|---|---|---|---|---|---|---|
| 2369 | 144/1.2 | 41.3/40.5 | 58-B | 78 | 52.7 | 10.7 | $4.5 \times 10^7$ |
| 2370 | 146/1.5 | 26.9/30.4 | 58-A | 51.3 | 84 | 12.7 | $5.7 \times 10^7$ |
| 2368 | 142/1.5 | 40.5/34.4 | 117-B | 87.3 | 41.3 | 13.4 | $3.9 \times 10^7$ |
| 2367 | 143/1.3 | 37.6/27.4 | 176-B | 93.3 | 41.3 | 13.4 | $3.0 \times 10^7$ |
| 2379 | 110/1.2 | 38.3/37.6 | 58-B | 61 | 42 | 6 | $3.2 \times 10^7$ |
| 2380 | 109/1.1 | 40.5/40.9 | 0 | 43 | 49 | 9 | --- |
| 2360 | 124/1.3 | 31.1/31.5 | 0 | 62 | 56 | 5.7 | --- |
| 2534 | 151/1.5 | 37.6/43.4 | 0 | -- | -- | 6.0 | --- |

The subscripts A and B denote the two laser channels. $E_L$ is laser energy. $T_D$ is the time delay between channels (and the delayed channel). X is the wall displacement and $r_o$ is the radius of the inner ring in the pinhole picture. $V_{imp}$ is the implosion velocity.

The implosion velocities computed in this way varied from $3 \times 10^7$ cm/sec to $5.7 \times 10^7$ cm/sec for the time-delayed shots.

The illumination pattern on the pellet when the paraxial focus coincides with the pellet conter provides illumination with each beam over a solid angle subtending $144^\circ$ about the laser axis. There is approximately a 40% increase in the intensity for the marginal rays over the paraxial rays. The remainder of the pellet is not illuminated, and thus thermal conduction is required to obtain good implosions. This effect becomes important for the larger pellets where peripheral smoothing by thermal conductivity becomes less effective.

In another series of target shots the paraxial focal position was changed to study the effect on implosion symmetry and neutron yield. The procedure was first to align the pellet in the usual manner so that both paraxial focal positions were located in the center of the pellet. The paraxial focal positions were then shifted by moving both lenses a pre-determined amount. These shifts in the focal position change the amount of surface illuminated by the laser and change the angle of incidence of the laser beam at the pellet surface. These shifts also compensate for possible differences between the paraxial focal position of the low power (YAG) and the focal position at full power. Results obtained with these configurations are shown in Table III and corresponding X-ray photographs are shown in Figure 4. The X-ray pinhole pictures of shots 2442 (99-μm-diameter pellet) and 2460 (100-μm-diameter target) illustrate the improvement in symmetry for 100-μm targets. Shot 2442 was aligned by the conventional (low power YAG) technique putting the paraxial focal position of each beam at the pellet center. For shot 2460 both $80^\circ$ lenses were moved in for a paraxial focal shift of approximately 10 μm measured from the center of the pellet. The X-radiation symmetry is greatly improved. When the $80^\circ$ lenses are moved enough to shift each paraxial focal position approximately 20 μm, measured from the center of the pellet, an asymmetry appears in the central emission region as observed on shot 2456. A similar set of experiments was conducted using 145-μm-diameter targets. Shot 2540 was aligned with no paraxial focal shift from the center of the pellet and shot 2541 was aligned with a displacement of 20 μ for each of the two paraxial foci from the center of the pellet. On the basis of this limited set of experiments, it has been demonstrated that by changing the illumination on the target it is possible to improve the symmetry of the implosion. Although there were some variations as a result of changes in the laser characteristics, the target shot in this series with the most symmetrical pinhole photograph had the largest yield of neutrons ($6.7 \times 10^6$).

TABLE III

| Shot | Target OD/Wall (μm) | $E_{LA}/E_{LB}$ (J) | Pulselength (ps) | Paraxial Focal Shift (μm) | $ID_v$ (μm) | $ID_H$ (μm) | Neutron Yield |
|---|---|---|---|---|---|---|---|
| 2442 | 99/0.9 | 19.6/16.8 | 164 | 0 | 40.0 | 31.3 | $3.7 \times 10^6$ |
| 2460 | 100/0.8 | 33.7/32.3 | 192 | 10 | 44.4 | 46.3 | $6.7 \times 10^6$ |
| 2456 | 100/0.9 | 38.0/33.5 | 190 | 20 | 30.0 | 36.7 | $1.6 \times 10^6$ |
| 2540 | 145/1.3 | 34.6/41.9 | 213 | 0 | 39.3 | 32.7 | $2.9 \times 10^6$ |
| 2541 | 145/1.5 | 48.5/60.8 | 308 | 20 | 38.7 | 43.3 | $6.0 \times 10^5$ |

The subscripts A and B denote the two laser channels. $E_L$ is laser energy. $ID_v$ is the vertical halfwidth of the implosion spike. $ID_H$ is the horizontal halfwidth of the implosion spike.

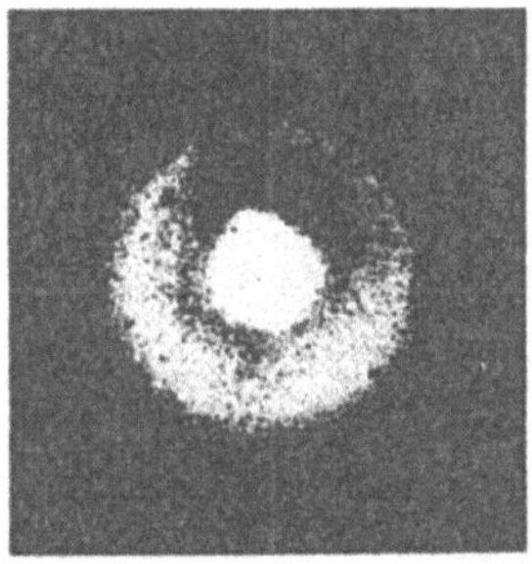

99 μm Diameter Target Shot 2442

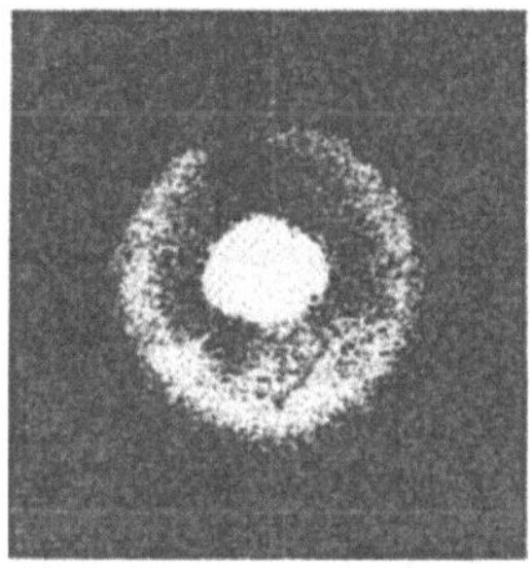

100 μm Diameter Target Shot 2460

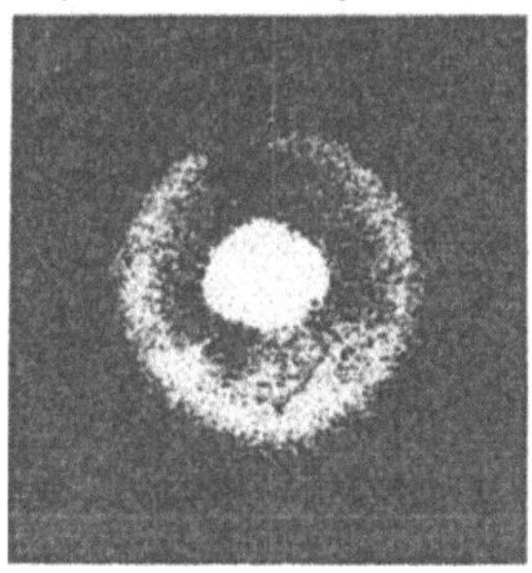

100 μm Diameter Target Shot 2456

Figure 4 X-Ray Pinhole Photographs Showing Changes in Illumination Pattern. A - Normal Focus. B - 10 μm Paraxial Focal Shift. C - 20 μm Paraxial Focal Shift.

The neutron yield was measured as a function of pellet diameter for these target shots in which the illumination was changed. The results of these measurements are shown in Figure 5 together with 50 μm target shots which are included for reference. As expected, the larger yields occured with those shots which exhibited the best implosion symmetry, as evidenced by the X-ray photographs, and for which laser power was a maximum. The solid line shows the dependence of the reciprocal mass on pellet diameter for these target shots. The aspect ratio was essentially 40:1 for these measurements. The results indicate that the neutron yield increases approximately as the inverse mass of the pellet.

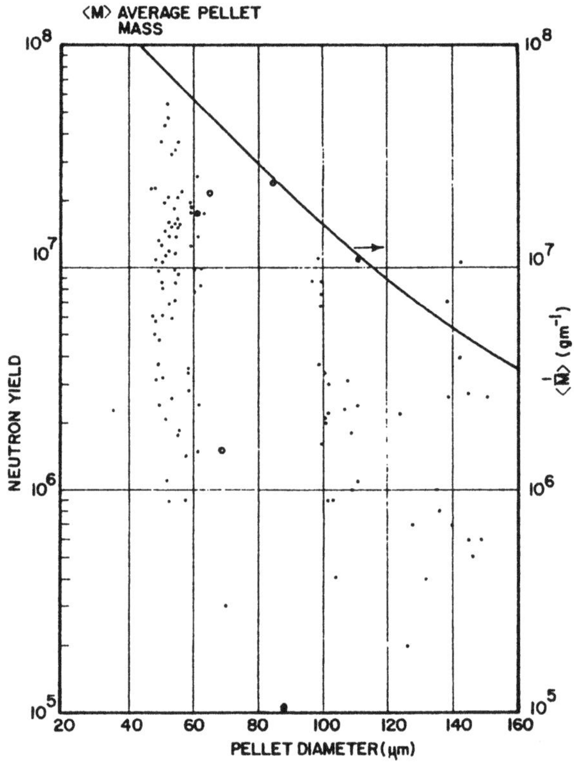


Figure 5 Neutron Yield and Reciprocal Mass as a Function of Target Diameter. • Glass Shell Pellets ° Polymeric Pellets Aspect Ratio 40:1.

## DISCUSSION OF IMPLOSION MEASUREMENTS

A simple similarity calculation of an adiabatic compression and expansion of the contained D-T gas shows that the neutron yield is consistent with a peak D-T gas temperature of between ½ and 2 keV[14]. The neutron yield calculated from this model is

$$N_{DT} = \lambda(1 - \lambda)p^2D^4C^{2/3}\ F_{DT}(\theta_o,C) \tag{16}$$

in terms of total gas fill pressure p, in atmospheres, compression C, the initial diameter D in hundreds of micrometers, $\theta_o$ in keV, and $\lambda = n_T/(n_D + N_T)$ the fraction of tritium molecules in the gas. The integral $F_{DT}(\theta,C)$ has been evaluated numerically for compressions larger than 10, where no dependence on C remains. The results of these calculations are shown in Figure 6. Also included in Figure 6 are the results of similar calculations assuming a

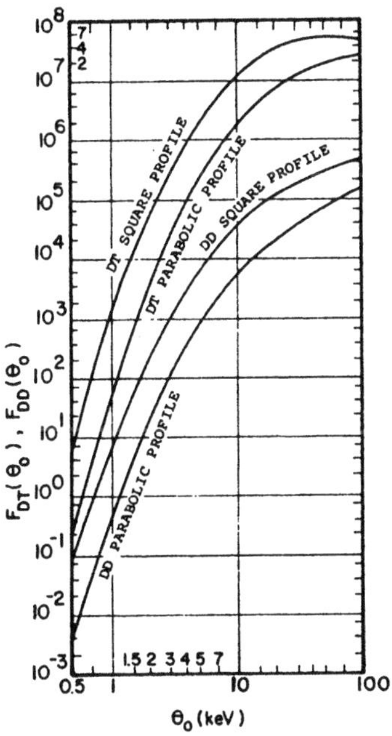


Figure 6 The Functions $F_{DT}$ and $F_{DD}$ as a Function of Peak Ion Temperature.

spatially uniform temperature profile, i.e., a square profile. This square profile is no longer a self-similar solution of the hydro equations, but may be usefully compared to the result for the parabolic profile obtained by solving the hydro equations. The result for D-D reactions is obtained by setting $\lambda(1-\lambda)$ equal to unity in Equation (16).

Measurements[12] of the tamper electron temperature at peak compression using foil filtration techniques on multiple-pinhole X-ray photographs gives electron temperatures in the ½ to 2-keV region.

It has been found that all target shots producing more than $10^6$ neutrons are associated with strongly compressed targets as determined from the X-ray pinhole photographs. There have been pellets with 10 atmospheres of D-T fuel which have compressed the fuel by a factor of 2000. Measurements of the electron temperature in the tamper using foil filtration techniques in the central region indicate a temperature range from 0.5 to 2.0 keV with the lower temperatures corresponding to larger diameter targets and higher gas pressures. The results of these experiments are shown in Figure 7. The higher temperatures are correlated with the higher

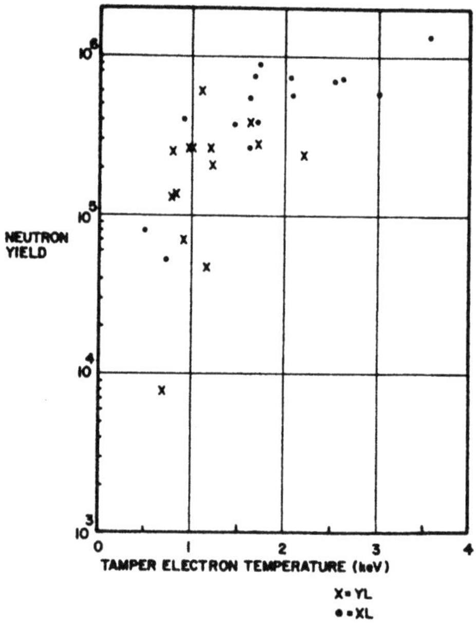


Figure 7 Neutron Yield as a Function of Tamper Electron Temperature. × $\frac{R}{\delta}$ = 40; 10 atmospheres D-T gas, • $\frac{R}{\delta}$ = 20; 10 atmospheres D-T gas, $\delta$ = wall thickness.

neutron yields which have been achieved using small diameter targets.

As shown in Table II, the parameters for shot 2360 are D = 124 μm, P = 10 atmospheres, $\lambda$ = 0.42, C = 1320 and $N_{DT} = 2.2 \times 10^6$. Solving for the ion temperature gives $\theta_o$ = 0.95 keV.

The present measurements can be used to estimate some core hydrodynamic quantities at peak compression. For shot 2534 in Table II, the central ring had a diameter of 12 μm corresponding to a compression of 1990 from which the peak average density $\sim 4.6\ g/cm^3$ (liquid D-T density $\sim .07\ g/cm^3$), peak pressure $\sim 5 \times 10^9$ atmospheres, internal energy $\sim$ 50 millijoules, and D-T fuel burn fraction $\sim 10^{-7}$ for the observed yield of $2.6 \times 10^6$ neutrons.

The present neutron yield levels are not sufficient to allow a measurement of the ion thermal broadening of the D-T neutron line[15]. With present technology it is estimated that yields of the order of $10^8$ neutrons are necessary to measure ion temperatures of 1 keV.

## CONCLUSION

These experiments have shown that it is possible to achieve high compression from spherical targets using high-energy pulsed laser systems. Compression of the fuel in the target has been measured to be approximately 20 times solid and neutron generation greater than $10^7$ was observed.

A marked increase in absorption and neutron yield was observed on going from the single pass to the circulating double pass laser configuration. The data on absorption by low- and high-Z targets obtained in this set of laser-target experiments does not permit an unambiguous determination as to whether the classical inverse bremsstrahlung mechanism is dominant and whether radiative preheat is reduced. In addition to a determination of the Z-dependence of the absorption, a concomitant measurement of the plasma density scale length is required in order to specify this currently unknown parameter. Greater insight into the nature of the absorption process can also be obtained by studying the variation of absorption with incident laser wavelength. The inverse bremsstrahlung model predicts that absorption varies approximately as $\lambda^{-2}$ where $\lambda$ is the wavelength. Hence, for fixed $ZL/\phi_0$, going from 1.06 μm to 0.53 μm illumination should give a significant increase in absorption.

## ACKNOWLEDGMENTS

The author would like to express his appreciation to the Experimental Group at KMSF and particularly to F.J. Mayer, R.R. Goforth, and G. Charatis in acquiring data and for many useful discussions. Thanks are also extended to P. Hammerling for his assistance with the analysis on the low-Z absorption measurements.

## REFERENCES

1. K.A. Brueckner, P.M. Campbell, and R.A. Grandey, Nuclear Fusion 15, 471 (1975).
2. G.S. Fraley and R.J. Mason, Phys. Rev. Lett. 35, 520 (1975).
3. C.E. Thomas and L. Siebert, Appl. Opt. 15, 462 (1976).
4. K.A. Brueckner, S. Jorna and K. Moncur, Appl. Opt. 13, 2183 (1974).
5. D.E. Solomon and T.M. Henderson, J. Physics D: Applied Physics 8, L85 (1975).
6. C.E. Thomas, Appl. Opt. 14, 1267 (1975).
7. G. Charatis, J. Downward, R. Goforth, B. Guscott, T. Henderson, S. Hildum, R. Johnson, K. Moncur, T. Leonard, F. Mayer, S. Segall, L. Siebert, D. Solomon, C. Thomas, Plasma Physics

and Controlled Nuclear Fusion Research (Proc. Int. Conf., Tokyo, 1974) IAEA, Vol. II, 317, Vienna (1975).
8. F.J. Mayer, G.R. Montry and E. Benn, 23rd Annual Conf. for Appl. of X-Ray Analysis, August 7-9, 1974, Denver, Colorado.
9. S.B. Segall, F.J. Mayer, R.R. Johnson and J.S. Hildum, Bull. Am. Phys. Soc. 19, 909 (1974).
10. R.R. Goforth, Rev. Sci. Int. (to be published).
11. R.J. Lanter and D.E. Bannerman, LASL Report No. LA-3498-MS, July 16, 1966; F.J. Mayer and H. Brysk, Nucl. Inst. Meth. 125, 323 (1975).
12. G. Charatis, R.R. Johnson and D. Sullivan, Bull. Am. Phys. Soc. 19, 909 (1974).
13. P.M. Campbell, G. Charatis and G.R. Montry, Phys. Rev. Lett. 34, 74 (1975).
14. F.J. Mayer and R.K. Osborn, KMSF Report No. TC-131 (1974); F.J. Mayer, R.R. Johnson and D.J. Tanner, KMSF Report FE-489 (1975).
15. G. Lehner and F. Pohl, Zeit. Für Physik 207, 83 (1969); H. Brysk, Plasma Phys. 15, 611 (1973).

This research was supported by the United States Energy Research and Development Administration under Contract E(11-1)-2709.

# COMPRESSION OF LASER-IRRADIATED HOLLOW MICROSPHERES

N. G. Basov, Yu. A. Zakharenkov, N. N. Zorev,
A. A. Koligrivov, O. N. Krokhin, A. A. Rupasov,
G. V. Sklizkov, A. S. Shikanov

Lebedev Physics Institute
Academy of Sciences of the USSR
Moscow, USSR

## INTRODUCTION

Recent experiments with solid and shell targets proved the existence of high compression of the substance in the center of the target[1-7]. The ultimate compression and temperature are specified by the mechanisms of absorption and conversion of high-power laser radiation energy to the plasma energy, by the dynamics and stability of compression, and by the processes occurring in a plasma corona[8-11].

In this work we discuss experimental results on the interaction of nanosecond laser radiation with hollow microspheres. The plasma-absorbed energy, spectral distribution, and the energy of plasma-scattered radiation were determined by various methods. The influence of initial perturbations on the microsphere compression stability was studied. By using pinhole cameras one succeeded in registering a 200-fold bulk compression of shells with uniform walls.

## EXPERIMENTAL SETUP

### Laser Arrangement and the Focusing System

A schematic representation of the laser arrangement "Calmar" and diagnostic intruments is shown in Fig. 1.

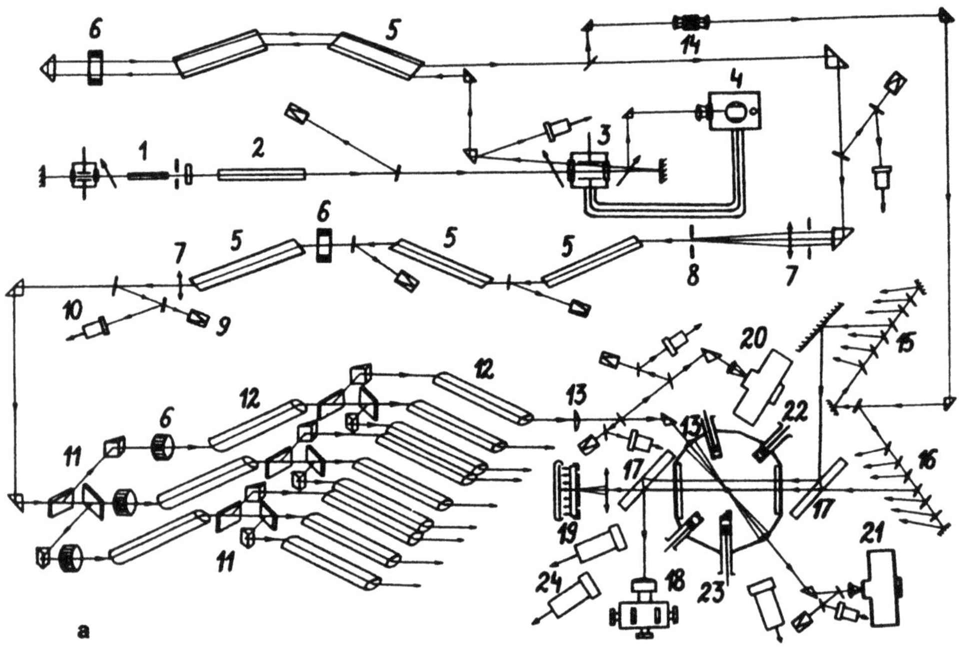


a

b

Fig. 1. Scheme of experimental setup (a) and microsphere photograph (b).
(1) Oscillator; (2) preamplifier; (3) Kerr cell; (4) laser-triggered spark gap; (5) two-passage amplifiers; (6) nonlinear absorbers; (7) lens; (8) diaphragm; (9) calorimeters; (10) photodiode; (11) splitting systems; (12) powerful amplifiers; (13) focusing systems; (14) KDP crystal; (15-16) optical delay systems; (17) Jamin interferometer; (18) interferometric camera; (19) multiframe schlieren system; (20-21) spectrographs; (22-23) multichannel X-ray film detector.

Unlike previous experiments[12] the YAG master oscillator was used in the nine-channel Nd-glass laser. As a result, the output radiation linewidth (FWHM) was reduced to ∿ 10Å ($\lambda_0$ = 10640 Å).

In the experiments described the laser energy reached 150 J at a pulse duration $\tau \approx 2.5$ nsec in the base and $\tau_f \lesssim 0.5$ nsec in the leading front. Since the energy consumed for the evaporation of the utilized microspheres was $10^{-3}$ J, the required value of contrast was about $10^6$.

At the output of amplifying cascades the laser radiation was focused onto the target by using nine two-lens systems (with effective focal lenght F = 20 cm). The degree of linear polarization of each beam in the vicinity of the target surface was not high, which is a result of the depolarization effect in the prisms of the beam-directing system. The optical paths for each beam were equalized in both the splitting and the beam-directing systems. As a result, the optical paths of the "fastest" and "slowest" beams differed by less than 2 cm, i.e., the lag time was less than $10^{-10}$s.

Because of Fresnel losses in the beam-directing system, and on parts of the diagnostic apparatus, the light energy in the target vicinity was E ∿ 100-120 J. The focal plane in each of the focusing systems was located somewhat farther from the target surface, and hence the beam diameter was d ≈ 400 microns in the target plane. The targets were solid microspheres of glass ($SiO_2$) or tin, or $SiO_2$, $Al_2O_3$ or polystyrene $(C_8H_8)_n$ hollow shells 100-350 microns in diameter, with a wall thickness of about 2-8 microns. The target was suspended on a rubber cement thread[13], the thickness of which reached 0.5 micron. The thread was tightened on a special holder. The beam spot dimension on the target remained unchanged regardless of the target diameter.

## Diagnostic Devices

The energy of the incident and plasma-scattered radiation was measured via a multichannel calorimeter connected with a computer having an output printer. In three of the nine beams there were calorimeters behind the targets, which made it possible to estimate the fraction of light energy transmitted through the target. The solid angle in which the transmitted energy was measured was greater by a factor of six than that of the radiation converging on the target. The shape of the incident, reflected, and transmitted pulses was determined by means of a coaxial photodiode and by a streak camera (not shown in Fig. 1) with time resolution better than $10^{-10}$ sec.

The spectral distribution of the reflected and plasma-scattered radiation emitted in various directions was investigated with spectrographs in the optical range. To study plasma luminosity in the X-ray spectral range ($\lambda \sim$ 2-10 Å), multichannel pinhole cameras were placed at different orientations relative to the plasma. The spatial resolution of these experiments was $\sim$ 20-25 microns at image magnifications from 1.1 to 2. Pinhole photographs were made with high-sensitivity X-ray photofilms of the UF-VR type.

The spectral distribution of X-ray quanta transmitted through beryllium and aluminum foils of different thicknesses was investigated by using detectors[14] in the region of soft quanta ($h\nu$ < 10 keV), and by means of scintillators and photomultipliers in the region of $h\nu$ < 10 keV.

The dynamics of plasma expansion during the first 20 nsec after beginning of the heating was investigated by a high-speed multiframe interferometric system. The formation of several spatially and temporally separated pulsed light sources was realized by a conventional scheme[1] of light delay using plane mirrors. The light beams produced by the light delay went, at small angles to each other, to a Jamin interferometer tuned for zero-order interference. In each shot the interference pictures corresponding to different moments were separated and recorded by a special multiframe camera, in the focal objective plane of which the prism splitter was mounted. The dynamics of shock waves propagating in the residual gas (hydrogen, deuterium, or air at pressures of $10^{-2}$ to 10 torr) was investigated by multiframe schlieren photography. It should be noted that the technique of multiframe schlieren photography has better spatial resolution than the shadow method[15] we have previously employed. As is known, the resolution of the shadow method is determined by the product of the refraction angle in plasma and the distance to the film, whereas in the schlieren method only the optical system aberration and diffraction at the optical diaphragm affect the resolution. Thus, the improved spatial resolution of this registration method results in a decrease of error in determining the shock wave energy, since

$E_{SW} \sim R^5$. This scheme had a spatial resolution of about 10 lines/mm.

Unlike Ref. 16, the holes in the screen at the focal plane of the schlieren objective (before the prism splitter) were used as optical diaphragms. By this scheme of registration the test radiation deviated by inhomogeneities was stopped by the diaphragm, and the dark zones of the image were the regions with a large gradient of refractive index. For the illumination of both high-speed optical diagnostic systems a fraction of the laser radiation, after the two-passage amplifier, was deflected to the nonlinear KDP crystal, thus doubling the laser radiation frequency. So the schlieren and interferometric images were recorded at the wavelength $\lambda = 0.532$ micron with the exposure time $\tau \gtrsim 10^{-9}$ sec.

The thickness and uniformity of the microsphere walls were measured with an interferometric technique allowing us to determine distortion of the radiation wave front penetrating the shell. The accuracy of measurement of the shell thickness was $\sim 0.1$ micron.

## EXPERIMENTAL RESULTS AND DISCUSSION

For obtaining the maximum temperature and substance compression it is important to determine the influence of thickness inhomogeneity and nonuniformity of hollow-microsphere irradiation on the dynamics of expansion and stability of pellet compression. By using pinhole cameras one could study plasma emissivity from hollow microspheres with various degrees of wall homogeneity. Figure 2a shows a pinhole photograph of a shell glass target with $R_o = 70$ micron. The wall of the shell was highly inhomogeneous; its thickness range was $\Delta R_o \sim 3$-7 microns. Figure 2b shows a pinhole photograph of a hollow microsphere of the same radius but of a higher homogeneity of the wall. In contrast to the experimental picture of Fig. 2a one can see the appearance of an external emitting region having a ring configuration. The mean thickness of the wall was $\Delta R_o \sim 3.5$ micron with a maximum deviation from this value amounting to 0.5 micron. Figure 2d shows a pinhole photograph and the microphotometer trace along a diameter for the shot with a glass target having $R_o = 70$ micron, $\Delta R_o = 4$ micron, and fluctuations of wall thickness of less than 0.3 micron. In this case the outer emitting plasma region acquires a definitely spherical shape with another emitting region in the center having a size of the order of

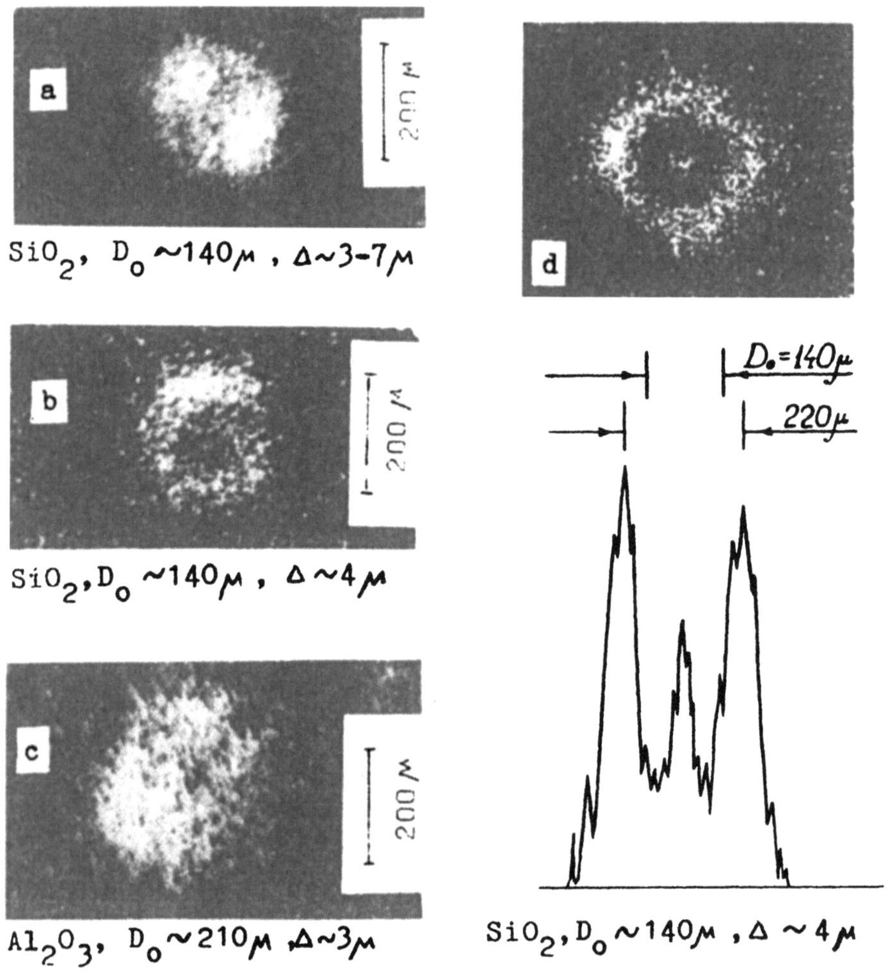


Fig. 2. X-ray pinhole plasma photographs of various targets (a,b,c); d) X-ray pinhole photograph and its microphotometer trace for $SiO_2$ shell with compression; Ø 140 μ, $\Delta R_o = 4$ μ.

the spatial resolution. The origin of these emitting regions seems to be quite different. The appearance of the outer emitting ring should be associated with the radiation of the expanding plasma corona, and it is only logical to assume that the strongest emission corresponds to the region near the critical density, where the electron temperature $T_e$ is maximum. Since the registration in pinhole photography is integral in time, the observed ring corresponds to the plasma region where the plasma emissivity is a maximum, or to the region with the minimum velocity of motion. Figure 3 illustrates a processed pinhole photograph similar to that of Fig. 2d, and shows the dependence of the volume plasma emissivity on the coordinate (the plasma was assumed optically thin). We equate volume emissivity with the number of quanta emitted from unit volume during registration and transmitted through the filters shielding the pinhole camera (in this case the Be foil thickness was 300 microns). The transition from the photographic density $S_z(x)$ to the volume emissivity of plasma $n_z(r)$ was realized by using the Abel transformation:

$$n_z(r) = -\frac{16b^2}{\pi\delta^2}\, j \int_r^{R_z} \frac{[10^{S_z(x)/\gamma}]'}{[x^2 - r^2]^{1/2}}\, dx ,$$

where the geometry and pinhole camera parameters were taken into account. Here $R_z$ equals the value of x at which $S_z(x) = 0.1$; b is the distance between the pinhole camera and film; $\delta$ is the diameter of the hole; j is the inertia of the film, and $\gamma$ is the film contrast factor determined in each shot. Note that in measuring the volume emissivity of the plasma central region the size of it was assumed equal to the spatial resolution

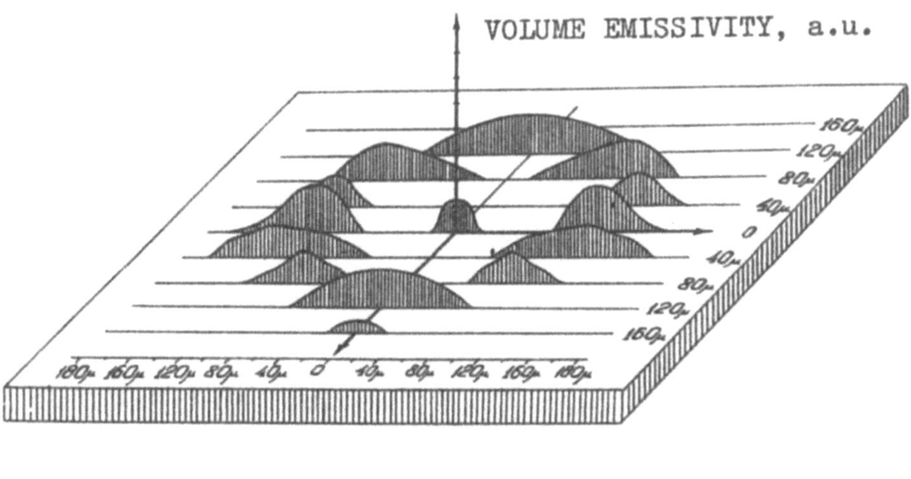


Fig. 3. Plasma volume emissivity distribution for a shot similar to Fig. 2d.

of the pinhole camera. As a result, in Fig. 3 only a minimum emissivity of the plasma central region is plotted, whereas the real value of $n_Z(0)$ seems to be considerably higher (radiation absorption from the center in the compressed and low-temperature shell was not taken into account). The emissivity distribution obtained differs from the experiments of Refs. 3 and 4, in which the diameter of the emitting corona was approximately the same as the initial diameter of the shell (the results of Refs. 3 and 4 are interpreted in Ref. 17). It should be noted that the results discussed in this paper were obtained at smaller flux densities than in Refs. 3 and 4. The increase in the diameter of the outer emission region is apparently caused by the fact that initially the critical-density plasma moves from the target surface just as in the case of "sharp" focusing of radiation onto a plane target[12,18]. Figure 4 shows a streak photograph of 5300 Å plasma radiation, i.e., the second harmonic of infrared radiation ($q \sim 5.10^{14}$ W/cm$^2$) emitted from the plasma region with an electron density of about $10^{21}$ cm$^{-3}$ (the critical density $N_c$ for Nd-laser radiation).

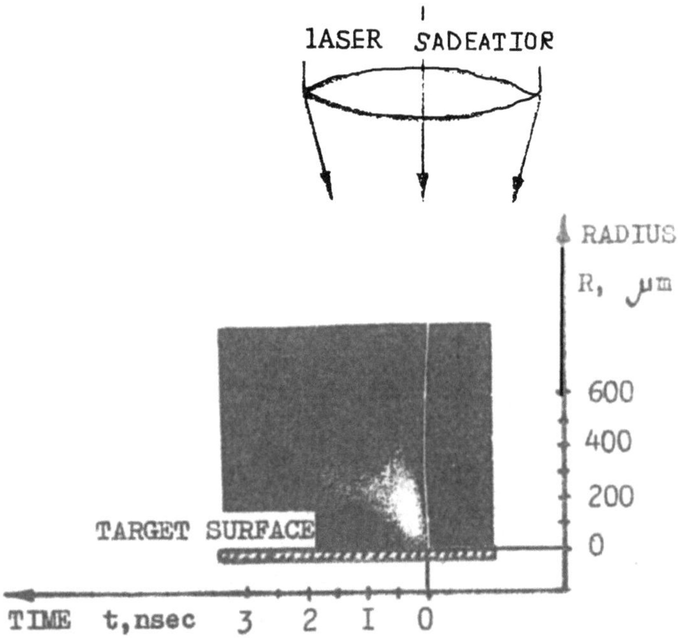


Fig. 4. Streak photograph of plasma radiation at the second harmonic radiation wavelength ($\lambda_0/2$ = 5300 Å).

It is seen that initially this region propagates from the target surface to a distance of ∿ 300 μ with high velocity. Another possible explanation is increased radiation absorption in the plasma region with one-fourth the critical density caused by the development of a parametric instability such as conversion into two plasmons[19]. As a result, the maximum region of plasma emissivity may be shifted toward smaller densities.

The appearance of the central emitting region is associated with the fact that due to deceleration, the unevaporated part of the shell target is compressed and heated up to high temperatures, which are determined by the kinetic energy of the imploding substance. The processes of heating and compression of the laser-irradiated glass microspheres were numerically calculated in Ref. 11. For the laser shot of Fig. 2d it follows that at the final stage of compression the target is a kernel with a compressed (by tenfold) and relatively cold periphery. The center of the target is highly heated ($T_e \sim 600$ eV) and is of smaller density. Note that since the lifetime of the compressed kernel is $\tau \sim 10^{-11}$ sec the parameter $n\tau$ amounts to about $5.10^{13}$. The volume reduction of the residual gas inside the target is about 200, which is negligibly less than the maximum estimated value[11]. The result obtained indicates the stability of a compressed shell with a homogeneous wall. The experiments with inhomogeneously thick shells (Fig. 2a, b) show no substantial compression. In the first case (Fig. 2a), when the initial perturbation is about ± 60%, the growth of perturbations leads, apparently, to a turbulent motion and smearing of the integrally recorded emissivity. In the case of inhomogeneities of the order of 10-15% of the initial thickness there appears an outer emitting region, whereas the central emitting kernel disappears due to the absence of appreciable compression. Figure 2b illustrates an extreme case with an imploding shell, when the wall of the shell is probably broken. In the case of $Al_2O_3$ shell targets having a sufficiently high homogeneity of wall thickness no compression was observed (Fig. 2c), which may be due to a characteristic microstructure of the $Al_2O_3$.

The electron "temperature" of the plasma was determined from the number of X-ray quanta transmitted through Be-filters of various thicknesses [20]. The results of measurements showed that for the solid glass microspheres 100-200 microns in diameter these ratios are in good agreement with those for the equilibrium plasma with $T_e \sim$ 250-350 eV. For the hollow microspheres of the same diameter the electron temperature was about 300 eV for 300- and 500-micron-thick filters, and $T_e \sim$ 600-700 eV for 300- and 700-micron-thick filters. The X-ray energy transmitted through the 300-micron-thick Be-filters ranged from $10^{-6}$ to $10^{-5}$ J (recalculated for 4π ster. solid angle) as shown by the experiments of Figs. 2b, d. To find the total X-ray energy one should determine accurately the plasma temperature, since there is a strong dependence of the filter transmission coefficient on the X-ray spectral distribution. Evaluations show that not more than $10^{-3}$ -$10^{-2}$ J is reradiated by a plasma in the X-ray range, and this value does not affect the energy balance.

The fraction of light energy absorbed by the microsphere was measured from the dynamics of the shock wave[1] propagating in the residual gas (hydrogen or deuterium). Figure 5 shows a seven-frame shock wave interferogram in deuterium (p = 10 torr) taken at the zero-order interference during the first 20 nsec of expansion.

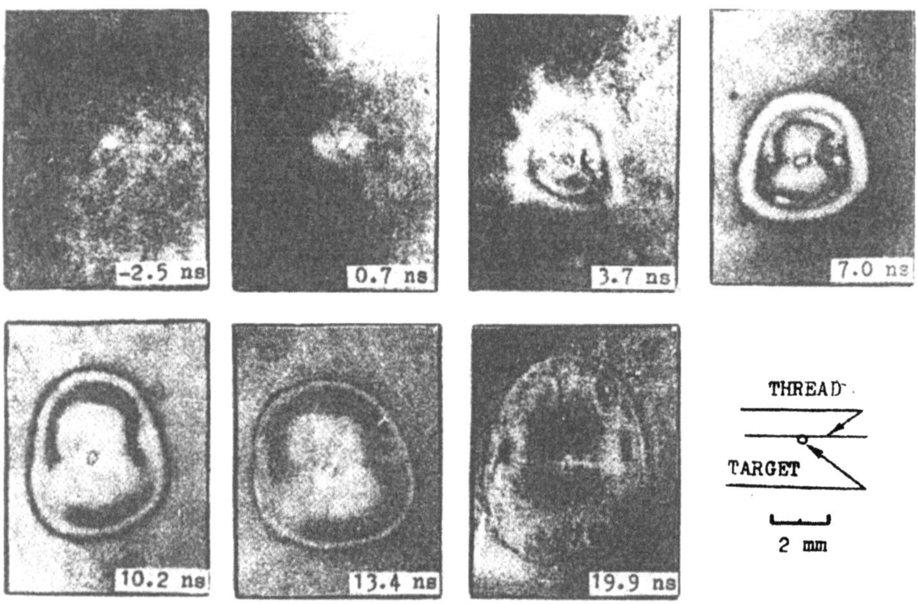


Fig. 5. Laser plasma interferograms corresponding to different times for a glass shell similar to that of Fig. 2a.

The target was a glass shell with an inhomogeneous wall. During the first 300 nsec the dynamics of the shock wave motion was studied by means of multiframe schlieren photography. Figure 6 shows a typical eight-frame schlieren photograph of the shock wave for a polystyrene hollow microsphere about 300 microns in diameter. Processing results of the interferometric and schlieren photographs show that the fraction of absorbed energy determined from the shock wave dynamics was 6% to 30-40% of the laser energy, depending on the size and type of microsphere.

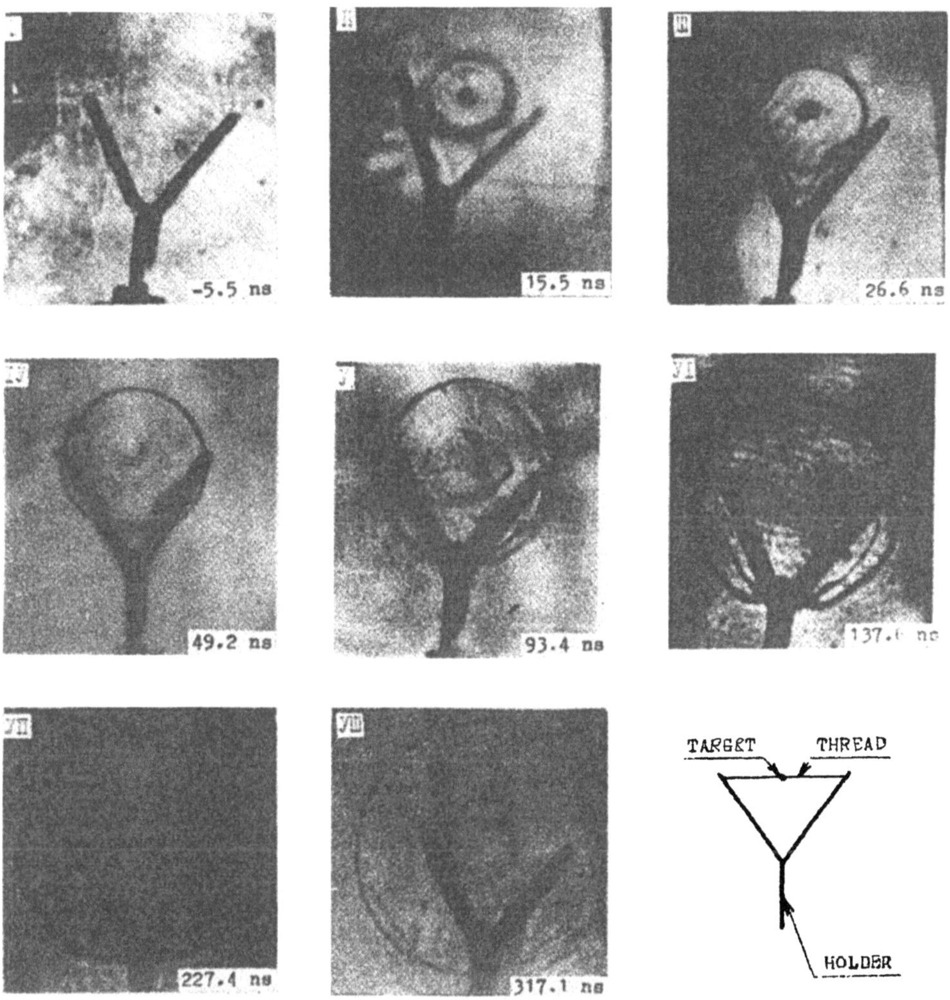


Fig. 6. Typical eight-frame schlieren photographs of the shock wave in the residual gas.; $(C_8H_8)_n$ shell, Ø 300 μ, $\Delta R_o = 5$ μ.

The target radiation was calorimetrically measured in three beams and gave a value for the "absorbed" energy 2-3 times greater than that determined by the shock wave. Thus, for instance, Fig. 2d shows a shot for which the absorbed energy measured by means of the shock wave was about 6 J, whereas the calorimetric measurements gave a value of 15 J.

Such a discrepancy may have two causes. First, perhaps the total light energy absorbed by the microsphere is not transformed into shock wave energy. The experimental results of Ref. 15 indicate that the speed of shock wave motion, and hence its energy, decreases with an increase in the atomic number of the target substance at the same absorbed energy.

The second reason is that the light beam normally incident on the spherical surface can be highly refracted in the regions of plasma corona with $N_e < N_c$. And since $nr \sin\theta = \text{const}$ (where n is the refractive index, θ is the angle of incidence of radiation onto the plasma layer) the beam deviation can be unlimitedly large because n varies from 1 in the vacuum to 0 in the region with critical density. As a result, the infrared radiation does not reach the critical-density regions where the absorption coefficient is maximum. The estimations of Refs. 21 and 22 show that the losses caused by refraction can be several tens of percent of the incident light energy, and depend strongly on the method of irradiating spherical targets and on the shape of the plasma corona density profile. At the present time the role of the parametric effect[23], which ensures laser absorption at large flux densities in the case of plane targets[19], is still not clear, since a part of the rays, due to refraction, do not reach the densities of $N_c$ and $\frac{1}{4}$ $N_c$ at which the absorbing instability develops. The results of the present experiment make it possible to assume that the greater value of the "absorbed" energy is obtained by using the calorimetric technique because a part of the rays does not go to the calorimeters located behind the target due to refraction. One may assume that the maximum absorbed energy can be obtained by means of calorimetric techniques, whereas the minimum value is attainable through optical methods. The plasma-reflected energy at the infrared radiation frequency was measured in the solid angles of the three-beam focusing systems using coaxial photodiodes and calorimeters. The reflection coefficient, i.e., the ratio of plasma-reflected energy to the energy of a given beam, varied in different observation channels by not more than 30-40%.

For the glass shell 140-microns in diameter this value was less than 0.05%, and it hardly depended on the shell thickness. Assuming that scattering is isotropic in the $4\pi$ ster. solid angle, the total reflection coefficient $R_{4\pi} \sim 2\%$ can be obtained. It should be, however, noted that by heating the spherical targets with a multichannel focusing system one cannot separate the scattered emission from the reflected rays of neighboring beams in the measured reflected radiation.

Figure 7 shows the spectrograms and their microphotometer traces for the incident radiation (a), and

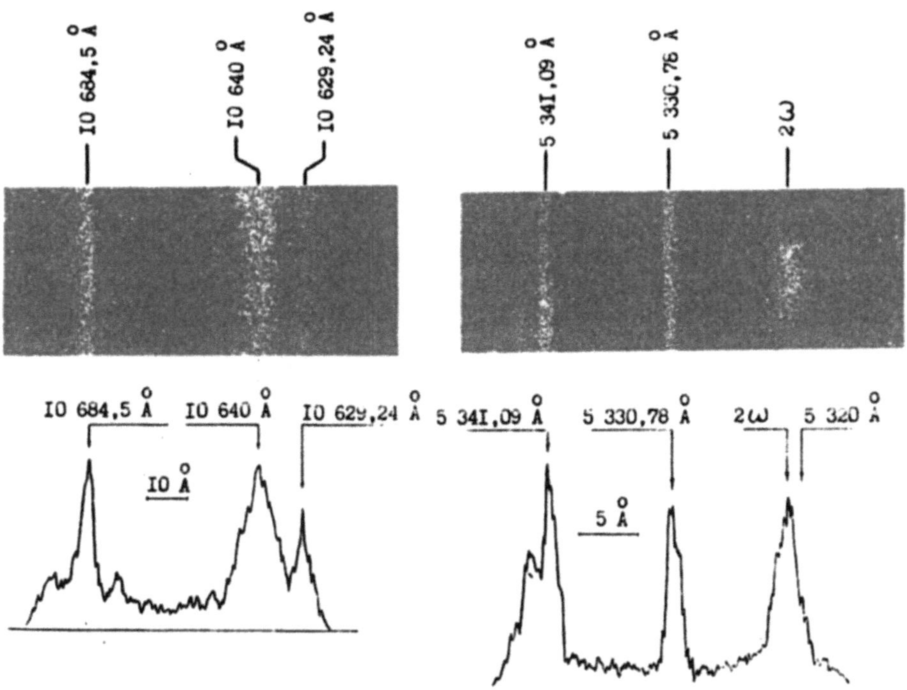


Fig. 7. Spectrograms and microphotograms: a) incident radiation; 10684.5 Å, 10629.24 Å - calibration lines; b) second harmonic radiation; $SiO_2$ shell, Ø 140 μ, $\Delta R_0 = 4$ μ; Ne 5341.09 Å, Ne 5330.78 Å - calibration lines.

the second harmonic radiation (b), which was observed at the solid angle of one of the focusing systems. The spectral maximum of the incident radiation corresponds to $\lambda_0 = 10640$ Å, and the spectral width (FWHM) is 10 Å. The spectral width (FWHM) of the second harmonic amounts to 5-6 Å but the maximum is shifted up to 1.7 Å toward longer wavelengths from the nominal value $\lambda_0/2 = 5320$ Å.

In Ref. 24 a method is proposed for the calculation of the electron temperature $T_e$ in the region of critical density by using the value of the second harmonic shift from the nominal value, provided the line is generated due to the decay instability. For the flux densities in question the relationship between the shift of the spectral maximum and the electron temperature is as follows ($\Delta\lambda$ is in Å, and $T_e$ in keV):

$$\Delta\lambda = (20\ Z/A\ T_e)^{1/2}.$$

For $\Delta\lambda \approx 1.5$ Å, $A = 20$, and $Z \approx 8$ we obtain $T_e \approx 280$ eV, which is in a good agreement with the X-ray data.

At flux densities $q \sim 10^{14}$ W/cm$^2$ (the case of sharper focusing but with similar spectral parameters for the incident radiation) the structure of the second harmonic becomes more complicated. A typical spectrogram of the second harmonic radiation and its microphotogram are shown in Fig. 8a. Asymmetric red broadening of the second harmonic up to 12 Å (FWHM) is accompanied by a shift in the maximum of the main component of up to 2.5 Å. Moreover, we observed the appearance of the red-shifted background with the spectral width up to 95 Å at the 0.1 $I_{max}$ level, which usually looks like an additional peak red-shifted by 25-30 Å from $\lambda_0/2 = 5320$ Å. The ratio of the additional peak intensity to the main peak of the second harmonic reaches 0.45.

We have also recorded the three-halves harmonic generation with flux densities $q \sim 10^{14}$ W/cm$^2$. Figure 8b shows the spectrogram and the microphotogram of the region near $(3/2)\omega_0$ obtained from the glass hollow microsphere irradiation and observed at the solid angle of one of the focusing systems. The radiation near $(3/2)\omega_0$ was comprised of two evident components, namely, the red component shifted from the nominal value $(2/3)\lambda_0 = 7093.3$ Å by 15-32 Å, and the blue one, normally of less intensity, which is shifted from the nominal value by 7-13 Å. The distance between these two compo-

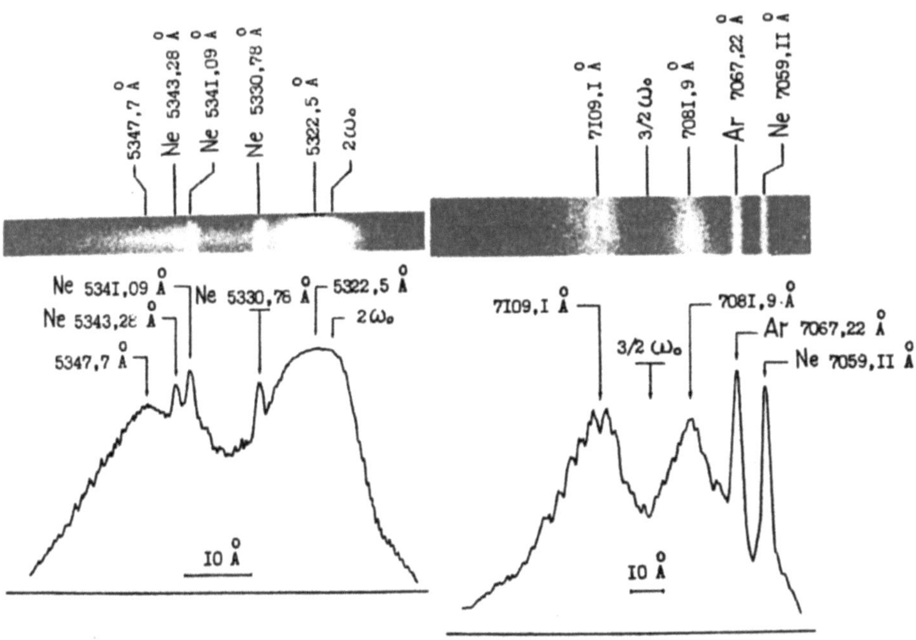


Fig. 8. Spectrograms and microphotograms; $SiO_2$ shell, Ø 83 μ, $\Delta R_o$ = 3.6 μ: a) second harmonic radiation; Ne 5343.28 Å, Ne 5341.09 Å, Ne 5330.78 Å-calibration lines; b) three-halves harmonic radiation; Ar 7067.22Å, Ne 7059.11 Å-calibration lines.

nents varied in different shots from 27 Å to 42.5 Å. The minimum between the peaks corresponds to the nominal value of $(3/2)\omega_o$ with an accuracy up to 2-3 Å. The ratio of the blue-component intensity to the red one varied in different shots from 0.35 to 0.95. The spectrogram represented in Fig. 8b shows that the red component spectral width (FWHM) is about 33 Å, and for the blue one it is about 24 Å. The total spectral width of three-halves harmonic generation (0.1 $I_{max}$) is 80 Å. Almost all the $(3/2)\omega_o$ spectrograms obtained reveal the modulations of the spectral intensity distribution with the distance between the peaks ~ 3.5 - 8.5 Å.

## ACKNOWLEDGMENTS

The authors are indebted to Yu. S. Leonov, F. I. Matveeva, Yu. A. Merkuliev, L. M. Ochkaeva, and

E. R. Rychkova for their help in the fabrication and selection of the targets, and to A. A. Erokhin, B. D. Makarov, and N. V. Novikov for assistance in the experiment.

REFERENCES

1. N. G. Basov, O. N. Krokhin, G. V. Sklizkov, S. I. Fedotov, A. S. Shikanov. JETP, 62, 203 (1972).
2. N. G. Basov, E. G. Gamaly, O. N. Krokhin, Yu. A. Mikhailov, G. V. Sklizkov, S. I. Fedotov. "Laser Interaction and Related Plasma Phenomena," Vol. 3, Plenum Press, New York (1974).
N. G. Basov, O. N. Krokhin, G. V. Sklizkov, Trudy FIAN, 76, 146 (1974).
3. P. M. Campbell, G. Charatis, G. R. Montry. Phys. Rev. Letters, 34, 74 (1975).
4. B. R. Guscott, G. Charatis, J. S. Hildum, R. R. Johnson, J. J. Mayer, N. K. Moncur, P. E. Solomon, C. E. Thomas. Preprint KMSF-U346, 7th European Conference on Controlled Fusion, Lausanne, Switzerland, Sept. 1-5.
5. G. H. McCall, R. L. Morse. Laser Focus, December, 1974, p. 40.
6. J. H. Nuckolls, J. L. Emmett, H. G. Ahlstrom, C. D. Hendricks, L. W. Coleman, J. A. Blaze, J. H. Holzrichter, D. G. Dahlbacka. Preprint UCRL-76957, 7th European Conference on Controlled Fusion, Lausanne, Switzerland, Sept. 1-5 (1975).
7. N. G. Basov, A. A. Kologrivov, O. N. Krokhin, A. A. Rupasov, G. V. Sklizkov, A. S. Shikanov. JETP Letters, 23, 474 (1976).
8. Yu. V. Alfanasiev, N. G. Basov, P. P. Volosevich, E. G. Gamaly, O. N. Krokhin, S. P. Kurdiumov, E. I. Levanov, V. B. Rozanov, A. A. Samarsky, A. N. Tikhonov. JETP Letters, 21, 150 (1975).
9. J. Nuckolls, J. Lindl, W. Mead, A. Thiessen, L. Wood, G. Zimmerman. 5th IAEA Conference on Plasma Physics and Controlled Nuclear Fusion Research, Tokyo, Japan, Nov. 11-15 (1974), paper IAEA-CN-33/F5-4.
10. J. S. Fraley, W. P. Gula, D. B. Henderson, R. L. McCrory, R. C. Malone, R. J. Mason, R. L. Morse. 5th IAEA Conference on Plasma Physics and Controlled Nuclear Fusion Research, Tokyo, Japan, Nov. 11-15 (1974), paper IAEA-CN-33/F5-5.
11. Yu. V. Afanasiev, P. P. Volosevich, E. G. Gamaly, O. N. Krokhin, S. P. Kurdiumov, E. I. Levanov, V. B. Rozanov. JETP Letters, 23, 470 (1976).

12. Yu. A. Zakharenkov, N. N. Zorev, O. N. Krokhin, Yu. A. Mikhailov, A. A. Rupasov, G. V. Sklizkov, A. S. Shikanov. JETP 70, 547 (1976).
13. E. G. Gamaly, A. I. Isakov, Yu. A. Merkuliev, A. I. Nikitenko, E. R. Rychkova, G. V. Sklizkov. Kvantovaya Elektronika, 2, 1043 (1975). E.R. Rychkova, Kvantovaya Elektronika, 2, 1048 (1975).
14. O. N. Krokhin, Yu. A. Mikhailov, V. V. Pustovalov, A. A. Rupasov, V. P. Silin, G. V. Sklizkov, A. S. Shikanov. JETP Letters, 20, 239 (1974).
15. O. N. Krokhin, G. V. Sklizkov, A. S. Shikanov. Trudy FIAN, 85, 143 (1975).
16. A. S. Shikanov, Yu. A. Zakharenkov. Proc. of the 11th Congress on High Speed Photography, Sept. 1974.
17. K. A. Brueckner, P. M. Campbell, R. A. Grandey. Nuclear Fusion, 15, 471 (1975).
18. O. N. Krokhin, Yu. A. Mikhailov, A. A. Rupasov, G. V. Sklizkov, A. S. Shikanov, Yu. A. Zakharenkov, N. N. Zorev. Proc. of the 12th International Conference on Phenomena in Ionized Gases, Eindhoven, The Netherlands, August 18-22, 1975.
19. N. G. Basov, O. N. Krokhin, V. V. Pustovalov, A. A. Rupasov, V. P. Silin, G. V. Sklizkov, V. T. Tikhonchuk, A. S. Shikanov. JETP, 67, 118 (1974).
20. F. C. Jahoda, E. M. Little, W. E. Quinn, G. A. Sawyer, T. F. Stratton. Phys. Rev., 119, 843 (1960).
21. K. Brueckner, R. Gover, J. Howard, F. J. Mayer, D. Mitrovich. Preprint KMSF-U176.
22. M. Lubin, E. Goldman, J. Soures, L. Goldman, W. Friedman, S. Letzring, J. Albritton, P. Koch, B. Yaakobi. Proc. of the Fuji Seminar on Laser Interaction with Plasma, p. 1 (1975).
23. V. P. Silin. Parametric Effect of High-Power Radiation on Plasma. Nauka Publishing House, Moscow, 1973 (in Russian).
24. O. N. Krokhin, V. V. Pustovalov, A. A. Rupasov, V. P. Silin, G. V. Sklizkov, A. N. Starodub, V. T. Tikhonchuk, A. S. Shikanov. JETP Letters, 22, 47 (1975).

# COLLECTIVE BEHAVIOR IN RECENT LASER-PLASMA EXPERIMENTS*

W. L. Kruer, R. A. Haas, W. C. Mead, D. W. Phillion, and V. C. Rupert

University of California, Lawrence Livermore Laboratory

Livermore, California 94550 U.S.A.

## ABSTRACT

The absorption of 1.06 μ light focused on small $C_8H_8$ discs has been measured in the intensity range of $10^{15}$ - $10^{17}$ W/cm . The data confirms the importance of collective plasma effects in determining the absorption of intense light. The measured absorption efficiencies are in the range of 30-40%, and the scattered light has a polarization dependence. The measured absorptions are shown to be inconsistent with classical inverse bremsstrahlung. Both the magnitude of the absorption and the observed polarization dependence of the scattered light are shown to be consistent with recent calculations of light absorption via collective processes. Comparisons are made, and improved models of the light absorption are discussed with reference to the data. In addition, the heated electron energies deduced from the X-ray data are consistent with those expected via collective processes.

## INTRODUCTION

The interaction of intense laser light with plasmas is a problem rich in nonlinear effects. For example, for Nd laser light with an intensity of $10^{16}$ W/cm$^2$, the energy of oscillation of electrons in the free-space electric field of the light is ∿ 4 keV, a value comparable to the thermal energy of the plasma.

*Research performed under the auspices of the U.S. Energy Research and Development Administration, Contract No. W-7405-Eng-48.

Under such conditions, classical inverse bremsstrahlung (collisional) absorption is quite weak, but many collective mechanisms are operative. Some of these processes enhance the absorption by coupling the incident light into plasma waves which are then damped. Such processes include resonance absorption,[1-4] parametric instabilities both near the critical density[5,6] and near one-fourth the critical density,[7,8] and quasi-resonant coupling on ion density fluctuations[9,10] driven up by subsidiary processes such as by two-stream instabilities in the plasma expansion. On the other hand, there is in general also stimulated scattering of the incident light in the plasma with density less than critical via the Raman and Brillouin instabilities.[11-13]

All these processes are operative in a highly inhomogeneous, expanding plasma and their evolution is strongly modified by the reaction of the turbulence back on the density and velocity gradients.[3,4,13] In addition, the absorbed energy fluxes can be far above those for which classical theory is valid, and the energy transport can be determined by heat-transport instabilities[14] or self-generated DC magnetic fields.[15-18] Although the understanding of the competition between these many effects is of course far from complete, it is encouraging that correlations with experiments are increasing. It is the purpose of this paper to discuss a number of these correlations by relating recent calculations with some recent experimental results. We will show that there is considerable evidence that collective plasma processes play an important role in determining the absorption of intense light.

## MEASUREMENTS OF LIGHT ABSORPTION

There have been many previously reported experiments[19-31] on the irradiation of targets of various types with intense laser light. In the experiments discussed here, small $C_8H_8$ discs (150μ diameter by 10μ thick) were irradiated with 1.06μ radiation using the Janus laser facility.[32] The energies on target ranged from 5-15 J in pulses with a full-width at half-intensity ranging from 60-150 ps. The light was focused to a target spot with diameters of approximately 90, 30 and 10μ. The resulting irradiation intensities ranged from $10^{15}$ - $10^{17}$ W/cm$^2$. Extensive diagnostics were taken, including measurements of optical energy balance, the angular distribution and time-dependence of the reflected light, the X-ray energy spectrum from ∿ 1-100 keV, the ion expansion energy, and the spatial distribution of the X-ray emission. Details of these experiments and numerical simulations of them using a 2-D fluid-dynamics code with multi-group electron transport (LASNEX)[33] are discussed in Ref. 17 and 18. We will here concentrate on the role of collective plasma effects on the light absorption, as evidenced by the data.

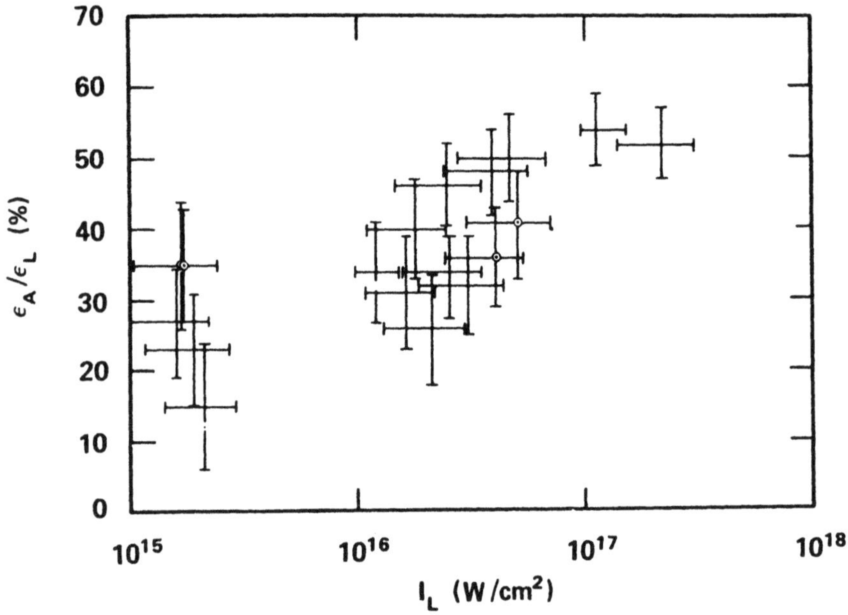


Fig. 1

The fraction of the incident laser light absorbed as a function of incident intensity $I_L$. The crosses denote measurements (with estimated error bars) determined by optical energy balance using an array of photodiodes distributed in angle around the target augmented by calorimeters to measure the light reflected and transmitted through the lenses. The points denote measurements in which the photodiodes are replaced by a box calorimeter enclosing the target.

Fig. 1 shows the measured absorption as a function of intensity. As indicated, on most shots the absorption was determined from measurements of the scattered light using $\sim$ 22 calibrated photo-diodes which were distributed in angle around the target (and calorimeters for the light scattered into the lens). The absorption measured by this technique ranges from .27 ± .09 at a nominal intensity of 2 x $10^{15}$ W/cm$^2$ to .41 ± .08 at 3-5 x $10^{16}$ W/cm$^2$. On a few shots, the photo-diodes were replaced with a calorimeter which enclosed the target. The absorption determined by this alternative technique ranges from .35 to .39 for the above intensities. The magnitudes of the absorption given by the different techniques are in reasonable agreement, but clearly the trend for the absorption to increase with intensity must be considered tentative; i.e. within the uncertainty of the measurements.

Fig. 2 shows the fraction of the incident light energy reflected back through the lens. This fraction decreases from ∿ 35% to ∿ 15% as the intensity increases in the range of $10^{15}$-$10^{17}$ W/cm². Fig. 3 shows some scattered light distributions averaged over the photodiodes in and out of the plane of polarization. Note that the scattered light peaks in the backwards direction ($\theta = 180^{\circ}$) and that very little light is refracting around the target. As the incident intensity increases, the scattered light distribution remains peaked in the backwards direction but becomes somewhat broader. Lastly, these measurements show that the scattered light has a polarization dependence. As shown in Fig. 4, photodiodes perpendicular to the plane of polarization collect on the average more reflected light than do photodiodes in the plane of polarization. The asymmetry ranges from a factor of ∿ 1.2 to a factor of ∿ 3 depending on scattering angle and intensity.

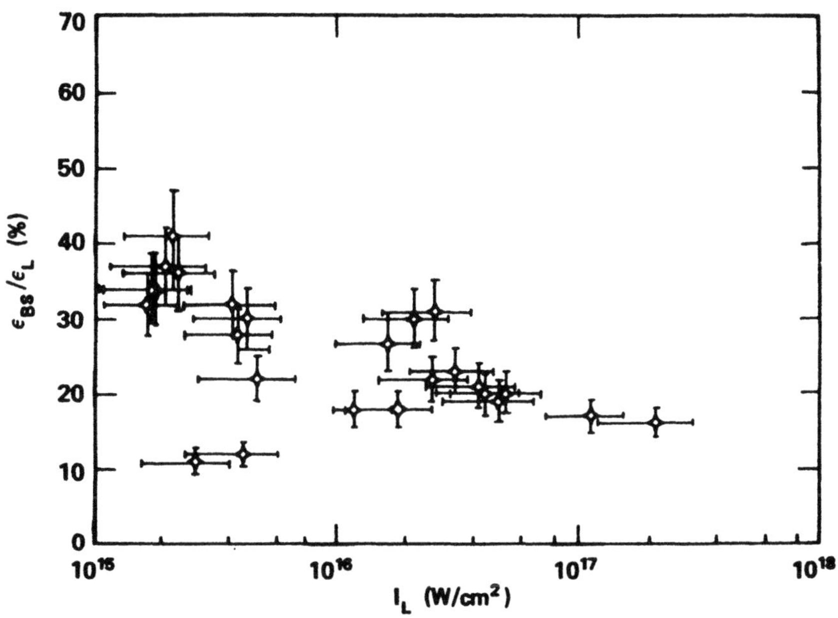


Fig. 2

The fraction of the incident light energy reflected through the focusing lens as a function of incident intensity. The measurements are made with a calorimeter, and the estimated error bars are shown.

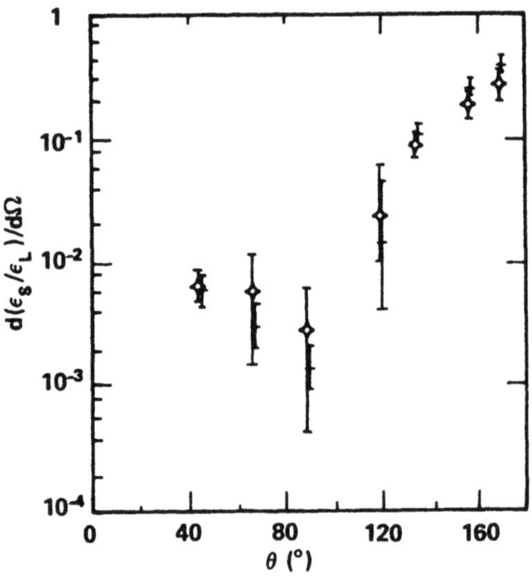


Fig. 3

The azimuthally averaged fraction of the incident laser light energy scattered per unit solid angle as a function of the scattering angle θ. Direct backscatter corresponds to θ = 180°. The circles denote the average value measured with photodiodes in experiments with a 30μ diameter target spot, and the crosses denote the average value in experiments with a 90μ diameter spot.

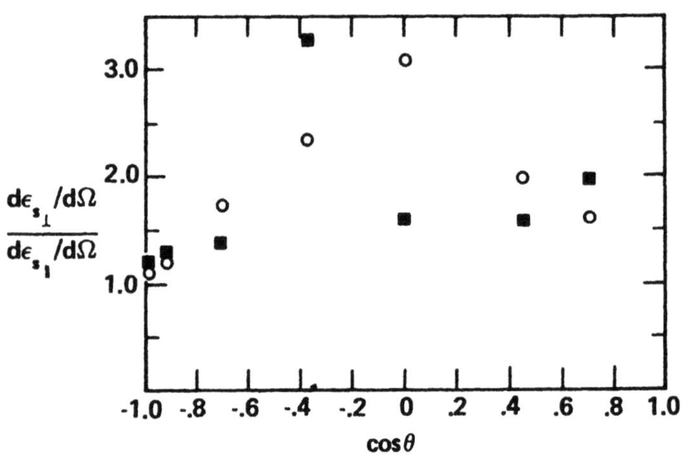


Fig. 4

The ratio of the fraction of the light energy scattered per unit solid angle perpendicular to the plane of polarization to the fraction scattered parallel to the plane as a function of the cosine of the scattering angle $\theta$. The circles denote averages over several experiments with intensities in the range $1.9 \pm .2 \times 10^{15}$ W/cm$^2$, and the squares denote averages over several experiments with intensities in the range $2.0 \pm .4 \times 10^{16}$ W/cm$^2$. The measurements are made with photodiodes.

As we will discuss, the measured absorption (and its polarization dependence) does not agree with inverse bremsstrahlung absorption, but is in semi-quantitative agreement with that expected via collective processes in the plasma. Let us first address the inverse bremsstrahlung and show that it is rather small for the higher intensities investigated. Then we will relate recent calculations[3] of light absorption via collective processes to these experiments.

## CLASSICAL INVERSE BREMSSTRAHLUNG

It is appropriate to first examine the classical collisional absorption. The absorbed intensity in a linear density profile is then[34]

$$I_{ABS} = I_{INC} \left[1 - \exp\left(- \frac{32}{15} k_o L \frac{\nu_{cr}}{\omega_o}\right)\right] .$$

Here $I_{INC}$ is the incident intensity, $k_o$ the free-space wave number, L the gradient length, and $\omega_o$ the light frequency. $\nu_{cr}$ is the electron-ion collision frequency evaluated at the critical density:

$$\nu_{cr} = 3.4 \; \langle Z \rangle \, \ell n \, \Lambda \, \frac{n_{cr} e^4}{m_e^{1/2} \theta_e^{3/2}} ,$$

where $n_{cr}$ is the critical density, $\theta_e$ the electron temperature, e the electron charge, $m_e$ the electron mass, <Z> the average charge state of the ions, and $\Lambda$ the usual ratio of the maximum to the minimum impact parameter. It is sufficient for these estimates to express the absorbed energy flux as

$$I_{ABS} = f \, n_{cr} \, \theta_e \, v_e ,$$

that is, as a factor f times the free-streaming value. Note that if we take f = 1, we are finding the minimum temperature consistent with a given flux of absorbed energy, and hence the maximum classical absorption.[35]

Putting these equations together gives $1 - \varepsilon = e^{-B/\varepsilon}$, where $\varepsilon$ is the absorption efficiency ($I_{ABS}/I_{INC}$) and

$$B = 2 \times 10^{10} \frac{\langle Z \rangle \, \ell n \Lambda \, f}{I_{INC}} \, k_o L \quad .$$

$I_{INC}$ is in W/cm$^2$. If $B \gg 1$, then $\varepsilon \simeq 1$. If $B \ll 1$, then $\varepsilon = \sqrt{B}$.

To illustrate the numbers, consider some parameters characteristic of these short-pulse experiments: $k_o L \simeq 10^2$, $\langle Z \rangle \simeq 3$, $\ell n \Lambda \simeq 5$ and $f = 1$ (to find the maximum inverse bremsstrahlung). Then

$$B = \frac{3 \times 10^{13} \text{ W/cm}^2}{I_{INC}} \quad .$$

For $I_{INC} = 10^{16}$ W/cm$^2$, $B \ll 1$ and $\varepsilon \simeq .06$. Note also that in this regime, $\varepsilon \propto I_{INC}^{-1/2}$.

As a matter of fact, there are a number of reasons why even this small value is an over-estimate of the inverse bremsstrahlung absorption. First, as already noted, the condition $f = 1$ assumes essentially the most efficient energy transport possible (and hence the lowest electron temperature consistent with the collisional absorption). As discussed in Ref. 17 and 18, the transport appears to occur at a rate much less than this free-streaming value; i.e. $f \ll 1$. Furthermore, we have neglected in this simple calculation several nonlinear effects which reduce the collisional absorption. These nonlinear effects are the intensity corrections to the collision frequency and the local steepening of the density profile near the critical density due to ponderomotive forces. Our principal point is that neither the magnitude nor the intensity dependence of the measured absorption is consistent with collisional absorption.

## COLLECTIVE ABSORPTION PROCESSES

On the other hand, the measured absorptions are consistent with those found in recent calculations of laser light absorption via the generation of electron plasma waves. To illustrate this point, we first refer to some two-dimensional simulations[3] of laser light absorption in inhomogeneous plasma which were carried out with a 2-D particle code. In these simulations uniform plane waves were propagated from vacuum into an inhomogeneous plasma slab, and the absorption was computed as a function of angle of incidence.

Some representative results are shown in Fig. 5, which is a plot of absorption efficiency versus angle of incidence. For these results the plasma slab had an initially linear density profile with scale length $k_0L = 10$ and an initial electron temperature of 4 keV. The incident light had an intensity of $10^{16}$W/cm$^2$(1.06 μ radiation) and was p-polarized; i.e., had its electric vector in the plane of incidence. Note that the fractional absorption computed in these short-term simulations (the points in Fig. 5) ranges from ≃ 15% at normal incidence to ≃ 60% at an angle of incidence of ≃ 20°.

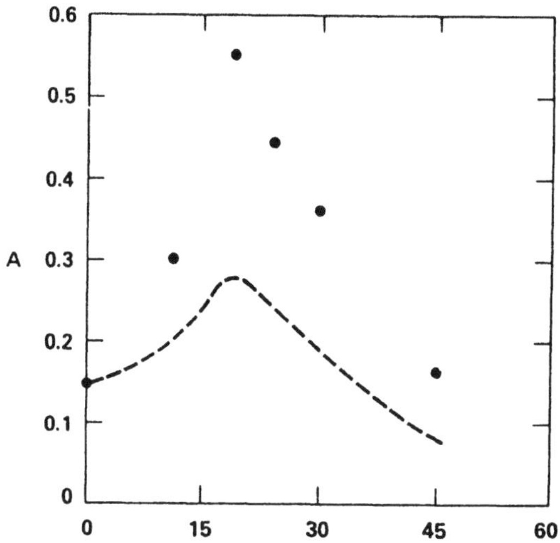


Fig. 5

The absorption efficiency as a function of angle of incidence θ. The points are the results of 2-D particle simulations[3] of p-polarized light incident on a plasma slab. The dashed line is an estimate, based on this data, of the net absorption efficiency for light incident with an equal mix of polarizations. The simulation results are for a plane wave with an intensity of $10^{16}$ W/cm$^2$ (1.06μ light) incident on a plasma slab with an initial density profile varying linearly from 0 to 1.7 $n_{cr}$ in a distance of 3 vacuum wavelengths. These results should only be considered as representative.

These absorptions are consistent with those expected in the nonlinearly steepened density profiles which are self-consistently generated in the simulations by strong ponderomotive forces near the critical density. An important effect of this strong local steepening is to reduce the efficiency of parametric instabilities and make resonance absorption effective over a rather broad range of angles. A few estimates will suffice to make these points clearer. The absorption efficiency due to parametric instabilities near the critical density in the steep profile is $\varepsilon_{pi} \sim 2\nu^*/\omega_0 \, k_0 \ell$, where $\nu^*$ is the effective damping rate and $\ell$ is the size of the region of plasma accessible to the instability excitation. At high intensities[6] $\nu^* \sim .05\ \omega_0$, and $\ell$ is $\sim$ few $c/\omega_0$ in the nonlinearly steepened profile. Hence instability generation in the steep profile is expected to provide at most a rather modest absorption of order 10-20%, consistent with the simulations.

The absorption efficiency due to resonance absorption of p-polarized light can be estimated by computing the energy damped by the resonantly-excited electron plasma waves. For a locally linear profile, the result is $\varepsilon_{ra} \sim |\phi(\tau)|^2/2$. Here $\tau = (k_0 L)^{1/3} \sin\theta$, $\theta$ is the angle of incidence, and $\phi(\tau)$ is the well-known resonance function which has a maximum value of $\simeq 1$ at $\tau = .8$. Of course, the details of both the resonance absorption and the instability generation depend on the actual shape of the nonlinearly steepened profile, but the qualitative features of the calculated absorption are consistent with these simple estimates.

In the simulations, the incident light is p-polarized. In experiments, generally only part of the incident light ( $\sim$ 1/2) has this polarization. The dashed line in Fig. 5 is our estimate, based on this simulation data, of the average absorption which results when one assumes an equal mix of polarizations for the incident light. Hence this estimate takes into account the fact that resonance absorption is only operative on one polarization. The resulting prediction is that the absorption via plasma wave generation is (1) polarization-dependent and (2) has a net efficiency of $\sim$ 15-30% as a function of angle of incidence.

Both these results compare favorably with the data. Since the light was focused with an f/1 lens, the angle of incidence averaged over energy on a plane target is $\simeq 15^0$. We then estimate from Fig. 5 about 25% net absorption due to plasma wave generation. Since p-polarized light is absorbed more efficiently than is s-polarized light (the electric vector out of the plane of incidence), we also estimate that diodes out of the plane of polarization collect $\sim$ 1.5 times as much reflected (nonabsorbed) light as do diodes in the perpendicular plane. Note from Fig. 4 that the observed asymmetry is not large (1.1-1.5) for most of the scattered light, which is reflected into angles within $\sim$ 30-$40^0$ of the direction of incidence. The principal point of these comparisons is simply that the observations are consistent with light absorption via collective mechanisms.

## IMPROVED MODELS OF THE LIGHT ABSORPTION

Although these correlations are sufficient to establish the importance of collective mechanisms, there are still some considerable uncertainties in our quantitative understanding of light absorption via such processes. Short-term simulations of the simple model discussed above are too ideal to predict in detail the magnitude of the absorption and its dependences on angle of incidence and intensity. The magnitude of the measured absorption ranges up to nearly twice as large at that estimated on the basis of Fig. 5. The angular spread of the scattered light is larger than the spread due to reflection from a plane critical surface, and the polarization asymmetry of this scattered light is sometimes as large as a factor of 3 for large scattering angles. Time-dependences in the back-reflected light are observed in the higher intensity experiments ($I \gtrsim 10^{16}$ W/cm$^2$). Finally, the absorption efficiencies measured in some more recent experiments using a f/2.5 lens are approximately the same as those measured with the f/1 lens. This suggests a weaker dependence of the absorption on angle of incidence than is apparent from Fig. 5.

Improved models of the light absorption must include a number of additional effects not included in the simple model. Many of these effects both enhance the absorption and in particular reduce its sensitivity to angle of incidence. First, if the transport is severely limited, the plasma near the critical density surface can be at an extremely high temperature. This would reduce the region over which the profile is strongly steepened and allow for more instability-generated absorption. In addition, the model discussed above treated a uniform plane wave incident on an infinite plasma slab. In practice, the spatial intensity profile of a light beam can be (and is) strongly modulated on spatial scales of 10-100μ. Such intensity modulations can crater the critical density surface either directly via variations in the ponderomotive pressure or indirectly by generating $\nabla n \times \nabla \theta$ magnetic fields which inhibit the transport and modify the fluid dynamics. Recent calculations show that a cratered critical surface can also be produced in other ways, for example, by a pinch-type instability[36] of light which is initially self-trapped in channels along the surface. Clearly a strongly rippled surface both reduces the sensitivity of the absorption to angles of incidence and broadens the angular distribution of the scattered light.

In addition to these relatively long wavelength density variations (compared to $\lambda_0$, the laser light wavelength), there are also additional sources of short wavelength ion turbulence. For example, electron-ion instabilities induced by heat transport to higher densities[14] and/or ion-ion instabilities driven by the expansion of different ion species can produce ion density fluctuations in the underdense plasma with a wavelength of $\sim 5 - 10\ \lambda_{De}$ (electron Debye lengths). These fluctuations enhance the absorption by coupling the light into electron plasma waves. The physics of the coupling is similar to that in absorption via parametric instabilities near the critical density, but now the ion fluctuations are generated by subsidiary processes rather than directly by the electric vector of the light. One important feature is that this mechanism operates over a broader range of densities and is less sensitive to a local profile steepening than are parametric instabilities.

It is instructive to give some simple estimates[10] of this absorption for these experiments. The effective collision frequency due to the ion turbulence is

$$\frac{\nu_{eff}}{\omega_o} \simeq 1/4 \left|\frac{\Delta n}{n_p}\right|^2 \frac{n_{cr}}{n_p} \left\langle \mathrm{Im} \frac{1}{\varepsilon(k,\omega_o)} \right\rangle$$

where $\Delta n$ is the mean amplitude of the fluctuations which are assumed to be isotropic in angle, $n_p$ the local plasma density, $n_{cr}$ the critical density, $\omega_o$ the light frequency, and $\varepsilon$ the plasma dielectric function evaluated at $\omega_o$ and at the wave number k of the fluctuations. The brackets denote an average over a range of wave numbers. To illustrate the numbers, take a range of wave numbers uniformly distributed from $k\lambda_{De} = .5$ to $k\lambda_{De} = 1.$, use the well-known dielectric function for a plasma with a Maxwellian distribution of velocities, and estimate the mean fluctuation level as $\Delta n = .1\ n_p$. This latter estimate is based on the amplitude of an ion wave at which ion trapping onsets in a plasma with $\theta_e = 10\ \theta_i$ ($\theta_i$ is the ion temperature). Then near the critical density $\nu_{eff} \simeq 1/800\ \omega_o$. In a linear density profile with $k_oL = 10^2$, the estimated absorption is $\simeq 20\%$, which is competitive with the net absorption estimated from Fig. 5. It is also interesting that this mechanism produces relatively soft heated electron distributions. The largest uncertainty at present is determining when such turbulence is actually present in the underdense plasma.

It is appropriate to briefly mention a few other collective mechanisms for absorption. For example, self-generated magnetic fields can introduce resonance absorption of even normally incident light.[37] In this case, the oscillation velocity of electrons in the electric field of the incident light resonantly drives electrostatic waves near the critical density via the Lorentz force. Recent calculations indicate that this mechanism can introduce a modest amount of absorption ($\sim$ 10%) if the size of the self-generated magnetic fields is of order 5 MG. In addition, there is also the two-plasmon decay instability which operates near one-fourth the critical density. However, present calculations[8] indicate that this instability is also strongly limited by a local profile steepening. Clearly a quantitative understanding of the many different absorption mechanisms is far from complete.

## STIMULATED SCATTERING

More quantitative calculations of the absorption must also include an improved understanding of the role of stimulated scattering in the plasma with density less than critical. The ideal model referred to for the "zeroth-order" estimates included a rather small region of underdense plasma (of size $\lesssim$ 5 free-space laser light wavelengths) and so can underestimate this scattering. Several features of the data indicate that stimulated scattering, particularly via the Brillouin instability, is playing some role in these experiments. For example, stimulated side-scattering preferentially scatters light out of the plane of polarization and so also leads to a polarization dependence to the scattered light. Hence this scattering can account for the large asymmetry with respect to the plane of polarization observed in the light scattered at large angles. However, the light scattered into such angles is a small fraction of the incident energy.

In these experiments the intensity is increased by reducing the diameter of the target spot. Hence the size of the underdense plasma decreases as the intensity increases. For the experiments with $I \gtrsim 10^{16}$ W/cm$^2$, the underdense plasma is then rather small. For example, LASNEX calculations indicate that the density scale lengths are $\sim$ 10-30$\mu$.

Because of the small mass and heat capacity of the underdense plasma, the stimulated reflectivity of high intensity light can rapidly self-limit when account is taken of the momentum and energy deposition into the plasma via the electrostatic waves.[13] One very simple example of this effect is shown in Fig. 6. This is a plot of back-reflection via the Brillouin instability vs. time as computed with a 1-D hybrid code (fluid electrons and particle ions). In this example, 1.06$\mu$ radiation with an intensity of $10^{16}$ W/cm$^2$ is incident on a 13.5$\mu$ slab of initially uniform plasma with a density $n = 1/3\ n_{cr}$ and with an electron temperature of 1 keV. The net amount of plasma is motivated by LASNEX simulations of how much plasma is in the underdense region when the laser pulse reaches high intensity in the experiments with $I \sim 10^{16}$ W/cm$^2$. Note that the stimulated reflectivity turns off

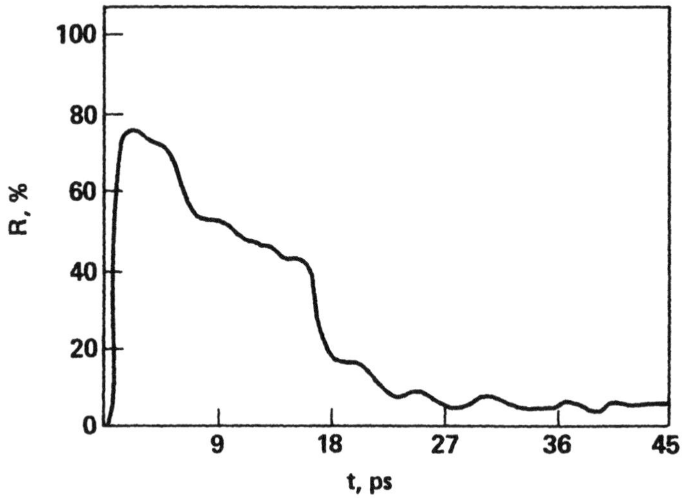


Fig. 6

An example of the back-reflection due to the Brillouin instability vs. time as computed with a 1-D code. Plasma conditions and light intensity are given in the text. The boundary conditions model contact with a reservoir of plasma at the initial temperature as detailed in Ref. 13. The time is scaled from an ion-electron mass ratio of 100 to a ratio of 3600 assuming a square root dependence. $\theta_e = 5\theta_i$ initially.

in a time of $\lesssim$ 20 ps. This is in agreement with estimates of how long it takes for the reflecting light to push the underdense plasma back, modifying its size and gradients enough to turn off the large reflection. Although this model is one-dimensional, clearly a similar behavior is expected for side-scattered light.

As shown in Fig. 7, the back reflection in the experiments with $I \geq 10^{16}$ W/cm$^2$ is time-dependent[21] (at least in most of the shots). Note that a sizeable reflectivity persists for $\sim$ 20 ps, consistent with the simple model. The largest uncertainty in these calculations is just how much relatively cold plasma ($\theta_e \sim 1$ keV) exists in the underdense plasma when the light pulse reaches a high intensity. This depends on both the transport and the form of the heated distributions, both of which are poorly known in detail.

Alternative explanations for a time-dependent reflectivity can be constructed. For example, Brillouin scatter can be small if the entire underdense plasma is at an extremely high temperature. The time-dependence could then be due to a break-up of the

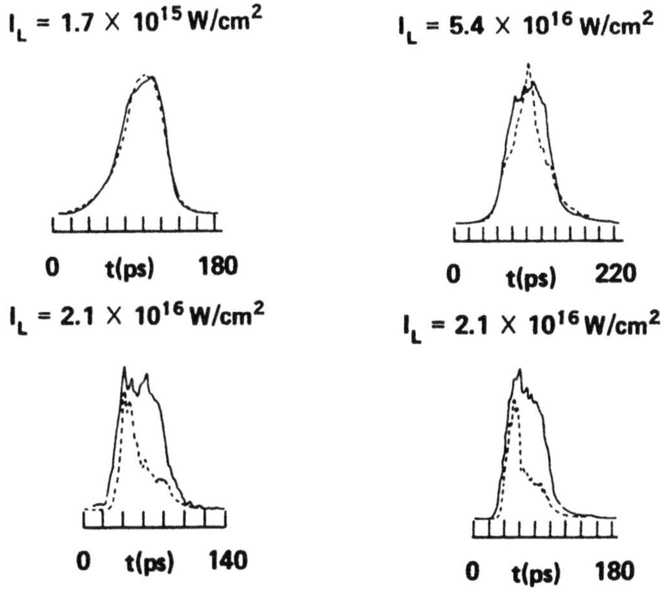


Fig. 7

The time dependence of the reflection back through the focusing lens measured in several experiments with different intensities. The solid line denotes the incident pulse and the dashed line denotes the reflected pulse. These pulses are not plotted on the same intensity scale.

critical surface on a hydrodynamic time scale, reflecting light into a larger solid angle. An improved understanding of the role of induced scattering awaits future experiments with more extensive regions of underdense plasma in which the temperatures and gradients are better characterized.

## HEATED ELECTRONS

We will conclude with a few comments on the heated electron energies. The measured X-ray spectra show a hard component. To calculate this spectrum, the LASNEX simulations[17,18] (and simple estimates) show that a large fraction of the absorbed energy must be put into rather hot electrons. For example, the hot electron energy employed in these calculations are $\simeq$ 10 keV for $I = 10^{15}$ W/cm$^2$ and $\simeq$ 40 keV for $I = 10^{16}$ W/cm$^2$. The 2-D particle simulations show that such energies are approximately what is expected from the collective absorption mechanisms we have discussed.

An estimate will help to clarify this point. From the cold wave breaking condition, the mean energy of the fast electrons produced by the resonantly-excited waves is [2-4]

$$\langle\varepsilon_e\rangle \underset{\sim}{\sim} 1/2\ mc^2\ (\frac{v_{os}}{c})\ \sqrt{k_o L}\ \ \phi(\tau)$$

where $v_{os}$ is the quiver velocity of an electron in the free-space value of the laser field and $\phi(\tau)$ is the resonance function previously mentioned. For $\phi(\tau) \sim 1$ and taking $k_o L \sim 2$ (locally steepened profile), $\langle\varepsilon_e\rangle \sim 350\ v_{os}/c$ (keV). For $I = 10^{15}$ W/cm², $\langle\varepsilon_e\rangle \sim 35$ keV. For $I = 10^{16}$ W/cm², $\langle\varepsilon_e\rangle \sim 12$ keV. Note in particular that these predictions would be much larger if one neglected the nonlinear profile steepening which is self-consistently included in the particle simulations.

Finally, we emphasize that relating the X-ray spectrum to heated distributions is rather indirect, depending not only on the heated electron spectra but also upon the space and energy dependent transport inhibition. LASNEX calculations show that different energy X-rays dominantly come from quite different regions of the plasma. Hence more detailed comparisons between calculations and experiment depends on an improved understanding of the heat transport (and the various mechanisms for its inhibition). This would enable, for example, the use of improved boundary conditions in the particle simulations. In view of such uncertainties it is premature to deduce details of the shape of the heated distributions from the data.

## SUMMARY

In conclusion, we have discussed some recent measurements of the absorption of 1.06μ radiation with focused intensities in the range $10^{15} - 10^{17}$ W/cm². The magnitude of the measured absorption efficiency is in the range of 30-40%, and the scattered light has an asymmetry with respect to the polarization vector of the incident light. The data was shown to be generally inconsistent with classical inverse bremsstrahlung. The magnitude of the absorption, the polarization dependence of the scattered light, and the heated electron energies were shown to compare favorably with recent calculations of light absorption via plasma wave generation. Improved models necessary for more quantitative comparisons were briefly discussed with reference to details of the data. Although the quantitative understanding of collective plasma effects on light absorption is at an early stage, correlations between the calculations and the data confirm that such processes play an important role.

## ACKNOWLEDGMENTS

We are grateful to J. Emmett, H. Ahlstrom, C. Hendricks, J. Holzrichter, J. Nuckolls, and R. Speck for encouragement and support and to M. Boyle, K. Estabrook, L. Koppel, H. Kornblum, B. Langdon, B. Lasinski, R. Lerche, J. Lindl, D. MacQuigg, K. Pettipiece, H. Shay, V. Slivinsky, K. Tirsell, E. Valeo and G. Zimmerman for valuable contributions and discussions.

## REFERENCES

1. V. L. Ginzburg, The Properties of Electromagnetic Waves in Plasmas (Pergamon, New York, 1964).

2. J. P. Freidberg, R. W. Mitchell, R. L. Morse and L. F. Rudsinski, Phys. Rev. Letters 28, 795 (1972).

3. K. G. Estabrook, E. J. Valeo and W. L. Kruer, Phys. Fluids 18, 1151 (1975); E. J. Valeo and W. L. Kruer, Phys. Rev. Letters 33, 750 (1974).

4. D. W. Forslund, J. M. Kindel, K. Lee, E. L. Lindman and R. L. Morse, Phys. Rev. A 11, 679 (1975); J. M. Kindel, K. Lee, and E. L. Lindman, Phys. Rev. Letters 34, 134 (1975).

5. D. F. Dubois and M. V. Goldman, Phys. Rev. 164, 207 (1967); V. P. Silin, Sov. Phys. JETP 21, 1127 (1965); K. Nishikawa, J. Phys. Soc. Japan 24, 916, 1152 (1968); P. K. Kaw and J. M. Dawson, Phys. Fluids 12, 2586 (1969).

6. W. L. Kruer and J. M. Dawson, Phys. Fluids 15, 446 (1972); J. S. DeGroot and J. I. Katz, ibid 16, 401 (1973).

7. E. A. Jackson, Phys. Rev. 153, 235 (1967); M. V. Goldman, Ann. Phys. 38, 117 (1966); W. L. Kruer and J. M. Dawson, Phys. Fl. 14, 1003 (1971).

8. A. B. Langdon and B. F. Lasinski, in Methods in Computational Physics, edited by J. Killen, B. Alder, S. Fernback and M. Rotenberg (Academic, New York, 1976) Vol. 16, (in press).

9. J. M. Dawson and C. Oberman, Phys. Fluids 5, 517 (1962) and 6, 394 (1963).

10. R. J. Faehl and W. L. Kruer, Phys. Fluids (to be published); W. L. Kruer, Phys. Fluids 15, 2423 (1972).

11. C. S. Liu, M. N. Rosenbluth and R. B. White, Phys. Fluids 17,

1211 (1974); and many references therein.

12. D. W. Forslund, J. M. Kindel and E. Lindman, Phys. Rev. Letters 30, 739 (1973).

13. W. L. Kruer, E. J. Valeo and K. G. Estabrook, Phys. Rev. Letters 35, 1076 (1975).

14. D. W. Forslund, J. Geophys. Res. 75, 17 (1970); R. J. Bickerton, Nuclear Fusion 13, 457 (1973); R. C. Malone, R. L. McCrory and R. L. Morse, Phys. Rev. Letters 34, 721 (1975).

15. J. A. Stamper, K. Papadopoulous, R. Sudan, S. Dean, E. McClean and J. Dawson, Phys. Rev. Letters 26, 1012 (1971); J. B. Chase, J. M. LeBlanc and J. R. Wilson; Phys. Fluids 16, 1142 (1973); J. J. Thomson, C. E. Max and K. G. Estabrook, Phys. Rev. Letters 35, 663 (1975).

16. J. A. Stamper and B. H. Ripin, Phys. Rev. Letters 34, 138 (1975).

17. W. Mead, R. Haas, W. Kruer, D. Phillion, H. Kornblum, J. Lindl, D. MacQuigg and V. Rupert, Lawrence Livermore Laboratory UCRL-78106 (1976), submitted to Phys. Rev. Letters.

18. R. Haas, W. Mead, W. Kruer, D. Phillion, H. Kornblum, J. Lindl, D. MacQuigg and V. Rupert, submitted to Phys. Fluids.

19. H. G. Ahlstrom, et al., in Plasma Physics and Controlled Nuclear Fusion Research, (International Atomic Energy Agency, Vienna, 1975) Vol II, p. 375.

20. G. Charatis, et al., ibid, p. 317.

21. M. Lubin, et al., ibid, p. 459.

22. C. Yamanaka, et al., ibid, p. 421.

23. D. Schirmann, et al., ibid, p. 449.

24. K. Eidmann, et al., ibid, p. 357.

25. M. Galanti, et al., ibid, p. 405.

26. C. Fabre, et al., ibid, p. 435.

27. N. G. Basov, et al., Sov. Phys. JETP 40, 61 (1975).

28. B. H. Ripin, et al., Phys. Rev. Letters 34, 1313 (1975).

29. J. F. Kephart, R. P. Godwin and G. H. McCall, Appl. Phys. Letters 25, 108 (1974).

30. T. P. Donaldson and I. J. Spalding, Phys. Rev. Letters 36, 467 (1976).

31. R. A. Haas, M. J. Boyle, K. R. Manes and J. E. Swain, J. Appl. Phys. 47, 1318 (1976).

32. J. F. Holzrichter and D. R. Speck, J. Appl. Phys., (to be published).

33. G. B. Zimmerman, Lawrence Livermore Laboratory UCRL-74811 (1973).

34. J. Dawson, P. Kaw and B. Green, Phys. Fluids 12, 875 (1969).

35. W. L. Kruer, in Progress in Lasers and Laser Fusion, edited by B. Kursunoglu, A. Perlmutter and S. Widmayer (Plenum Press, New York, 1975) p. 5-26.

36. E. J. Valeo and K. G. Estabrook, Phys. Rev. Letters 34, 1008 (1975); K. G. Estabrook, Phys. Fluids, (in press).

37. W. L. Kruer and K. G. Estabrook, Lawrence Livermore Laboratory UCRL-77717 (1976).

# EXPERIMENTS ON PARAMETRIC INSTABILITIES IN LASER-PLASMA INTERACTIONS

Francis F. Chen

University of California

Los Angeles, California

Recent experiments in laser-fusion laboratories have indicated the importance of non-classical processes in the absorption of intense laser radiation by solid targets. For instance, backscattered light typically ranges from 10% to 50% of the incident intensity, larger than can be explained by classical reflection at the critical layer; and the excess has often been attributed to a parametric instability such as stimulated Brillouin or Raman scattering. Absorption of light--typically 40%--is lower than classical inverse bremsstrahlung in some experiments and higher in others. The parametric decay and OTS (oscillating two-stream) instabilities are supposed to occur near the critical layer to enhance the absorption. Even when classical processes are sufficient to explain the observed absorption, anomalous effects are indicated by the appearance of suprathermal electrons, as revealed by x-ray spectra. An even more serious effect is the generation of fast ions, which have in some instances accounted for as much as 80% of the absorbed energy. Acceleration of ions by electric fields generated in nonlinear processes has been conjectured. Finally, experiments on transmission through and fast ions from plastic foils have shown that heat conductivity is anomalously low, thus raising the possibility of large, self-generated magnetic fields.

Because of the small size and short time scale of the experiments, confirmation of theoretically predicted plasma phenomena must necessarily rely on indirect measurements. It would be well-nigh impossible, for instance, to verify the absorption of light via the parametric decay instability by detecting the ion acoustic and electron plasma waves which are generated in the corona of an imploding pellet. Instead, one normally uses a computer code to study the nonlinear development of a particular mechanism and to predict the macroscopic observables. In this example, the observables could include the net energy absorption and the non-thermal tail of the electron distribution (which, in turn, is measured indirectly from the x-ray spectrum). Needless to say, even if the measurements should agree with the computer simulations, the sequence of events would not be uniquely identified; in fact, even the existence of the initiating instability would not be proved.

Just as computer simulation has brought theory closer to experiment, experimental simulation in larger, less dense plasmas can bring the complex phenomena of laser-fusion closer to theory. A prime example of this is the work done with microwaves in large, tenuous plasmas. The experimental discovery by Kim, Stenzel, and Wong[1] of density depressions--"cavitons"--formed by self-trapped rf energy has led to computational predictions of solitons, profile modification, and ion acceleration. A second example is the verification of Brillouin backscattering in the $CO_2$-laser experiments reviewed below. Without proof that the linear effect really occurs in nature, computer extrapolations to nonlinear saturation and calculations of the resultant reflection coefficient would be on shaky ground indeed. Finally, we cite the recent investigation by Obenschain, Luhmann, and Greiling[2] of the effect of finite bandwidth pumps on parametric thresholds. In this case, theory preceded experiment; but by using microwaves, the authors were able to demonstrate control of parametric instabilities well before the technology was in hand to do so with lasers.

In toroidal-confinement fusion, the understanding of micro-instabilities in stellarators and tokamaks was impeded by the simultaneous occurrence of many phenomena in the complex magnetic geometry. Progress was made only after plasma sources such as Q-machines and multipoles were devised so that the plasma could be controlled to exhibit one effect at a time. The development of suitable plasma sources is, similarly, needed for experimental unraveling of complex laser-fusion phenomena. When laser light forms a plasma from a solid target, many unnecessary complications make it difficult to study the physics of the laser-plasma interaction. For instance, a slight amount of prepulse can change the density gradient scale length and greatly affect instability thresholds. The uniformity of illumination is often poor at high intensities. Most solids, particularly tampered glass shells, have different atomic masses of high-Z atoms, giving rise to an unknown mixture of charge states that can interact. Spherical targets have refraction effects and contain a continuum of angles of incidence; even flat targets give rise to hemispherical plasmas. Expansion of the plasma causes Doppler shifts which obscure the intrinsic frequency shifts in scattered light; differential expansion can greatly affect thresholds. The density profile, important in determination of thresholds, cannot be measured except in a crude way. Worst of all, the critical ($\omega_o = \omega_p$, $n = n_c$), quarter-critical ($\omega_o = 2\omega_p$, $n = n_c/4$), and underdense ($\omega_o > 2\omega_p$, $n < n_c/4$) regions are all present at once; and each has its own set of parametric instabilities. To remove this unnecessary complication, an obvious procedure is to employ lasers incident on gas or plasma targets which are everywhere underdense. Only a few such experiments have been done so far; these are reviewed in the next section. A relatively large body of literature exists for microwave simulations; a review of these would be outside the scope of this paper. For the reasons given above, we do not regard experiments on solid targets to be sufficiently controllable for detailed studies of laser-plasma interactions.

## SURVEY OF RELEVANT EXPERIMENTS

### Experiments on Backscattering in Underdense Plasmas

Anomalous reflection of light from plasmas can occur through the parametric decay of an incident photon ($\omega_o$, $\underline{k}_o$) into a red-shifted photon ($\omega_2$, $\underline{k}_2$) and either a plasmon (electron plasma wave) or a phonon (ion acoustic wave) ($\omega_1$, $\underline{k}_1$). Thresholds for these processes, called respectively stimulated Raman scattering (SRS) and stimulated Brillouin scattering (SBS), were calculated for homogeneous plasmas by Drake et al.[3]. These thresholds are easily exceeded, but in a series of papers Rosenbluth, Liu, and others[4] pointed out that plasma inhomogeneity, finite interaction length, and differential expansion impose considerably higher thresholds. In another series of papers, Forslund, Biskamp, Kruer, and others[5] give similar results and in addition investigated the nonlinear development of SBS and SRS through computer simulation.

Relevant results have been summarized by Chen[6]. In the $CO_2$ experiments discussed below, the intensity of $\simeq 10^{11}$ W/cm$^2$ far exceeds the homogeneous thresholds for both SRS and SBS. However, the instability is convective, and the relevant threshold is set by the finite length $\ell$ of the interaction region--either the plasma length or the depth of focus of the laser light. The instability grows exponentially at the rate $\gamma_o$, but the number of e-foldings N above thermal noise is limited by the rate of energy loss of the decay waves by convection or damping. The threshold is set by the value of N required for detectability. It turns out that SRS has a threshold of order $c^2/v_e^2$ larger than SBS ($v_e$ being the electron thermal speed), amounting to about $10^{13}$ W/cm$^2$ for the experiments in question, and is therefore expected to be absent. Brillouin scattering has the threshold

$$N = \frac{2\gamma_o^2 \ell}{c\gamma_i} >> 1 \,, \qquad \text{where} \qquad \gamma_o = \frac{1}{2}\,\frac{v_o}{c}\left(\frac{\omega}{\omega_i}\right)^{\frac{1}{2}} \omega_{pi} \,, \tag{1}$$

$\gamma_i$ being the ion wave damping rate and $v_o = eE_o/m\omega_o$ the electron quiver velocity in the field $E_o$ of the incident wave. According to numerical computations[5], SBS would dominate over SRS even under laser-fusion conditions because of the tendency of SRS to saturate at low levels. In the absence of critical-layer phenomena and SRS, the only competing parametric processes are filamentation and stimulated sidescatter. The latter has a lower intrinsic growth rate than backscatter and is important only in solid-target experiments where the plasma is more inhomogeneous in the beam direction than perpendicular to it. Filamentation has a growth length longer than is available in the $CO_2$ experiments.

Eq. (1) is valid when ion Landau damping $\gamma_i$, rather than convection, limits the growth of the ion wave. Thus, the time required to set up an ion wave of the proper amplitude is of order $\gamma_i^{-1} \simeq 0.1$ nsec; otherwise, it would be of order $\ell/c_s \simeq 150$ nsec, longer than the $\simeq 50$ nsec pulse lengths used. The simulation experiments, therefore, have sufficient duration to test the validity of (1). The dependence of Eq. (1) on $\ell$ is particularly important in the light of recent explanations[7] of the observed decrease of back reflection at very large intensities[8]. These explanations involve a steepening of the density profile.

In the experiment of Turechek and Chen[9], the target was a pulsed, high-current arc in 1-15 torr of $H_2$, He, A, or He + A, a pressure low enough to ensure that the plasma density was every-

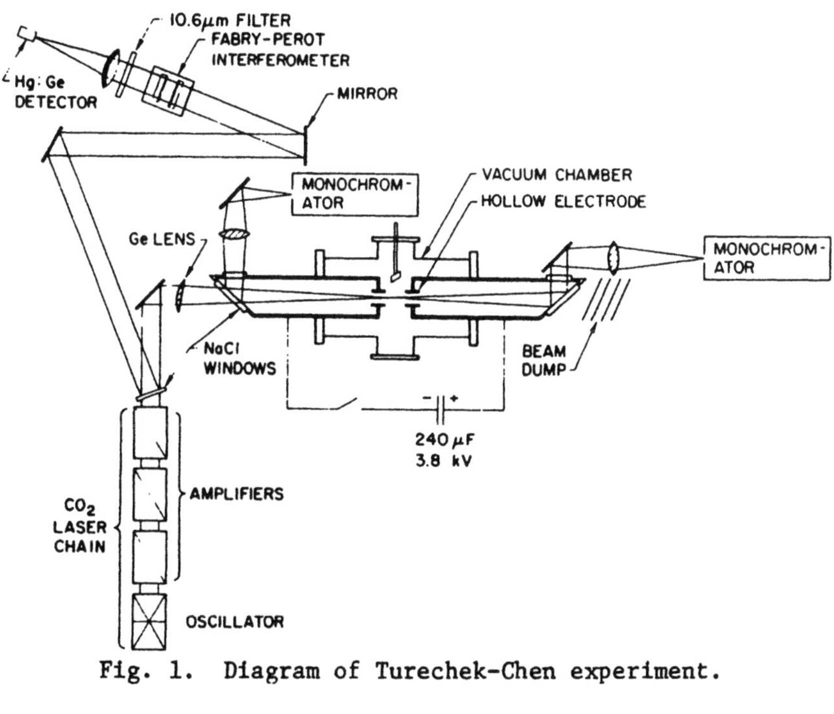


Fig. 1. Diagram of Turechek-Chen experiment.

where below $n_c/4$ for 10.6 μm. Typically, $n \simeq .02\ n_c$ and $n_c/4 = 2.5 \times 10^{18}$ cm$^{3}$. The $CO_2$ laser beam entered and left the plasma axially through tubular electrodes (see Fig. 1). The beam from a 45-J, 0.5-GW double-discharge system was focussed by a 75-cm focal length, f/20 lens to provide a uniformly illuminated region about 1 mm in diameter and 3 cm long, with peak intensity $\simeq 5 \times 10^{10}$ W/cm$^2$. The plasma was about 2 cm in diameter and 10 cm long, so that its inhomogeneity was weaker than that of the pump field. Backscattered light was extremely weak--about $10^{-7}$ of the incident intensity, or 9 e-foldings above Thomson scattering. SBS appeared as short, random spikes after the peak of the incident pulse, often occurring well into the tail, when the intensity had fallen to less than 10% of its peak value (see Fig. 2). Multiple-shot spectra of the backscattered light were taken with a low-resolution Fabry-Perot interferometer of large area, as required by the low intensity. These are shown for various gases in Fig. 3.

There are several points of particular interest in this experiment. First, observable backscatter was seen only during a short time interval during the rise of the arc current. At this time, the plasma is only 5% ionized, with $n_e \simeq 1 \times 10^{16}$ cm$^{-3}$ and and $KT_e \simeq 2$ eV. Heating by the initial peak of the laser pulse heats the plasma, completes the ionization, and brings $n_e$ to

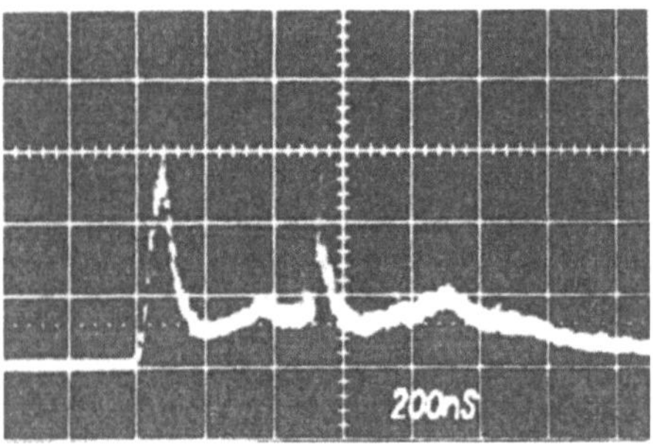


Fig. 2a

Backscattered signal (Initial peak is stray light)

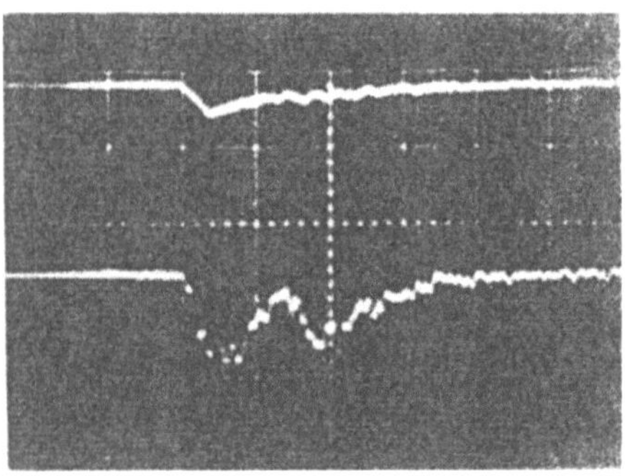

Fig. 2b

upper: Continuum intensity

lower: Intensity of forbidden line

(Turechek and Chen)

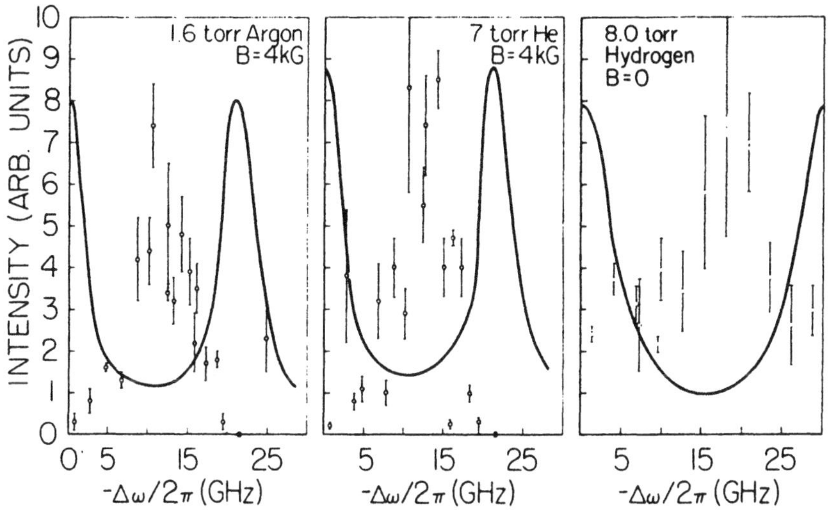


Fig. 3 Spectra of red-shifted backscattered light. (Turechek and Chen)

$2 \times 10^{17}$ cm$^{-3}$ or above. At later times, there is insufficient neutral density for this to happen, and the arc itself cannot raise $n_e$ above $3 \times 10^{16}$ cm$^{-3}$ because of pressure balance. In spite of the attempt to create a plasma target independently, the experimental conditions are greatly affected by the pump field itself.

Second, the red-shifts $f_i$ shown in Fig. 3 do not scale simply with ion mass. Since the shift $\omega_i = 2\pi f_i$ is equal to $2k_o c_s$ when $\omega_o >> \omega_p$, one would have expected an $M^{-\frac{1}{2}}$ dependence. However, inverse bremsstrahlung absorption varies as Z, so that $T_e$ varies from gas to gas. The measured shifts show that $c_s \simeq (ZKT_e/M)^{\frac{1}{2}}$ remains approximately constant. To verify the change of Z and $T_e$ with M, exhaustive spectroscopic measurements were made of the laser-heated focal region. A lower limit to $T_e$ was determined from line ratios using a time-dependent coronal model. The value of Z was 1, 2, and 8 for H, He, and A, respectively. Densities were measured by line broadening, mostly of He II 4686 A, introduced, if necessary, as a diagnostic impurity. In addition, numerical computations were made for the evolution of $n_e$ and $T_e$, taking into account ionization and recombination, absorption by electron-ion and electron-neutral collisions, multiple ion charge species, nonlinear radial heat conduction, and electron-ion temperature relaxation. Results for $T_e$ in various gases are shown on Fig. 4, which verifies that $\omega_i$ varies properly with ion mass when the difference in $T_e$ is taken into account.

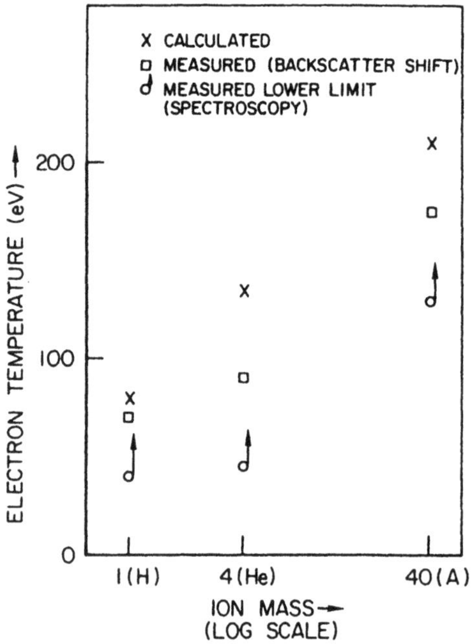


Fig. 4 Computed and measured electron temperatures, as compared with those from SBS red-shift. (Turechek and Chen)

The magnitude of the threshold, Eq. (1), was also verified. From the backscatter intensity, the number N of e-foldings above thermal noise was determined to be about 10. Plotting the required incident intensity $I_o$ as a function of $n_e$ and $T_e$ for $\ell$ = 10 cm, one finds that the initial laser spike is below threshold in the arc-generated plasma. After $n_e$ and $T_e$ are raised by laser heating, ion Landau damping is reduced so that even the intensity in the tail of the pulse is above threshold, as observed.

Finally, a third point of this experiment is that the low-frequency fluctuations in SBS were detected directly by using the time-dependent Stark effect in helium. The line 6671 A of He I is accompanied by a forbidden line at 6632 A. The forbidden transition can be induced by a large electric field. This is seen in synchronism with backscatter. In Fig. 2b, the continuum and forbidden-line intensities are shown. It is seen that the continuum has an increase associated with laser heating by the initial spike; the 6632 A line shows this also, but in addition has peaks correlated with the occurrence of SBS. It turns out that the ion-wave field is too weak to cause the forbidden line increase, but the beat-wave field at $\omega_i$ between the incident and reflected light is of the right order of magnitude. Apparently, the nonlinearity generating the low-frequency field stems from the quantum mechanical processes in the atom. Although this effect is far from understood, it is an example of the type of diagnostics needed for careful studies of parametric instabilities in laser-plasma interactions.

The second experiment reviewed here is that of Offenberger et al.[10], whose apparatus is shown in Fig. 5. Here again, the idea was to verify the existence of SBS in a plasma known to be underdense everywhere. The plasma was generated by the $CO_2$ laser pulse itself the technique is well known among those studying laser heating of plasma confined in linear solenoids. The laser light was focussed by a 52-cm focal length lens through a small aperture into a chamber filled, typically, with 18 Torr of $H_2$. To prevent breakdown between the aperture and the lens, the intervening space was pumped out just before each pulse. Laser breakdown created plasma which expands radially and axially, the density at the time of backscatter being lower than the filling density of $1.2 \times 10^{18}$ $cm^{-3}$ on axis and higher than this in the shock front. A 110-kG pulsed magnetic field could be used to retard the radial expansion and increase $T_e$. At the time of backscatter (≈20 nsec after the beginning of the pulse) the plasma conditions were computed to be $T_e \simeq 70$ eV, $T_i \simeq 10$ eV, $n_e \simeq 4 \times 10^{17}$ $cm^{-3}$ (an average of the B = 0 and B ≠ 0 cases). The column length ℓ was measured by streak photography to be 4.4 mm. The double-discharge $CO_2$ laser produced 125 J in a 40-nsec, 1-GW spike plus a 1-μsec, 200 MW tail. The focussed intensity was $10^{11}$ W/cm$^2$.

Reflected light occurred during the peak of the pulse and was sufficiently intense (0.2%) to be analyzed by a monochromator and a high resolution (3.5 A) Fabry-Perot interferometer in series. Fig. 6a shows the SBS spectrum for B = 0, taken by averaging shots in a 10-A window, corresponding to 2.7 GHz resolution. Because of the large reflected intensity relative to stray light, the high-finesse interferometer, and better reproducibility because $I_o$ was well above threshold, this spectrum is considerably more accurate than those of Turechek and Chen. Both a red-shift (≈17 GHz) and a width (≈5 GHz FWHM) could be determined. Furthermore, the scattered intensity vs. incident power and initial pressure could be measured, as shown in Figs. 6b and 6c. The latter is purported to represent variation with interaction length ℓ.

From the slope of Fig. 6b, the known value of ℓ, and the computed values of $T_e$, and $T_i$, and n, one can calculate the value of $\gamma_o$ to be $10^{11}$ $sec^{-1}$. From the second of Eqs. (1), one obtains $\gamma_o \simeq 5 \times 10^{10}$ $sec^{-1}$, which is excellent agreement. However, the authors prefer to use the random-phase theory of Tsytovich, which predicts a growth rate $\gamma_k$ of $5 \times 10^{11}$ $sec^{-1}$, as compared with $6 \times 10^{11}$ $sec^{-1}$ from Fig. 6b. This agreement is surely fortuitous, and the method itself is suspect because random-phase theory predicted a faster growth than coherent-phase theory.

Offenberger et al. also compared the width of the spectrum with that due to ion Landau damping and found that it was larger than expected. Inclusion of collisional damping, however, is not

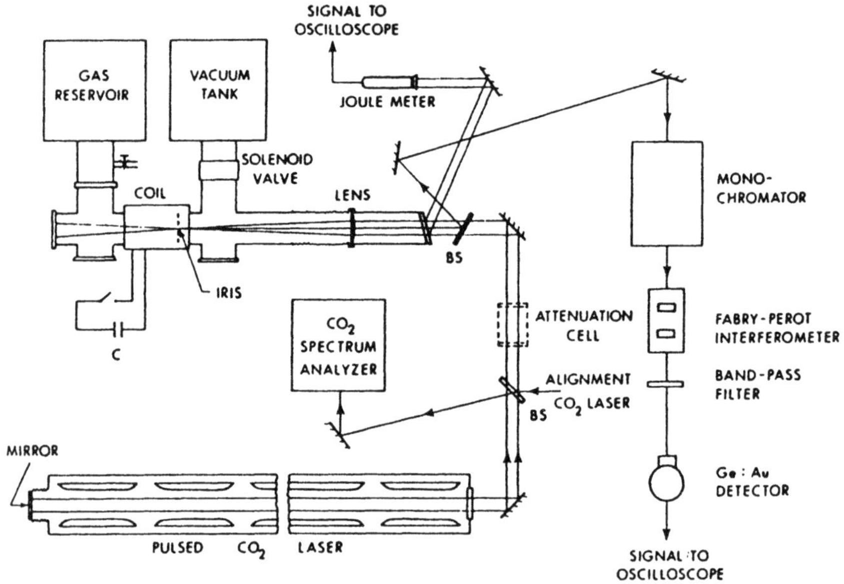


Fig. 5 Diagram of experiment of Offenberger et al.

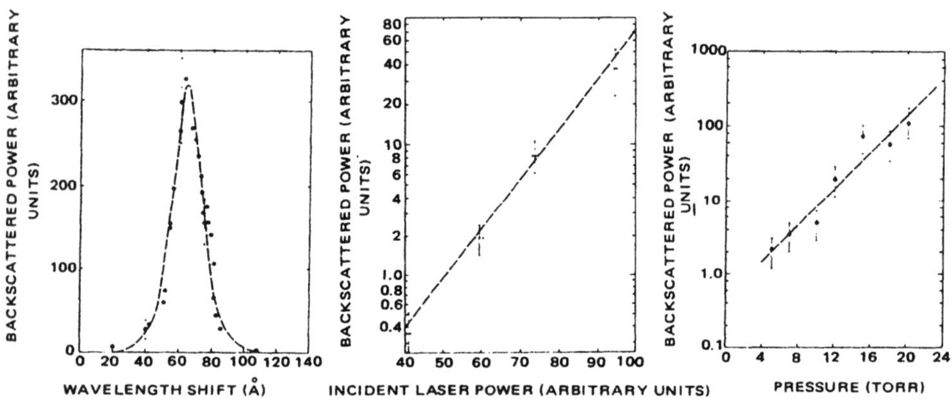


Fig. 6 Data of Offenberger, Cervenan, Yam, and Pasternak.
(a) SBS spectrum, (b) Power dependence, (c) Pressure dependence

straightforward, since the latter can decrease the total damping by disturbing the resonant particles. Furthermore, the width also depends on the number of exponentiations and the duration of the spiky signals. In checking the mass dependence between H and He, the authors failed to account for the difference in $T_e$. They also pointed out that the product of $I_o$ and $\ell$ was larger in the tail of the pulse than at the peak, but no SBS was observed there. Finally, computing N from $\gamma_o$ and Eq.1, one finds N = 8.7, about the same as in the Turechek-Chen experiment. Since $I_s/I_o$ was $\approx 10^{-6}$ in that case for 9 e-foldings above thermal noise, the value of $I_s/I_o$ = $2 \times 10^{-3}$ in Offenberger's experiment suggests an initial level of ion acoustic turbulence far above thermal. In summary, this experiment has by far the best data obtained so far, but there are many questions still left unanswered and many diagnostic checks to be made.

The third experiment along this line is that of Massey, Berggren and Pietrzyk.[11] Again, laser breakdown was used to produce the plasma in a large (100-kG) magnetic field (Fig. 7). The plasma density was measured as a function of distance along $\underline{B}$ by ruby interferometry (Fig. 8). For a typical pressure of 28 Torr of $H_2$, the maximum density is $1.8 \times 10^{18}$ cm$^{-3}$; however, much of the data was taken at higher pressures, where $n > n_c/4$. The e-beam sustained $CO_2$ laser produced 125 J in a 2-μs pulse, with 150 MW in the tail and about twice as much at the peak. The intensity during

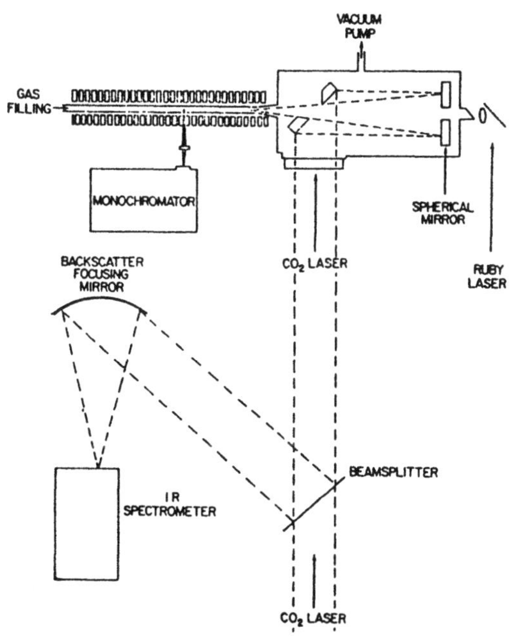


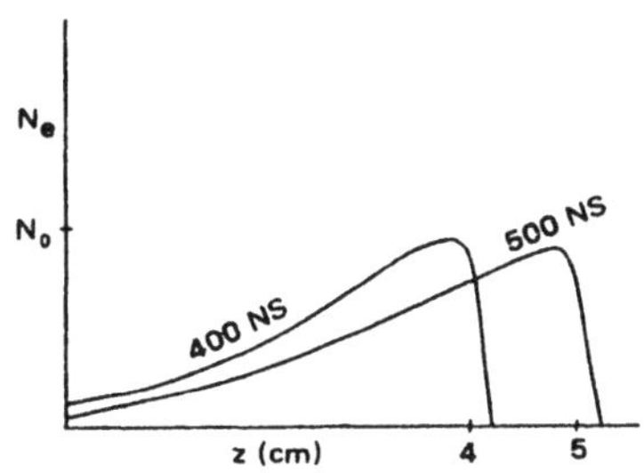


Fig. 8 Axial density variation.

Fig. 7 Diagram of experiment of Massey, Pietrzyk, and Scudder.

backscatter was about $5 \times 10^9$ W/cm$^3$. In spite of the lower intensity, higher backscatter (1 - 5%) was observed than in Offenberger's experiment. The beam was focussed by an 0.5-m focal length mirror, which resulted in backscatter during the gain-switched peak. Earlier work with a 1-m f.l. mirror produced late backscatter, presumably because the plasma, at low $I_o$, was too short at early times.

Data on spectrum, $I_o$ dependence, and $n_e$ dependence are shown in Fig. 9. These data were taken with pyroelectric detectors and a grating spectrometer. The scattered power, as large as 9 J, could be measured calorimetrically. Single-shot spectra were obtained with liquid crystal detectors. The electron temperature was measured by C V impurity line emission, using a time-dependent coronal model, yielding $KT_e = 80 \pm 25$ eV.

In interpreting the data, the density profile of Fig. 8 was approximated by a square profile, 1.5 - 2 mm long, at the peak density. From Fig. 9b, extrapolation to zero power would give the initial turbulence level if $T_e$ were constant. If the variation of $T_e$ with power P is taken into account by $T_e \propto P^{2/7}$, the resulting number of exponentiations is N = 4.8, indicating a very high level of initial turbulence. Using this value of N, and the measured

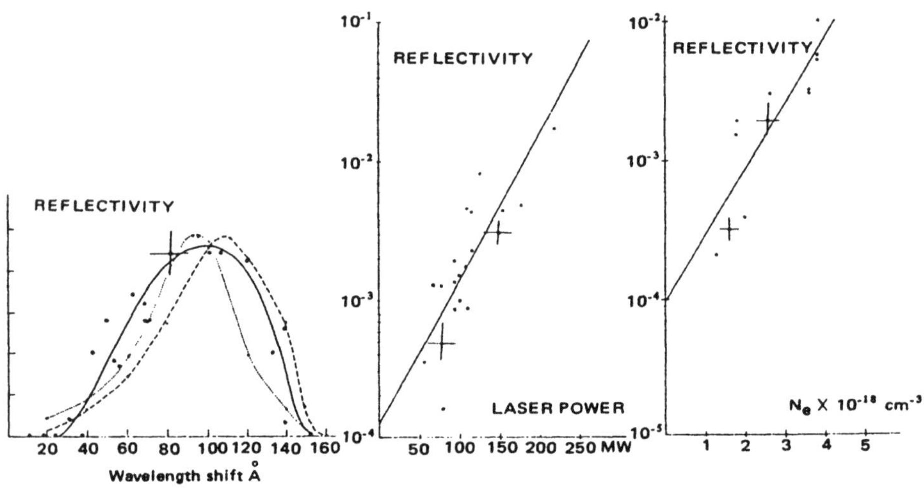


(a) (b) (c)

Fig. 9 Data of Massey et al.: Backscattered power vs. (a) red-shift, (b) laser power, and (c) plasma density. The dashed curve in (a) is theory for $T_e = T_i = 80$ eV, and the dotted curve, for 6C eV.

values of $T_e$ and $n_e$, one can compute the spectrum from theory; good agreement is shown in Fig. 9a. In Fig. 9c, the pressure dependence is interpreted as a dependence on $n_e$, whereas Offenberger et al. interpret it as a dependence on $\ell$. This point is unresolved.

These three backscattering experiments are compared in Table I (hydrogen data only). A, B, and C refer to Refs. 9, 10, and 11, respectively. F. L. is focal length; R the backscattered power divided by the incident power at that time; N the number of e-e-foldings; $f_i$ the ion-acoustic frequency shift (measured); $\Delta f$ the frequency resolution; and FWHM the measured width of the spectrum.

Each of the experiments had its particular advantages. Experiment A was the only one in which $n_e$ was definitely below $n_c/4$, and in which SBS grew from thermal noise, presumably because the arc current was well below the threshold of the ion acoustic instability. Furthermore, A had the forbidden-line diagnostic. The small signals did not permit varying $I_o$. In B and C, the initial turbulence level was not understood, but it gave large signals permitting variation of $I_o$ and, in the case of C, a single-shot spectrum. Between these, B had the higher $I_o$ and the better spectral resolution, but C had better diagnostics of the plasma. All three experiments revealed the spiky nature of SBS; which of many reasons is responsible for the fast decay of SBS once it starts is unknown. It is clear that much more work needs to be done, including checking the inhomogeneous

TABLE I

| Expt. | $\ell$(cm) | $KT_e$(eV) | $n_e$(cm$^{-3}$) | W(J) | P(GW) (peak) | $I_o$(GW/cm$^2$) (peak) | $I_o$(GW/cm$^2$) (tail) |
|---|---|---|---|---|---|---|---|
| A | 3 | 45 | $2 \times 10^{17}$ | 45 | 0.5 | 50 | 3 |
| B | 0.44 | 70 | $4 \times 10^{17}$ | 125 | 1.0 | 100 | 20 |
| C | 0.2 | 80 | $2 \times 10^{18}$ | 125 | 0.3 | 10 | 5 |

| Expt. | F.L.(cm) | R(%) | N | $f_i$(GHz) | $\Delta f$(GHz) | FWHM(GHz) |
|---|---|---|---|---|---|---|
| A | 75 | $10^{-4}$ | 9.2 | 18 | 4.3 | 10 |
| B | 52 | 0.2 | 8.7 | 17 | 2.7 | 5.3 |
| C | 50 | 1-5 | 4.8 | 25 | 6.7 | 21 |

threshold, measuring the initial level of turbulence, confirming the existence of ion waves, and finding the angular distribution of scattering.

### Other Experiments on Plasma Targets

A number of other experiments have been reported concerning nonlinear laser-plasma effects in underdense plasmas. Using a dc magnetically stablized arc, Marhic[12] was the first to detect the radial ponderomotive force applied by a laser beam to a plasma well below critical density. The plasma was almost fully ionized, with $n_e = 3 \times 10^{14}$ $cm^{-3}$ and $T_e = 1.4$ eV. The first demonstration of the beat-frequency mixing of two laser beams with $\Delta\omega \simeq \omega_p$ was done by Stansfield, Nodwell, and Meyer[13] in a partially ionized atmospheric arc with a density of $2 \times 10^{16}$ $cm^{-3}$ and $KT_e = 1.5$ eV. Fabre[14] has developed a technique of using a ruby laser to produce a plasma from a polyethylene foil, and then doing a $CO_2$-laser experiment on the plasma. The plasma target has $n_e \simeq 5 \times 10^{18}$ $cm^{-3}$, $T_e \simeq 9$ eV, r = 1 mm, L = 1.5 cm, and lasts 50-100 nsec. Finally, plasma foci[15] can produce plasmas of density $2 \times 10^{19}$ $cm^{-3}$, temperature 7 keV, raidus 1 mm, and length 1 cm, lasting 70 nsec.

## COMPARISON OF LASER-PLASMA SOURCE COMBINATIONS

For future experiments on parametric effects, we wish to consider the relative merits of various available radiation and plasma sources.

1. In the microwave range, high powers are easily available in the 3-30 cm wavelength regime; for instance, 100 kW in 20-μsec pulses with a klystron or traveling wave tube is common. Even 1 MW in 3-μs pulses with a magnetron is not an unusually cumbersome apparatus. We shall choose this in a 10-cm wavelength as our reference microwave source. The cutoff density being about $10^{11}$ $cm^{-3}$, the ideal plasma target for work in the underdense region up to $n = n_c$ is a low-pressure filament discharge of the type developed at UCLA. These have $KT_e \simeq 3$ eV and characteristic dimension $L \simeq 100$ cm $\gg \lambda_o$.

2. In the 4-8 mm microwave range, a reasonably high power would be 40 kW in a 1-μsec pulse from a magnetron. At 8 mm, $n_e$ is $2 \times 10^{13}$ $cm^{-3}$. Possible plasma targets at these higher densities are pulsed, partially ionized plasmas produced by ultraviolet radiation from sparks or by laser irradiation of solids. However, we shall take as a reference system a rf-generated plasma produced, say, by a Lisitano coil; the temperature would be about 8 eV.

3. Recent development of far-infrared lasers has yielded about 300 W at 337 μm with the HCN laser and 200 kW at 496μm (≃0.5 mm) with the $CH_3F$ (methyl fluoride) laser, optically pumped by a 9.55-μm $CO_2$ laser. For reference, we shall take a 1-MW $CH_3F$ laser (which will soon be achieved) with a 50-nsec pulse length. To go up to the critical density of $5 \times 10^{15}$ $cm^{-3}$, one can use a high current arc, which has $KT_e \simeq 5$ eV.

4. Moving to 10.6-μm $CO_2$ lasers, we find that 1 GW (peak) in a 50-nsec pulse is achievable with relative ease and flexibility with TEA, double-discharge or e-beam sustained lasers. It is not easy to reach the critical density of $10^{19}$ cm$^{-3}$ except with a plasma focus, but in the underdense regime of $n \simeq 10^{16-18}$ one can use pulsed, high-current arcs, z-pinches, θ-pinches, or laser breakdown plasmas. Whatever the initial temperature, laser heating will raise $kT_e$ to ≃ 100 eV; hence, we shall take this for the reference temperature.

5. At the highest frequencies, we have ruby lasers or dye lasers with $\lambda_o \simeq 0.7$ μm and $n_c \simeq 2 \times 10^{21}$ cm$^{3}$. We assume 1 GW in a 30-psec pulse and a plasma focus with $KT_e = 1$ keV.

6. The system that all these are to simulate is a near-term Nd-glass laser irradiating a microballoon. For the reference situation, we take $\lambda_o = 1.06$ μm, P = 1 TW in a 100 psec pulse, $n_c = 10^{21}$ cm$^{-3}$, and a heated corona with $KT_e \simeq 2$ keV.

These characteristics are summarized in Table II. In addition, we need to compute the incident intensity $I_o$. The diffraction limited spot size is approximately $f\lambda_o$, where f is the f-number of the focussing optics. The diffraction limit is expected to be reached in microwave and F.I.R. systems, where mode purity is easily controllable. For multi-mode lasers, we assume that the spot diameter exceeds the diffraction limit by a spoilage factor S, which is normally higher for shorter wavelength. Thus, $I_o = P/(\pi/4)(S\lambda_o f)^2$. To obtain the f-number, we assume a typical beam diameter and a focal length compatible with the size of the plasma. The pulse length is τ.

TABLE II

| $\lambda_o$ | P | f | S | $I_o(W/cm^2)$ | $n_c(cm^{-3})$ | $T_e$(eV) | τ |
|---|---|---|---|---|---|---|---|
| 10 cm | 1 MW | 1 | 2 | $3 \times 10^3$ | $10^{11}$ | 3 | 3 μs |
| 8 mm | 40 kW | 1 | 2 | $2 \times 10^4$ | $2 \times 10^{13}$ | 8 | 1 μs |
| 0.5 mm | 1 MW | 1 | 3 | $6 \times 10^7$ | $5 \times 10^{15}$ | 5 | 50 ns |
| 10.6 μm | 1 GW | 15 | 5 | $2 \times 10^{11}$ | $10^{19}$ | 100 | 50 ns |
| 0.7 μm | 1 GW | 25 | 50 | $2 \times 10^{11}$ | $2 \times 10^{21}$ | 1000 | 30 ps |
| 1.06 μm | 1 TW | 1.5 | 20 | $1 \times 10^{17}$ | $10^{21}$ | 2000 | 100 ps |

These are typical, not maximum, values.

Suitability of these systems for simulating effects in pellet irradiations (last line, Table II) can be evaluated by several criteria.

1. Ponderomotive force. Profile modifications and parmetric instabilities are driven by the force[6].

$$F_{NL} = -\frac{\omega_p^2}{\omega^2}\ \nabla \frac{\langle E^2 \rangle}{8\pi}$$

The ratio of $F_{NL}$ to the $\nabla p$ force is a measure of the importance of parametric effects. Hence, we define a quantity G as

$$G \equiv \frac{F_{NL}}{\nabla p} = -\frac{\omega_p^2}{\omega^2}\frac{\nabla \langle E^2 \rangle / 8\pi}{\nabla p} = -\frac{\omega_p^2}{\omega^2}\frac{\langle E^2 \rangle / 8\pi}{nKT} = \frac{n}{n_c}\frac{4\pi I/c}{8\pi nKT} = \frac{I}{2cn_cKT}$$

where $T = T_e + T_i$ and $n_c = m\omega^2/4\pi e^2$.

2. Exponentiation. The growth rate of instabilities is scaled to the ion plasma frequency $\Omega_p$. The number of exponentiations in the pulse length $\tau$ is, therefore, some fraction of $\Omega_p\tau$. In calculating $\Omega_p\tau$, we assume hydrogen and a density of 0.1 $n_c$.

3. Classical absorption. In an experiment in which the critical density is reached, the amount of beam energy absorbed classically by the plasma depends on the density scale length L. For a linear gradient $\nabla n \simeq n_c/L$ in plane geometry, the energy loss after reflection is

$$\frac{\Delta I}{I_o} = 1 - \exp\left[-\frac{32}{15}(7.8 \times 10^{-9})\frac{Z \ln\Lambda\, n_c^2\, L\lambda_o^2}{c^2 T_{ev}^{3/2}}\right]$$

for Z = 1 and ln $\Lambda$ = 5, this becomes

$$\frac{\Delta I}{I_o} = 1 - \exp\,(-1.16 \times 10^{-2} L/\lambda_o^2 T_{ev}^{3/2})$$

The effect of $\Delta I$ on the plasma is measured by the ratio Q of $\tau\Delta I$ to the plasma energy $\frac{3}{2}\bar{n}KTL$ per $cm^2$, where $\bar{n} = \frac{1}{2}n_c$.

Thus $$Q \equiv \tau\Delta I/0.75\, n_cKTL \approx \tau\Delta I/n_cKT_eL,$$

where $T_i = T_e/3$ has been assumed for convenience. In Table III, both $\Delta I/I_o$ and Q are given for a typical value of L.

4. Plasma heating. Even in underdense regions, inverse bremsstrahlung absorption causes pressure-gradient and thermo-electric forces which compete with radiation pressure in driving instabilities, self-focussing, and spontaneous magnetic fields. The rate of absorption is

$$q = \frac{3.1 \times 10^{-7} Zn_e^2 \ln\Lambda\ I}{T_{ev}^{3/2}\omega^2(1 - \omega_p^2/\omega^2)^{1/2}} \quad ergs/cm^3/sec$$

If thermal conduction is neglected, the change in temperature is given by

$$\frac{d}{dt}\left(\frac{3}{2}\, nKT\right) = q(T)$$

Integrating over time and assuming $Z = 1$, $\ell n\ \Lambda = 5$, and $n/n_c = 0.1$, we find

$$T_{ev}^{5/2} - T_o^{5/2} = 1.7 \times 10^{13}\ nI\tau/\omega^2 = 4.8 \times 10^{-11}\ \lambda_o^2 n_c I\tau$$

where I is in $W/cm^2/sec$. The value of $\Delta T/T$ is shown in Table IV.

The relative effect of heating is measured by the ratio H of $\Delta(nKT)$ to the pondermotive potential $(\omega_p^2/\omega^2)\langle E^2\rangle/8\pi$. H is given by

$$H \approx 10^{-8}\ \frac{n_cT_{ev}}{I_o}\,\frac{\Delta T}{T}$$

and is given in Table IV for $n = 0.1\ n_c$.

On the other hand, thermal conduction can limit the temperature rise. The latter then depends on the beam power rather than its intensity. We have considered radial conduction from a long, thin, heated cylinder of radius a, with the temperature at radius $b \gg a$ held constant at the initial temperature $T_o$. If there is no magnetic field, the approximate steady-state solution for the temperature $T_{ev}$ of the heated region is

$$T_{ev}^{3/2}\ (T_{ev}^{7/2} - T_o^{7/2}) = 1.9 \times 10^{-30}\ \lambda_o^2 n^2\ P\ \ell n\ (b/a).$$

TABLE III

| $\lambda_o$ | G | $\Omega_p\tau$ | L(cm) | $\Delta I/I_o$ | Q |
|---|---|---|---|---|---|
| 10 cm | 0.9 | 400 | 100 | .002 | 0.18 |
| 8 mm | .02 | 1700 | 2 | .002 | 0.72 |
| 0.5 mm | 0.28 | 1400 | 1 | 0.34 | 280 |
| 10.6 μm | .02 | $7 \times 10^4$ | 0.3 | .96† | 200† |
| 0.7 μm | $5 \times 10^{-6}$ | 600 | 0.5 | .98† | $4 \times 10^{-4}$† |
| 1.06 μm | 5.2 | 1300 | 20(μm) | 0.02 | 360 |

TABLE IV

| $\lambda_o$ | $\Delta T_e/T_e$ (heat capacity) | H (heat capacity) | $\Delta T_e/T_e$ (heat conduction) | H (heat conduction) | A |
|---|---|---|---|---|---|
| 10 cm | 0.10 | 0.10 | $5 \times 10^{-5}$ | $5 \times 10^{-5}$ | 15 |
| 8 mm | .003 | 2.1 | $4 \times 10^{-6}$ | $3 \times 10^{-4}$ | 9.5 |
| 0.5 mm | 3.1 | 12.7 | 0.16 | 0.67 | 19 |
| 10.6 μm | 4.0 | 200 | 0.11 | 5.5 | 46† |
| 0.7 μm | $4 \times 10^{-5}$† | 3.6† | $2 \times 10^{-4}$† | 25† | 135† |
| 1.06 μm | 3.0 | 0.6 | .004 | $8 \times 10^{-4}$ | 18.6 |

†These values pertain to plasma densities not attainable with the quoted sources.

Here, P is the beam power in watts, and we have taken Z = 1, $n = 0.1\ n_c$, and reasonable $\ell n\ \Lambda$ factors. $\Delta T/T$ and H are also given in Table IV for conduction-limited temperature rise, with b/a = 10.

5. Other effects. Finally, we consider swelling and refraction. The intensity of the light-wave field increases near the cutoff layer because of the decrease in group velocity and the constructive interference with the reflected wave. For perpendicular incidence, this amplification factor is given by[16].

$$\left(\frac{E_{max}}{E_o}\right)^2 = 3.74\ (k_oL)^{1/3} \simeq 7\ (L/\lambda_o)^{1/3} \equiv A$$

This quantity is also shown on Table IV. We have assumed a scale length L appropriate to radial incidence. The amplification can, of course, be increased with axial incidence on a long, thin plasma; but in practice collisional absorption limits this. Refraction has been estimated by considering the geometrical-optics deflection of a ray incident on a spherical plasma of radius a with parabolic density profile and maximum density n. The worst case has impact parameter $p = 3^{-\frac{1}{2}}$ a and deflection $\Delta r \simeq 0.2$ a $(n/n_c)$. Therefore, refraction can affect the alignment of optics if a is large compared to the spot size and if $n/n_c$ is not very small. Two other effects, the damping of waves and the generation of magnetic fields, cannot easily be compared among experimental systems. An idea of collisional damping (except for neutral collisions) can be obtained from the heating results, but Landau damping depends critically on $T_e/T_i$. Magnetic fields can arise from many effects, including heating and ponderomotive force, but the effect is strongly geometry-dependent.

Tables II - IV show that different wavelength ranges have different advantages. All the systems have sufficiently large G and $\Omega_p\tau$ to generate parametric instabilities. Only the 10-cm and FIR systems have enough power to approach the nonlinear regime of laser-fusion. Microwave experiments are excellent for isolating parametric phenomena from heating, since Q and H are small. Moreover, the long-wavelength experiments in large, tenuous plasmas offer the possibility of detailed probing of the plasma. On the other hand, only the laser simulations can reproduce the large heating effects that occur in laser fusion, such as anomalous conduction, thermal self-focussing, and thermoelectric generation of magnetic fields. Daignostics of $CO_2$ and FIR experiments, though easier than for pellet experiments is still a considerable problem.

The results of Tables II to IV can, of course, be adjusted by changing, say, the initial $T_e$, or $\tau$, or L, or $n/n_c$. The lower of the values of $\Delta T/T$ in the tables should be taken, except for the Nd-glass case, where the boundary conditions do not resemble those used in the calculation. Although there is some freedom in designing an experiment, it is clear that one cannot always scale every parameter properly or completely isolate a given phenomenon. Nonetheless, well diagnosed and analyzed experiments under less extreme conditions than those in a pellet implosion can provide the missing link between theory (including computation) and reality.

## FUTURE EXPERIMENTS

In the remaining space we can only mention a few of the important problems that need scrutiny. Backscattering and sidescattering are by no means well documented. Resonant absorption has never been observed directly; it is not known whether the absorption occurs by collisions or by an instability such as the radiative decay instability. The two-plasmon decay and other parametric processes at the $n_c/4$ layer need to be studied. Filamentation in a plasma has not been detected directly. The mechanism of anomalous heat conduction--perhaps a thermal instability involving spontaneous magnetic fields--needs to be determined. And, of course, it would be nice to know how laser-plasma interactions produce such efficient fast-ion accelerators.

ACKNOWLEDGMENTS

We are indebted to R. Massey and J. Turechek for access to unpublished data. This work was supported by the U.S. Energy Research and Development Administration, Contract E(04-3)-34, P.A. 236.

References

1. H. C. Kim, R. Stenzel, and A. Y. Wong, Phys. Rev. Letters 33, 886 (1974).
2. S. P. Obenschain, N. C. Luhmann, Jr., and P. T. Greiling, Phys. Rev. Letters 36, 1309 (1976).
3. J. F. Drake, P. K. Kaw, Y. C. Lee, G. Schmidt, C. S. Liu, and M. N. Rosenbluth, Phys. Fluids 17, 778 (1974).
4. M. N. Rosenbluth, Phys. Rev. Letters 29, 565 (1972); M. N. M. N. Rosenbluth and R. Z. Sagdeev, Comments on Plasma Physics and Controlled Fusion 1, 129 (1972); C. S. Liu, M. N. Rosenbluth, and R. B. White, Phys. Rev. Letters 31, 697 (1973) and Phys. Fluids 17, 1211 (1974); M. N. Rosenbluth, R. B. White, and C. S. Liu, Phys. Rev. Letters 31, 1190 (1973); D. Pesme, G. Laval, and R. Pellat, Phys. Rev. Letters 31, 203 (1973); R. W. Harvey and G. Schmidt, Phys. Fluids 18, 1395 (1975).
5. D. W. Forslund, J. M. Kindel, and E. L. Lindman, Phys. Rev. Letters 30, 739 (1973), Phys. Fluids 18, 1002 (1975) and 18, 1017 (1975); D. F. DuBois, D. W. Forslund, and E. A. Williams, Phys. Rev. Letters 33, 1013 (1974); D. Biskamp and H. Welter, Phys. Rev. Letters 34, 312 (1975); W. L. Kruer, K. G. Estabrook, and K. H. Sinz, Nuclear Fusion 13, 779 (1973).
6. F. F. Chen, Laser Interaction and Related Plasma Phenomena, ed. by H. J. Schwarz and H. Hora (Plenum Press, New York, 1973), vol. 3A, 291.
7. W. L. Kruer, E. J. Valeo, and K. G. Estabrook, Phys. Rev. Letters 35, 1076 (1975).
8. M. Lubin, E. Goldman, J. Soures, L. Goldman, W. Friedman, S. Letzring, J. Albritton, P. Koch, and B. Yaakobi, Plasma Physics and Controlled Nuclear Fusion Research 1974 (Int'l Atomic Energy Agency, Vienna, 1975) II, 459; C. Yamanaka, T. Yamanaka, T. Sasaki, J. Mizui, and H. B. Kang, Phys. Rev. Letters 32, 1038 (1974); J. Martineau, S. Repoux, M. RAbeau, G. Nierat, and M. Rostaing, Opt. Commun. 12, 307 (1974); K. Dick and H. Pepin, Opt. Commun. 13, 289 (1975).
9. J. J. Turechek and F. F. Chen, Phys. Rev. Letters 36, 720 (1976).
10. A. A. Offenberger, M. R. Cervenan, A. M. Yam, and A. W. Pasternak, Jour. Appl. Phys. 47, 1451 (1976).
11. R. Massey, K. Berggren, and Z. A. Pietrzyk, Phys. Rev. Letters 36, 963 (1976). The data given here are from more recent work: R. Massey, Z. A. Pietrzyk, and D. W. Scudder, Sixth Annual

Anomalous Absorption Conference, May 10-21, 1976, Vancouver, B.C., Canada, Abstract 57.
12. M. E. Marhic, Phys. Fluids 18, 837 (1975).
13. B. L. Stansfield, R. Nodwell, and J. Meyer, Phys. Rev. Letters 26, 1219 (1971).
14. E. Fabre and C. Stenz, Phys. Rev. Letters 32, 323 (1974).
15. Y. Kitagawa, A. Thein, E. Setoyama, C. Yamabe, and M. Yokoyama, J. Phys. Soc. Japan (to be published); J. W. Mather, Methods of Experimental Physics 9B, ed. by H. R. Griem and R. H. Lovberg, Academic Press, New York & London (1971).
16. R. W. White and F. F. Chen, Plasma Phys. 16, 565 (1974).

# MULTIPLE WAVE INTERACTION AND HARMONIC GENERATION IN LASER CREATED PLASMAS

J.L. Bobin

C.E.A. - Centre d'Etudes de Limeil

Boîte Postale n° 27
94190 - Villeneuve-Saint-Georges - FRANCE

ABSTRACT

The mechanisms leading to harmonic generation in inhomogeneous laser created plasmas are reviewed. After a description of classical and three-quanta processes, the role of Raman up-conversion is emphasized. The limitations due to the selection rule on k vectors are stated. In many cases the only way of getting rid of them comes from four wave processes. These have also the advantage of accounting for the asymetric broadening of the $2\omega_L$ and $3\omega_L/2$ harmonics. Theoretical results are compared to experimental data. Finally, the use of harmonic and subharmonic generation for laser plasma interaction and laser induced plasma flow diagnostics is discussed.

## I. INTRODUCTION AND SCOPE

The observation of second harmonic generation from laser created plasmas[1] was no surprise. Indeed, the effect was predicted long ago by the theorists of non linear optics[2]. More intriguing was the occurence of a line with frequency 3/2 of the incoming one $\omega_L$. The existence of a single sideband on the low frequency (Stokes) side was a further puzzle. Since there seems to be a definite threshold and the one sided broadening has the order of magnitude of ion-acoustic frequencies, the connexion with some sort of parametric interaction of laser light with the plasma is quite clear[3]. However, the zoology of such mechanisms is as rich as the manifold of theoretical approaches to study them. An enormous amount of litterature dealing with non linear effects in plasmas, has been published during the last 10 years[4]. Most of it is theory. Since it is very difficult to isolate a single process in a laboratory experiment, conclusive measurements are rather scarce. If, as in atomic (or nuclear) physics, charactersitic lines and broadenings could be identified and accurately related to well defined processes, then the experimental situation would become far better.

It is the aim of the present paper to review those non linear processes which might be involved in harmonic and subharmonic generation. The theory is grounded on cold plasma fluid equations and a quantum mechanical interpretation. Non linearities in plasmas are associated with resonances whose occurence will be discussed first, together with the kind of wave-wave coupling they imply. Actually, the effects we are intersted in, are due to the coupling of different sources of non linearities : currents, gradients, $\vec{U} \times \vec{B}$ force. Of importance are the frequency and wave-vector matchings. They correspond to restrictive conditions in the case of three-quanta processes. However, the coupling of Raman up conversion to the parametric decay is an important exception.

Scattering will be the main process dealt with. The Compton effect leads to harmonic generation when taking into account the $\vec{U} \times \vec{B}$ term and a cross section can be evaluated in any order. In a plasma, electron motions are collective rather than individual provided the Debye radius has a sense and the wave number is smaller than its reciprocal. Thus we are to consider the interaction of waves with the plasma modes.

All considered effects are stimulated. Spontaneous emission of photons or quasi particles will be ignored throughout. This implies high fluxes of incoming laser radiation : well over threshold values. In the equations therefore, damping terms will be, as a rule, neglected except in the discussion of line broadening. Furthermore, as usual, energy conservation is disregarded. We are not interested here in the zero frequency modes associated with the energy equation, although they may play a role in the line shape of scattered light[5].

## II - RESONANCES IN THE PROPAGATION OF E.M. WAVES IN A PLASMA

In principle, non linearities in which electron can be accelerated up to energies in the MeV range should be dealt with relativistically. The formulation is then quite complicated. However, simple results of interest to the topic of this paper were recently derived by A. Decoster[6]. The approach uses relativistic cold plasma equations coupled to Maxwell's equations. Let $\vec{p}$ be the macroscopic momentum of the fluid (ions or electrons).

Setting $p_o = (mc^2 + p^2)^{1/2}$, one has with self evident notations ( $\sigma$ stands for e or i) :

$$(2.1)\begin{cases} \partial n / \partial t + \vec{\nabla} . (n_\sigma c\vec{p} / p_{o\sigma}) = 0 \\ \{\partial/\partial t + (c\vec{p}_\sigma/p_{o\sigma}) . \vec{\nabla}\} \vec{p}_\sigma = q_\sigma\{\vec{E} + (\vec{p}_\sigma/p_{o\sigma}) \times \vec{B}\} \\ \vec{\nabla}.\vec{E} = 4\pi \Sigma n_\sigma q_\sigma \\ \partial\vec{B}/\partial t = c\vec{\nabla} \times \vec{E} \\ c \vec{\nabla} \times \vec{B} - \partial\vec{E}/\partial t = 4 \pi\vec{j} \end{cases}$$

in which the current density $\vec{j}$ is the product of a particle density by a velocity. Note that in the above system of equations, energy conservation is ignored.

Now, consider an electromagnetic wave propagating in an uniform plasma with a given phase velocity $\omega/k$ ($>c$). Following Winkles & Eldridge[7], it can be shown that there exists a Lorentz transform such that in the new frame, fields are no longer space dependant. Then we may set $\vec{B} = 0$ and get the system :

$$(2.2)\begin{cases} \partial n/\partial t + \vec{\nabla} . (nc\vec{p}/p_o) = o \\ \partial/\partial t + ( c\vec{p}/p_o) . \vec{\nabla}) \vec{p} = q \vec{E} \\ \Sigma n_\sigma q_\sigma = o \quad , \quad - \partial\vec{E}/\partial t = 4\pi\vec{j} \; . \end{cases}$$

Let the ions have the mass $m_i$ and the charge Z. The equations can be solved by using a dimensionless auxiliary variable :

$$(2.3)\vec{a} = \vec{p}_e/m_e c - \vec{u}_o \text{ with } \vec{u}_o = (Z\vec{p}_e^o + \vec{p}_i^o)/ m_i c(1+Zm_e/m_i)$$

in which $\vec{p}_\sigma^{\,0}$ are constant vectors. By substitution one gets a second order differential equation for $\vec{a}$, which is formally identical to the equation of motion for a non relativistic particle with unit mass in a scalar potential :

$$(2.4)\quad d^2\vec{a}/dt^2 = -\omega_p^2 \; (\vec{a} + \vec{u}_o)\,\{1 + (\vec{a} + \vec{u}_o)^2\}^{-1/2}$$

$$+ (\vec{a} - m_i\vec{u}_o/Zm)\{(m_i^2/Z^2m_e^2 + (\vec{a} - m_i\vec{u}_o/Zm_e^2)\}^{-1/2}$$

where $\omega_p^2 = 4\pi \; n_e \; e^2/m_e$ is the usual electron plasma frequency squared. In order to solve this equation, one has to make a choice about $\vec{u}_o$ . The condition a = o defines a situation in which ions and electrons have the same bulk velocity. The velocity can be set equal to zero in the laboratory frame. Hence :

$$(2.5)\qquad \vec{u}_o = - \,\mathrm{sh}\,\psi\,\vec{U}_k$$

in which $\psi$ is the argument of the coefficients in the above mentionned Lorentz transform and $\vec{U}_k$ is the unit vector in the direction of propagation of the E.M. wave. Then the densities in both frames are constants and linked by the relationship :

$$(2.6)\qquad n = n_L \;\mathrm{ch}\,\psi$$

(The subscript $_L$ refers to the Laboratory). Assume for a choice, that this property is true for any value of a. One is then able to solve the equation (2.4)or rather its first integral :

$$(2.7)\quad (1/2)\;(d\vec{a}/dt)^2 + \omega_p^2\;\mathrm{ch}\,\psi\{(1 + (\vec{a} - \mathrm{sh}\psi\;\vec{U}_k)^2)^{1/2}$$

$$+ (m_i^2/Z^2\;m_e^2 + (\vec{a} + m_i\;\mathrm{sh}\,\psi\,\vec{U}_k/Zm_e)^2)^{1/2}\} = \mathrm{const.}$$

In this equation, the potential is first expanded as a power series of $\vec{a}$ exp $(-\psi)$. The vector $\vec{a}$ has two components : $a_k$ in the direction of $U_k$ and $a_\perp$ which may include a precession angle. Then setting :

$$(2.8)\qquad \begin{aligned} a_k &= a_k^{(0)} + a_k^{(1)} + a_k^{(2)} + \ldots, \\ a_\perp &= a_\perp^{(0)} + a_\perp^{(1)} + a_\perp^{(2)} + \ldots, \\ \omega_k &= \Omega_k + \Omega_k^{(1)} + \ldots, \\ \Omega_\perp &= \Omega + \Omega^{(1)} + \ldots., \end{aligned}$$

the latter being the longitudinal and transverse frequencies respectively, one is able to solve the equations up to any order. Since the expressions are very long, they are not to be given here. We restrict ourselves to the relevant results : in the $(2n + m - 2)$th order of perturbation, a resonance occurs for $2n\,\Omega \sim m\,\Omega_k$ in which longitudinal and transverse oscillations are coupled. Accordingly ch $\psi \sim m/2n$. In order of apperance, such resonances are :

| PERTURBATIVE ORDER | 2 | 3 | 4 | 5 | 6 | 7 | 8 |
|---|---|---|---|---|---|---|---|
| RESONANCE FOR $\Omega / \Omega_K$ = | 1 | 3/2 | 2 | 5/2 | 1&3 | 7/2&5/4 | 4&3/2 |

Of course the theory is no longer valid in the vicinity of such poles. The important point is however the possibility of resonant coupling for such associations of longitudinal and transverse frequencies.

## III. PARAMETRIC RESONANCES

Another and more usual way of evidencing resonances is grounded on the cold plasma equations in the so called dipole approximation : the incoming E.M. wave has an infinite wavelength i.e. $k_L = o$. Let it be : $E_L \sin \omega_L t$ (L now for laser). The non relativistic fluid equations are :

$$(3.1)\quad \begin{cases} \partial n/\partial t + \vec{\nabla}.(n\,\vec{\upsilon}) = 0 \\ \partial\vec{\upsilon}/\partial t + (\vec{\upsilon}.\vec{\nabla})\vec{\upsilon} = q\vec{E}/m. \end{cases}$$

Putting in the right-hand side the high frequency transverse field only, one gets an "Equilibrium" solution thereafter denoted by the index o. The equations are then linearized. The perturbed longitudinal field is given by the Poisson equation. Fourier transforming in space the fluid equations and setting for the perturbed densities

$$(3.2)\quad \nu_\sigma = n_\sigma \exp - i \int(\vec{k}.\,\vec{\upsilon}_o)\,dt,$$

one gets for a plasma of electrons and ions with charge Z :[8]

$$(3.3)\quad \begin{cases} \partial^2\nu_e/\partial^2 + \omega_{p\,e}^{\,2}\,(\nu_e + Z\nu_i \exp{-ia} \sin\omega_L t) = 0 \\ \partial^2\nu_e/\partial_t^2 + \omega_{p\,i}^{\,2}\,\{\nu_i + (\nu_e/Z)\exp{-ia} \sin\omega_L t\} = 0 \end{cases}$$

where

(3.4) $a = (e/m_e - Ze/m_i)\,(\vec{k}.\vec{E}_L)\,/\,\omega_L^{\,2}$

and $\omega_{p\,\sigma}^{2}$ is the square of the plasma frequency of species $\sigma$. As usual one uses a Bessel expansion for the exponential and looks for a solution through the multiple time scale method. Let t, be a slow time and

(3.5) $\nu_e = \nu_1\,(t_1)\;\nu\,(t).$

If $\partial\nu_1/\partial t$ and $\partial\nu_i/\partial t_1$ are zero, the equations reduce to

(3.6) $\partial^2\nu/\partial t^2 - \omega^2_{pe}\,\{\nu + \sum_l J_l(a)\,\exp\, il\omega_L t\} = 0$

For small values of the argument a, it can be approximated by a Mathieu equation whose cases of instability are well known viz

(3.7) $\omega_L = \pm\, 2\,\omega_{pe}/n.$

If $\partial\nu_1/\partial t_1$ and $\partial\nu_i/\partial t_1$ are small but non zero, one gets setting

(3.8) $\nu_e = \nu_{e+}(t_1)\exp\, i\omega_L t + \nu_{e-}(t_1)\exp\, -i\omega_L t,$

(3.9) $$\begin{cases} Z\nu_i(t) = \nu_{e+}\exp i\omega_L t\;\;\{\sum_l J_l(a)\,\omega_{pe}^{\,2}\,\exp il\omega_L t\}/(l\omega_1+\omega_{pe})^2 \\ +\nu_{e-}\exp -i\omega_L t\;\;\{\sum_l J_l(a)\,\omega_{pe}^{\,2}\exp\, il\omega_L t\}\,/\,(l\omega_1-\omega_{pe})^2 \\ \partial\nu_{e\pm}\,/\,\partial t_1 = \pm i\,(\nu_{e\pm}/2)\,\omega_{pe}\,\omega^2_{pi}\{\sum_l J^2_l(a)\}\,/\,(l\omega_L-\omega_{pe})^2 \end{cases}$$

which exhibits resonances for $\omega_L = \pm\,\omega_p/n$. Since one should have $\omega_L > \omega_{pe}$ in order that the E.M. wave propagates through the plasma, all found resonances but two are non physical. Exceptions are

(3.10) $\omega_L = 2\,\omega_{pe}\;,\quad \omega_L = \omega_{pe}.$

The former involves coupling with electron plasma waves only and as shown by E.A. Jackson[9] the instability requires $k_L \neq 0$. The latter implies coupling with electron plasma waves and ion acoustic (low frequency) waves. The corresponding instabilities : parametric decay and oscillating two stream have been extensively studied by theoreticians.

## IV. SECOND HARMONIC GENERATION

This is the simplest case of harmonic generation whose properties were predicted quite long ago[10]. Consider a plasma in which motionless ions are just a neutralizing background. We again use a non relativistic cold fluid description. Let N be the equilibrium electron density and n and $\vec{U}$ be small perturbations. $\varepsilon$ is an order of magnitude indicator (eventually set equal to 1).

The fluid equations are

$$(4.1)\quad \begin{cases} \partial n/\partial t + \vec{\nabla}.(N.\vec{U}) = 0 \\ \partial\vec{U}/\partial t + \varepsilon(\vec{U}.\vec{\nabla})\vec{U} = -e\{\vec{E}+\varepsilon(\vec{U}/c)x\vec{B}\}/m \end{cases}$$

coupled to Maxwell's equations. All small terms labelled $\varepsilon$ are sources of non linearities including the current in the third Maxwell's equation. Let $\vec{U}$ and n be expanded according to

$$(4.2)\qquad n = n_1 + \varepsilon\, n_2\ , \qquad \vec{U} = \vec{U}_1 + \varepsilon\vec{U}_2 .$$

One then gets assuming $E_L$, $B_L \propto \exp i\,\omega_L t$

$$(4.3)\quad \begin{cases} \vec{U}_1 = -ie\vec{E}_L/m\omega_L \qquad n_1 = e\vec{\nabla}\cdot(N\vec{E}_L)/m\omega_L^{\ 2} \\ \vec{U}_2 = ie^2\{(\vec{E}_L.\vec{\nabla})\vec{E}_L + i\omega\ (\vec{E}_L x\vec{B}_L)/c\}/2m^2\omega^3 \\ \vec{J}_2 = N\vec{U}_2 + n_1\vec{U}_1 . \end{cases}$$

Obviously both terms in $\vec{J}_2$ have the frequency $2\omega_L$ More precisely since

$$(4.4)\quad \begin{cases} i\omega_L\vec{B}_L/c = \vec{\nabla}x\vec{E}_L,\ \ \vec{E}_L x\vec{\nabla}x\vec{E}_L = \vec{\nabla}(\vec{E}_L.\vec{E}_L)/2 - (\vec{E}_L.\vec{\nabla})\vec{E}_L\ , \\ \vec{\nabla}.(N\vec{E}_L) = \vec{\nabla}N.\vec{E}_L + N\vec{\nabla}.\vec{E}_L = \vec{\nabla}N.\vec{E}_L + 4\pi Nen_1 = \\ \vec{\nabla}N.\vec{E}_L + \omega_p^{\ 2}\vec{\nabla}.(N\vec{E}_L)/\omega^2_{\ L} = (\vec{\nabla}N.\vec{E}_L)(1+\omega_p^{\ 2}/\omega_L^{\ 2} + \\ \omega_p^{\ 4}/\omega_L^{\ 4} + \ldots.) = (\vec{\nabla}N.\vec{E}_L)/(1-\omega_p^{\ 2}/\omega_L^{\ 2})\ , \end{cases}$$

one finally gets :

$$(4.5)\quad \vec{J}_2 = ie^2\{N\vec{\nabla}(\vec{E}_L.\vec{E}_L)/4 + (\vec{\nabla}N.\vec{E}_L)\vec{E}_L/(1-\omega_p^{\ 2}/\omega_L^{\ 2})\}\ /m^2\omega_L^3 .$$

In a spatially uniform plasma ($\vec{\nabla} N = 0$), only the magnetic dipole term $\vec{E}_L \times \vec{B}_L$ contributes to $\vec{J}_2$. Since it is a longitudinal component, no transverse wave at second harmonic frequency can be generated. In an inhomogeous plasma, there is a contribution of the $\vec{\nabla} N$ term which can give rise to second harmonic generation. As seen in equation (4.5) this is the more so at the cut-off density $\omega_L \simeq \omega_p$. Also note that the incoming radiation should be obliquely incident onto the density gradient with the electric vector in the plane of incidence. The resonant second harmonic generation is thus associated with resonant absorption.

Quantum mechanically, this effect is a two photon absorption followed by a single photon emission. It can be also depicted as a three wave process in which only transverse oscillations are involved. The gradient provides the non linear coupling.

## V. EFFECT OF DENSITY MODULATION

We now investigate the reflexion of laser light impinging onto an inhomogeneous plasma in which longitudinal oscillations have been triggered. According to the results of section 3, they occur near the critical density ($\omega_L = \omega_p$) where electron plasma waves and ion acoustic low frequency waves are generated, and also near the quarter critical density ($\omega_L = 2\,\omega_p$) where electron plasma waves are induced, and independantly, the Raman backscattering which also produces an electron plasma wave, is an absolute instability. Now, assume that a plasma has a linear density profile of length L. Abcissa $z = 0$ corresponds to the critical density $N_c$. In the absence of perturbations the electron plasma frequency along the profile is

$$\omega_p^{\,2} = \omega_L^{\,2}\,(1 + z/L). \tag{5.1}$$

Consider the case in which the light travels from $Z = -\infty$ to the critical density and back. If $\eta_o(z)$ is the refractive index, the phase $\phi_0$ of the reflected light can be written :

$$\phi_0 = -\,(2\,\omega_L\,L/c)\int_{-L}^{0} \eta_0(z)dz = -\,4\omega_L L/3c. \tag{5.2}$$

Now, when the parametric decay takes place at cut-off, the phase is to be calculated taking into account electron motion under the influence of electron ($\omega_e$) and ion acoustic ($\omega_a = \omega_L - \omega_e \ll \omega_L$) waves[11].

Let

(5.3) $\tilde{n}_e = n_e \cos \omega_e t$ and $\tilde{n}_a = n_a \cos \omega_a t$

be the corresponding density perturbations. Setting $\alpha_\sigma = n_\sigma / N_c$, the modified refractive index near the critical density is

(5.4) $\tilde{\eta} = \{\alpha_e \cos\omega_e t + \alpha_a \cos\omega_a t - z(1 - \alpha_e \cos\omega_e t - \alpha_a \cos\omega_a t)/L\}^{1/2}$

The phase of the back reflected radiation is

(5.5) $\tilde{\phi} \sim 2\omega_L / c \int_{-L}^{1} \tilde{\eta}\, dz$

where 1 is the time dependant limit of integration due to the motion of the critical density :

(5.6) $1 = 1_e + 1_a = (\alpha_e \cos\omega_e t + \alpha_a \cos\omega_a t)L$

On integrating, one gets an expression for $\tilde{\phi}$ as a function of $\phi_0$, electron density fluctuations and time. Using the Bessel's identity, the out-going wave has the form :

(5.7) $\exp(i\omega_L t + \tilde{\phi}) = \exp\{(i(\omega_L t - |\phi_0|)\} x \{\sum_{q=-\infty}^{\infty} J_q(|\phi_0| \alpha_e) x$
$\exp\{-iq(\omega_L - \omega_a)t - q\pi/2)\}\{\sum_{m=-\infty}^{\infty} J_m(|\phi_0| \alpha_a)\ x$
$\exp(-i|m\omega_a t - m\pi/2|)\}.$

Thus, one gets the harmonics of the laser frequency with sidebands due to ion acoustic waves. Since a whole spectrum of ion acoustic frequencies is actually permitted, one expects an almost continuous satellite structure with holes corresponding to the ion plasma frequency at cut-off.

This simple minded approach can also be used for Raman backscattering at the quarter critical density. Since the two electron plasma wave resonance, and an absolute Raman instability may occur, electrons oscillate with the frequency $\omega_L/2$. Then setting :

(5.8) $1'_e = (4n_e/N_c)\cos(\omega_L/2)t,$

$\eta' = \{\alpha_e \cos\omega_L t/2 \quad - (2z/L)(1 - \alpha'_e \cos\omega_L t/2)\}^{1/2}$

the phase of the backscattered radiation is

(5.9) $\tilde{\phi} \sim -(\omega_L/c) \int_{-L}^{1'} \eta\, dz - (\omega_L/c) \int_{1'}^{L} \eta\, dz.$

since, here, the terms corresponding to the ingoing and the out going wave, which have different frequencies, should be separated. By the use of the same Bessel expansion as before, one gets a set of harmonic and subharmonic lines separated by $\omega_L/2$ i.e. $\omega_L/2$, $3\omega_L/2$ $2\omega_L$, $5\omega_L/2$, ... with intensities depending on the Bessel's functions arguments.

## VI. MORE ON SUBHARMONIC GENERATION

In the previous sections, we have seen how sources of non linearities, independantly or not, act :

- non linear currents : resonances in which longitudinal and transverse waves are coupled

- $\vec{E}\times\vec{B}$ term : harmonic scattering by a single electron

- $\vec{E}\times\vec{B}$ coupled with gradient : second harmonic generation

- gradient coupled with non linear current : harmonic and subharmonic generation.

It is also possible to investigate coupling between sources of non linearities through a double expansion when the gradient is not too steep and can be represented by the small vector :

(6.1) $$\vec{\eta} = \vec{\nabla} N/N .$$

Another small parameter is of course $\upsilon/c$. The cold plasma equations are then solved expanding $\vec{\upsilon}$ and n up to the first order in $\eta$ and to any order in $\upsilon/c$. Such calculations are rather long and tedious. Since details are given elsewhere[12], we are to skip them and state only the relevant result. Doing so, there is no basic difference with what was done for second harmonic generation. This effect is again found in second order together with a new result : subharmonic generation at frequency $\omega_L/2$ [13], whose connexion to Raman scattering is not very clear. Now, if in the above procedure, one takes into account density modulations due to parametric decay at cut-off, setting

(6.2) $$N = N\{1+n_a(\omega_a) + n_e(\omega_L-\omega_a)\}$$

one finds among the frequencies occuring in second order :

(6.3) $\omega = (3\omega_L - \omega_a)/2$ , $\omega = 2\omega_L \pm \omega_a$

with wave vectors

(6.4) $\vec{k}\,\{(3\omega_L-\omega_a)/2\} = (2\vec{k}_L+\vec{k}_e)/2$, $\vec{k}(2\omega_L \pm \omega_a) = 2\vec{k}_L \pm \vec{k}_a$.

Note that due to the plasma dispersion relation one cannot have simultaneously the $(3\,\omega_L - \omega_a)/2$ frequency and $2\,\omega_L + \omega_a$. So one has to retain $2\,\omega_L - \omega_a$ only and both lines exhibit an asymetric broadening with a width corresponding to the ion acoustic spectrum.

The occurence of the $3\,\omega_L/2$ line can be physically interpreted as the parametric excitation of transverse waves due to the coupling of a Langmuir wave to the longitudinal oscillation of electrons at second harmonic frequency under the influence of the Lorentz force in presence of a gradient. The occurence of the $2\,\omega_L$ line is a Raman up-conversion.

## VII. QUANTUM MECHANICAL INTERPRETATION

In most of the cases reviewed so far, resonances and harmonic generation are associated with the coupling of longitudinal and transverse oscillations. Electromagnetic (transverse) waves can be described equivalently by photons. Longitudinal oscillations can be also represented by quasi particles : plasmons (for electron plasma waves) and phonons (for ion acoustic waves). Three quanta processes are more likely. They are listed in the table 1.

The effects below are directly involved in harmonic and subharmonic generation :

- Raman scattering with subharmonic $\omega_L/2$ at quarter critical density
- Raman upconversion with
  - harmonic $3\,\omega_L/2$ at quarter critical density
  - harmonic $2\omega_L$ at critical density
- Three photon with harmonic $2\omega_L$ in a gradient.

TABLE 1

(  = Photon, —— = Plasmon, ---- = Phonon)

| PROCESS | SCHEME | SELECTION RULES |
|---|---|---|
| Parametric decay |  | $\omega_L=\omega_e+\omega_a$, $\vec{k}_L=\vec{k}_e+\vec{k}_a$, resonant for $\omega_L\sim\omega_p$ |
| Parametric resonance |  | $\omega_L=2\omega_e$, $\vec{k}_L=\vec{k}_{e1}+\vec{k}_{e2}$ |
| Brillouin scattering |  | $\omega_L=\omega_L+\omega_a$, $\vec{k}_L=\vec{k}_s+\vec{k}_a$ |
| Raman scattering |  | $\omega_L=\omega_S+\omega_e$, $\vec{k}_L=\vec{k}_s+\vec{k}_e$, resonant at $\omega_s=\omega_e=\omega_L/2$ |
| Raman upconversion |  | $\omega_L+\omega_e=\omega_u$, $\vec{k}_L+\vec{k}_e=\vec{k}_u$ |
| Three photons |  | Obvious, resonance for $\omega_L=\omega_p$ |

The Raman upconversion is actually coupled to any other process producing electron plasma waves : parametric decay for second harmonic generation, Raman scattering or parametric resonance for 3 $\omega_L/2$ generation. The coupling schemes are :

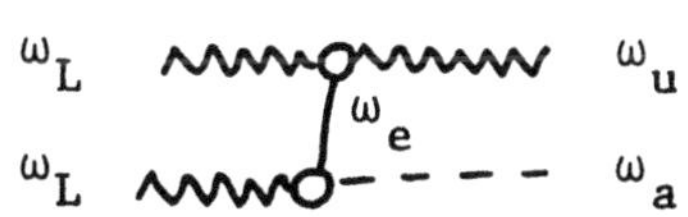


for second harmonic generation. This coupling at the critical density was found in section 6.

A shift is predicted for the $\omega_u$ line whose frequency is $2\omega_L - \omega_a$. Furthermore one has :

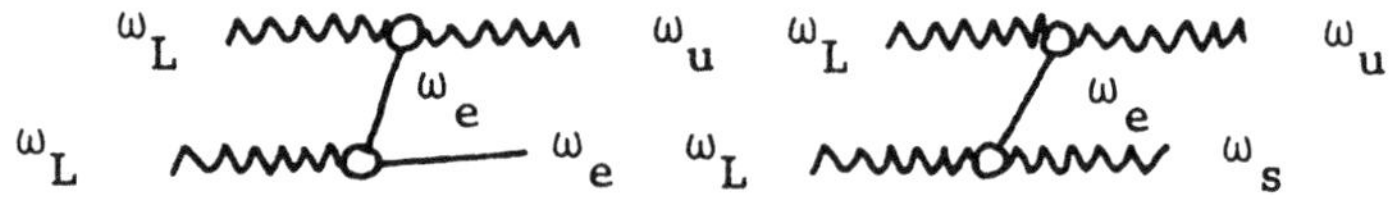


for $3\omega_L/2$ generation at the quarter critical density.

The case of $3\omega_L/2$ generation at the critical density is slightly more complicated. Indeed two quanta $\omega_L$ combine with one plasmon in order to give two photons with frequency $(3\omega_L - \omega_a)/2$ according to the coupling scheme :

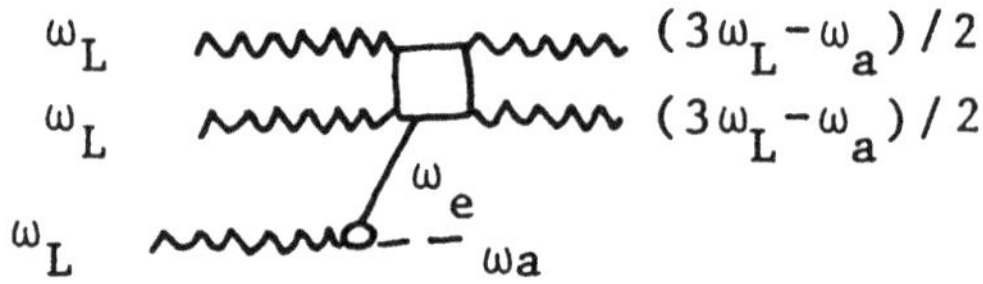


The probability of such a mechanism is expected to be small. Note that it is an example of a stimulated two photon emission process coupled with a two photon absorption.

## 8. COUPLING OF THREE WAVE INTERACTIONS

In order to get $2\omega_L$, $3\omega_L/2$ and $\omega_L/2$, the Raman up-conversion is coupled to a parametric decay or to Raman back-scattering. A plasmon can be created or deleted by either one of two mechanisms. In the case of two plasmon coupled with back-scattering, the density is quarter critical for the incoming E.M. wave and nearly critical for the out going one ($\omega_R$, $\vec{k}_R$). Thus $\vec{k}_R \sim 0$which implies $k_L \sim k_1 = k_2$ and the diagram of Fig. 1a. In order to experimentally achieve such a situation, one should have either two beams 60° apart or a large angular aperture focusing lens. The same condition on $k_R$ holds for the coupling of back-scattering with up-conversion which therfore takes place according to the diagram 1b. Coupling between the two plasmon resonance and the up-conversion involves no restriction on $k_e$ other than $k_{e1} = k_{e2}$. Hence the scheme 1c. Finally the coupling of the up-conversion to the parametric decay at the critical density takes place according to the figure 1d with $k_L \sim 0$, $k_u \sim k_e$.

Assume first that the linear dispersion relations still hold i.e.

(8.1) $$\omega^2 = \omega_p + k^2 c^2, \; \omega_e^2 = \omega_p^2 + 3\, k_e^2 \upsilon_e^2$$

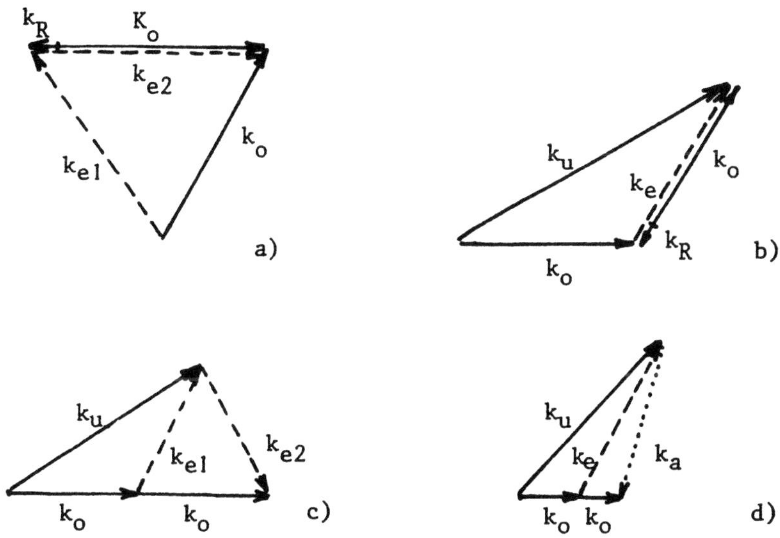


Fig. 1 - Coupling of threewave processes

for E.M. waves and electron plasma waves respectively, $\upsilon_e$ being the thermal velocity of the electrons. Now, for the Raman back-scattering, setting $\omega_L = 2\omega_e = 2\omega_R$, one gets, by using the dispersion relations :

$$k_R^2 = 3\, k_e^2 \upsilon_e^2 / c^2 .$$

Hence :

(8.2) $$\omega_L^2 = 4\omega_p^2 \{1-2(\upsilon_e/c)\sqrt{3}\}\{1-2(\upsilon_e/c)\sqrt{3}-9(\upsilon_e/c)^2\}^{-1}$$

which means that the corresponding density is slightly smaller than quarter critical. This relationship holds whether or not there is a coupling with either the two plasmon resonance or the up conversion. In the latter case, the angle between the two $k_L$ should accordingly have a specified value which if $\upsilon_e/c$ is small is about 70°.

For the coupling of up conversion with the two plasmon decay one has $\omega_u = 3\omega_L/2, \omega_1 = 2\omega_e$, $k_u^2 = 2\, k_L^2 + k_e^2$. Substituting the dispersion relations yield :

(8.3) $$\omega_L^2 = 4\omega_p^2 \, (c^2+3\upsilon_e^2)\, (c^2 - 3\upsilon_e^2)^{-1}$$

which also corresponds to a density smaller than quarter critial, but different from the previous one. It does not depend on the angle between $k_L$ and $k_e$. However, due to the dispersion relation for E.M. waves, the angle between $k_L$ and $k_u$ is fixed (about 23°).

Finally, at cut-off, the coupling of the up-conversion to the parametric decay takes place for the common k vector satisfying $k^2(c^2-12v_e^2) = 3\omega_p^2$. No angular condition is involved.

## 9. SUPERELASTIC SCATTERING AND OTHER SCHEMES

One way of relaxing the severe conditions on k matching stated in section 8, is to introduce a fourth quantum in the up-conversion process. In order, that it carries the extra momentum without perturbing too much the $\omega$ matching, it should be a phonon. The corresponding schemes are

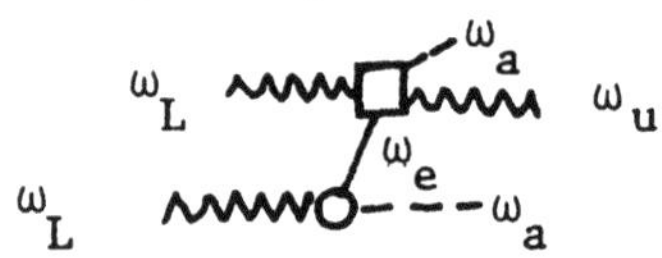


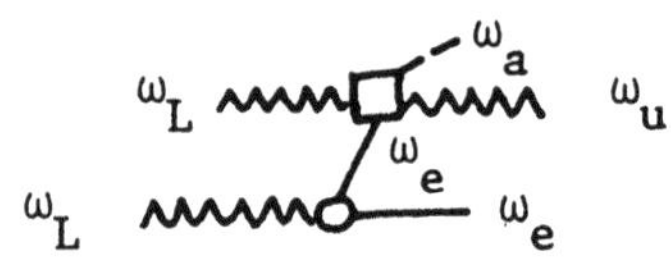


for the second harmonic and the three halves harmonic respectively. These mechanisms account for the asymetric broadening of both lines.[14,15]. The sideband is expected to extend up to the ion-plasma frequency at cut-off for $3\omega_L/2$ and to twice this value for $2\omega_L$.

Other schemes may also be considered. For instance forward Brillouin scattering generates a $(\omega_L-\omega_a)$ wave which in a density gradient is able to create a $2(\omega_L-\omega_a)$ wave through the mechanism of section 4 [15]. Finally an infinite set of cascade process leads to higher order harmonics. The $3\omega_L/2$ wave may interact with itself in a gradient to create $3\omega_L$, or recombines with a plasmon $(\omega_L-\omega_a)$ to give $5\omega_L/2$, or with a plasmon $\omega_L/2$ to contribute to the $2\omega_L$ line.....

## 10. LINEWIDTHS AND GROWTH RATES

In real life, dispersion relations are non linear. The selection rules for the $\omega$ are only approximately satisfied. The resonance is actually obtained with a frequency mismatch depending on the growth and damping rates of the involved waves. This is an alternative way of relaxing the restrictions on k matching. Let us consider, for instance, the two plasmon decay. Setting :

(10.1) $\omega_e = \omega_p + \delta - i\Gamma, \qquad \omega_L = 2\omega_e + \Delta,$

and solving E.A. Jackson's dispersion relaxation

(10.2) $1-\alpha^2\chi_e(\omega)\chi_e(\omega-\omega_L)\ \{1+\chi_e(\omega)\}^{-1}\ \{1+\chi_e(\omega-\omega_L)\}^{-1} = 0$

where $\chi_e$ is the electron susceptibility, and

(10.3) $\alpha = (\vec{k}_L.\vec{k}_e)\ (e\vec{k}_e.\vec{E}_L)/2k_e^2\ m_e^2\ \omega_L^2,$

one gets the growth rate

(10.4) $\gamma = (\alpha^2\omega_p^2 - 4\Gamma^2 - 4\ \delta^2)/4\Gamma$

and the mismatch

(10.5) $\Delta = 2\ \Gamma\ \gamma/\omega_p.$

The frequency of the antistokes line is $3\omega_L/2 - \Delta/2$ which implies a broadening depending on the angles, the incoming flux and the damping rate of the electron plasma waves. This broadening combines with the width $\Delta_0$ of the incoming line to give an overall width

(10.6) $\Delta(3/2) = (\Delta_0^2 + \langle\Delta^2\rangle /4)^{1/2}$

where the average is taken over the k vectors. The line as a whole is red-shifted by $\langle\Delta\rangle/2 < \omega_a$. The growth rate of $3\omega_L/2$ line is about the same as the one of the two plasmon resonance as shown in ref. 16.

The situation is quite similar in the case of $2\omega_L$ generation by up-conversion coupled to the parametric decay. The frequency shift of the latter combines with the ion acoustic spectrum

We have already seen that the superelastic scattering implies a red-shift of the generated transverse wave ($\omega_u$, $\vec{k}_u$). Its amplitude together with that of the acompanying ion-acoustic waves are given by the coupled equations :

(10.7) $$\begin{cases} \partial E_u/\partial t + \Gamma_u\ E_u = i\ C_u E_L E_e E_a^* \\ \partial E_a/\partial t + \Gamma_a\ E_a = i\ C_a E_L E_e E_u^* \end{cases}$$

where the star denotes the complex conjugate and $\Gamma_u$ and $\Gamma_a$ are damping rates (phenomenological). The coupling coefficients $C_u\ C_a$ were calculated in the ref. 17. The instability threshold is given by

(10.8) $E_L\ E_e = \Gamma_u\ \Gamma_a/C_u\ C_a$

and the growth rate is

(10.9) $$\begin{cases} \gamma = (1/2)\ \{-\ (\Gamma_u+\Gamma_a) + |(\Gamma_u-\Gamma_a)^2 + 4\ C_u C_a\ E_L^2\ E_e^2|^{\ 1/2}\} \\ \sim C_u^{1/2}\ C_a^{1/2}\ E_L\ E_e \end{cases}$$

far from threshold. Note that it depends on the amplitude of an electron plasma wave which itself is exponentially generated in a decay instability. Such a behaviour was also found in ref. 3 for the mechanism dealt with in section 6.

## 11. COMPARISON WITH EXPERIMENTAL RESULTS

Harmonics and subharmonics are observed when a low Z target is irradied by a high flux ( $> 10^{13}$ W/cm$^2$) of Nd glass laser radiation ( $\lambda = 1.06$ µm, $\omega_L = 1.8\ 10^{15}$ S$^{-1}$). A striking feature of the $2\omega_L$ and $3\omega_L/2$ lines is a sideband on the Stokes (low frequency) side only. They are observed in the light either backscattered (it is actually meant : within the angular aperture of the focusing lens) or side scattered (90° or 135°). Their intensity is about 1 % of the incoming intensity.

In the experiments performed at Limeil, the $3\omega_L/2$ line was correlated with a strong reflectivity at $\omega_0$ and the occurence of high energy X rays consistent with suprathermal electrons. On the contrary the $2\omega_L$ line was correlated with neutron emission. Maximum intensities of backscattered $2\omega_L$ and $3\omega_L/2$ lines are obtained for different focusing conditions of a laser beam with given intensity and power[18]. It was shown using the second harmonic emission that the vacuum caustics is not perturbed by the plasma[19], and the cut-off density of both harmonics is readily plotted versus the flux density at cut-off. However the $3\omega_L/2$ line comes mainly from a zone outside the target where the flux density is smaller Assuming that the scale length of the density gradient is 100µm and correcting accordingly for the geometry, one gets the solid line of fig. 2 which as for the second harmonic before saturation has a unit slope. Since the energy flux on the target is a constant this behaviour is typical of mechanisms in which two impinging photons combine (note that the process described in section 6 implies a power 1/2 dependance on the flux density at cut-off, whereas the actual dependance is a 0.6 power law). The threshold is consistent with the detailed measurements by Pant and al[20].

Time resolved line shapes were recorded in the case of light scattered at 135°[21]. They are reproduced on fig. 3 with an ordinate scale linear instead of logarithmic. The main line and the different components are thus more easily distinguishable. They are tentatively identified as indicated. The second harmonic has 4 components: a sharp line which exists even at lower flux densities, a rather broad sideband which peaks at $\omega_{aM} \sim 0.2\omega_{ac}$ (ion plasma frequency at cut-off) and two satellite lines, one blue shifted by $\Delta\omega_B \sim 0.25\ \omega_{ac}$, the other one red shifted by $\Delta\omega_R \sim 1.25\omega_{ac}$. According to Nishikawa[22], the highest growth rates of the parametric decay are obtained for $k_a \sim 0.1$-$0.2\ k_D$.

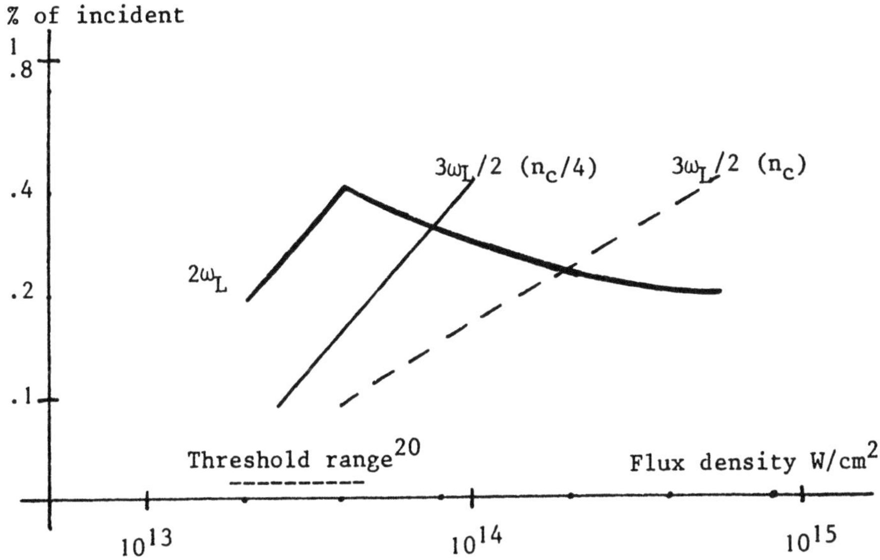


Fig. 2 - Intensity of harmonic lines

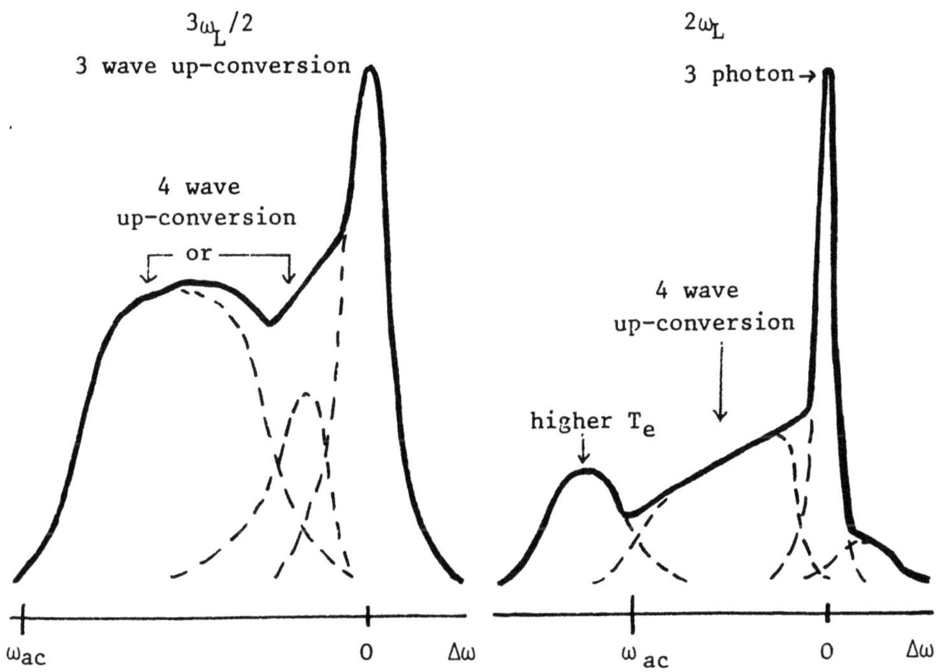


Fig. 3 - Shape of harmonic lines

Furthermore, assuming a reasonable electron temperature (0.5keV) one expects a large value of $k_a$ ($>4k_u$). Accordingly the most probable up-conversion process is four wave back-scattering, which may thus account for the side band. The satellite on the low frequency side would require a higher phase velocity of the ion acoustic wave together with a Doppler shift. They could be related to a two temperature electron distribution due to the parametric decay instability. Therefore a smaller $k_a$ could be colinear with $k_u$ and undergo a large Doppler shift due to the plasma flow at cut-off. In the same conditions the other satellite could be related to antistokes Brillouin scattering.

In the $3\,\omega_L/2$ line, the peak is associated with the coupling of Raman up conversion to the two plasmon decay with a broadening depending on the growth rates, while the red shifted satellite is attributed to superelastic scattering. A rather weak sideband in between might be due to the mechanism described in section 6. It could be due as well to 4 wave up-conversion whereas the satellite could be related to the higher temperature component of the electron distribution.

Different line shapes were recorded in back reflected light both with Nd glass[14,20] and $CO_2$[23] laser irradiation. In the experiment with $CO_2$, the second harmonic has a low frequency sideband peaking at $0.1\,\omega_{ac}$ which might also come from four wave scattering.

## 12. CONCLUSION

Obviously, the study of harmonic scattering in laser created plasmas is a source of information about the way the impinging E.M. waves couple to an inhomogeneous plasma. Although the connexions are not yet crystal clear, existence, broadening, global shift[24], sidebands, satellites of harmonic lines may be eventually ascribed, sometimes through a one to one correspondence, to definite instabilities and other plasma properties. As shown in the theoretical part of this paper, the theory of the processes leading to harmonic generation is mainly phenomenological and includes various approaches. The interpretation of comparatively scarce experimental data is not firmly established and remains controversial. We need both a better theoretical frame occasionally backed by numerical simulations[25] and more accurate experimental results with very detailed time and frequency resolution. It should be noted finally that the occurence of harmonics is also a diagnostic for the plasma flows. Localization of the emitting zones leads to the evaluation of the scale lenght of the density gradients[19,26].

I am indebted to A. Decoster for permission of quoting his results prior to publication.

REFERENCES

1. M. Decroisette, B. Meyer, G. Piar, J. de Phys. 32, Colloque C 5b (1971) 119
2. e. g. N. Bloembergen. Non Linear Optics Benjamin (1965)
3. J.L. Bobin, M. Decroisette, B. Meyer, Y. Vitel, Phys. Rev. Lett 30 (1973) 594
4. e. g. The Review by T.W. Johnston, Canadian Summer School on Plasmas, Banff (1975) which contains a pertinent bibliography
5. L. Landau, G. Placzek, Phys. Z. Soviet 5 (1934) 172. This approach has been recently taken up by R. Balescu and J. Paiva Veretennicoff. Physica to be published.
6. A. Decoster. to be published
7. BB. Winkles, O. Eldridge, Phys. Fluids 15 (1972) 1790
8. Yu. M. Aliev, VP Silin, Soviet Phys. J.E.T.P. 21 (1965) 601
9. E.A. Jackson. Phys. rev. 153 (1967) 235
10. Y.R. Shen. Rev. Mod. Phys. 48 (1976) 1
11. J. Martineau, J.L. Bobin, Phys. Lett. 47A (1974) 43
12. J.L. Bobin in Laser interaction and related plasma phenomena vol. 3B, H. Schwarz, H. Hora, Ed. Plenum N.Y. (1974) p. 591
13. C.S. Chen, J. Plasma. Phys. 5 (1971) 107
14. C. Yamanaka et al. Phys. Rev. Lett. 32 (1974) 1038
15. P.L. Mascheroni. Rept. F.R.C.R. n° 84 (1975) U. of Texas, Preprint
16. J.L. Bobin. Optics Comm. 14 (1975) 339
17. Nguyen Duc Long, KJ. Parbhakar, T.W. Johnston, Proceeding of the 7th European Conference on Plasma Physics and Nuclear Fusion. Lausanne 1 (1975) 73.
18. A. Salères, M. Decroisette, F. Floux, J.L. Bobin. Phys. Lett. 45A (1973) 454
19. A. Salères, M. Decroisette, C. Patou, Optics Comm. 13 (1975) 321
20. H.C. Pant, K. Eidmann, P. Sachsenmaier, R. Sigel. Optics Comm. 16 (1976) 396
21. M. Decroisette, B. Meyer, Y. Vitel. Phys. Lett. 45A (1973) 443
22. K. Nishikawa. J. Phys. Soc. Japan 24 (1968) 1152
23. E. Fabre, C. Popovics, C. Stenz. Proceedings of the 7th European Conference on Plasma Physics and Controled Fusion Lausanne 1 (1975) 80
24. ON. Krokin, VV. Pustovalov et al... JETP Lett. 22 (1975) 21
25. D. Biskamp, H. Welter Phys. Rev. Lett. 34 (1975) 312
26. S. Jackel, J. Albritton, E. Goldman. Phys. Rev. Lett. 35 (1975) 514 and private communications.

# COUPLING SATURATION IN THE NONLINEAR THEORY OF PARAMETRIC DECAY INSTABILITIES

D. F. DuBois and B. Bezzerides

Theoretical Division, Los Alamos Scientific Laboratory

P. O. Box 1663, Los Alamos, New Mexico, 87545

An understanding of the nonlinear stage of parametric instabilities is needed to predict the anomalous absorption of the pump mode in laser or microwave interactions with a plasma. The waves excited in the linear stage of the instability grow up from thermal noise until some nonlinear mechanism limits their growth resulting in a saturated (perhaps quasi-steady state) level of waves. Considerable theoretical analysis [1,2] has been devoted to the "ladder-decay" mechanism whereby the secondary waves excited by the pump excite a ladder of further decay instabilities which spread the wave spectrum out of the primary unstable region of wave-vector space ($\vec{k}$) into regions where the waves are stabilized by linear dissipative processes. The transfer of waves in $\vec{k}$ space arises formally as a nonlinear, i.e.,wave intensity dependent, correction to the wave damping (e.g. nonlinear induced scattering of Langmuir waves from ions). These analyses employ the standard techniques of weak turbulence theory [3] including the random-phase approximation which assumes that waves of different $\vec{k}$ are uncorrelated. This assumption is clearly questionable for parametric instabilities in which a coherent pump mode induces a definite phase and amplitude correlation between the unstable daughter waves.

In this paper we wish to describe some recent work which attempts to partially remove the random-phase approximation by retaining the correlations induced by the coherent pump mode. We find that wave-intensity dependent corrections to the mode coupling coefficients now arise in addition to the modification of the wave frequency and damping familiar in ordinary weak turbulence theory. This new effect arising from pump induced correlations occurs at the same level of the weak turbulence expansion as the familiar nonlinear induced-scattering corrections. We find in the examples treated that the mode-coupling coefficients are reduced, resulting in a strong saturation mechanism for decay instabilities.

In section I we derive the nonlinear equations describing decay instabilities starting from the susceptibility expansion of the nonlinear Poisson equation. In section II we apply the theory to the $2\omega_{pe}$ (electron-electron) decay instability. We find that coupling saturation dominates the mechanism of induced scattering from ions producing a much lower wave level than previous estimates [2]. In section III the electron-ion decay instability is considered for $T_e \gg T_i$. The coupling saturation mechanism strongly limits the wave energy in the most favorably matched mode for strong pump amplitudes. This mode in turn drives further decay instabilities spreading the spectrum to lower $|\vec{k}|$. The total wave energy in a one-dimensional model increases as $E_0^{4/3}$ for sufficiently strong pump intensities $E_0$, whereas if coupling saturation is neglected the power law is $E_0^4$.

## I. NONLINEAR DISPERSION RELATION

We will base our derivation on the nonlinear Poisson equation

$$\varepsilon(k)\ E(k) = \frac{1}{2!}\int\frac{d^4k'}{(2\pi)^4}\int\frac{d^4k''}{(2\pi)^4}\ X_{(2)}(k,k',k'')E(k')E(k'')(2\pi)^4\delta^4(k-k'-k'')$$

$$+\frac{1}{3!}\int\frac{d^4k'}{(2\pi)^4}\int\frac{d^4k''}{(2\pi)^4}\int\frac{d^4k'''}{(2\pi)^4}\ X_{(3)}(k,k',k'',k''')E(k')E(k'')E(k''')$$

$$\times\ (2\pi)^4\delta^4(k-k'-k''-k''') + O(E^4) \tag{1.1}$$

where $\varepsilon(k)$ is the linear dielectric function for a homogeneous, isotropic plasma, and $X_{(2)}$ and $X_{(3)}$ are the second and third order nonlinear susceptibilities, respectively. We employ the notation $(\vec{k},\omega) = k$. As is well-known, a monochromatic pump wave

$$E_0(k) = E_0(2\pi)^4\left[\delta^4(k-k_0) + \delta^4(k+k_0)\right]$$

linearly excites a pair of growing decay modes $(\vec{k}_1,\ \omega_1)$ and $(\vec{k}_2,\ \omega_2)$, where $\omega_0 = \omega_1 + \omega_2$ and $\vec{k}_0 = \vec{k}_1 + \vec{k}_2$ for those cases where we may neglect anti-Stokes coupling.[4] We assume then that nonlinearly the saturated spectrum is composed of a collection of paired waves $\vec{k}_{n1},\ \omega_{n1}$ and $\vec{k}_{n2},\ \omega_{n2}$ where

$$k_0 = k_{n_1} + k_{n_2}\ . \tag{1.2}$$

This pairing, or correlation, exists in the linear theory. It is a direct result of the sinusoidal modulation of the system by the coherent pump in time and space which couples together modes whose frequencies and wave vectors differ by integral multiples of $\omega_0$ and $\vec{k}_0$, respectively. Since this coupling arises from the pump imposed symmetry, it must also exist in the nonlinear theory. In a more detailed and rigorous treatment, we examine the response of the nonlinear or turbulent plasma [4] to a small perturbation in the presence of the coherent pump modulation and show that the dispersion relation to be derived below follows from the marginal stability requirement of the nonlinear system. Therefore, we postulate that the total field satisfying Eq. (1.1) may be written

$$E(k) = E_0(k) + \sum_{s=1}^{2} \sum_{n_s} (2\pi)^4 \left[ E_{n_s} \delta^4(k - k_{n_s}) + E^*_{n_s} \delta^4(k + k_{n_s}) \right] + \delta E(k) \ . \tag{1.3}$$

Here, $E_{n_1}$ and $E_{n_2}$ are the complex amplitudes of the $n^{th}$ correlated pair associated with a given frequency $\omega_{n_1}$ or $\omega_{n_2}$, and $\delta E(k)$ denotes the presumably weak contribution to the field from nonresonant beats of the excited pairs.

If we substitute (1.3) into (1.1), we obtain a set of coupled equations for $E_{n_1}$, $E_{n_2}$, and $\delta E$. To lowest (linear) order in $\delta E$, we find for $E_{n_1}$

$$\begin{aligned} \varepsilon(k_{n_1}) E_{n_1} &= X_{(2)}(k_{n_1}, k_0, -k_{n_2}) E_0 E^*_{n_2} + \frac{1}{(2\pi)^4} \frac{1}{\delta^4(0)} \\ &\times \sum_s \sum_{n_s'} X_{(2)}(k_{n_1}, k_{n_s'}, k_{n_1} - k_{n_s'}) E_{n_s'} \, \delta E(k_{n_1} - k_{n_s'}) \\ &+ \sum_s \sum_{n_s'} X_{(3)}(k_{n_1}, k_{n_1}, k_{n_s'}, -k_{n_s'}) E_{n_1} |E_{n_s'}|^2 \\ &+ \sum_{n'} X_{(3)}(k_{n_1}, -k_{n_2}, k_{n_1'}, k_{n_2'}) E^*_{n_2} E_{n_1'} E_{n_2'} \end{aligned} \tag{1.4}$$

To obtain this we have used the invariance of $X_{(3)}(k,k',k'',k''')$ to permutations of the last three indices. There are six identical contributions which cancels the 3! in (1.1).

A similar equation can be written for $E_{n_2}$. The beat terms $\delta E$ are given by

$$\varepsilon(k_{n_1}-k_{n_1'})\delta E(k_{n_1}-k_{n_1'}) = (2\pi)^4\delta^4(0)\Big[X_{(2)}(k_{n_1}-k_{n_1'},k_{n_1},-k_{n_1'})E_{n_1}E^*_{n_1'}$$

$$+X_{(2)}(k_{n_1}-k_{n_1'},k_{n_2'},-k_{n_2})E^*_{n_2}E_{n_2'}\Big] \tag{1.5a}$$

and

$$\varepsilon(k_{n_1}-k_{n_2'})\delta E(k_{n_1}-k_{n_2'}) = (2\pi)^4\delta^4(0)\Big[X_{(2)}(k_{n_1}-k_{n_2'},k_{n_1},-k_{n'_2})E^*_{n_2'}E_{n_1}$$

$$+X_{(2)}(k_{n_1}-k_{n_2'},-k_{n_2},k_{n_1'})E_{n_1'}E^*_{n_2}\Big], \tag{1.5b}$$

where we have used $k_{n_1}-k_{n_1'} = -k_{n_2} + k_{n_2'}$ which follows from (1.2) We will show that all the nonlinear terms in (1.4) and the corresponding equation for $E_{n_2}$ are consistent with a renormalized set of coupled equations,

$$\tilde{\varepsilon}(k_{n_1})E_{n_1} = \tilde{\chi}(k_{n_1},k_0,-k_{n_2})E_0E^*_{n_2}, \tag{1.6a}$$

$$\tilde{\varepsilon}(k_{n_2})E_{n_2} = \tilde{\chi}(k_{n_2},k_0,-k_{n_1})E_0E^*_{n_1}, \tag{1.6b}$$

a result identical <u>in form</u> with that derived from the linear theory, except that the tilded quantities are now functions of the field intensities $|E_{n_1}|^2$ and $|E_{n_2}|^2$.

Let us pause to comment on the linear theory of decay instabilities. In this case coupled equations of the form (1.6a) and (1.6b) arise but with $\tilde{\varepsilon}(k) \to \varepsilon(k)$ and $\tilde{\chi} \to X_{(2)}$. The solubility condition for these equations produces a dispersion relation

$$\tilde{\varepsilon}(k_{n_1})\tilde{\varepsilon}(k_{n_2})^* - \tilde{\chi}(k_{n_1},k_0,-k_{n_2})\tilde{\chi}^*(k_{n_2},k_0,+k_{n_1})|E_0|^2 = 0 \tag{1.7}$$

which determines $\omega_{n_1}$ given $\vec{k}_{n_1}$, and $\vec{k}_0$. If this dispersion relation has a solution $\omega_{n_1}(\vec{k}_{n_1})$ then it can be shown to also have a solution if $\omega_{n_1}(\vec{k}_{n_1})$ is replaced by $\omega_{n_2}(\vec{k}_0-\vec{k}_{n_1}) = \omega_{n_2}(\vec{k}_{n_2}) = \omega_0-\omega^*_{n_1}(\vec{k}_{n_1})$and $\vec{k}_{n_1}$ is replaced by $\vec{k}_{n_2} = \vec{k}_0 - \vec{k}_{n_1}$. Neither Re $\omega_{n_1}(k_{n_1})$, or Re $\omega_{n_2}(k_{n_2})$ <u>necessarily</u> correspond to the dispersion curves of uncoupled modes such as Langmuir or ion acoustic modes. The imaginary parts are equal, Im $\omega_{n_1}(k_{n_1})$= Im $\omega_{n_2}(k_{n_2})$ and provide the growth rate of the parametric instability.[2] The same relation between $\omega_{n_1}(k_{n_1})$ and $\omega_{n_2}(k_{n_2})$ exists in the nonlinear theory but here we consider only the marginally stable modes Im $\omega_{n_1}$ = Im $\omega_{n_2}$ = 0 for which nonlinear saturation effects balance the growth rate. These marginally stable modes form the highly excited part of the wave spectrum. To complete the nonlinear theory we must compute the renormalized $\tilde{\varepsilon}$ and $\tilde{\chi}$.

If we substitute the expressions of (1.5) for $\delta E$ into (1.4), we find terms either proportional to $E_{n_1}$, and therefore contributing to $\tilde{\varepsilon}$, comparing with (1.6a), or terms proportional to $E^*_{n_2}$ contributing to the right hand side of (1.6b). Consideration of terms proportional to $E_{n_1}$ leads to

$$\tilde{\varepsilon}(k_{n_1}) = \tilde{\varepsilon}(k_{n_1}) - \sum_s \sum_{n'_s} \left[ \frac{X_{(2)}(k_{n_1}, k_{n'_s}, k_{n_1} - k_{n'_s}) X_{(2)}(k_{n_1} - k_{n'_s}, -k_{n'_s}, k_{n_1})}{\varepsilon(k_{n_1} - k_{n'_s})} + (k_{n_1} \leftrightarrows k_{n_2}) + X_{(3)}(k_{n_1}, k_{n_1}, k_{n'_s}, -k_{n'_s}) \right] |E_{n'_s}|^2 \tag{1.8}$$

The terms proportional to $E^*_{n_2}$ occur proportional to products such as $E_{n'_1} E_{n'_2} E^*_{n_2}$. We can use the ansatz (1.6b) to write the cross product as

$$E_{n_1} E_{n_2} = \frac{\tilde{\chi}(k_{n_2}, k_0, -k_{n_1})}{\tilde{\varepsilon}(k_{n_2})} E_0 |E_{n_1}|^2 \tag{1.9}$$

Therefore, the terms proportional to $E^*_{n_2}$ are also proportional to $E_0$ and we find the following equation for $\tilde{\chi}$ :

$$\tilde{\chi}(k_{n_1}, k_0, -k_{n_2}) = X_{(2)}(k_{n_1}, k_o, -k_{n_2}) + \sum_{n'_1} \left[ \frac{X_{(2)}(k_{n_1}, k_{n'_1}, k_{n_1} - k_{n'_1}) \, X_{(2)}(k_{n_1} - k_{n'_1}, k_{n'_2}, -k_{n_2}) + (k_{n'_2} \rightleftarrows k_{n'_1})}{\varepsilon(k_{n_1} - k_{n'_1})} + X_{(3)}(k_{n_1}, -k_{n_2}, k_{n'_1}, k_{n'_2}) \right] \frac{\tilde{\chi}(k_{n'_2}, k_o, -k_{n'_1})}{\varepsilon(k_{n'_2})} |E_{n'_1}|^2 \tag{1.10}$$

It is clear that to terms of order $|E_{n_s}|^2$ the nonlinearities higher than the third make no contribution to these equations. Similarly we find

$$\tilde{\chi}(k_{n_2},k_o,-k_{n_1}) = X_{(2)}(k_{n_2},k_o,-k_{n_1})$$

$$+\sum_{n_1'}\left[\frac{X_{(2)}(k_{n_2},k_{n_2'},k_{n_2}-k_{n_2'})\;X_{(2)}(k_{n_2}-k_{n_2'},k_{n_1'},-k_{n_1})}{\varepsilon(k_{n_2}-k_{n_2'})}\right.$$

$$\left.+\,(k_{n_2'}\rightleftarrows k_{n_1'})+X_{(3)}(k_{n_1},-k_{n_2},k_{n_1'},k_{n_2'})\right]\frac{\tilde{\chi}(k_{n_2'},k_o,-k_{n_1'})}{\tilde{\varepsilon}\,(k_{n_2'})}\,|E_{n_1'}|^2 \qquad (1.11)$$

The Eqs. (1.6), (1.8), (1.10), and (1.11) form a complete set for the determination of $E_{n_s}$. The nonlinear terms of $\tilde{\varepsilon}$ are familiar from the usual form of weak turbulence theory, resulting in the modification of wave damping and frequency[1]. The contribution to Im $\tilde{\varepsilon}(\vec{k}_n,\omega_n)$ from the beating of high frequency waves, $|E_{n'}|^2$ includes the process of induced scattering from ion fluctuations [1,2]. The $X_{(3)}$ correction to the $\tilde{\varepsilon}$ has the well-known effect of weakening the nonlinear induced scattering of Langmuir waves from electrons but plays no role in the nonlinear induced scattering from ions which was the only effect in Im$\tilde{\varepsilon}$ considered in [1,2]. The $X_{(3)}$ contribution to the renormalized $\tilde{\chi}$ appears to be quite important. The nonlinear terms of $\tilde{\chi}$, on the other hand, are novel for several reasons:

(1) They change the coupling to the pump of the excited fields. In many situations of interest, we find that $\tilde{\chi}$ is significantly reduced from its linear value, leading to saturated energies substantially lower than those from just $\tilde{\varepsilon}$ ;

(2) from the derivation above it is clear that these terms result from the correlated product $E_{n_1}E_{n_2}$. In the RPA, this correlation vanishes; however, it is intrinsic to the parametric coupling that this correlation is strong;

(3) we see that these terms lead to an integral equation for $\tilde{\chi}$ as compared with $\tilde{\varepsilon}$ where the nonlinearity is simply additive.

Since the solution to this integral equation implies that $\tilde{\chi}$ contains contributions to all orders in $|E_{n_s'}|^2$ it is important to understand its relation to the weak turbulence expansion. First we remark that if higher order terms in the nonlinear Poisson equation (1.1) are taken into account as well as higher iterations of the terms already explicit in (1.1) we find that in the square brackets of (1.11) additional terms in powers of $|E_{n_s'}|^2$ appear. In this sense we have a weak turbulence expansion of the kernel of the integral Eqn.(1.11) for $\tilde{\chi}$; we keep only the leading term of this expansion. In the cases of interest discussed in Section II and III we see that the iterative solution of (1.11) for $\tilde{\chi}$ in powers of $|E_{n_s'}|^2$ does not converge because of the singular nature of the factor $\tilde{\varepsilon}^{-1}(k_{n_2'})$ for the most perfectly matched modes in the sum over $n'$.

Up to this point we have not made any assumptions concerning the sources of the fields $E_{n_s}$. We can treat the case where the fields are driven up from thermal noise by adding fluctuating discrete particle density source $4\pi\delta\rho_{n_1}/(i|k_{n_1}|)$, $4\pi\delta\rho_{n_2}/(i|k_{n_2}|)$ to the right hand sides, respectively, of Eqns. (1.6a) and (1.6b). Multiply (1.6a) by $E^*_{n_1}$ and (1.6b) by $E^*_{n_2}$ and take the ensemble average, denoted by $\langle\ldots\rangle$, of both equations. As in usual weak turbulence theory we assume that the $E_{n_s}$ are gaussian random variables. If we also assume as a result of the random sources that modes with different pair indices n are uncorrelated, i.e. $\langle E_{n_s}E_{n_s'}\rangle = \langle E_{n_s}E^*_{n_s'}\rangle = 0$ for $n \neq n'$ we can decompose the average of a product of four amplitudes $E_{n_s}$ into products of averages of pairs of amplitudes $E_{n_s}$ to write

$$\langle \tilde{\varepsilon}(k_{n_1})|E_{n_1}|^2\rangle = \langle \tilde{\varepsilon}(k_{n_1})\rangle\langle|E_{n_1}|^2\rangle$$

$$\langle \tilde{\chi}(k_{n_1},k_{n_1},-k_{n_2})E^*_{n_1}E^*_{n_2}\rangle = \langle \tilde{\chi}(k_{n_1},k_{n_o},-k_{n_2})\rangle\langle E^*_{n_1}E^*_{n_2}\rangle$$

Note, however, we do retain the correlation between $E_{n_1}$ and $E_{n_2}$ - members of a given correlated pair. The RPA is replaced by the condition that no correlation exists between members of different correlated pairs. The result of this analysis is that in the nonlinear dispersion relation for the resonant excitations of the turbulent, noise-driven system is the same as (1.7) with $\tilde{\varepsilon}$ and $\tilde{\chi}$ replaced by their ensemble averages[4]. In this case we replace $|E_{n_s}|^2$ in (1.8), (1.10) and (1.11) by $\langle|E_{n_s}|^2\rangle$.

The same results are derived from a more complete and rigorous formulation of weak turbulence theory in a longer publication[4]. It is important to emphasize that CS is distinct from pump depletion. In the present work, the pump parameters are held fixed.

## II. NONLINEAR SATURATION OF THE $2\omega_{pe}$ DECAY INSTABILITY

We now consider the nonlinear saturation of the electron-electron decay instability (EED or $2\omega_{pe}$) to illustrate the relative significance of the coupling saturation (CS) mechanism. Postovalov, Silin, and Tikhonchuk (PST)[2] have recently calculated the nonlinear saturation of this instability, taking as the dominant saturation mechanism the spreading of the spectrum in angle due to nonlinear induced scattering of waves from ions. We merely take over the results of their calculation showing how the CS mechanism enters. The solubility condition of Eq. (1.7) is equivalent to

$$(\omega_{n_1} - \tilde{\omega}_1 + i\tilde{\gamma}_1)\ (\omega_0 - \omega_{n_1} - \tilde{\omega}_2 + i\tilde{\gamma}_2) + \tilde{\Gamma}(k_{n_1})^2 = 0 \tag{2.1}$$

$$\tilde{\Gamma}(k_{n_1})^2 = \frac{\tilde{\chi}(k_{n_1},k_0,-k_{n_2})\ \tilde{\chi}^*(k_{n_2},k_0,-k_{n_1})}{\frac{\partial}{\partial\omega_{n_1}}\ \mathrm{Re}\varepsilon(k_{n_1})\frac{\partial}{\partial\omega_{n_2}}\ \mathrm{Re}\varepsilon\ (k_0-k_{n_1})}\ |E_0|^2, \tag{2.2}$$

where we assume that the frequency, $\omega_{n_1}$, is near one of the roots of $\tilde{\varepsilon}(k)$ and $\tilde{\varepsilon}(k-k_0)$, a condition which allows us to use resonance approximations for the $\varepsilon$'s. In Eq. (2.1), $\omega_1 = \omega_\ell(\vec{k})$, $\omega_2 = \omega_\ell(\vec{k}_0-\vec{k})$, the unperturbed Langmuir frequencies for the decay plasmons (the non-linear shift in the frequencies is small), and, since $\vec{k}_0-\vec{k} \simeq \vec{k}$, $\tilde{\gamma}_1 \simeq \tilde{\gamma}_2 = \tilde{\gamma}_\ell$, the corresponding damping, including the nonlinear enhancement [2].

From the real and imaginary parts of Eq. (2.1), we have

$$\omega_{n1} = \frac{\omega_0}{2} + \frac{\omega_\ell^o(\vec{k}_n) - \omega_\ell^o(\vec{k}_0 - \vec{k}_n)}{2} \tag{2.3}$$

and the marginal stability condition

$$\tilde{\gamma}_\ell = \left[\tilde{\Gamma}^2 - \left(\frac{\Delta\omega}{2}\right)^2\right]^{1/2} \tag{2.4}$$

where $\Delta\omega = \omega_\ell^o(\vec{k}) + \omega_\ell^o(\vec{k}_0 - \vec{k}) - \omega_0$. The solution of (12) obtained by PST with $\tilde{\Gamma}^2 = \Gamma^2$, i.e., in the absence of CS, is localized in k about $k_m$ where $\Delta\omega(k_m) = 0$, but spread in angle about the fastest growing mode direction by nonlinear induced scattering from ions. They obtain, for the total energy normalized to the particle energy density,

$$W = \sum_{s=1}\sum_{n_s} \frac{|E_{n_s}|^2}{4\pi n\ T_e} = \rho\,\frac{(\sqrt{P} - 1)^2}{\sqrt{P}} \tag{2.5}$$

where $\rho = \frac{243}{\pi\sqrt{2}}\ \frac{\gamma_\ell}{\omega_{pe}}\ \frac{m_i}{m_e}\ \frac{v_e^2}{c^2}\left(1 + \frac{T_i}{T_e}\right)$, and $P = \frac{\Gamma^2}{\gamma_\ell^2}$, the ratio of the pump energy to the threshold energy. To include CS, we turn to Eq. (1.11).

We have computed $X_{(3)}$ by taking the third functional derivative of the density using the techniques of reference [6]. The result is:

$$X_{(3)}(k_{n_1},-k_{n_2},k_{n_1'},k_{n_2'}) = \frac{e^2}{m^2\,\omega_{pe}{}^4}$$

$$\times\Big[(\hat{k}_{n_1}\cdot\hat{k}_{n_1'})(\hat{k}_{n_2}\cdot\hat{k}_{n_2'})\;\chi_e(k_{n_1}-k_{n_1'})(\vec{k}_{n_1}-\vec{k}_{n_1'})^2$$
$$+(\hat{k}_{n_1}\cdot\hat{k}_{n_2'})(\hat{k}_{n_2}\cdot\hat{k}_{n_1'})\;\chi_e(k_{n_1}-k_{n_2'})(\vec{k}_{n_1}-\vec{k}_{n_2'})^2 \tag{2.6}$$
$$+(\hat{k}_{n_1}\cdot\hat{k}_{n_2})(\hat{k}_{n_1'}\cdot\hat{k}_{n_2'})\;\chi_e(k_{n_1}+k_{n_2})(\vec{k}_{n_1}+\vec{k}_{n_2})^2\Big]$$

where we have assumed $\omega_{n_s} \approx \omega_{pe} >> |\vec{k}_{n_s}|v_e$, $\omega_{n_s'} \approx \omega_{pe} >> |\vec{k}_{n_s'}|v_e$ (s=1,2). Here $\chi_e(k)$ is the familiar collisionless electron susceptibility. The terms for which the frequency arguments of $\chi_e$ are small dominate. For the $2\omega_{pe}$ instability we can have

$$\omega_{n_1} - \omega_{n_1'} << |\vec{k}_{n_1}-\vec{k}_{n_1'}|v_e$$

$$\omega_{n_1} - \omega_{n_2} << |\vec{k}_{n_1}-\vec{k}_{n_2'}|v_e$$

since $\omega_{n_1}$, $\omega_{n_2} \simeq \omega_{pe}$ but in the last term $\omega_{n_1} + \omega_{n_2} \simeq 2\omega_{pe} >> |\vec{k}_{n_1}+\vec{k}_{n_2}|v_e$ so this term is smaller by terms of order $|\vec{k}_{n_1}+\vec{k}_{n_2}|^2/k_{D_e}^2$. The well-known result for $X_{(2)}$ can be written to leading terms as

$$X_{(2)}(k_{n_2},k_{n_2'},k_{n_2}-k_{n_2'}) = i\frac{e}{m\omega_{pe}^2}(\hat{k}_{n_2}\cdot\hat{k}_{n_2'})\chi_e(k_{n_2}-k_{n_2'})|\vec{k}_{n_2}-\vec{k}_{n_2'}| \tag{2.7}$$

The square bracketed expression in (1.11) can then be written on substitution of (2.6) and (2.7) as

$$-\frac{e^2}{m^2}\frac{(\hat{k}_{n_2}\cdot\hat{k}_{n_2'})(\hat{k}_{n_1}\cdot\hat{k}_{n_1'})}{\omega_{pe}{}^4}\frac{\chi_e(k_{n_1}-k_{n_2'})[\chi_i(k_{n_1}-k_{n_1'})+1]}{\varepsilon(k_{n_1}-k_{n_1'})} + (k_{n_1} \rightleftarrows k_{n_2}) \tag{2.8}$$

where we have used $\varepsilon(k) = 1 + \chi_e(k) + \chi_i(k)$ and $\chi_i(k)$ is the collisionless ion susceptibility.

In (1.11) we use $k_{n_1} = k$, $k_{n_2} = k_o - k$ and define $\tilde{\chi}(\vec{k}) \equiv \tilde{\chi}(k_{n_2},k_{n_o},-k_{n_1})$ which then obeys (changing sums to integrals)

$$\tilde{\chi}_-(\vec{k}) = \chi(\vec{k}) + \frac{e^2k_{D_e}^2(T_e/T_i)}{m^2\omega_{pe}^3(1+T_e/T_i)}\int\frac{d^3k'}{(2\pi)^3}\;\frac{(\hat{k}\cdot\hat{k}')^2\;|E(\vec{k}')|^2\;\tilde{\chi}_-(\vec{k}')}{\Delta\omega(\vec{k}') - 2i\gamma_\ell(\vec{k}')} \tag{2.9}$$

where we have taken the second term in (1.11) equal to the first since $|\vec{k}_o - \vec{k}| \simeq |\vec{k}|$, employed a resonant approximation for $\varepsilon(k-k_o)$ as in Eq. (10), and used the static limit for $|\vec{k}-\vec{k}'|^2 \, \varepsilon(\vec{k}-\vec{k}', \omega-\omega') \simeq k_{D_e} + k_{D_i}$, and for $\chi_e = k_{D_e}/(\vec{k}-\vec{k}')^2$ and $\chi_i = k_{D_i}/(\vec{k}-\vec{k}')^2$. Since the spectrum, as derived by PST, is narrowly confined to $k' \simeq k_m$, for which $\Delta\omega = 0$, and assuming that $\tilde{\chi}$ is slowly varying in angle compared to variation of $|E|^2$, we find the $\tilde{\chi}$ for the most unstable vector $\vec{k}_m$ is given by

$$\tilde{\chi}_-(\vec{k}_m) \simeq \chi(\vec{k}_m) \left[ 1 + \frac{i\omega_{pe} W}{2\tilde{\gamma}_\ell(\vec{k}_m)\left(1 + \frac{T_e}{T_i}\right)} \right]^{-1} \tag{2.10}$$

A similar calculation for $\tilde{\chi}(k_{n_2}, k_o, -k_{n_1}) = \tilde{\chi}_+$ in (9) yields $\tilde{\chi}_+ = \tilde{\chi}_-$. It follows from (2.4) that when $\Delta\omega = 0$, $\gamma_\ell = \tilde{\Gamma}$. If we combine this with (2.10) and identify $\tilde{\Gamma}^2$ as in (2.2), we find

$$\tilde{P} = \frac{\tilde{\Gamma}^2(\vec{k}_m)}{\gamma_\ell^2} = P \left[ 1 + \frac{\omega_{pe}^2 W^2}{4\gamma_\ell^2 \tilde{P} \left(1 + \frac{T_i}{T_e}\right)^2} \right]^{-1}, \text{ or } \tilde{P} = P - \frac{1}{4} \frac{\omega_{pe}^2}{\gamma_\ell^2} \frac{W^2}{\left(1 + \frac{T_i}{T_e}\right)^2} \tag{2.11}$$

which we now use to replace P in Eq. (2.5) to obtain W.

To solve the nonlinear equation for W resulting from this substitution we note that $\rho/W >> 1$, generally, and we find

$$W = 2(\gamma_\ell/\omega_{pe}) \quad (P - 1)^{1/2} \, (1 + T_i/T_e) \tag{2.12}$$

Comparing with the value for $W = W_{PST}$ in the absence of CS we have $W/W_{PST} \sim \gamma_\ell/\omega_{pe}\rho <<1$, if $P >> 1$. Thus CS greatly reduces the excited wave energy in this case. We do not regard (17) as a complete formula for the wave energy, however, since the resulting spectrum localized near $k_m$ is probably unstable to decay into plasmons and ion waves which spreads the plasmon spectrum to lower values of k. The point we wish to make here is that CS greatly reduces the wave-energy predicted by previous theories which used only induced scattering from ions as the saturation mechanism.

## III. SATURATION OF THE ELECTRON-ION DECAY INSTABILITY FOR $T_e >> T_i$

The nonlinear saturation of the decay instability of an electromagnetic pump wave into a Langmuir wave and an ion acoustic wave has been treated by a number of workers in the case of equal electron and ion temperatures[1]. In this case ($T_e \simeq T_i$), since the ion acoustic wave is heavily damped the instability is best viewed as nonlinear scattering of the pump wave into a Langmuir wave by ion density fluctuations. The secondary Langmuir wave can undergo another induced scattering process to produce a tertiary Langmuir wave of still lower energy and wave number. This cascade process continues to spread Langmuir wave energy out of the parametrically unstable region of k space until the system is stabilized by collisional damping. For strong pumps this may ultimately result in the modulational "collapse" of the Langmuir spectrum as wave energy piles up at low wave number [5]. This saturation theory at $T_e = T_i$ also ignores the OTS instability which has a nearly equal threshold. We have shown elsewhere, that a weak turbulence theory, such as the one considered here, cannot be valid for the OTS instability [4].

For unequal temperatures $T_e \gg T_i$ the instability is a resonant decay process involving a Langmuir wave and ion acoustic wave. A cascade of decay processes can again stabilize the system. We will see that coupling saturation is a competing and often dominant saturation mechanism. The OTS instability in this case has a threshold which is higher by roughly $(m_i/m_e)^{1/2}$ and so can be rigorously suppressed.

The marginal stability condition $\mathrm{Im}\omega_{n_1} = 0$ as determined from the nonlinear dispersion relation (2.1) can be written generally as

$$\tilde{\gamma}_1 \tilde{\gamma}_2 = \tilde{\Gamma}^2 \left[1 + (\Delta\tilde{\omega})^2/(\tilde{\gamma}_1 + \tilde{\gamma}_2)^2\right]^{-1} \tag{3.1}$$

where $\Delta\tilde{\omega} = \tilde{\omega}_1 + \tilde{\omega}_2 - \omega_o$. For the EID instability we identify mode $1 \equiv \ell$ as the Langmuir mode as mode $2 \equiv a$ as the ion acoustic mode. In the present work we will neglect any nonlinearities in the ion wave damping or frequency, setting $\tilde{\omega}_a = \omega_a$, $\tilde{\gamma}_a = \gamma_a$. In the context of the present coupled mode theory there appear to be no nonlinearities in $\gamma_a$ in one dimension and only weak nonlinearities in two dimensions. From (1.8) we can compute the Langmuir wave damping $\tilde{\gamma}_\ell = (\partial \mathrm{Re}\varepsilon/\partial\omega_\ell)^{-1} \cdot \mathrm{Im}\,\tilde{\varepsilon}(k, \omega_\ell)$ and find the familiar result

$$\frac{\tilde{\gamma}_\ell}{\gamma_\ell} = 1 + \sum_{n_1'} M(k_{n_1}, k_{n_1'}) W_{n_1'} \; ; \; W_{n_1} = \frac{|E_{n_1}|^2}{4\pi n T_e} \tag{3.2}$$

where

$$M(k,k') = \frac{-\omega_p}{8\gamma_\ell} \frac{(\vec{k} \cdot \vec{k}')^2}{|\vec{k}|^2 |\vec{k}'|^2} \frac{k_{D_e}^2}{|\vec{k}''|^2} \mathrm{Im} \frac{1}{\tilde{\varepsilon}(\vec{k}'', \omega'')} \tag{3.3}$$

This result is valid if $\omega'' = \omega - \omega' \ll |\vec{k}''| v_e$ where $\vec{k}'' = \vec{k} - \vec{k}'$. In distinction to the case for $T_e = T_i$, $\mathrm{Im}\, 1/\varepsilon(\vec{k}'', \omega'')$ is dominated at low frequencies by sharp ion-acoustic wave resonances:

$$M(k,k') \cong \frac{\omega_p}{16\gamma_\ell} \frac{(\vec{k}\cdot\vec{k}')^2}{|\vec{k}|^2|\vec{k}'|^2} \left[ \frac{\gamma''_a \, \omega''_a}{(\omega''-\omega''_a)^2 + \gamma''^2_a} - \frac{\gamma''_a \, \omega''_a}{(\omega''+\omega''_a)^2 + \gamma''^2_a} \right] \tag{3.4}$$

To make the problem analytically tractable we will consider only the one dimensional problem here. In the limiting case of small spontaneous noise sources the spectrum in one dimension becomes a discrete set of marginally stable modes located at $k = k_n$ as discussed by Kruer and Valeo [1] for $T_e$=$T_1$. The spacing of the modes is determined by $M(k_n,k_m)$ which clearly localizes them at the resonant peaks of M where $\omega'' = \omega_n - \omega_m = \pm\omega_a(\vec{k}_n - \vec{k}_m)$ where $m = n \pm 1$. At these values

$$M(k_n,k_m)_{max} \equiv \begin{matrix} W_T^{-1} & \text{for } k_n > k_m \\ -W_T^{-1} & \text{for } k_n < k_m \end{matrix} \quad ; \quad W_T = 16\,\frac{\gamma_a}{\omega_a}\,\frac{\gamma_\ell}{\omega_p} \tag{3.5}$$

The quantity $W_T$ is recognized as the threshold value of $E_o^2/4\pi nT_e$ for the decay instability. There is then a set of discrete modes spaced at frequency intervals equal to the acoustic frequency $\omega_a(2k_n)$. A Langmuir wave with $\vec{k}$ decays into a Langmuir wave $\vec{k}'$ in the opposite direction where $|\vec{k}-\vec{k}'| \simeq 2|\vec{k}|$.

We label the modes so that n = 1 corresponds to the "matched" or fastest growing mode excited by the pump for which $\Delta\omega(k_1) = 0$ and which has the largest wave number (i.e. $k_1 > k_n (n > 1)$). For all other modes at $n > 1$, $\Delta\omega(\vec{k}_n) \gtrsim \omega_a(2k_n)$. If we impose the condition

$$\frac{\tilde{P}(\tilde{\gamma}_\ell + \gamma_a)^2}{\Delta\omega(\vec{k}_n)} < \frac{\tilde{P}(\tilde{\gamma}_\ell + \gamma_a)^2}{4\omega_a^2} << 1 \tag{3.6}$$

where

$$\tilde{P} = \tilde{\Gamma}^2/\gamma_a\gamma_\ell \tag{3.7}$$

then the marginal stability condition (3.1) can be written

$$\frac{\tilde{\gamma}_\ell(k_n)}{\gamma_\ell} = 0 \quad \text{for } n > 1 \tag{3.8a}$$

$$\frac{\tilde{\gamma}_\ell(k_1)}{\gamma_\ell} = \tilde{P} \quad \text{for } n = 1 \tag{3.8b}$$

On substitution of (3.5) into (3.3) we can write (3.8a) explicitly as

$$1 + \frac{W_{n+1}}{W_T} - \frac{W_{n-1}}{W_T} = 0 \tag{3.9}$$

This has the solution ($n > 1$)

$$W_n = C - \frac{n}{2} W_T \tag{3.10}$$

The constant C is determined by matching to the $n = 1$ equation (note $W_n = 0$ for $n < 1$)

$$1 + \frac{W_2}{W_T} = 1 + \frac{(C - W_T)}{W_T} = \tilde{P}$$

$$C = \tilde{P}\, W_T \tag{3.11}$$

Thus we have

$$W_n = (\tilde{P} - \frac{n}{2}) W_T \tag{3.12}$$

The total energy is easily summed

$$W_{TOT} = \sum_{n=1}^{n\,max} (\tilde{P} - \frac{n}{2}) W_T \cong \tilde{P}^2 W_T \tag{3.13}$$

where $n_{max} = 2\tilde{P}$. This equation does not determine $W_{TOT}$ since $\tilde{P}$ is itself determined by the wave spectrum.

In the equation (1.11) for the renormalized $\tilde{\chi}$ we focus our attention on the factor

$$\tilde{\varepsilon}^{-1}(k_{n_2'}) = \tilde{\varepsilon}^{-1}(k_o - k_{n_1'}) \simeq (\partial\varepsilon/\partial\omega_2)^{-1}\left[\omega_o - \omega_{n_1'} - \omega_a(k_o - k_{n_1'}) + i\gamma_a(k_o - k_{n_1'})\right]^{-1}$$

From the nonlinear dispersion relation (2.1) we find the real part of $\omega_{n_1}$ at marginal stability to be

$$\mathrm{Re}\,\omega_{n_1} = \omega_o - \tilde{\omega}_2(k_{n_2}) - \Delta\tilde{\omega}\,\tilde{\gamma}_2(\tilde{\gamma}_1 + \tilde{\gamma}_2)^{-1} \quad . \tag{3.14}$$

With this relation we find

$$\varepsilon^{-1}(k_{n_2}) = (\partial \mathrm{Re}\varepsilon/\partial\omega_2)^{-1} \cdot \left[\Delta\omega\,\gamma_{\bar{a}}(\gamma_\ell + \gamma_a)^{-1} + i\,\gamma_a\right]^{-1} .$$

This factor is large only for $\Delta\omega = 0$ when $\tilde{\varepsilon}^{-1}(k_{n_2}) = \frac{-i}{2}(k_1^2/k_{D_e}^2)$ x $(\omega_a/\gamma_a)$. When the condition (3.6) is satisfied only the $k_{n_1} = k_1$ ($\Delta\omega = 0$) term to the sum in (1.11) contributes significantly and we can obtain a simple algebraic equation for $\tilde{\chi}(k_{n_2}, k_o, -k_{n_1})$ for $\vec{k}_{n_1} = \vec{k}_1$.

$$\tilde{\chi}(k_{n_2},k_o,-k_{n_1}) = \frac{X_{(2)}(k_{n_2},k_o,-k_{n_1})}{1 - i\xi W_1} \quad \text{for } k_{n_1} = k_1$$

We also find $\tilde{\chi}(k_{n_1},k_{n_o},-k_{n_2}) = \tilde{\chi}(k_{n_2},k_{n_o},-k_{n_1})$ which means that $\Gamma^2(k_{n_1})$ defined in (2.2) is a real quantity. Using (2.2) we can write the following relation for P

$$\tilde{P} = \frac{\tilde{\Gamma}^2(k_{n_1})}{\gamma_a\gamma_\ell} = \frac{P}{1 + \xi^2 W_1{}^2} \tag{3.15}$$

where

$$\xi = \left[k_{n_1} \simeq k_{n_1'} = k_1\right] \frac{\omega_a}{2\gamma_a} \frac{k_1{}^2}{k_{D_e}{}^2} \tag{3.16}$$

The factor in square braces is the corresponding factor in (1.11) with $\vec{k}_{n_1}$, $\vec{k}_{n_1'}$ set equal to the matched wave vector $\vec{k}_1$. To evaluate this quantity it is necessary to compute the various $X_{(2)}$'s and $X_{(3)}$ which appear. Space does not permit us to pursue the details here. We merely list the pertinent results. For $X_{(2)}(k_{n_2}-k_{n_2'},k_{n_1'},-k_{n_1})$ and $X_{(2)}(k_{n_2},k_{n_1'},k_{n_2}-k_{n_1'})$, $X_{(2)}(k_{n_2}-k_{n_1'},k_{n_2'},-k_{n_1})$ (the latter two occuring in the $(k_{n_2'} \rightleftarrows k_{n_1'})$ term in (1.11)) we have expressions of the form of (2.7) since these three susceptibilities have two high frequency ($\sim\omega_{pe}$) arguments and one low frequency argument. However, now

$$X_{(2)}(k_{n_2}\,,\, k_{n_2'},\, k_{n_2}-k_{n_2'}) = \frac{i}{|\vec{k}_{n_1}-\vec{k}_{n_1'}|} \frac{e}{T_e} \frac{k_{D_e}^2}{|\vec{k}_{n_2}||\vec{k}_{n_2'}|} \tag{3.17}$$

since all three arguments $k_{n_2}-k_{n_2'},k_{n_2},k_{n_2'}$ satisfy relations on the form $\omega \ll |\vec{k}|v_e$ . We also find by functional differentiation[6] the third order susceptibility

$$X_{(3)}(k_{n_1},k_{n_1'},k_{n_2'},-k_{n_2}) = \frac{e^2}{T_e{}^2} \frac{(\hat{k}_{n_1} \cdot \hat{k}_{n_1'})}{|\vec{k}_{n_2}||\vec{k}_{n_2'}|} \tag{3.18}$$

Using these results we find for the square brackets

$$\left[k_{n_1} = k_1\right] = \lim_{k_{n_1'}\to k_{n_1}\to k_1} \frac{(\hat{k}_{n_1}\cdot\hat{k}_{n_1'})}{|\vec{k}_{n_2}||\vec{k}_{n_2'}|} \frac{e^2}{T_e^2} \left\{ \frac{-k_{D_e}^{\;2}}{(\vec{k}_{n_1}-\vec{k}_{n_1'})} \frac{1}{\varepsilon(k_{n_1}-k_{n_1'})} + \frac{1}{\varepsilon(k_{n_1}+k_{n_1'}-k_o)} + 1 \right\} \tag{3.19}$$

To calculate the quantity $\xi$ defined in (3.16) we need the limit of this as $k_{n_1} \to k_{n_1'} \to k_1$. This limit depends sensitively on the dispersion $\omega_n(\vec{k}_n)$ of the marginally stable mode frequencies as given by Eqn. (3.14) which we substitute for $\omega_{n_1}=\omega_{n_1}(\vec{k}_{n_1})$ and $\omega_{n_1'}=\omega_{n_1}'(k_{n_1'})$ in $\varepsilon(k_{n_1}-k_{n_1'})$. Two limiting cases will be considered here. (The linear damping $\gamma_\ell$ is taken as the collisional damping $\gamma_c$).

i.) $\tilde{\gamma}_\ell(k_1) = \tilde{P}\gamma_c \ll \gamma_a$: $\omega_{n_1}(k) = \omega_\ell(k)$

$$\lim_{\vec{k}_{n_1}\to\vec{k}_{n_1'}\to\vec{k}_1} (\vec{k}_{n_1}-\vec{k}_{n_1'})^2\, \varepsilon(k_{n_1}-k_{n_1'}) = k_{D_e}^{\;2} \tag{3.20}$$

if $m_e/m_i \ll 9k_1^2/k_{D_e}^2$

$$\xi = \frac{1}{72}\,\frac{\omega_a}{\gamma_a}\,\frac{k_{D_e}^2}{k_1^{\;2}}\;; \quad \xi W_T = \frac{2}{9}\,\frac{\gamma_c}{\omega_p}\,\frac{k_{D_e}^{\;2}}{k_1^{\;2}}$$

ii.) $\tilde{\gamma}_\ell(k_1) = \tilde{P}\gamma_c \gg \gamma_a$ : $\omega_{n_1}(k) = \omega_o-\omega_a(k)$

$$\lim_{\vec{k}_{n_1}\to\vec{k}_{n_1'}\to\vec{k}_1} (\vec{k}_{n_1}-\vec{k}_{n_1'})^2\, \varepsilon(k_{n_1}-k_{n_1'}) = 3|\vec{k}_{n_1}||\vec{k}_{n_1'}| = 3k_1^{\;2} \tag{3.21}$$

$$\xi = \frac{-5}{72}\,\frac{\omega_a}{\gamma_a}\,\frac{k_{D_e}^{\;2}}{k_1^{\;2}}\;; \qquad \xi W_T = \frac{-10}{9}\,\frac{\gamma_c}{\omega_p}\,\frac{k_{D_e}^{\;2}}{k_1^{\;2}}$$

From (3.12) we find $W_1 = (\tilde{P} - 1/2)W_T$ where $\tilde{P}$ is a function of $W_1$ given by (3.15). Elimination of $\tilde{P}$ gives the nonlinear equation determining $W_1$.

$$W_1 = \left(\frac{P}{1 + \xi^2 W_1^{\;2}} - \frac{1}{2}\right) W_T \tag{3.22}$$

When $\xi W_1 \gg 1$, $\tilde{P} \gg 1$ we find

$$W_1 = \left(\frac{P}{\xi^2 W_T{}^2}\right)^{1/3} W_T \qquad ; \quad \xi^2 W_1{}^2 = (\xi W_T)^{2/3} P^{2/3} >> 1 \tag{3.23}$$

$$\tilde{P} = \left(\frac{P}{\xi^2 W_T{}^2}\right)^{1/3} \tag{3.24}$$

Thus for $P >> (\xi W_T)^{-1}$ the effective pump parameter increases as only the third root of the unrenormalized parameter P. In this limit the total wave energy increases as $P^{2/3}$ :

$$W_{TOT} = \left(\frac{P}{\xi^2 W_T{}^2}\right)^{2/3} W_T \tag{3.25}$$

For $P << (\xi W_T)^{-1}$ we have the much stronger P dependence

$$W_{TOT} = P^2 W_T \tag{3.26}$$

The conditions $P = (E_o{}^2/4\pi n T_e) W_T{}^{-1} > |\xi W_T|^{-1}$ and $\tilde{\gamma}_\ell = \tilde{P}\gamma_c < \gamma_a$ require $(9/10)(k_1/k_{De}) < \sqrt{\pi/8}\ (m_e/m_i)$ which usually cannot be satisfied for reasonable parameters. Therefore only case (ii) above applies in practice where

$$\frac{E_o{}^2}{4\pi n T_e} > 16\,\frac{\gamma_a}{\omega_a}\frac{\gamma_c}{\omega_{pe}}$$

$$\text{and} \tag{3.27}$$

$$\frac{E_o{}^2}{4\pi n T_e} > \frac{1}{|\xi|} = \frac{72}{5}\frac{\gamma_a}{\omega_a}\frac{k_1{}^2}{k_{De}{}^2}$$

The first condition assures that the pump intensity is above the linear instability threshold and the second condition $[P > (\xi W_T)^{-1}]$ is that for strong coupling saturation of the wave-energy; $W_{TOT} \sim P^{2/3}$ .

The results above were derived for a one-dimensional model, however, we expect qualitatively similar results for the total wave energy in two dimensions.

The effective damping of the pump wave can be computed as usual [1] by balancing the effective energy dissipation of the pump against the collisional energy dissipation of the excited Langmuir waves. We have

$$\gamma_o^{NL} \frac{E_o^{\,2}}{4\pi n T_e} = \sum_{n=1}^{n\,max} \gamma_c W_n = \gamma_c W_{TOT} \tag{3.28}$$

where $\gamma_o^{NL}$ is the effective nonlinear damping seen by the pump which is often written as an effective collision frequency $\gamma_o^{NL} = \frac{1}{2}\nu_{eff}$ . a factor of 2 which arises to include the particle kinetic energy associated with the Langmuir waves ($k_n/k_o << 1$) and electromagnetic waves cancels on both sides of (3.28). [This formula is not correct for the n = 1 mode since its dispersion formula may be strongly altered from the ordinary Langmuir dispersion by the parametric coupling. However, for $P >> 1$ this mode makes a negligible contribution.] We can write this as

$$\gamma_o^{NL} = \frac{1}{2}\nu_{eff} = \gamma_c\, \tilde{P}^2 (W_T/W_o) = \gamma_c \frac{\tilde{P}^2}{P} \tag{3.29}$$

Here we have used the definition of the pump ratio $P$ to write $P = W_o/W_T$ where $W_o = E_o^{\,2}/4\pi n T_e$ .

In the weak pump regime we use (3.26) to write with $\gamma_c = \nu_{ei}/2$

$$2\gamma_o^{NL} = \nu_{eff} = \nu_{ei}\, P \quad ; \quad P << |\xi W_T|^{-1} \tag{3.30}$$

In this case $\nu_{eff}$ increases linearly with $P$ or with $E_o^{\,2}$ and because of the definition of $P$ is independent of $\nu_{ei}$ . In the strong pump regime we use (3.25) to find

$$\nu_{eff} = \nu_{ei}(\xi^2 W_T^{\,2} P)^{-1/3} \quad ; \quad P >> |\xi W_T|^{-1} \tag{3.31}$$

The saturation due to CS therefore results in a $\nu_{eff}$ which ultimately decreases as $P^{-1/3}$ or $E_o^{-2/3}$ for sufficiently strong pumping. Thus the anomalous absorption of the pump becomes less efficient for strong pumps.

## IV. SUMMARY AND CONCLUSIONS

In treating the nonlinear wave spectrum generated by parametric decay instabilities we have attempted to go one step beyond the usual RPA and include pump induced correlations between modes of a given parametric pair ($k_{n1}$, $k_{n2}$) . We still assume that modes corresponding to different parametric pairs ($n_s \neq n_s'$) are uncorrelated since the waves are excited from random particle noise. When this added correlation ($k_{n1}$, $k_{n2}$) is taken into account in the nonlinear theory we find that the effective mode coupling coefficients are reduced from their linear values ($X_2$) when the wave energy in the nearly perfectly matched modes ($\Delta\omega \approx 0$) becomes sufficiently great. Thus the nonlinear wave-wave interactions reduce the mode correlations which are responsible for the instability. This effect appears to be a dominant saturation mechanism for both the EED and EID (for $T_e \gg T_i$) decay instabilities.

Finally we wish to make a number of observations and conclusions based on the above results:

1) The EED nonlinear spectrum calculated in Section II is narrowly confined because of coupling saturation (C.S.) to the region in $\vec{k}$ space around the most favorably matched mode $\vec{k}m$, which lies in a direction which is in the plane of, and bisects the 90° angle between, $\vec{E}_o$ and $\vec{k}_o$. This spectrum is unstable to further decay into Langmuir and ion waves. We can estimate the competition of C.S. with further decay or spreading of the spectrum by using the results of Section III. We adopt a one dimensional model in which all wave vectors lie parallel or anti-parallel to the direction of $\vec{k}m$. Such waves will be the most strongly coupled. Then $W_1$ is again the energy in the matched waves [where $|\vec{k}m| = k_1$ and $\Delta\omega(k_1) = 0$] which we denoted by W in Section II and we have a ladder of modes which $|\vec{k}_n| < k_1$ with energies $W_n$. For $\tilde{P} \gg 1$ we again have $W_1 = \tilde{P}W_T$ [see (3.12)] where now P is given by (2.11) with W replaced by $W_1$. The resulting quadratic equation for $W_1$ has two limiting solutions:

$$\left.\begin{aligned} W_1 &= PW_T \\ W_{TOT} &= P^2 W_T \end{aligned}\right\} \quad 1 < P \ll P_c \tag{4.1a}$$

$$\left.\begin{aligned} W_1 &= 2(\gamma_c/\omega_p)P^{1/2} \\ W_{TOT} &= 16\, P_c P W_T \end{aligned}\right\} \quad P \gg P_c \tag{4.1b}$$

where

$$P_c = (1 + T_i/T_e)^2(\omega_a/16\gamma_a)^2 \tag{4.2}$$

Eqn. (4.1b) for $W_1$ agrees with (2.12) for $P \gg 1$ but now since the energy can spread to modes of lower k the total wave energy increases as P (or $E_o^2$). The critical pump parameter $P_c$ is small for $T_e \tilde{} T_i$ where $\gamma_a/\omega_a \sim 1$ which inhibits further decay processes but for $T_e/T_i \gg 1$ $P_c \sim 16$ for a hydrogen plasma and so the cascade process dominates C.S. except for higher P. These one-dimensional estimates are clearly crude. However, we may conclude that because of C.S. the Langmuir spectrum for the EED instability will spread to lower $|\vec{k}|$ rather than spread in angle as assumed by PST[2].

2) The condition for the validity of the weak turbulence theory is [5]

$$W_{TOT} \ll k_1^2/k_{D_e}^2 \tag{4.3}$$

which ensures that the spectrum is modulationally stable. If we apply this to the EID instability by using Eqn. (3.25) for $W_{TOT}$ (with the restriction $P > |\xi W_T|^{-1}$) we find that the simultaneous satisfaction of these two inequalities requires

$$|\xi W_T|^{-1} \ll P \ll [(5/72)(\omega_a/\gamma_a)]^{3/2} \; |\xi W_T|^{1/2} \tag{4.4}$$

The outer members of this inequality impose a restriction on $|\xi W_T|$ which, in view of (3.21), can be written

$$1 \gg \frac{72}{5} \; \frac{\gamma_a}{\omega_a} \frac{\omega_p}{\gamma_c} \; \frac{k_1^2}{k_D^2} \tag{4.5}$$

For the set [7] of experimental parameters $\gamma_c/\omega_p \sim 10^{-3}$, $k_1/k_D \sim .05 - .1$ and $\sqrt{m_e/m_i} \simeq 4 \times 10^{-3}$ this is easily satisfied. However, we note that for a sufficiently collisionless plasma this might not be satisfied.

If we apply the same inequality to the EED instability using (4.1b) for $W_{TOT}$ we arrive at the inequality

$$P_c \ll P \ll \frac{k_m^2}{k_{D_e}^2} \left( \frac{\gamma_a}{\omega_a} \; \frac{\omega_p}{\gamma_c} \right) \frac{1}{(1+T_i/T_e)^2} \tag{4.6}$$

when $P_c$ is given by (4.2). This is most readily satisfied if $T_e \simeq T_i$ so that $\gamma_a/\omega_a \sim 1$ and by <u>smaller</u> $\gamma_c/\omega_{p_e}$ .

3) The increase of wave-energy as $(|E_o|^2)^{2/3}$ for the EID instability may be consistent with the recent experiments of Mizuno and DeGroot [7] whose measured wave energy is consistent with a curve proportional to $|E_o|^2$. A power law of $(|E_o|^2)^{2/3}$ would appear to fit their data in the higher range of $|E_o|$ as well as does $|E_o|^2$. Moreover, for small $|E_o|$, $[P \sim O(1)]$, a power law $|E_o{}^2|^2$ as predicted by (3.26) would appear to fit the data. Any detailed agreement may be fortuitous because of our crude estimates. However, it is clear that without coupling saturation an energy proportional to $(|E_o|^2)^2$ would be predicted <u>for all $|E_o|$</u> in gross disagreement with the experiment.

The application of this type of theory to more realistic 2 or 3 dimensional models, the time evolution of the spectrum and the application to other decay instabilities remain for future work.

This work was performed under the auspices of U.S. Energy Research and Development Administration. A partial account of the work in this paper has appeared in B. Bezzerides and D. F. DuBois, Phys. Rev. Lett. 36, 729(1976).

## REFERENCES

1. D. F. DuBois and M. V. Goldman, Phys. Fluids 15, 419 (1972) Phys. Rev. Lett. 28, 218 (1972); E. Valeo, F. Perkins, and C. Oberman, Phys. Rev. Lett. 28, 340 (1972); Y. Kuo and J. A. Fejer, Phys. Rev. Lett. 29, 1667 (1973); W. L. Kruer and E. Valeo, Phys. Fluids 16, 675 (1973).

2. V. V. Postovalov, V. P. Silin, and V. T. Tikhonchuk Zh. Eksp. Teor. Fiz. 65, 1880 (1973) [Sov. Phys. JETP, 38 938 (1974).]

   The nonlinear enhancement of $\tilde{\gamma}_\ell$ is given by PST as $-\frac{1}{2}\int d^3k' Q(k,k')\, W(k')$ with Q defined in their Eqn. (1.9). To avoid confusion, the reader should note that $\tilde{\gamma}(k)$ as defined by PST is not the same as our $\tilde{\gamma}_\ell$.

3. B. B. Kadomtsev, "Plasma Turbulence" (Academic Press, N.Y. 1965); R. Z. Sagdeev and A. A. Galeev, "Lectures on Nonlinear Plasma Theory" (Benjamin, N.Y., 1969); V. N. Tsytovich, "Nonlinear Effects in Plasma" (Plenum, N.Y., 1970).

4. D. F. DuBois and B. Bezzerides, "Nonlinear Theory of Parametric Instabilities ", Los Alamos Report LA-UR-75-727 (to be published November, 1976). It is shown here that the present type of weak turbulence theory does not apply to the modulational or OTS instabilities.

5. A. A. Vedenov, A. V. Gordeev and L. I. Rudakov, Plasma Physics 9, 719 (1967); V. E. Zakharov, Zh. Eksp. Teor. Fiz. 62, 1745 (1972), [Sov. Phys. JETP 35, 908(1972).]

6. D. F. DuBois, Phys. Fluids (to be published) "Space-Time Formulation of Plasma Turbulence Theory", Los Alamos Scientific Laboratory Report #LA-UR-75-726. To be published, Phys. Rev. A, November, 1976.

7. K. Mizuno and J. S. DeGroot, Phys. Rev. Lett. 35, 219 (1975).

# VARIATION PRINCIPLE FOR NONLINEAR WAVE PROPAGATION*

T.Watanabe†, Y.C.Lee†† and Kyoji Nishikawa

Faculty of Science, Hiroshima University

Hiroshima, Japan

## §1. Introduction

It is now widely recognized that one of the most important nonlinear effects which arise associated with the propagation of a moderately strong high-frequency wave is a modification of the density profile of the plasma. Such a profile modification occurs as a result of the action of the ponderomotive force of the high-frequency fields. When the plasma is confined by a static magnetic field, the force acting along the magnetic field is most important and the balance of the ponderomotive force and the pressure gradient force in this direction determines the stationary density profile. The profile modification in turn affects the properties of the wave propagation, and the entire problem has to be solved self-consistently. In general, this problem requires a solution of highly complicated partial differential equations with appropriate boundary conditions.

In this paper, we show that under quite general conditions, this problem can be reduced to solving a simple variation principle for the high-frequency electric field. An advantage of the variation principle is that it contains only low-order differential operation, usually of second order, whereas the original differential equations often contain higher-order differentials. Therefore, the variation principle will be suited for numerical analyses of

---

* Supported in part by the Scientific Research Fund of the Ministry of Education in Japan.

† Institute of Plasma Physics, Nagoya University, Nagoya, Japan.

†† Department of Physics, University of California, Los Angeles, Ca., USA.

the problem.

The following are the basic assumptions and/or approximations used in the present formalism:

1) We consider a stationary state and assume the local charge neutrality for the static component of the density.
2) The intensity of the high-frequency wave is sufficiently small, so that only the second order nonlinear effect causing the local density-profile modification will be taken into account. This approximation will be justified when the excursion length due to the oscillating field is much less than its effective wavelength in the direction of the excursion.
3) We neglect the static flow of the particles. This may be justified when the flow velocity is much less than the excursion velocity due to the oscillation field.
4) For simplicity, we use the collisionless fluid model with isotropic pressure $\Pi$ whose space-time dependence takes place only through the local density. Generalization including the kinetic effects can be made to a certain extent (see §5).

Under these assumptions and/or approximations, our theory is applicable to any waves, either electrostatic or electromagnetic, and to arbitrary configuration of the magnetic field.

We first derive in §2 a simple but general expression for the ponderomotive force acting along the magnetic field. Using the charge neutrality for the static part of the density, we then derive in §3 a relation which determines the self-consistent density profile in terms of the oscillating variables. The variation principle is derived in §4 from Maxwell's equations and the relation derived in §3. Finally in §5, a brief discussion is given on the kinetic effects and on the validity of the present formalism.

## §2. Ponderomotive Potential along the Magnetic Field

We start from the equation of motion in the collisionless fluid model,

$$\dot{\vec{v}} + \vec{v}\cdot\vec{\nabla}\vec{v} = \frac{e}{m}\left\{\vec{E} + \vec{v}\times\frac{(\vec{B}_0+\vec{B}')}{c} - \frac{\vec{\nabla}H}{e}\right\}, \qquad (1)$$

where $\vec{B}_0$ and $\vec{B}'$ are, respectively, the static and the oscillating components of the magnetic field, $\vec{\nabla}H$ stands for the pressure gradient force which we assume to be parallel to the local density gradient $\vec{\nabla}n$, i.e. $H=\int^n d\Pi/n$, and the rest of the notation is standard. We take the time average, denoted by a bar; neglecting the static flow of the particles, i.e. $\bar{\vec{v}}=0$, we obtain,

$$m\overline{\vec{v}\cdot\vec{\nabla}\vec{v}} - \frac{q}{c}\overline{\vec{v}\times\vec{B}'} = -\vec{\nabla}\{q\Psi + \overline{H}\}, \tag{2}$$

where the static electric field is represented by $-\vec{\nabla}\Psi$. The left--hand side of this equation is the ponderomotive force which we shall denote by $\vec{G}$. Introducing the oscillating displacement vector $\vec{r}$ by the relation $\dot{\vec{r}}=\vec{v}$ with $\overline{\vec{r}}=0$, we can rewrite $\vec{G}$ in the form,

$$\vec{G} = m\left[\frac{1}{2}\vec{\nabla}\,\overline{|\vec{v}|^2} - \overline{\vec{v}\times(\vec{\nabla}\times\vec{v})}\right] - q\,\overline{\vec{r}\times(\vec{\nabla}\times\vec{E})}, \tag{3}$$

where we used the relation

$$\overline{\dot{X}Y} = -\overline{X\dot{Y}}, \tag{4}$$

as well as Faraday's induction law, $\dot{\vec{B}}'=-c\vec{\nabla}\times\vec{E}$.

We shall calculate $\vec{G}$ in the approximation second-order with respect to the wave field. For this purpose, we can use for $\vec{v}$, $\vec{r}$ and $\vec{E}$ entering in (3) the linear relation

$$\dot{\vec{v}} = \frac{q}{m}\left[\vec{E} + \frac{\vec{v}\times\vec{B}_0}{c} - \frac{\vec{\nabla}H}{q}\right]. \tag{5}$$

We then show that the component of $\vec{G}$ parallel to the static magnetic field can be written in the form of the following potential force:

$$\vec{b}\cdot\vec{G} = -\vec{b}\cdot\vec{\nabla}\{\overline{\vec{P}\cdot\vec{F}/2}\}, \tag{6}$$

where $\vec{P}(=q\vec{r})$ is the polarization vector per particle, $\vec{b}(=\vec{B}_0/B_0)$ is the unit vector along the static magnetic field, and $\vec{F}$ is given by the relation

$$\vec{F} = \vec{E} - \frac{\vec{\nabla}H}{q}. \tag{7}$$

To prove the relation (6), we first take the scalor product of (5) with $\vec{r}$ and average over the time; using (4) we obtain

$$-\frac{m}{2}\overline{|\vec{v}|^2} = \frac{q}{2}\left\{\overline{\vec{r}\cdot\vec{F}} + \overline{\vec{r}\cdot(\vec{v}\times\vec{B}_0)}/c\right\}. \tag{8}$$

Next we take the curl of (5), then take its vector product with $\vec{r}$ and average over the time, obtaining

$$-m\overline{\vec{v}\times(\vec{\nabla}\times\vec{v})} = q\left[\overline{\vec{r}\times(\vec{\nabla}\times\vec{E})} + \overline{\vec{r}\times\vec{\nabla}\times(\vec{v}\times\vec{B}_0)}/c\right]. \tag{9}$$

Substituting (8) and (9) into (3), we find

$$\vec{b}\cdot\vec{G} = -\vec{b}\cdot\vec{\nabla}\left[\frac{1}{2}\overline{\vec{P}\cdot\vec{F}}\right] + qS/c,$$

where

$$S = -\vec{b}\cdot\vec{\nabla}\,\overline{\vec{r}\cdot(\vec{v}\times\vec{B}_0)}/2 + \vec{b}\cdot\left\{\overline{\vec{r}\times\vec{\nabla}\times(\vec{v}\times\vec{B}_0)}\right\}. \tag{10}$$

Our problem is to show that S vanishes identically. This can be shown by rewriting the last term as follows:

$$\vec{b}\cdot\left\{\overline{\vec{r}\times\vec{\nabla}\times(\vec{v}\times\vec{B}_0)}\right\} = -\vec{b}\cdot\left[\overline{\vec{v}\times\vec{\nabla}\times(\vec{r}\times\vec{B}_0)}\right]$$

$$= \frac{\vec{B}_0\cdot}{2B_0}\left\{\overline{\vec{r}\times\vec{\nabla}\times(\vec{v}\times\vec{B}_0)} - \overline{\vec{v}\times\vec{\nabla}\times(\vec{r}\times\vec{B}_0)}\right\}$$

$$= \frac{1}{2B_0}\left\{\overline{(\vec{B}_0\times\vec{r})\cdot[\vec{\nabla}\times(\vec{v}\times\vec{B}_0)]} - \overline{(\vec{v}\times\vec{B}_0)\cdot[\vec{\nabla}\times(\vec{B}_0\times\vec{r})]}\right\}$$

$$= \frac{1}{2B_0}\vec{\nabla}\cdot\left\{\overline{(\vec{v}\times\vec{B}_0)\times(\vec{B}_0\times\vec{r})}\right\}$$

$$= \frac{1}{2B_0}\vec{\nabla}\cdot\left\{\overline{\vec{r}\cdot(\vec{v}\times\vec{B}_0)\cdot\vec{B}_0}\right\},$$

where in the first line we again used (4) and the rest of the calculation is straightforward. Since $\vec{\nabla}\cdot\vec{B}_0=0$, the last expression just cancels the first term on the right-hand side of (10).

## §3. Self-Consistent Density Profile

We first integrate (2) along the $\vec{B}_0$-field. Using (6), we find

$$q\Psi + \overline{H} - \overline{\vec{P}\cdot\vec{F}}/2 = f(\psi), \tag{11}$$

where $\psi$=const. represents the magnetic surface and the right-hand side is an arbitrary function of $\psi$. This equation is written for each species of the particle. We eliminate $\Psi$ by combining the equations (11) for the electron and the ion. Assuming for simplicity that the ions are singly ionized, we obtain

$$\sum\{\overline{H} - \overline{\vec{P}\cdot\vec{F}}/2\} = f(\psi), \tag{12}$$

where the summation is for the electron and the ion and the right-hand side is again an arbitrary function of $\psi$.

We now calculate $\overline{H}$. Since H is a function of the density alone, we can expand it in the form

$$H = H_0 + H_0' n_1 + H_0'' n_1^2/2 + \cdots, \tag{13}$$

where the dash denotes the derivative with respect to n, $n_1$ is the oscillating part of the density and the subscript depicts the value evaluated at the time-averaged density $n_0$. Correct to the second order, one can then write as $\overline{H}=H_0+H_0''\,\overline{n_1^2}/2$ with $\overline{H}''$ being treated as a constant. Substituting this result into (12), we obtain

$$\sum H_0 = \sum\{\overline{\vec{P}\cdot\vec{F}} - H_0''\,\overline{n_1^2}\}/2 + f(\psi). \tag{14}$$

The right-hand side has only explicit dependence on the oscillating variables through $\vec{P}$, $\vec{F}$ and $n_1$. If we assume the local charge neutrality for $n_0$, i.e. $n_{0e}=n_{0i}=n_0$, then this equation determines the self-consistent density profile $n_0$ in terms of the oscillating variables. In the particular case when $\Pi_0=n_0T$ with a uniform temperature T, we have $H_0$=T log $n_0$, so that eq.(14) gives

$$n_0 = N(\psi) \exp\left\{ \frac{(\overline{\vec{P}\cdot\vec{F}} - H_0'' \overline{n_1^2})}{2(T_e + T_i)} \right\}, \tag{15}$$

where $N(\psi)$ is na arbitrary positive function of $\psi$. This result can be regarded as a generalized Boltzmann distribution.

## §4. Variation Principle

We start from Maxwell's two equations, i.e. those representing Faraday's induction law and Ampère's law. Eliminating the magnetic field, we get

$$\frac{\partial^2}{\partial t^2}\vec{E} + c^2\vec{\nabla}\times(\vec{\nabla}\times\vec{E}) = -4\pi n_0 \sum \frac{\partial^2}{\partial t^2}\vec{P}, \tag{16}$$

where on the right-hand side we retained only those nonlinear effects that arise from $n_0$. We replace $\partial^2/\partial t^2$ by $-\omega^2$, where $\omega$ is the frequency of the high-frequency wave. We then take the scalar product of (16) with the variation $\delta\vec{E}$ and integrate over the entire space. We use integration by parts and neglect the surface integral, assuming either the vanishing of the perturbation at the boundary or a periodic boundary condition. The result can be written in the form

$$\delta\int d\vec{x}\, \{|\vec{E}|^2 - |\vec{B}'|^2\} = -8\pi \int d\vec{x}\; n_0 \sum \delta\vec{E}\cdot\vec{P}, \tag{17}$$

where the relation $c(\vec{\nabla}\times\vec{E}) = -\dot{\vec{B}}'$ is used.

On the other hand, if we take the variation of (12) with respect to $\vec{E}$, we have

$$\sum \delta\widetilde{H} = \sum \frac{1}{2}\left[\overline{\delta\vec{P}\cdot\vec{F}} + \overline{\vec{P}\cdot\delta\vec{F}}\right]. \tag{18}$$

In the present approximation, we can use the linear relation between $\vec{P}$ and $\vec{F}$ to calculate the right-hand side. As can be seen from (5), the linear relation is written in terms of the Fourier transform in time in the form

$$\vec{P}(\omega) = \overleftrightarrow{\chi}(\omega)\cdot\vec{F}(\omega), \tag{19}$$

where $\overleftrightarrow{\chi}(\omega)$ is a hermitian tensor, $\overleftrightarrow{\chi}^T(\omega)=\overleftrightarrow{\chi}(-\omega)$, and its spatial dependence can arise only from that of the static magnetic field. We can then show the relation $\overline{\delta\vec{p}\cdot\vec{F}}=\overline{\vec{p}\cdot\delta\vec{F}}$ as follows:

$$\begin{aligned}\overline{\delta\vec{p}\cdot\vec{F}} &= \delta\vec{p}(\omega)\cdot\vec{F}(-\omega) + \delta\vec{p}(-\omega)\cdot\vec{F}(\omega)\\ &= \{\overleftrightarrow{\chi}(\omega)\cdot\delta\vec{F}(\omega)\}\cdot\vec{F}(-\omega) + \{\overleftrightarrow{\chi}(-\omega)\cdot\delta\vec{F}(-\omega)\}\cdot\vec{F}(\omega)\\ &= \delta\vec{F}(\omega)\cdot\overleftrightarrow{\chi}(-\omega)\cdot\vec{F}(-\omega) + \delta\vec{F}(-\omega)\cdot\overleftrightarrow{\chi}(\omega)\cdot\vec{F}(\omega)\\ &= \delta\vec{F}(\omega)\cdot\vec{P}(-\omega) + \delta\vec{F}(-\omega)\cdot\vec{P}(\omega) = \overline{\delta\vec{F}\cdot\vec{P}}.\end{aligned}$$

Using this relation, we find from (18) and (7)

$$\Sigma\,\delta\overline{H} = \Sigma\,\overline{\vec{P}\cdot\delta\vec{F}} = \Sigma\,\{\overline{\vec{P}\cdot\delta\vec{E}} - \frac{1}{q}\,\overline{\vec{P}\cdot\delta\vec{\nabla}H}\}. \qquad (20)$$

The last term can be calculated as follows:

$$\begin{aligned}-\Sigma\frac{1}{q}\,\overline{\vec{P}\cdot\vec{\nabla}\delta H} &= -\Sigma\,\overline{\vec{r}\cdot\vec{\nabla}\,\delta(H_0'n_1)}\\ &= -\Sigma\,n_0^{-1}\,\{\vec{\nabla}\cdot(n_0\overline{\vec{r}\,\delta[H_0'n_1]}) - \delta\,\overline{(H_0'n_1)\vec{\nabla}\cdot(n_0\vec{r})}\}\\ &= -n_0^{-1}\,\Sigma\,\{\vec{\nabla}\,(n_0\overline{\vec{r}\,\delta[H_0'n_1]}) + \overline{n_1\,\delta[H_0'n_1]}\},\end{aligned} \qquad (21)$$

where in the last line we used the continuity relation, $n_1+\vec{\nabla}\cdot(n_0\vec{r})=0$. On the other hand, the left-hand side of (20) can be written as as

$$\begin{aligned}\Sigma\,\delta\overline{H} = \Sigma\,\overline{n^{-1}\delta\Pi} &= n_0^{-1}\Sigma\,\delta\overline{\Pi} - n_0^{-2}\Sigma\,\overline{n_1\,\delta[\Pi_0'n_1]}\\ &= n_0^{-1}\,\{\Sigma\,\delta\overline{\Pi} - \Sigma\,\overline{n_1\delta\,[H_0'n_1]}\},\end{aligned} \qquad (22)$$

where we used the approximation, $n_0^{-1}\delta[\Pi_0'n_1] \doteqdot \delta[n_0^{-1}\Pi_0'n_1] \doteqdot \delta[H_0'\,n_1]$. Using (21) and (22) in (20), we obtain

$$\delta\,\Sigma\,\overline{\Pi} = \Sigma\,\{n_0\overline{\vec{P}\cdot\delta\vec{E}} - \vec{\nabla}\,(n_0\overline{\vec{r}\,\delta\,[H_0'n_1]})\}. \qquad (23)$$

If we integrate this expression over the entire space, the contribution of the last term vanishes and we get

$$\delta \int \Sigma \overline{\Pi}\, d^3\vec{x} = \int d^3\vec{x}\, \Sigma\, n_0 \overline{\vec{P}\cdot\delta\vec{E}} \quad . \qquad (24)$$

Apart from the numerical factor $-8\pi$, the right-hand side of this equation is identical with the time average of the right-hand side of (17). Therefore, we finally arrive at the following variation principle:

$$\delta \int d^3\vec{x} \left\{ \frac{\overline{|\vec{E}|^2}}{8\pi} - \frac{\overline{|\vec{B}'|^2}}{8\pi} + \overline{\Pi_e} + \overline{\Pi_i} \right\} = 0, \qquad (25)$$

where the variation is taken with respect to $\vec{E}$. This equation, together with (14), determines the self-consistent profiles for both the oscillating electric field and the static density. For actual calculations, we need explicit expressions for all the variables appearing in (14) and (25) in terms of the electric field. This can be done by assuming a relation between the pressure and the density, say the adiabatic relation for the high-frequency component, and by using the linear relations among the oscillating variables.

## §5. Discussions

Let us first discuss the generalization of the above formalism to include some of the kinetic effects which are disregarded in the fluid model. There are both dissipative and nondissipative effects, of which we shall consider only the nondissipative effects. They include finite Larmour radius corrections, deviation from the isotropy of the pressure gradient force, etc. These can be taken into account in our formalism by a slight generalization of the preceding analysis, provided the following two conditions are satisfied:

i) These kinetic effects yield only minor correction to the propagation characteristics of the high-frequency wave;

ii) The effective wavelength of the high-frequency wave is much shorter than the gradient scale-lengths (in the propagation direction) of all the static quantities.

Condition i) admits us to use the fluid model to calculate the nonlinear effects; the kinetic effects are to be taken into account only in the calculation of the linear terms. Condition ii) then allows us to use the <u>local</u> linear relation to calculate the kinetic effects.

Because of condition i), we can use eq.(12) or eq.(14) with $\vec{P}$ and $n_i$ being calculated in the fluid approximation. In eq.(16), however, we have to use the complete polarization vector, including the kinetic effects, on the right-hand side. Let us denote by $\Delta\vec{P}$ the deviation of the polarization vector from the one calculated by the fluid model described in the previous sections. Then the term which has to be added to the right-hand side of (17) is

$$-8\pi \int d^3\vec{x}\, n_o \sum \delta\vec{E}\cdot\Delta\vec{P} \tag{26}$$

We make the Fourier decomposition of $\Delta\vec{P}$, then because of condition ii) we can write in the form

$$\begin{aligned}\Delta\vec{P}(\omega) &= \int \frac{d^3\vec{k}}{(2\pi)^3}\, \Delta\overleftrightarrow{\chi}(\vec{k},\omega)\cdot\vec{E}(\vec{k},\omega)\exp[i\vec{k}\cdot\vec{r}] \\ &= \Delta\overleftrightarrow{\chi}(-i\vec{\nabla},\omega)\cdot\vec{E}(\omega),\end{aligned} \tag{27}$$

where $\Delta\overleftrightarrow{\chi}(\vec{k},\omega)$ denotes the kinetic correction to the polarizability. In the absence of the dissipative effect, the lowest order expression for the matrix element of $\Delta\overleftrightarrow{\chi}$ may be written in the form

$$\Delta\chi_{\alpha\beta}(\vec{k},\omega) = A_{\alpha\beta}(\omega)\, k_\alpha k_\beta \qquad (\alpha,\beta = x, y, z), \tag{28}$$

with $A_{\alpha\beta}(\omega)=A_{\beta\alpha}(-\omega)$. Substituting (28) into (27) and using the result in (26), we obtain

$$\begin{aligned}&-8\pi\int d^3\vec{x}\, n_o \overline{\sum \delta\vec{E}\;\Delta\vec{P}} \\ &= -8\pi\,\delta\int d^3\vec{x}\, n_o \sum\sum_{\alpha,\beta} A_{\alpha\beta}(\omega)\,\nabla_\alpha E_\alpha(\omega)\,\nabla_\beta E_\beta(-\omega),\end{aligned} \tag{29}$$

where in carrying out the partial integration we ignored the derivative of $n_o$ according to the condition ii). Choosing $\vec{E}(\omega)$ and $\vec{E}(-\omega)$, instead of $\vec{E}(t)$, as the variational variables, we can then write the variation principle in the following form:

$$\delta\int d^3\vec{x}\left[\frac{|\vec{E}(\omega)|^2}{4\pi} - \frac{|\vec{B}(\omega)|^2}{4\pi} + \overline{\Pi}_e + \overline{\Pi}_i + n_o\sum\sum_{\alpha,\beta} A_{\alpha\beta}(\omega)\,\nabla_\alpha E_\alpha(\omega)\,\nabla_\beta E_\beta(-\omega)\right] = 0$$

where in the last term on the left-hand side the first summation takes place over the species of the particle.

Let us finally discuss the validity of our formalism. First, the use of the fluid model as the zero-order approximation can be justified when $k\lambda_D \ll 1$ and $k_\perp\rho \ll 1$ where $k$ and $k_\perp$ are respectively the effective wavenumber and its component perpendicular to the $B_0$-field, $\lambda_D$ is the Debye length and $\rho$ is the Larmour radius. Assumption that $\nabla\Pi$ is parallel to $\vec{\nabla}n$ excludes the case when there exists a temperature gradient perpendicular to $\vec{\nabla}n$. The second-order approximation for the nonlinear effect can be justified when $|\vec{k}\cdot\vec{r}| \ll 1$. Finally, the neglect of the static flow of the particles is justified if $|\vec{\bar{v}}| \ll |\vec{v}|$. Since the static flow predominantly occurs along the magnetic field, this condition can alternatively be written as

$$|\vec{b}\cdot\vec{\bar{v}}| \ll |q\vec{b}\cdot\vec{E}/m\omega| \ll \omega/k_\parallel ,$$

where the second inequality arises from the condition $|\vec{k}\cdot\vec{r}| \ll 1$.

# THREE-WAVE COUPLING COEFFICIENTS FOR MAGNETIZED PLASMAS

L. Stenflo and J. Larsson

Department of Plasma Physics

Umeå University, S-901 87 Umeå, Sweden

## INTRODUCTION

In order to investigate the resonant interaction between three waves it is necessary to calculate the coupling coefficients. This has previously been done by many authors for different cases. Reviews [e.g. 1,2] thus contain numerous references to basic works. In the present paper we will reconsider a general expression, found in reference [3] , in order to demonstrate how it can be applied to particular situations. This is illustrated by means of four specific examples.

## BASIC EXPRESSIONS

Let us assume that the resonance conditions $\omega_3 = \omega_1 + \omega_2$ and $\underset{\sim}{k}_3 = \underset{\sim}{k}_1 + \underset{\sim}{k}_2$ are satisfied for three undamped waves which propagate in a spatially uniform plasma situated in a constant external magnetic field $B_0\hat{z}$. We choose to study the development in time of the amplitudes $E_j (j = 1,2,3)$ of the electric fields $\underset{\sim}{E}_j = E_j\hat{e}_j$ ($\hat{e}_j$ is the polarization vector of unit length) of the waves. One then derives coupled bilinear equations of the wellknown form

$$\frac{\partial E_1^*}{\partial t} = c_1 E_2 E_3^*$$

$$\frac{\partial E_2^*}{\partial t} = c_2 E_1 E_3^* \qquad (1)$$

$$\frac{\partial E_3}{\partial t} = c_3 E_1 E_2$$

(More generally one could in (1) replace $\partial/\partial t$ by $\frac{\partial}{\partial t} - \frac{\partial \omega_j}{\partial \underline{k}_j} \cdot \frac{\partial}{\partial \underline{r}}$ - Im $\omega_j$ respectively).

We write the coupling coefficients $c_j$ (which are constants) as

$$c_1 = \frac{i\omega_1}{W_1} \frac{E_1^* V}{E_2 E_3^*} ; \quad c_2 = \frac{i\omega_2}{W_2} \frac{E_2^* V}{E_1 E_3^*} ; \quad c_3 = \frac{-i\omega_3 E_3 V}{W_3 E_1 E_2} \qquad (2)$$

where

$$W = \varepsilon_0 \underline{E}^* \cdot \frac{1}{\omega} \frac{\partial}{\partial \omega} \omega^2 \underline{\underline{\varepsilon}} \cdot \underline{E} \qquad (3)$$

is the wave energy ($\underline{\underline{\varepsilon}}$ is the usual dielectric tensor) and V is a common factor which is proportional to $E_1 E_2 E_3^*$. As $W_j$ is proportional to $E_j E_j^*$ it is thus obvious that $c_j$ are independent of the amplitudes $E_{1,2,3}$. Eqs. (1) are sometimes written in the form

$$\frac{1}{2} \frac{\partial W_{1,2}}{\partial t} = - \omega_{1,2} \text{ Im } V ; \qquad \frac{1}{2} \frac{\partial W_3}{\partial t} = \omega_3 \text{ Im } V \qquad (4)$$

which shows that the total energy $W_1 + W_2 + W_3$ is conserved. The interaction coefficient V for a cold plasma is wellknown [e.g. 1]. Thus

$$V = \sum_\sigma m \Big[ n_1 \underset{\sim}{v}_2 \cdot \underset{\sim}{v}_3^* + n_2 \underset{\sim}{v}_1 \cdot \underset{\sim}{v}_3^* + n_3^* \underset{\sim}{v}_1 \cdot \underset{\sim}{v}_2 - \frac{i\omega_c N_0}{\omega_3 - \underset{\sim}{k}_3 \cdot \underset{\sim}{v}_d} \Big( \frac{k_{2z}}{\omega_2 - \underset{\sim}{k}_2 \cdot \underset{\sim}{v}_d} - \frac{k_{1z}}{\omega_1 - \underset{\sim}{k}_1 \cdot \underset{\sim}{v}_d} \Big) \underset{\sim}{v}_3^* \cdot (\underset{\sim}{v}_1 \times \underset{\sim}{v}_2) \Big] \tag{5}$$

where $\sum_\sigma$ stands for a summation over the particle species, m is the mass, $\underset{\sim}{v}_d$ the drift velocity, $\omega_c = qB_0/m$ the gyrofrequency, $n_j$ the density perturbation due to wave j, and $\underset{\sim}{v}_j$ the corresponding mean velocity perturbation. By means of the linear relations between $n_j$, $\underset{\sim}{v}_j$ and $E_j$ one then often writes out $c_j$ explicitly for different cases of interest. The calculation of the corresponding expression for a Vlasov plasma is much more complicated although the final result is formally equally simple. One finds the symmetric expression [3]

$$V = \sum_\sigma m \int d\underset{\sim}{v}\, F_0(\underset{\sim}{v}) \sum_{\substack{p_1+p_2=p_3 \\ p_j = 0, \pm 1, \pm 2, \ldots}} e^{i(p_1\theta_1 + p_2\theta_2 - p_3\theta_3)}$$

$$\Big[ \frac{\underset{\sim}{k}_1 \cdot \underset{\sim}{u}_{1p_1}}{\omega_1 - k_{1z}v_z - p_1\omega_c} \underset{\sim}{u}_{2p_2} \cdot \underset{\sim}{u}_{3p_3}^* + \frac{\underset{\sim}{k}_2 \cdot \underset{\sim}{u}_{2p_2}}{\omega_2 - k_{2z}v_z - p_2\omega_c} \underset{\sim}{u}_{1p_1} \cdot \underset{\sim}{u}_{3p_3}^* + \tag{6}$$

$$+ \frac{\underset{\sim}{k}_3 \cdot \underset{\sim}{u}_{3p_3}^*}{\omega_3 - k_{3z}v_z - p_3\omega_c} \underset{\sim}{u}_{1p_1} \cdot \underset{\sim}{u}_{2p_2} - \frac{i\omega_c}{\omega_3 - k_{3z}v_z - p_3\omega_c}$$

$$\Big( \frac{k_{2z}}{\omega_2 - k_{2z}v_z - p_2\omega_c} - \frac{k_{1z}}{\omega_1 - k_{1z}v_z - p_1\omega_c} \Big) \underset{\sim}{u}_{3p_3}^* \cdot (\underset{\sim}{u}_{1p_1} \times \underset{\sim}{u}_{2p_2}) \Big]$$

where the angles $\theta_j$ and velocity vectors $\underset{\sim}{u}_j$ are defined from $e^{i\theta_j} =$ $= (k_{jx} + ik_{jy})/k_{j\perp}$ and

$$(\omega_j - k_{jz}v_z - p\omega_c)\, \underset{\sim}{u}_{jp} + i\omega_c \hat{z} \times \underset{\sim}{u}_{jp} = \frac{iq}{m\omega_j} \{(\omega_j - k_{jz}v_z - p\omega_c) J_p(j) \underset{\sim}{E}_j +$$

$$+ \left[ (v_z E_{jz} + \frac{p\omega_c}{k_{j\perp}^2} \underset{\sim}{k}_{j\perp} \cdot \underset{\sim}{E}_j)\, J_p(j) + \frac{iv_\perp \omega_c}{k_{j\perp}^2} (\hat{z} \times \underset{\sim}{k}_j) \cdot \underset{\sim}{E}_j \frac{d}{dv_\perp} J_p(j) \right] \underset{\sim}{k}_j \} \tag{7}$$

where $J_p(j) = J_p(k_{j\perp}v_\perp/\omega_c)$ denotes a Bessel function of order p.

## SPECIFIC EXAMPLES

Let us now illustrate the use of (6) and (7) by means of four examples.

a) If all waves are longitudinal, it is possible to simplify (7). Thus

$$\underset{\sim}{u}_p = \frac{iq}{m\omega_d} \frac{E}{(1-\omega_c^2/\omega_d^2)k} (\underset{\sim}{k} - \frac{i\omega_c}{\omega_d} \hat{z} \times \underset{\sim}{k} - \frac{\omega_c^2}{\omega_d^2} k_z \hat{z}) J_p \tag{8}$$

where $\omega_d = \omega - k_z v_z - p\omega_c$

The wave energy (3) reduces to

$$W = \varepsilon_0 \omega \frac{\partial \varepsilon_\ell}{\partial \omega} E\, E^* \tag{9}$$

where $\varepsilon_\ell(\omega, \underset{\sim}{k}) = 1 + \sum_{\sigma,p} \frac{q^2}{m\varepsilon_0 k^2} \int \frac{d\underset{\sim}{v}}{\omega_d} (\frac{p\omega_c}{v_\perp} \frac{\partial F_0}{\partial v_\perp} + k_z \frac{\partial F_0}{\partial v_z}) J_p^2$

is the usual dielectric constant. Inserting (8) in (6) and then (6) and (9) in (2) we thus have comparatively simple expressions for the coupling constants which, although written in a different form, agree with those of previous works (see references in [1] and [2]).

b) For the case [e.g. 4] where two transverse waves (1 and 3) and one longitudinal wave (2) propagate parallel to $\underset{\sim}{B}_0$ we easily (as

$k_\perp = 0$) obtain from (6), (7), (2) and (3)

$$c_1 = -\frac{\omega_1^2}{(\partial\,\omega^2\varepsilon_\pm/\partial\,\omega)_1}C_\pm\ ,\quad c_3 = \frac{\omega_3^2}{(\partial\,\omega^2\varepsilon_\pm/\partial\,\omega)_3}C_\pm \quad\text{and}\quad c_2 = -\frac{C_\pm}{(\partial\,\varepsilon_\ell/\partial\,\omega)_2} \tag{10}$$

where

$$\varepsilon_\pm = 1 - \sum_\sigma \frac{q^2}{m\varepsilon_0\omega^2}\int d\underset{\sim}{v}\,\frac{(\omega-k_zv_z)F_0-(k_zv_\perp^2/2)\,\partial F_0/\partial v_z}{\omega-k_zv_z\mp\omega_c} \tag{11}$$

and

$$C_\pm = \sum_\sigma \frac{q^3}{m^2\varepsilon_0\omega_1\omega_3}\int d\underset{\sim}{v}\left\{\frac{F_0}{(\omega_2-k_{2z}v_z)^2}\left[k_{2z}\mp\right.\right.$$

$$\left.\omega_c\left(\frac{k_{1z}}{\omega_3-k_{3z}v_z\mp\omega_c}-\frac{k_{3z}}{\omega_1-k_{1z}v_z\mp\omega_c}\right)\right] - \tag{12}$$

$$\left.-\frac{k_{1z}k_{3z}}{2}v_\perp^2\frac{\partial F_0}{\partial v_z}\frac{1}{(\omega_1-k_{1z}v_z\mp\omega_c)(\omega_2-k_{2z}v_z)(\omega_3-k_{3z}v_z\mp\omega_c)}\right\}$$

c) Let us now assume that two high frequency, almost longitudinal waves (1 and 3) interact with a low frequency magnetosonic wave (2). The high frequency waves have to satisfy the dispersion relation

$$1 - \frac{k_\perp^2}{k^2}\frac{\omega_{pe}^2}{\omega^2-\omega_{ce}^2} - \frac{k_z^2}{k^2}\frac{\omega_{pe}^2}{\omega^2} \approx 0 \tag{13}$$

whereas the low frequency wave is described by $\omega_2 = k_2v_A$ and $W_2 = 2\varepsilon_0|E_2|^2c^2/v_A^2$, where $v_A$ is the Alfvén velocity.

If $|\omega_{ce}| \gg \omega_{pe}$ [5] we then obtain (for most angles, or $(k_{1\perp}/k_1)^2 \gg$ $\gg \omega_2/\omega_{ci} \ll 1$)

$$c_1 = - c_3 = \frac{ik_{2\perp}}{B_o} \frac{|\omega_{ce}|}{\omega_2} e^{i\phi}$$

and (14)

$$c_2 = \frac{ik_{2\perp}}{B_o} \frac{m_e}{m_i} \left(\frac{\omega_{ce}}{\omega_{pe}}\right)^4 \frac{k_1^2}{k_{1\perp}^2} e^{i\phi}$$

where $\phi$ is the angle between $\underset{\sim}{k}_{1\perp}$ and $\underset{\sim}{k}_{3\perp}$. We have here used the fact that $k_{1\perp}/k_1 \approx k_{3\perp}/k_3$ as $\omega_1 \approx \omega_3 \approx |\omega_{ce}| >> \omega_2$.

d) Let us finally consider the case where two high frequency ($\omega_{1,3} >> |\omega_{ce}|$ and $k_{1,3}v_{te}$) transverse waves interact with a magneto-sonic wave ($\omega_2 << \omega_{ci}$). Considering one of the transverse waves ($\omega_3, \underset{\sim}{k}_3$) as a pump wave one then finds the growth rate (well above threshold, i.e. $\gamma >>$ Im $\omega_2$)

$$\gamma \approx \sqrt{c_1 c_2^*} \; |E_3| = \frac{k_{2\perp}}{\sqrt{\omega_2 \omega_3}} \frac{\omega_{pi}}{\omega_3} \frac{q_e |E_3|}{2m_e} |\hat{e}_1 \cdot \hat{e}_3^*| \tag{15}$$

which is essentially (apart from the angular factor $k_{2\perp}/k_2$) the same as that of the decay into a transverse wave and an ion sound wave (stimulated Brillouin scattering). As $\omega_2$ is small one ought to extend the analysis to include oscillations at $\omega_3 + \omega_2$ as well as $\omega_3 - \omega_2$ and $\omega_2$ in order to replace (15) by the dispersion relation

$$\omega^2 - k^2 V_A^2 = \frac{q^2 \omega_{pi}^2}{m_e^2 \omega_3^2} k_\perp^2 \left[ \frac{|(\underset{\sim}{k}+\underset{\sim}{k}_3) \times \underset{\sim}{E}_3|^2}{(\underset{\sim}{k}+\underset{\sim}{k}_3)^2((\omega+\omega_3)^2-(\underset{\sim}{k}+\underset{\sim}{k}_3)^2c^2-\omega_{pe}^2)} + \right.$$
$$\left. + \frac{|(\underset{\sim}{k}-\underset{\sim}{k}_3) \times \underset{\sim}{E}_3|^2}{(\underset{\sim}{k}-\underset{\sim}{k}_3)^2((\omega-\omega_3)^2-(\underset{\sim}{k}-\underset{\sim}{k}_3)^2c^2-\omega_{pe}^2)} \right] \tag{16}$$

which can be analysed by means of well developed methods [6] .

## REFERENCES


1 Tsytovich, V.N., Nonlinear Effects in Plasma, Plenum Press, New York (1970);
Pustovalov, V.V. and Silin, V.P. in Proceedings of the P.N. Lebedev Physics Institute (Ed. by D.V. Skobeltsyn) 61, 37 (1975).

2 Kaw, P.K. in Advances in Plasma Physics (Ed. by A. Simon and W.B. Thompson) 6, 179 and 207 (1976);
Porkolab, M., Physica 82C, 86 (1976).

3 Larsson, J., J.Plasma Phys. 14, 467 (1975);
Larsson, J. and Stenflo, L., Beitr. aus der Plasmaphysik, 16, 79 (1976).

4 Sjölund, A. and Stenflo, L., Z.Phys. 204, 211 (1967);
Wilhelmsson, H., J.Plasma Phys. 3, 215 (1969).

5 Kaufman, A.N. and Stenflo, L., Phys.Scripta 11, 269 (1975).

6 Drake, J.F., Kaw, P.K., Lee, Y.C., Schmidt, G., Liu, C.S. and Rosenbluth, M.N., Phys.Fluids 17, 778 (1974);
Manheimer, W.M. and Ott, E., Phys. Fluids 17, 1413 (1974).

# GENERALIZED PONDEROMOTIVE FORCES AND THREE-WAVE INTERACTION

Shayne Johnston and Allan N. Kaufman

Department of Physics and Lawrence Berkeley Laboratory

University of California, Berkeley, California 94720

The subject of this article is a unified Hamiltonian approach to the theory of nonlinear interactions among waves and particles. The unifying feature of the approach is a generalization of the concept of "ponderomotive force". The formulation can be said to retain the conceptual simplicity of the familiar ponderomotive-potential method [1-3], but to remove the approximations [4]. The essence of the approach is to replace the usual method of time-averaging by the performance of a canonical transformation. The transformation is designed to eliminate the terms in the Hamiltonian of a particle which are linear in the wave potentials, replacing them with bilinear terms at combination frequencies. The new entity (the "oscillation center") thus has no first order jittering motion. The transformation formalism leads to explicit expressions for the required nonlinear currents, which can be decomposed into the current of oscillation centers and the "polarization" corrections [4]. The oscillation-center representation is thus quite analogous to the more familiar guiding-center representation in strong magnetic fields.

Such an approach has previously been applied to the theory of induced scattering of waves by resonant particles [5]. The useful extension of this point of view to other nonlinear processes is advocated here. We shall demonstrate the oscillation-center approach by deriving a compact general formula for the three-wave coupling coefficient in collisionless plasma. We wish here to emphasize ideas rather than the details of the formalism [6]. Accordingly, our subsequent discussion will be fairly schematic.

We consider a collisionless, nonuniform and possibly relativistic plasma, confined by spatially inhomogeneous electric and magnetic fields. We treat the linear normal modes of the configuration as fully electromagnetic, assuming that their eigenfrequencies are nearly real. We interpret the confinement as due to the adiabatic invariants of each particle, and shall implicitly assume the existence of the action-angle variables $(\underline{I}, \underline{\theta})$ associated with the unperturbed Hamiltonian $H_o(\underline{I})$. Accordingly, we separate all the plasma particles into two categories: the vast majority which comprise the nonresonant particles, and the small subset of "resonant" particles which satisfy [7,8]

$$\omega_a \approx \underline{\ell}_a \cdot \partial H_o / \partial \underline{I} \equiv \underline{\ell}_a \cdot d\underline{\theta}/dt \ . \tag{1}$$

In Eq. (1), $\omega_a$ denotes the real part of the eigenfrequency for normal mode $a$ , and the vector $\underline{\ell}_a$ represents a set of three integers. We wish to concentrate on the nonresonant particles in this article, and so we decompose (in some mathematically smooth way) the unperturbed phase-space distribution function in the form

$$f_o(\underline{r},\underline{p}) = \tilde{f}_o(\underline{r},\underline{p}) + f_o^{res}(\underline{r},\underline{p}) \ , \tag{2}$$

where $\tilde{f}_o$ represents the nonresonant distribution.

We choose to work with conjugate variables $(\underline{r},\underline{p})$, where $\underline{r}$ denotes the Cartesian position vector in physical space. The unperturbed Hamiltonian for a plasma particle can be written

$$H_o(\underline{r},\underline{p}) = e\Phi_o(\underline{r}) + \left\{[\underline{p}c - e\underline{A}_o(\underline{r})]^2 + m^2c^4\right\}^{\frac{1}{2}} \ . \tag{3}$$

Adopting the radiation gauge $\phi' = 0$, we consider a set of three perturbing normal modes,

$$\underline{A}'(\underline{x},t) = \sum_{a=1}^{3} \underline{A}_a(\underline{x}) e^{-i\omega_a t} + \text{c.c.} , \quad (4)$$

whose (positive) frequencies satisfy the resonant matching condition $\omega_3 = \omega_1 + \omega_2$. A small frequency mismatch $\Delta\omega \ll \omega_a$ can be taken into account in the usual way [9]. Representative components of the perturbed Hamiltonian are

$$H_a' = - e/c \, \underline{A}_a(\underline{r}) \cdot \underline{\partial} \, H_o(\underline{r},\underline{p}) \, e^{-i\omega_a t} , \quad (5)$$

$$H_3'' = e^2/c^2 \, \underline{A}_1(\underline{r}) \, \underline{A}_2(\underline{r}) : \underline{\partial} \, \underline{\partial} \, H_o(\underline{r},\underline{p}) \, e^{-i\omega_3 t} , \quad (6)$$

$$H_o''' = - e^3/c^3 \, \underline{A}_1(\underline{r}) \, \underline{A}_2(\underline{r}) \, \underline{A}_3^*(\underline{r}) \; \vdots \; \underline{\partial} \, \underline{\partial} \, \underline{\partial} \, H_o(\underline{r},\underline{p}) , \quad (7)$$

where the primes refer to the order in the perturbation, the subscripts identify the time dependence, and $\underline{\partial} \equiv \partial/\partial\underline{p}$. Let us devise a canonical transformation to eliminate $H'$, i.e., all first-order terms in the perturbed Hamiltonian:

$$(\underline{r},\underline{p},H) \rightarrow (\underline{R},\underline{P},K) , \quad (8)$$

$$K = H_o + K'' + \cdots . \quad (9)$$

Such a transformation corresponds to simple Hamilton-Jacobi perturbation theory. If the particle were in resonance with any of the three primary modes, then a two-time-scale refinement of the transformation [5,10] would be required. The forces derivable from the oscillation-center Hamiltonian $K''$ may be viewed as generalized ponderomotive forces. Note, however, that we have not ordered frequencies and averaged over time. We have simply performed a canonical transformation.

The perturbative generating function for the transformation, $S(\underline{r},\underline{P},t)$, is determined by the equations [5]

$$D_t S_a(\underline{r},\underline{p},t) = - H_a'(\underline{r},\underline{p},t) , \quad (10)$$

$$D_t = \partial/\partial t + [\;\; ,H_o] . \quad (11)$$

The resultant Hamiltonian $K''(\underline{r},\underline{p},t)$ is then a sum of $H''$ and a known bilinear functional [5] of first-order quantities. The nonresonant phase-space distribution function can now be decomposed in the form

$$\tilde{f}(\underline{r},\underline{p},t) = F(\underline{r},\underline{p},t) + \Delta(\underline{r},\underline{p},t) , \tag{12}$$

$$F = \tilde{f}_o + F'' + \cdots , \quad \Delta = \Delta' + \Delta'' + \cdots , \tag{13}$$

where $F$ denotes the solution of the Vlasov equation for oscillation centers, and $\Delta$ represents the difference between $F$ and $\tilde{f}$ at the same phase point $(\underline{r},\underline{p})$. Again, we have explicit formulas for $\Delta'$ and $\Delta''$ in terms of S, H and $\tilde{f}_o$ [5].

Let us apply these ideas to calculate the slow evolution of the amplitudes of the interacting normal modes. Our concern is therefore with coupled equations of the form

$$\underset{\sim}{D}(\omega_a + i\partial/\partial t)\cdot\underline{E}_a(\underline{x}) = (4\pi/i\omega_a)\underline{j}''_a(\underline{x}) , \tag{14}$$

where $\underset{\sim}{D}(\omega)$ denotes the linear dispersion operator (assumed nearly Hermitian), and $\underline{j}''_a$ represents the nonlinear current source at frequency $\omega_a$ due to the beating of the other two modes. Retaining only the nonresonant terms in Eq.(14), we write

$$\underset{\sim}{D}'(\omega_a + i\partial/\partial t)\cdot\underline{E}_a = (4\pi/i\omega_a)\tilde{\underline{j}}''_a , \tag{15}$$

where $\underset{\sim}{D}'$ denotes the Hermitian part of $\underset{\sim}{D}$. Since $\omega_a$ is a linear eigenfrequency, we have

$$\underset{\sim}{D}'(\omega_a + i\partial/\partial t)\cdot\underline{E}_a \rightarrow i\partial\underset{\sim}{D}'(\omega_a)/\partial\omega_a\cdot d\underline{E}_a/dt . \tag{16}$$

Now, the total energy of wave a can be written

$$W_a = (\omega_a/4\pi)\int d^3x\, \underline{E}_a^*\cdot\partial\underset{\sim}{D}'(\omega_a)/\partial\omega_a\cdot\underline{E}_a . \tag{17}$$

Combining relations (15) to (17), we thus obtain the action evolution equation

$$\omega_a^{-1}\,\tilde{d}W_a/dt = -2\,\mathrm{Im}\int d^3x\, c^{-1}\,\underline{A}_a^*\cdot\tilde{\underline{j}}''_a . \tag{18}$$

where the symbol $\tilde{d}$ denotes the evolution due to nonresonant currents.

We shall proceed to evaluate the right-hand side of Eq.(18) using the oscillation-center transformation. The physical current density $\tilde{\underline{j}}(\underline{x},t)$ of nonresonant particles can be written

$$\tilde{\underline{j}}(\underline{x},t) = e\int d^3r \int d^3p\ \delta(\underline{x}-\underline{r})\ \tilde{f}(\underline{r},\underline{p},t)\ \underline{\partial}H(\underline{r},\underline{p},t)\ . \quad (19)$$

Invoking the decomposition (12) of $\tilde{f}$, we can break up the second-order contributions to Eq.(19) in the form

$$\tilde{\underline{j}}''_3 = \underline{J}''_3 + \Delta\underline{j}''_3\ , \quad (20)$$

where

$$\underline{J}''_3 \equiv e\int d^3r \int d^3p\ \delta(\underline{x}-\underline{r})\ F''_3\ \underline{\partial}H_o\ , \quad (21)$$

and

$$\Delta\underline{j}''_3 \equiv e\int d^3r \int d^3p\ \delta(\underline{x}-\underline{r})$$
$$\times\ (\tilde{f}_o\underline{\partial}H''_3 + \Delta'_1\underline{\partial}H'_2 + \Delta'_2\underline{\partial}H'_1 + \Delta''_3\underline{\partial}H_o)\ . \quad (22)$$

We rewrite the currents (21) and (22) using our generating-function formulas, and substitute the results into Eq.(18). A crucial sequence in the subsequent manipulations is the following:

$$\begin{aligned} -\int d^3x\ c^{-1}\ \underline{A}^*_3\cdot\underline{J}''_3 &= +\int d^3r \int d^3p\ H'^*_3\ F''_3 \\ &= -\int d^3r \int d^3p (D_t\ S_3{}^*)\ D_t{}^{-1}\ [K''_3,\tilde{f}_o] \\ &= +\int d^3r \int d^3p\ S_3{}^*\ [K''_3,\tilde{f}_o] \\ &= +\int d^3r \int d^3p\ \tilde{f}_o\ [S_3{}^*,K''_3]\ . \end{aligned} \quad (23)$$

Judicious manipulation and partial integration lead finally to the following compact and general formula:

$$- \int d^3x\, c^{-1}\, \underline{A}_3^* \cdot \underline{\tilde{j}}_3'' = \int d^3r \int d^3p\, \tilde{f}_o(\underline{r},\underline{p}) \times \Big\{ H_o''' + [S_1, H_1''^*] + [S_2, H_2''^*] + [S_3^*, H_3''] + \frac{1}{3} \big( [S_2, [S_3^*, H_1']] + [S_3^*, [S_2, H_1']] + [S_3^*, [S_1, H_2']] + [S_1, [S_3^*, H_2']] + [S_1, [S_2, H_3'^*]] + [S_2, [S_1, H_3'^*]] \big) \Big\} . \tag{24}$$

This expression is to be inserted into Eq.(18).

Formula (24) is manifestly symmetric under interchange of the subscripts (1, 2, $3^*$). This symmetry implies conservation of action in the three-wave process (the Manley-Rowe relations [11]). We have presented here the essence of a purely classical and quite general proof of that conservation law. It is a consequence of the static nature of the equilibrium, and of separating out the dissipation to resonant particles. For purely electrostatic modes, a formula essentially equivalent to Eq. (24) has been derived by Laval and Pellat [9]. Only the triple Poisson bracket terms involving $H_a'$ survive in the electrostatic limit.

In the limiting case of a uniform, nonrelativistic, magnetized plasma, the expression (24) can be shown to reduce to the following:

$$c^{-1}\, \underline{A}_3^* \cdot \underline{\tilde{j}}''(\underline{k}_3, \omega_3) = m^{-2} \int d^3v\, \tilde{f}_o(v_\perp, v_z) \times \Big\{ (i\underline{k}_1 \cdot \underline{\partial} S_1)(D_t \underline{\partial} S_2) \cdot (D_t \underline{\partial} S_3^*) + (i\underline{k}_2 \cdot \underline{\partial} S_2)(D_t \underline{\partial} S_3^*) \cdot (D_t \underline{\partial} S_1) + (-i\underline{k}_3 \cdot \underline{\partial} S_3^*)(D_t \underline{\partial} S_1) \cdot (D_t \underline{\partial} S_2) + \Omega_B \underline{\partial} S_3^* \cdot [ik_{1z}(D_t \underline{\partial} S_2) \times \underline{\partial} S_1 + (1 \leftrightarrow 2)] \Big\} , \tag{25}$$

where, in this equation only, we have defined $\underline{\partial} \equiv \partial/\partial\underline{v}$. The generating function $S_a$ can now be written explicitly as an infinite sum over Bessel functions [5]. Formula (25) was recently derived by Larsson [12] using different notation and a different method.

This work was supported by the United States Energy Research and Development Administration under Contract W-7405-ENG-48.

References

1. A. G. Litvak and V. Yu. Trakhtengerts, Zh. Eksp. Teor. Fiz. 60, 1702 (1971) [Sov. Phys. - JETP 33, 921 (1971)].

2. J. F. Drake, P. K. Kaw, Y. C. Lee, G. Schmidt, C. S. Liu, and M. N. Rosenbluth, Phys. Fluids 17, 778 (1974).

3. W. M. Manheimer and E. Ott, Phys. Fluids 17, 1413 (1974).

4. S. Johnston, A. N. Kaufman, and G. L. Johnston, Bull. Am. Phys. Soc. 20, 1375 (1975) (to be published).

5. S. Johnston, Phys. Fluids 19, 93 (1976).

6. S. Johnston and A. N. Kaufman (to be published).

7. A. N. Kaufman, Phys. Fluids 14, 387 (1971).

8. A. N. Kaufman, Phys. Fluids 15, 1063 (1972).

9. G. Laval and R. Pellat, in Plasma Physics (Les Houches), edited by C. Dewitt and J. Peyraud (Gordon and Breach, New York, 1975), p. 261.

10. R. L. Dewar, Phys. Fluids 16, 1102 (1973).

11. A. N. Kaufman and L. Stenflo, Plasma Phys. 17, 403 (1975).

12. J. Larsson, J. Plasma Phys. 14, 467 (1975).

# MODULATION INSTABILITY AND STRONG LANGMUIR TURBULENCE

V.N.Tsytovich

P.N.Lebedev Physics Institute

Moscow USSR

ABSTRACT

The problems analyzed include the theoretical description of Langmuir condensation of strong nonlinear motions in nonequilibrium plasmas and the statistical theory of strong Langmuir turbulence.

## I. INTRODUCTION. LANGMUIR CONDENSATION: THE STATE OF THE ART

1. The problems of nonlinear interactions in plasmas are the most important problems of plasma physics at the present time since they are related to various applications, such as the powerful electron beam plasma and laser plasma interaction. The first approach [1] reported as far back as 1965 at the conference in Belgrade was simple and clear from the physics aspect. Further developments led to the interesting, important, and complicated problem facing plasma physics at the present time: Langmuir condensation. Let us follow the logic of the development of nonlinear plasma physics to clarify how this problem arose. The processes analyzed in [1] were later (in the seventies) rediscovered in the USA and called induced Raman and Brillouin scattering (see for example [2]). These are the so-called three-wave decay processes $t \rightleftarrows t + \ell$ (Fig. 1) which satisfy the conservation laws

$$\omega^t_{\vec{k}_1} = \omega^t_{\vec{k}_2} + \omega^\ell_{\vec{k}} \quad ; \quad \vec{k}_1 = \vec{k}_2 + \vec{k} \qquad (1)$$

In [1] it was pointed out that the decay process (1) can be regarded as induced emission of Langmuir waves by transverse waves (Fig.1), and is therefore similar to induced Cherenkov emission of Langmuir waves by particles (Fig.2). Thus for a beam of transverse waves the decay process (1) leads to an instability similar to the known beam plasma instability. The further development of this idea showed that the maximum of the growth rate of such an instability corresponds to the case $\omega^t \to \omega_{pe}$, where $\omega_{pe} = (4\pi n_o e^2/m_e)^{1/2}$ is the electron plasma frequency. For $\omega < 2\omega_{pe}$, process (1) is forbidden and the most important processes are those including the low-frequency ion-sound waves (denoted by the symbol "s")

$$t \rightleftarrows t' + s \tag{2a}$$

$$t \rightleftarrows \ell + s \tag{2b}$$

and the induced scattering processes

$$t + e \rightleftarrows \ell + e' \;;\quad t + i \rightleftarrows \ell + i' \tag{2c}$$

The importance of process (2a) in the problem of back-scattering of laser radiation was pointed out by M.Rosenbluth in 1972 [3]. The presence of low-frequency waves in processes (2a) and (2b) is of greatest importance as it leads to a new concept of a quasidecay [4]. The quasidecay corresponds to such a high intensity of electromagnetic waves that the growth rate of the decay instability becomes larger than the frequency of the low-frequency wave (s-wave). Then the low-frequency field is no longer a wave field but can be regarded as a field of virtual waves. It is necessary to mention that quasidecay is the limiting case for both processes (2b) and (2c), which can not be distinguished from each other at high intensities. The process (2c) occurs for sufficiently low intensities if the process (2b) is forbidden (for example, due to the strong damping of s-waves). As was

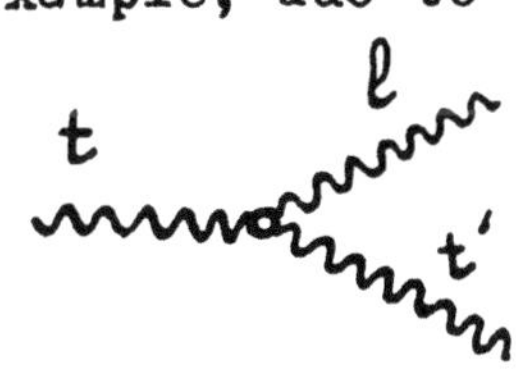


Fig. 1

Fig. 2

already pointed out in 1964 [5] there are two possible channels of the process (2c) (see Fig.3). The process shown in Fig.3b is a transitional scattering process [6] . The significance of these processes in the physics of any nonlinear wave conversion was pointed out only recently [7] . A process containing a virtual wave (Fig. 3b) is converted to quasidecay in the limit of high intensities. The nonlinear growth rate $\gamma^N$ should be accompanied by the nonlinear frequency shift $\Delta\omega^N$ according to the dispersion relation of the nonlinear plasma response. Thus the criteria of a quasidecay should be written in the form

$$|\Delta\omega^N + i\gamma^N| \gg |\omega^t - \omega^l| \tag{3}$$

Similar to the known Bloembergen dynamic equations [8] for the decay interactions, the dynamic equations for the two-wave interactions in the limit (3) have been found in [9] . The quasidecay process is converted to an aperiodic parametric instability [10,11] only in the limit when the intense transverse wave can be regarded as the given pump wave with $\kappa^t \to 0$. This instability is often called an oscillating two-stream instability. All the interrelations between the decay, quasidecay, induced scattering, and oscillating two-stream instabilities have been analyzed in detail.

2. Let us now discuss the development of research on the dynamics of the field of Langmuir oscillations. For the sake of simplicity the external magnetic field is assumed to be absent. Processes, similar to those of (2) for electromagnetic waves, are as follows for Langmuir waves:

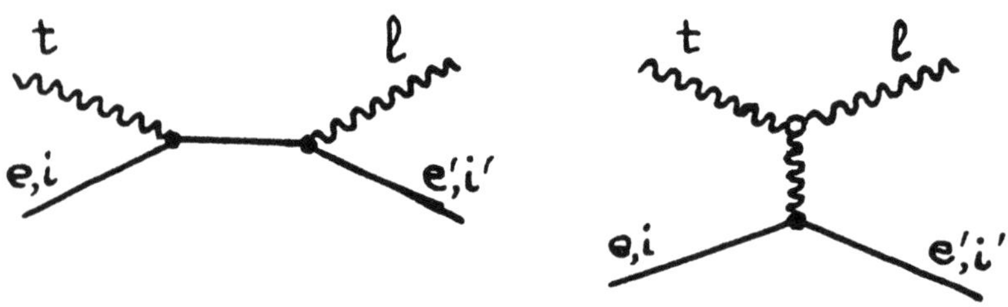


Fig. 3a

Fig. 3b

$$\ell \rightleftarrows \ell' + s \tag{4a}$$

$$e + \ell \rightleftarrows e' + \ell' \; ; \; i + \ell \rightleftarrows i' + \ell' \tag{4b}$$

The Langumuir waves excited by lasers or radiowaves in the processes described above, or by the beam plasma instability or some other source, then become independent of the source of excitation since the processes (4) remove them from the excitation region. The further dynamics of the Langmuir field is described by the same processes (4) and leads to a decrease in the frequency of Langmuir waves

$$\omega^{\ell} = \omega_{pe} + \frac{3k^2 V_{Te}^2}{2\omega_{pe}} \; ; \; \omega_{pe}^2 = \frac{4\pi n_o e^2}{m_e} \; ; \; V_{Te} = \sqrt{\frac{T_e}{m_e}} \tag{5}$$

and therefore to a decrease in the wave numbers. The waves are concentrated in the region of small k, forming the so-called Langmuir condensation [12], which is somewhat similar to a Bose-condensation [13]. Since the amount of energy accumulated in the Langmuir condensation increases in time for a constant source of waves, this leads to the Langmuir paradox. In the stationary turbulent state, the source of the waves should be compensated by the outflow of waves [14]. Thus the problem is to find the mechanism of energy loss in Langmuir condensation. Such a mechanism could be the binary collisions discussed in detail in [14]. But this mechanism does not work for a large power of excitation or for high temperatures. The solution of the Langmuir paradox is the following. The concentration of energy in the condensation is followed by an increase in the nonlinear frequency shift, thus converting the processes to the quasidecay type (see(3)). It is known [14] that the nonlinear frequency shift of Langmuir waves is largest when their group velocity is less than the ion thermal velocity:

$$V_g = \frac{3kV_{Te}^2}{\omega_{pe}} \ll V_{Ti} \; ; \; k \ll k_* = \frac{k_d V_{Ti}}{3V_{Te}} \; ; \; k_d = \frac{\omega_{pe}}{V_{Te}} \tag{6}$$

$$\Delta\omega^N(k \ll k_*) \cong - \frac{\omega_{pe} |E^{\ell}|^2}{8\pi n_o (T_e + T_i)} . \tag{7}$$

Substituting (7) and (5) in $\Delta\omega^N \gg |\omega_1^{\ell} - \omega_2^{\ell}|$ we obtain

$$\frac{W}{n_o(T_e + T_i)} \gg \frac{3}{1} \frac{k_o^2}{k_d^2} \; ; \; W = \frac{|E^{\ell}|^2}{4\pi} \tag{8}$$

where W is the energy density of Langmuir waves, $\kappa_o$ is the characteristic wave number of the packet of Langmuir waves and $\zeta$ is a constant of the order of unity. The decrease of $\kappa_o$ in the processes (4) and the increase of W in the condensation means that (8) will be satisfied at some moment in the dynamics of the Langmuir field. The quasidecay instability of Langmuir waves was first investigated in the limit of frequencies $\omega'$ and wave numbers $\kappa'$ of the perturbations

$$\kappa' \ll \kappa_o \; ; \; \omega' \ll \omega_{pe} \; ; \; v_{Ti} \ll \frac{\omega'}{\kappa'} \ll v_{Te} \qquad (9)$$

by A. Vedenov and L. Rudakov [15] in 1964. It was called the modulation instability, because in the case $\kappa' \ll \kappa_o$, the perturbations modulate the amplitude of the Langmuir wave. Independently, at the same time, this process was investigated by my student Gailitis [16] in the limit

$$\frac{\omega'}{\kappa'} \ll v_{Ti} \qquad (10)$$

by using the principle of minimum energy. The physics of modulation instability can be most easily understood by using this principle. The total energy of the thermal motions and Langmuir waves of a state with a local depletion in plasma density is proved to be less than the energy of a state with a homogeneous distribution of density. In density depletion the energy density of trapped Langmuir waves increases, thus pushing the plasma away from this region and increasing the depletion (see Fig.4). The threshold of this process is determined by the criteria (8). According to [16] $\zeta \simeq 3$. A description of the modulation instability can be found by considering the density depletion adiabatically, i.e. by substituting $4\pi e^2(n_o+\delta n)/m_e$ in (5) in place of $4\pi e^2 n_o/m_e$. This gives the equation for the envelope E of the field $\vec{E}\exp(-i\omega_{pe}t)$

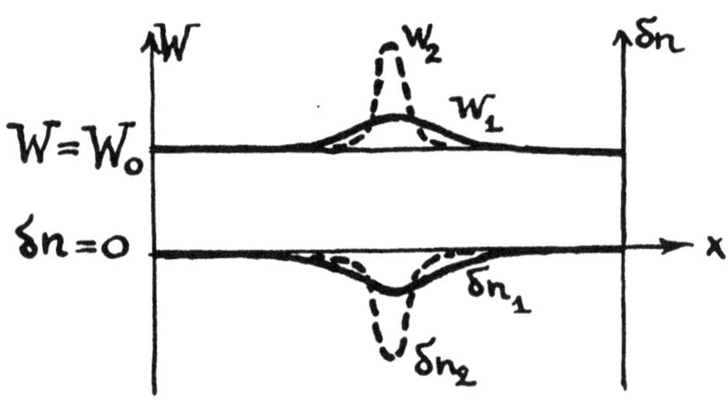


Fig. 4

$$\mathrm{div}\left(2i\omega_{pe}\frac{\partial \vec{E}}{\partial t}+\frac{3V_{Te}^2}{\omega_{pe}}\Delta\vec{E}\right)=\omega_{pe}^2\,\mathrm{div}\left(\frac{\delta n}{n_0}\vec{E}\right) \tag{11}$$

The equation for $\delta n$ can be found by combining the influence of the thermal pressure with the pressure of the electrostatic field of Langmuir oscillations:

$$\frac{\partial^2}{\partial t^2}\delta n=\Delta\left(V_s^2\delta n+\frac{1}{4\pi m_i}|\vec{E}|^2\right) \tag{12}$$

These equations were widely used both for investigations of the linear stage of modulation instability and for its nonlinear stage. The modulation instability was investigated both for the case of a strong regular pump wave [17] and for the case of a random Langmuir field [18] .In the first case, for a homogeneous pump wave ($E_0$ = Const), it was found that the growth rate [15] is maintained even for $k' \gg k_0$ and reaches the maximum value

$$\gamma_{max}\cong\omega_{pe}\sqrt{\mu E_0^2/4\pi n_0 T_e}\quad ;\ \mu=\frac{4m_e}{3m_i} \tag{13}$$

at the wave numbers

$$k_{max}\cong k_*\left(E_0^2/4\pi n_0 T_e\mu\right)^{1/4} \tag{14}$$

The modulation instability of a random field is essentially different [18] for $E_0^2/4\pi n_0 T_e\mu \ll 1$. But for $E_0^2/4\pi n_0 T\mu \gg 1$ the maximum growth rate and kmax is determined by (13) and (14), where $E_0^2/4\pi = W_*$ is the total turbulent energy density of waves for $k<k_*$ . Equations (11) and (12) were widely used for one-dimensional numerical computations [19], [20] , which had shown the existence of an energy flux from the condensation to larger $k$-s, forming a spectrum $\propto 1/k^2$ .Langmuir turbulence dominated by modulation instability has been called strong Langmuir turbulence. The reason is that in this case the nonlinear frequency shift is larger than the linear dispersion shift,both of them being less than $\omega_{pe}$. The belief exists that in the case of strong Langmuir turbulence, the energy flux changes its sign, and the energy flows finally to the large k, where it is absorbed by the tails of fast particles. Thus, the energy loss of Langmuir waves is simply due to Landau damping.

At the present time, there exists an extensive interest in the theory of strong Langmuir turbulence [21], [22], and several theoretical models were proposed: the model of weakly interacting solitons [23] , the model of noninteracting cavitons [17], and the dynamic model [24]. These models were mainly based on the equations (11) and (12) without investigations of the limits of their validity. They have played an important role as steps to a deeper understanding of the complicated processes of modulation (or quasidecay) interactions of Langmuir waves. It is obvious that at the present time the problem of modulation interactions is one of the most important, both in the development of fundamental theoretical research in plasma physics and for different applications

3. In the present report we give an analysis of the problems of modulation interaction and strong Langmuir turbulence from a general description using the nonlinear theory of a nonequilibrium plasma, We shall try to give answers to some of the problems that are still awaiting solution. Let us list some of them:

a) can the nonequilibrium particle distribution, usually present in the case of strong Langmuir turbulence, alter essentially the modulation instability?
b) what are the limits of validity of the simplest equations (11) and (12)? What are the additional nonlinear terms usually neglected? Can they be more important than those included in (11) and (12)?
c) what are the new regular nonlinear motions excited by modulation instability? Can the spikons be explained in the present theory?
d) is it possible to find a hierarchy of nonlinear motions similar to the eddies hierarchy in liquids?
e) what kind of interactions can exist between the new nonlinear motions, or are the interactions negligible?
f) on what can a proper statistical theory of strong Langmuir turbulence be based?
g) what are the spectra of turbulence and the spectra of fast particles?
h) what theoretical predictions can be made concerning the real three-dimensional turbulence?

We shall begin by developing a general nonlinear theory which includes the modulation interactions together with other nonlinear interactions, such as scattering and decays. We shall propose a simple treatment which can be used to develop the theory for any nonequilibrium particle distribution up to any order in a field, and will present the results of calculations up to the fifth order (sec.II). Then, we will discuss new nonlinear motions, such as solitons and cavitons, and propose a theory for spikons (sec.III). We shall also discuss problems of statistical theory which can give answers for the case of three-dimensional turbulence, where there is no hope of applying numerical methods to analyze the problem (sec.IV).

## II. THE KINETIC THEORY OF MODULATION INTERACTIONS

1. The general theory of nonlinear interactions can be developed by using the expansion of the plasma response in terms of the powers of the field strength. For the longitudinal field we have

$$\varepsilon_\kappa E = \int S_{1,2} E_1 E_2 d_{1,2} + \int \Sigma_{1,2,3} E_1 E_2 E_3 d_{1,2,3} + \\ + \int S'_{1,2,3,4} E_1 E_2 E_3 E_4 d_{1,2,3,4} + \int \Sigma'_{1,2,3,4,5} E_1 E_2 E_3 E_4 E_5 d_{1,2,3,4,5} \quad (15)$$

where $\varepsilon = \varepsilon_\kappa$ is the linear dielectric permeability; $E = E_\kappa$ is the Fourier component of the electric field strength; and $E_1 = E_{\kappa_1}, E_2 = E_{\kappa_2} \ldots; d_{1,2} = \delta(\kappa - \kappa_1 - \kappa_2) d\kappa_1 d\kappa_2$ $d_{1,2,3} = \delta(\kappa - \kappa_1 - \kappa_2 - \kappa_3) d\kappa_1 d\kappa_2 d\kappa_3 \ldots;\ \delta(\kappa) = \delta(\vec{\kappa})\delta(\omega);\ \kappa = \{\vec{\kappa}, \omega\}$ $d\kappa = d\vec{\kappa} d\omega;\ S_{1,2} = S_{\kappa_1,\kappa_2},\ \Sigma_{1,2,3} = \Sigma_{\kappa_1,\kappa_2,\kappa_3} \ldots$ are the matrix elements of nonlinear plasma response, calculated by using the kinetic approach. These matrix elements will be symmetrized with respect to the last two indexes.

A method can be proposed here for describing a Langmuir field which does not apply any averaging procedures (with respect to phases or fast time scale). We shall separate the Langmuir field from others by using the fact that both linear dispersion (5) and the nonlinear shifts (7) are small compared to plasma frequency $\omega_{pe}$. We divide the Langmuir field into the positive frequency part $E^+_\kappa$ and the negative frequency part $E^-_\kappa$. For the positive part

$$\alpha = \frac{|\omega - \omega_{pe}|}{\omega_{pe}} \ll 1 \quad (16)$$

and a similar relation with the opposite sign of $\omega$ can be written for $E^-_\kappa$. Thus the Langmuir field is considered as two sufficiently narrow lines near $\pm\omega_{pe}$. The problem is to find from the general expansion (15) the equation which contains only the Langmuir field. Besides the Langmuir field, the right-hand side of equation (15) for $E^+$ also contains the low-frequency field ($\omega \ll \omega_{pe}$), the fields with frequencies near $\pm 2\omega_{pe}$, $\pm 3\omega_{pe}$, etc. Equation (15) can be used to express these fields, called virtual fields in the following, in terms of the Langmuir field. After some calculations, the following equation is obtained:

$$\varepsilon_\kappa E^+_\kappa = \int \tilde{\Sigma}_{1,2,3} E^+_1 E^+_2 E^-_3 d_{1,2,3} + \\ + \int \tilde{\Sigma}'_{1,2,3,4,5} E^+_1 E^+_2 E^-_3 E^+_4 E^-_5 d_{1,2,3,4,5} + \ldots \quad (17)$$

2. We shall not write here the exact expression for the term of the fifth order in the field strength, as it is too long. The expression $\Sigma$ for the cubic nonlinearities is:

$$\tilde{\Sigma}_{1,2,3} = \tilde{\Sigma}^{L}_{1,2,3} + \tilde{\Sigma}^{++}_{1,2,3} \qquad (18)$$

where $\tilde{\Sigma}^{L}_{1,2,3}$ describes the processes with a virtual low-frequency field, while $\tilde{\Sigma}^{++}_{1,2,3}$ describes the processes with a virtual field having frequencies near $\pm 2\omega_{pe}$:

$$\tilde{\Sigma}^{L}_{1,2,3} = 2\left(\Sigma_{1,2,3} + 2 S_{1,2+3}\, \varepsilon^{-1}_{2+3}\, S_{2,3}\right) \qquad (19)$$

$$\tilde{\Sigma}^{++}_{1,2,3} = \Sigma_{3,1,2} + 2 S_{3,1+2}\, \varepsilon^{-1}_{1+2}\, S_{1,2} \qquad (20)$$

The graphs corresponding to these matrix elements are shown in Fig.5.

Next we shall use the expansion in terms of another parameter:

$$\beta = \frac{k^2 v_{Te}^2}{\omega_{pe}^2} \ll 1\,; \quad v_{Te}^2 = \frac{T_e}{m_e} = \int v^2 \Phi^{(e)} d\vec{v}\,; \quad \int \Phi^{(e)} d\vec{v} = 1 \qquad (21)$$

which is natural for the Langmuir field ( $\Phi^{(e)}$ is the normalized electron distribution function). The calculations show that $\tilde{\Sigma}^{++}_{1,2,3}$ is less than $\tilde{\Sigma}^{L}_{1,2,3}$ by a factor of $\beta$. In a first approximation in the expansion in terms of $\beta$ , $\tilde{\Sigma}^{L}_{1,2,3}$ has the form

$$\tilde{\Sigma}^{L}_{1,2,3} \cong \tilde{\Sigma}_0 \equiv -\frac{1}{4\pi n_0 T^{ef}_{2+3}} \cdot \frac{\varepsilon^{(i)}_{2+3}}{\varepsilon_{2+3}} \cdot \frac{(\vec{k}\vec{k}_1)}{|\vec{k}||\vec{k}_1|} \cdot \frac{(\vec{k}_2 \vec{k}_3)}{|\vec{k}_2||\vec{k}_3|} \qquad (22)$$

and

$$\frac{1}{T^{ef}_{2+3}} = \frac{(\vec{k}_2 + \vec{k}_3)^2}{\omega_{pe}^2 m_e}\left(\varepsilon^{(e)}_{2+3} - 1\right) \qquad (23)$$

Here, $\varepsilon^{(e)}$ and $\varepsilon^{(i)}$ are the electron and ion parts of the total dielectric permeability $\varepsilon = \varepsilon^{(e)} + \varepsilon^{(i)} - 1$. The appearance of $\varepsilon^{(i)}$ in (22) is due to the compensation of the two terms of (19). These compensation effects are well known in the nonlinear plasma theory of random fields [14]. Averaging equation (17) with respect to the wave phases leads to well-known results of weak turbulent interactions. For a Maxwellian distribution $T^{ef}$ is approximately equal to $T_e$ . Equations (11) and

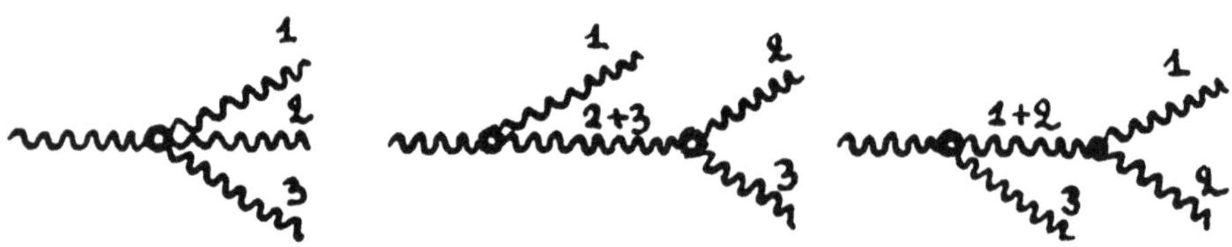


Fig. 5

(12) can be found from (22) for the special case when only the real parts of $\varepsilon^{(i)}_{2+3}$ and $\varepsilon^{(e)}_{2+3}$ are taken into account, and it is supposed that both of these values essentially exceed unity. If one denotes by $\delta n/n_0$ a value whose Fourier components are

$$\frac{\delta n_{\kappa'}}{n_0}=\frac{\varepsilon^{(i)}_{\kappa'}}{\varepsilon_{\kappa'}}\int\frac{(\vec{\kappa}_2\vec{\kappa}_3)}{|\vec{\kappa}_2||\vec{\kappa}_3|}\frac{E^{+}_{\kappa_2}E^{-}_{\kappa_3}}{4\pi n_0 T^{ef}}\delta(\kappa'-\kappa_2-\kappa_3)d\kappa_2 d\kappa_3 \qquad (24)$$

and introduces the envelope of the Langmuir field by

$$\vec{E}(\vec{r},t)=\int\frac{\vec{\kappa}}{|\vec{\kappa}|}E^{+}_{\kappa}\exp\left[-i\omega t+i\omega_{pe}t+i(\vec{\kappa}\vec{r})\right]d\kappa \qquad (25)$$

one finds that equation (17), with only the cubic term taken into account, is converted into equation (11),and that equation (24) is converted to equation (12) with $T^{ef}$ instead of $T_e$. Thus we obtain the generalization of equations (11) and (12) for the case of nonequilibrium plasma.But expression (22) contains much more information and takes into account the nonlinear interactions not taken into account by (11) and (12). Among them are processes of ion-sound emission, described by

$$Jm\ \varepsilon^{-1}_{2+3}\cong-\pi\,\delta(\varepsilon_{2+3})$$

and nonlinear Landau damping by electrons and ions described by $Jm\,\varepsilon^{(e)}_{2+3}$ and $Jm\,\varepsilon^{(i)}_{2+3}$, which contain $\delta(\omega-\omega_1-(\vec{\kappa}-\vec{\kappa}_1)\vec{v})$. These processes are the same as (2b) and (2c). It can be shown that the modulation interactions alone cannot produce the energy flux to higher wave numbers. Only the combination of all the nonlinear effects,including sound emission and nonlinear Landau damping, produces this flux. The proposed procedure to derive the equations for the Langmuir field has the important advantage of a regular procedure which can be used to find the necessary correction terms. The additional terms could even be dominating terms due to a strong compensation effect in (22).

3. One of the possible corrections is the set of higher nonlinear terms described by $\Sigma'$ in (17). In a general case, this coefficient contains 57 terms, each of which can be represented by the corresponding graph. There are now terms with virtual waves having frequencies up to $\pm 3\omega_{pe}$. Shown in Fig.6 are the graphs which contribute to $\Sigma'$ in the lowest order in the expansion with respect to the parameter $\beta$ (see (21)). We obtain

$$\tilde{\Sigma}'_{1,2,3,4,5}\cong\tilde{\Sigma}'_0\equiv\frac{1}{2(4\pi n_0 T'_{ef})^2}\cdot\frac{\varepsilon^{(i)}_{2+3+4+5}\,\varepsilon^{(i)}_{2+3}\,\varepsilon^{(i)}_{4+5}}{\varepsilon_{2+3+4+5}\,\varepsilon_{2+3}\,\varepsilon_{4+5}}\cdot\frac{(\vec{\kappa}\vec{\kappa}_1)(\vec{\kappa}_2\vec{\kappa}_3)(\vec{\kappa}_4\vec{\kappa}_5)}{|\vec{\kappa}_1||\vec{\kappa}||\vec{\kappa}_2||\vec{\kappa}_3||\vec{\kappa}_4||\vec{\kappa}_5|} \qquad (26)$$

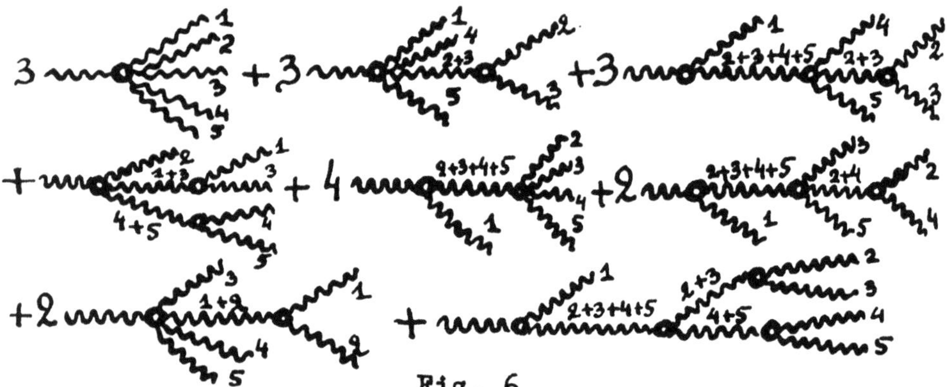

Fig. 6

where $$\frac{1}{(T'_{ef})^2} = \int \frac{\partial^2}{\partial H^2}\Phi^{(e)} d\vec{v} \; ; \quad H = \frac{m_e v^2}{2} \qquad (27)$$

In the one-dimensional (1D) case

$$\frac{1}{(T'_{ef})^2} = \frac{1}{m_e^2}\int \frac{1}{v}\frac{\partial}{\partial v}\frac{1}{v}\frac{\partial}{\partial v}\Phi^{(e)} dv \qquad (28)$$

To compare this result with the nonlinear term of the third order in the field, it is necessary to write the approximate expression for $T^{ef}$ in the 1D case:

$$1/T^{ef} = m_e^{-1}\int \frac{1}{v}\frac{\partial}{\partial v}\Phi^{(e)} dv \qquad (29)$$

According to (24), $k' = k_2 + k_3$, and then equation (23) indicates that $T^{ef}$ is a function of $\omega'/k'$, i.e. of the characteristic low-frequency phase velocity $T^{ef} = T^{ef}(\omega'/k')$. The same is true for $T'_{ef}$, since in the general case it is necessary to substitute $(v - \frac{\omega'}{k'})$ for $v$ in equation (28). Obviously $T_e$ is more dependent on the distribution of high-velocity particles, and both $T^{ef}$ and $T'_{ef}$ on the distribution of low-velocity particles (see (21) and (29)). It can be shown that the presence of a relatively small fraction ($n'/n_0 \ll 1$) of a very cold plasma with the density $n'$ can produce a negative $T^{ef}$ for sufficiently large values of $\omega'/k'$, thus stabilizing the fastest modulation disturbances.

The result (26) demonstrates the following properties of the nonlinear term of the fifth order in the fields:

a) $\tilde{\Sigma}'_0$ changes its sign when the motion changes from subsonic to supersonic, as is the case for $\tilde{\Sigma}_0$. It should be noted that in nonequilibrium plasma the sound velocity is determined by $T^{ef}$: $V_s = (T^{ef}/m_i)^{1/2}$

b) The sign of $\tilde{\Sigma}'_0$ is opposite to that of $\tilde{\Sigma}_0$.

c) The relative order of magnitude of fifth-power nonlinearities with respect to cubic nonlinearities is determined for subsonic motions by the factor $|E|^2 T^{ef}/4\pi n_0 (T'_{ef})^2$

4. To complete the analyses of the corrections for equations (11) and (12), it is necessary to analyze the so-called electron nonlinearities [25], i.e. all effects

arising from further expansion of the nonlinear matrix elements $S, \Sigma, S', \Sigma'$ with respect to the parameters $\alpha$ and $\beta$. Symbolically this expansion can be written in the form

$$\tilde{\Sigma} = \tilde{\Sigma}_0 + \tilde{\Sigma}_1\alpha + \tilde{\Sigma}^{(1)}\beta + \tilde{\Sigma}_2\alpha^2 + \tilde{\Sigma}_1^{(1)}\alpha\beta + \tilde{\Sigma}^{(2)}\beta^2 + \ldots \tag{30}$$

$$\tilde{\Sigma}' = \tilde{\Sigma}_0' + \tilde{\Sigma}_1'\alpha + \tilde{\Sigma}'^{(1)}\beta + \ldots \tag{31}$$

The direct calculations show that for the general collisionless case

$$\tilde{\Sigma}_1 = \tilde{\Sigma}_2 = \tilde{\Sigma}_1' = 0 \tag{32}$$

In expansion (30), for the term $\tilde{\Sigma}_0$ it is convenient to choose an expression which differs from (22) by substitution of the factor $(\varepsilon_{2+3}^{(i)} - 1)/(\varepsilon_{2+3} - 1)$ for the factor $\varepsilon_{2+3}^{(i)} / \varepsilon_{2+3}$. The reason for this is that unity in terms $\varepsilon_{2+3}^{(i)}$ and $\varepsilon_{2+3}$ gives terms of the same order of magnitude or a higher order of magnitude than other terms included in $\tilde{\Sigma}^{(1)}$ and because the factor $(\varepsilon_{2+3}^{(i)} - 1)/(\varepsilon_{2+3} - 1)$ leads to equations (11) and (12). The results of the calculation of $\tilde{\Sigma}^{(1)}$ in the 1D case and the 3D case differ:

$$\tilde{\Sigma}_{1D}^{(1)} = 0 \;;\; \tilde{\Sigma}_{3D}^{(1)} \neq 0 \tag{33}$$

In both the 1D and 3D cases $\tilde{\Sigma}_1^{(1)} \neq 0$ and $\tilde{\Sigma}^{(2)} \neq 0$. The terms $\tilde{\Sigma}^{(1)}$ and $\tilde{\Sigma}_1^{(1)}$ are not sensitive to the distribution of electrons because these matrix elements are proportional to the integral $\int \Phi^{(e)} d\vec{v}$. The term $\tilde{\Sigma}^{(2)}$ depends on $\int v^2 \Phi^{(e)} d\vec{v}$. The calculation of $\tilde{\Sigma}'^{(1)}$ should take into account 43 terms whose graphs are similar to those shown in Fig.6. In the 1D case, the result is $\tilde{\Sigma}'^{(1)} = 0$, but it is $\tilde{\Sigma}'^{(1)} \neq 0$ in the 3D case. These results are essential for the analysis of highly supersonic motions when the $\tilde{\Sigma}_0$ and $\tilde{\Sigma}_0'$ terms become small. In the 1D case, the $\tilde{\Sigma}_1^{(1)}$ and $\tilde{\Sigma}^{(2)}$ terms then determine the nonlinear response. They are both positive, and pure electron modulation instability becomes impossible. In the 3D case, the presence of terms $\tilde{\Sigma}^{(1)} \neq 0$ prevents supersonic collapse, i.e. prevents the existence of a modulation density depletion which develops to a size of the order of the Debye length $v_{Te}/\omega_{pe}$.

## III. NEW NONLINEAR MODES OF PLASMA MOTION CREATED BY MODULATION INTERACTIONS

1. The development of a regular Langmuir field due to modulation instability can lead to formation of new nonlinear dynamic modes: Langmuir solitons, cavitons, spikons, etc. An ensemble of such independent modes cannot describe the turbulence. But the nonlinear plasma modes can have some relation to plasma turbulence similar, in a sense, to the relation of a single eddy to the turbulence of an incompressible liquid. To find the possible self-consistent coherent nonlinear motions of a Langmuir field, it

is possible to use the equations with cubic and fifth-power nonlinear response found above. It is useful to introduce the dimensionless variables:

$$\vec{\mathcal{E}}(\vec{\xi},\tau)=\vec{E}(\vec{r},t)/\sqrt{4\pi n_0 T_e \mu}\ ;\ \nu=\delta n/n_0\mu\ ;\ \mu=4m_e/3m_i$$

$$\tau=t\,\omega_{pe}\mu/2\ ;\ \vec{\xi}=\vec{r}\sqrt{\mu}/\sqrt{3}\,r_d\ ;\ r_d=V_{Te}/\omega_{pe} \tag{34}$$

$$V_{Te}=\sqrt{T_e/m_e}\ ;\ T_e=m_e\int v^2\Phi^{(e)}d\vec{v}$$

Neglecting in the first approximation the imaginary parts of $\Sigma_0$ and $\Sigma_0'$, we have

$$\mathrm{div}\left(i\frac{\partial}{\partial\tau}+\Delta\right)\vec{\mathcal{E}}=\mathrm{div}\,(\nu+\rho)\vec{\mathcal{E}} \tag{35}$$

$$\left(\frac{\partial^2}{\partial\tau^2}-V_s^2\Delta\right)\nu=\Delta|\vec{\mathcal{E}}|^2 \tag{36}$$

$$\left(\frac{\partial^2}{\partial\tau^2}-V_s^2\Delta\right)\rho=-\chi\Delta\nu^2 \tag{37}$$

In the variables used here, the ion-sound velocity and the constant $\chi$ are

$$V_s=\sqrt{T^{ef}/T_e}\ ;\ \chi=\mu\left(T^{ef}\right)^3/2T_e\left(T'_{ef}\right)^2 \tag{38}$$

These equations are valid for characteristic velocities larger than the ion thermal velocity $V_{Ti}=\sqrt{T_i/T_e}$, and for $|\mathcal{E}|^2\ll\mu^{-1}$. The nonlinearities of the fifth order in the field strength become important for $\chi\gg\mu$. From the equations (35), (36), and (37), the following conservation laws can be found: conservation of number of Langmuir quanta N, conservation of energy H and momentum P. The expressions for them generalize the known expressions for cubic nonlinearities,

$$N=\int|\vec{\mathcal{E}}|^2d\vec{\xi}\ ; \tag{39}$$

$$H=\int\left\{|\mathrm{div}\vec{\mathcal{E}}|^2+(\nu+\rho)|\vec{\mathcal{E}}|^2+\frac{\nu^2+\vec{v}^2}{2}+\left(\nu\rho V_s^2+(\vec{v}\vec{V})\right)-\frac{\chi\nu^3}{3}\right\}d\vec{\xi}; \tag{40}$$

$$\vec{P}=\frac{1}{2}\int\left\{i\left(\vec{\mathcal{E}}\,\mathrm{div}\vec{\mathcal{E}}^*-\vec{\mathcal{E}}^*\mathrm{div}\vec{\mathcal{E}}\right)+2\left(\nu\vec{v}+\rho\vec{v}+\nu\vec{V}\right)\right\}d\vec{\xi} \tag{41}$$

where $\vec{v}$ and $\vec{V}$ satisfy the equations

$$\frac{\partial\nu}{\partial\tau}=-\mathrm{div}\,\vec{v}\ ;\ \frac{\partial\rho}{\partial\tau}=-\mathrm{div}\,\vec{V} \tag{42}$$

2. Let us first analyze the one-dimensional solitary pulses with a density depletion that moves at the constant velocity $u_0$. From (35), (36), and (37) we have:

$$\left(i\frac{\partial}{\partial\tau}+\frac{\partial^2}{\partial\xi^2}\right)\mathcal{E}=-\left(\frac{|\mathcal{E}|^2}{V_s^2-u_0^2}+\frac{\chi|\mathcal{E}|^4}{(u_0^2-V_s^2)^3}\right)\mathcal{E} \tag{43}$$

If only the cubic nonlinear term is taken into account, the solitons [23] exist only for subsonic velocities $u_0 < V_s$

$$\mathcal{E} = \frac{\mathcal{E}_0 \exp\left(-i\Omega\tau + i\frac{u_0}{2}(\xi - u_0\tau)\right)}{\cosh\left[\mathcal{E}_0(\xi - u_0\tau)/\sqrt{2(V_s^2 - u_0^2)}\right]}; \quad \Omega = -\frac{u_0^2}{4} - \frac{\mathcal{E}_0^2}{2(V_s^2 - u_0^2)} \quad (44)$$

The solitons do not form a complete set of nonlinear modes, since even for $u_0 < V_s$ periodic nonlinear waves can exist. In contrast to the case of cubic nonlinearities, the fifth-power nonlinear term in (43) describes the supersonic solitons [27] (which we call spikons). Neglecting the cubic nonlinear term, we find from (43)

$$\mathcal{E} = \frac{\mathcal{E}_0 \exp\left(-i\Omega\tau + i\frac{u_0}{2}(\xi - u_0\tau)\right)}{\left\{\cosh\left[2\sqrt{\chi}\,\mathcal{E}_0^2(\xi - u_0\tau)/\sqrt{3}(u_0^2 - V_s^2)^{3/2}\right]\right\}^{1/2}}; \quad \Omega = -\frac{u_0^2}{4} - \frac{\mathcal{E}_0^4 \chi}{3(u_0^2 - V_s^2)^3} \quad (45)$$

This solution exists for $\mathcal{E}_0^2 \chi \gg (u_0^2 - V_s^2)^2$.

3. It can be shown that the interactions between the nonlinear modes are necessarily connected with processes (4). The conservation laws (39), (40), and (41) show that merging of subsonic solitons is possible only if some energy is released. The numerical calculations [29] and the analytical theory [28] show that this release of energy corresponds to a sound emission process, which is actually process (4a). It has been proved [28] that only a small part of the energy of colliding subsonic solitons is converted into nonsoliton periodic Langmuir waves. If two equal solitons merge, the energy is doubled and the width diminished by a factor 2 (for spikons the width is diminished by a factor $\sqrt{2}$). The processes (4b) and (4c), i.e. the scattering on electrons and ions, also contribute essentially in the interactions of solitons and spikons. To describe these interactions it is necessary to take into account the imaginary parts of $\Sigma_0$ in (22) and $\Sigma_0'$ in (26), which gives the additional terms $\delta\nu$ and $\delta\rho$ in equations (35), (36), and (37). The energy H and momentum P cease to be constants of motion, but the number of quanta still is. Since $\delta\nu/\nu \ll 1$, $\delta\rho/\rho \ll 1$, the scattering processes lead to a slow variation of parameters $u_0$ and $\mathcal{E}_0$ in time. For subsonic solitons, the conservation of $N$, $\mathcal{E}_0\sqrt{V_s^2 - u_0^2} = \text{const}$, shows that the amplitude changes only if $u_0$ is comparable with $V_s$; for $u_0 \ll V_s$, $\mathcal{E}_0 = \text{const}$. The change in velocity $u_0$ leads to a braking of a soliton [30,26]:

$$(V_s^2 + 5u_0^2)\frac{du_0}{d\tau} = -\frac{9\sqrt{3}\,\zeta(3)}{8\pi^{5/2}}\mathcal{E}_0\sqrt{\mu}\,\frac{u_0\Psi(u_0)V_s^2}{\sqrt{V_s^2 - u_0^2}} \quad (46)$$

where $\zeta(3) \cong 1.08$

$$\Psi(u_0) = -\frac{\sqrt{2\pi}\, 4 V_{Te}}{3\mu} \cdot \frac{1}{u_0} \frac{\partial}{\partial u_0} \Phi^{(2)}\left(u_0 V_{Te}\sqrt{3\mu/2}\right). \quad (47)$$

For a Maxwellian distribution $\Psi(u_0) \cong 1$. Equation (46) determines the mean free path of a soliton (from initial $u_0$ up to $u \approx V_{Ti}$; but the mean free path from $V_{Ti}$ to $u_0 = 0$ is small as compared with that found from equation (46)). The braking of solitons discontinues their merging since there are no more collisions. For spikons, the conservation of number of quanta, $(u_0^2 - V_s^2)^{3/2} = \text{const}$, leads to the conservation of velocity. On the other hand, the scattering on electrons decreases the amplitude of the spikons:

$$\frac{d\mathcal{E}_0}{d\tau} = -\frac{21\zeta(3)\sqrt{\mu}}{8\sqrt{2}\,\pi^{5/2}} \cdot \mathcal{E}_0^3 \cdot \frac{u_0^2 \Psi(u_0) V_s^2 \sqrt{\gamma}}{(u_0^2 - V_s^2)^{3/2}} \quad (48)$$

The scattering also leads to a new effect - the presence of long-range interactions between the solitons and spikons. A soliton at a distance which essentially exceeds its width, $\sim \sqrt{2}/\mathcal{E}_0$, creates a density perturbation

$$\delta\nu = \sqrt{6\mu/\pi} \cdot \frac{u_0 \mathcal{E}_0}{(V_s^2 - u_0^2)} \cdot \frac{1}{\xi} \quad (49)$$

which decreases with the distance only as the first power of $\xi$.

This analysis shows how complicated the interactions actually are between the new nonlinear modes. The intrinsic interweaving of the modulation interactions, the processes of sound emission (4a), the processes of scattering (4b), (4c), and energy exchange between the soliton-type and periodic-type nonlinear modes create a complicated statistical picture. All the nonlinear interactions can take part in a final balance [31].

4. Let us now consider three-dimensional motions and so-called cavitons. It can be shown that the balance between thermal effects and nonlinear effects, which leads to a uniformly moving soliton in the 1D case, is no longer possible in the 3D case. Thus the nonlinear modes are always time- and space-dependent. The self-similar type of these motions will be called cavitons. By the Langmuir collapse, we imply such a self-similar caviton in which the relative density depletion $\delta n/n_0$ can reach a value of the order of unity ($\nu \approx \mu^{-1}$). The hypothesis of collapse was proposed in [17]. It was based on the cubic nonlinear equations (11) and (12). Up to the present time, no final proof exists that the collapse is possible.
For subsonic motions $\nu$ is usually of the order of unity and therefore $\delta n/n_0$ is small ($\approx \mu$). We shall prove

that supersonic collapse is impossible. According to [17], the supersonic caviton can be described by

$$\mathcal{E} \approx \frac{\mathcal{E}_0}{\tau} f(\vec{\xi}/\tau^{2/3}) ; \quad \nu \approx \frac{\nu_0}{\tau^{2/3}} f(\vec{\xi}/\tau^{2/3}), \quad (50)$$

where $\tau = 0$ and $\vec{\xi} = 0$ are the moment of time and the position of the field singularity. This solution was found from (11) and (12), which are valid only for $\tau \gg \sqrt{\mu}$ (see the analysis of sec.II). This means that $\nu \ll \mu^{-2/3}$ or $\delta n/n_0 \ll \mu^{1/3}$. For $\tau < \sqrt{\mu}$, the electron nonlinearities cannot be neglected. For the 3D case, $\Sigma^{(1)} \neq 0$. But for $\tau \approx \sqrt{\mu}$, the term $\Sigma'^{(1)}$ will be of the same order as the term $\Sigma^{(1)}$. This means that for $\tau < \sqrt{\mu}$ it is necessary to sum up the whole series of the perturbation theory in terms of the field strength. Since the nonlinear response is not larger than the linear dispersion term, the result will be that the increase of the density depletion for $\tau < \sqrt{\mu}$ is forbidden. In a nonequilibrium plasma, the solution (50) is restricted by the nonlinear term of the fifth order for $\tau \gg \chi^{3/2}$ if $\chi \gg \mu^{2/9}$. The self-similar solution for $\tau \ll \chi^{3/2}$ will be determined by the fifth-power nonlinear term and has the form:

$$\mathcal{E} \approx \frac{\mathcal{E}_0}{\chi^{3/20}\tau^{9/10}} f(\vec{\xi}/\tau^{3/5}); \quad \rho \approx \frac{\rho_0}{\chi^{1/5}\tau^{6/5}} f(\vec{\xi}/\tau^{3/5}). \quad (51)$$

The maximum density depletion is $\delta n/n_0 \approx \mu\rho \lesssim V_s^{1/3}\mu \ll \mu^{5/6}$ since $V_s \ll \mu^{-1/2}$. The value $\delta n_{max}/n_0$ is determined also by the electron nonlinearities and is even less in this case than for cubic nonlinearities.

5. The arguments given show that in the present theory supersonic collapse is absent if the supersonic self-similar motion can exist. But it is doubtful whether supersonic self-similar motion can exist at all. Solutions (50) and (51) do not take into account the emission of sound waves, and the question is: is the total energy released in sound waves for supersonic motion small or not as compared to the initial caviton energy? In other words, the question concerns the superimposition of process (4a). The solitons in the 1D case are stationary waves which emit sound waves during collisions only, while the cavitons are not stationary and can emit sound waves during the whole process of self-rarification. To estimate the rate of emission, it is possible as the first approximation to regard the motion (50) as given and to calculate the sound emission by a method proposed in [28]. The results of such calculations show [26] that the total energy emitted during the process (50) is 90.5% of the initial energy. One should also have in mind that this emission is due to the sonic tail of the supersonic mo-

tion. The result means that sound emission can never be neglected for the process (50), i.e. solution (50) is not self-consistent. Examples exist of 2D motion for which the sound emission forms a kind of barrier not allowing the motion to pass from the subsonic to the supersonic range [24]. One should also have in mind that if the motion reaches velocities close to sound the processes (4b) of scattering by electrons will be strengthened. This follows from (43), which shows that $\delta n/n_0$ tends formally to $\infty$ for $u \to V_s$. The scattering by electrons makes it finite ($\approx |\mathcal{E}|^2 \mu^{-1/2}$). The long-range interactions connected with a scattering are also strengthened for $u \to V_s$, and caviton (50) cannot be isolated from other cavitons. The numerical calculations of [32] show, on the other hand, that the interactions with resonant particles restrict the velocity of the process to velocities less than, but of the order of, the sound velocity. For subsonic motion, the cavitons usually are of the dipole type with a pancake shape [33]. The analysis and the arguments given show that the emission of sound waves and the scattering processes can play an important role in caviton problems.

6. It is also necessary to point out that the fundamental question of the role of initial conditions for the formation of cavitons is essentially related to the problem of long-range interactions of the cavitons even when such interaction is weak. Long-range interactions are strongest for subsonic motions and especially for $u < V_{Ti}$. The equation which takes into account long-range interaction has the form ($\vec{\mathcal{E}} = \vec{\mathcal{E}}(\vec{\xi},\tau)$, $\vec{\mathcal{E}}' = \vec{\mathcal{E}}(\vec{\xi}',\tau)$):

$$\operatorname{div}\left(i\frac{\partial}{\partial\tau} + \Delta\right)\vec{\mathcal{E}} = -\operatorname{div}\left\{\frac{|\vec{\mathcal{E}}|^2\vec{\mathcal{E}}}{V_s^2 + V_{Ti}^2} + \frac{\sqrt{\pi/2}\,V_{Ti}}{V_s^2 + V_{Ti}^2}\cdot\frac{\partial}{\partial\tau}\int\frac{d\vec{\xi}'\,|\vec{\mathcal{E}}'|^2\vec{\mathcal{E}}}{4\pi^2|\vec{\xi}-\vec{\xi}'|^2}\right\} \quad (52)$$

The second term of (52), which describes long-range interaction contains only the factor $V_{Ti} = \sqrt{T_i/T_e}$, which is not small compared with unity for $T_i \approx T_e$. This means that at the initial phase of field self-contraction, long-range interaction is of the same order of magnitude as that due to local nonlinearities. For $u \gg V_{Ti}$ the long-range interaction is smaller. The question arises: can the cavitons become isolated from each other? To answer this question it is necessary, at least, to have some information concerning the stability of cavitons and the dependence of the stability on the initial and boundary conditions. In connection with this problem it is necessary also to remember that the turbulent state arises when small deviations in the initial and boundary conditions lead to a drastic difference in the final macroscopic motions [14] .For modulation interactions it has been proved that small-scale disturbances develop on the background of large-scale disturbances, similar to the eddy hierarchy in liquids. If this is the case, even weak long-range interactions can very essentially change the development of the modulation field contraction. This case should correspond to strong turbulence, which is of the greatest interest.

## IV. STATISTICAL THEORY OF STRONG LANGMUIR TURBULENCE

1. The necessity of the use of a statistical description for strong turbulence is well known [14] . This approach is able to exclude the extra detail of the complicated dynamics of modulation interactions described above and to find the averaged values which are usually measured in experiments. The statistical theory for a strong Langumir turbulence was proposed in [12]. The advantages of a statistical theory as compared to different dynamic models are the following:

a) It is possible throughout to consider all the nonlinear interactions in their interrelationships.

b) A kinetic theory of simultaneous modulation, decay, and scattering interactions can be constructed.

c) The changes in distribution functions of resonant tail particles and subthermal particles can be found.

d) It is possible to develop a theory of isothermal plasmas ( $T_e = T_i$ ), when the sound emission processes (4a) are absent (the dynamic models for this case have not yet been developed).

e) Finally, the theory for real 3D distributions of particles, waves, and modulation disturbances can be constructed.

We give below the main features and main results obtained in a statistical theory, based on the general expression for a cubic nonlinear plasma response.

2. The starting point is the general equation (15). Let us consider a developed strong Langmuir turbulence in the stage when the different nonlinear modes have been interacting for a sufficiently long time. This state can arise as a result of modulation instability of Langmuir condensation, or as a result of subsequent development of modulation and other interactions of an initially regular strong pump. Due to interactions of modes, the correlation function of the Langmuir field is no longer constant in space and time [34] :

$$\langle E^{\ell}_{k} E^{\ell}_{k'} \rangle = I_{k,k'} \tag{53}$$

These space and time variations are described by the dependence of $I_{k,k'}$ on $k'$. Averaging over time scales which are larger than that of characteristic variations in time, and averaging over distances which are larger than that of the characteristic variations in space of the correlation function one should obtain $\overline{I_{k,k'}} = I_k \delta(k')$. $I_k$ is the value which describes the spectra of strong Langmuir turbulence. Besides the Langmuir field $E^l$, we introduce the low-frequency modulation field $E^m$, which is regular over the time scales and sizes of modulation disturbances. The same averaging of this field should give $\overline{E^m_{k'}} = 0$. Since the variations of $I_{k,k'}$ are due to modulation disturbances, the correlations between $I_{k,k'}$ and $E^m_{k''}$ should be significant:

$$\overline{I_{k,k'} E^m_{k''}} = U_{k,k'} \delta(k'+k'') \tag{54}$$

There also exists correlations of the modulation fields:

$$\overline{E^m_{k'} E^m_{k''}} = w_{k'} \delta(k'+k'') \tag{55}$$

and the following correlations:

$$\overline{I_{k,k'} I_{k_1,k''}} - \overline{I_{k,k'}} \cdot \overline{I_{k_1 k''}} = G_{k,k_1,k'} \delta(k'+k'') \tag{56}$$

In the absence of modulation disturbances equation (15) leads, after averaging, to the known equations [14] describing only processes (4). In the linear approximation of modulation disturbances, we find from (15) and (17) the known equation of the modulation instabilities discussed in detail in Chapter 8 of [14]. For the modulation instability of a random field the renormalization of the charge density should be taken into account [18]. To describe the nonlinear stage of modulation instability a self-consistent system of equations for the values $G$, $U$, $w$, $I_{k,k'}$, and $I_k$ was found [34]. Besides being influenced by processes (4) the modulation disturbances in the nonlinear stage essentially alter processes (4) to change the direction of turbulent energy flux (for proof see [34]). It is possible to divide the total k-space into the energy-containing region and the inertial region. In the energy-containing region, the modulation instability is stabilized by the nonlinear frequency shifts as will be proved below. In the inertial region, the energy is transferred from the energy-containing region to higher $k$.

3. Let us denote by the superscript $^{(0)}$ the values $G$, $U$, $w$, and $I$ which describe the energy-containing region. We will assume that the level of the turbulence in the inertial region is much lower than in

the energy-containing region, and will therefore neglect the influence of the inertial region on the energy-containing region. This allows the equations for $G^{(0)}$, $v^{(0)}$, $w^{(0)}$ and $I^{(0)}$ to be closed. The solution for $W^{(0)}/n_oT_e \ll 1$ and $k_o \ll k_*$, gives:

$$\tilde{\mathcal{E}}_k I_k^{(0)} = \int dk' \frac{|S_{k'}|^2 w_{k'}^{(0)}}{|1 + D_{k'}^{(0)}|^2} \left\{ \frac{I_{k-k'}^{(0)}}{\tilde{\mathcal{E}}_{-k}} + \frac{I_k^{(0)}}{\tilde{\mathcal{E}}_{k-k'}} \right\} \tag{57}$$

where

$$\tilde{\mathcal{E}}_k = \mathcal{E}_k + \mathcal{E}_k^N \; ; \quad \mathcal{E}_k^N = - \int \tilde{\Sigma}^{(0)}_{k,k_1} I_{k_1}^{(0)} dk_1 \tag{58}$$

and $\tilde{\Sigma}^{(0)}_{k,k_1}$ is the expression $\tilde{\Sigma}_0$ (22) for $k_2 = k$, $k_3 = -k_1$

$$S_{k'} = \frac{1}{4\pi n_o T^{ef}} \cdot \frac{\omega_{pe}^2}{k'^2 v_{Te}^2} \tag{59}$$

$$D_{k'}^{(0)} = \frac{\left(1 + \frac{T^{ef}}{T^{ef} + T^i}\right)}{8\pi n_o T^{ef}} \int \left\{ \frac{I_{k_1-k'}^{(0)}}{\tilde{\mathcal{E}}_{k_1}} + \frac{I_{k_1}^{(0)}}{\tilde{\mathcal{E}}_{k'-k_1}} \right\} dk_1 \tag{60}$$

For $w_{k'}^{(0)} = 0$ equation (57) is converted into the known equation [14] . A similar equation can be found for $w_{k'}^{(0)}$. In the limit $k' \ll k$ ,the equation (57) is converted into a dispersion relation from which we find that the finite level of modulation disturbances $w_{k'}^{(0)}$ produces a frequency shift for the Langmuir waves. This frequency shift completely stabilizes the modulation instability of the energy-containing region if

$$\int \frac{\omega_{pe}^2}{k'^2 v_{Te}^2} \cdot \frac{w_{k'}^{(0)}}{4\pi n_o T_e} dk' \cong \left(\frac{W^{(0)}}{n_o T_e}\right)^2 \tag{61}$$

for which $D_{k'}^{(0)}$ is of the order of unity. On the other hand, the same estimation (61) can be found as a result of nonlinear interactions of the modulation disturbances described by the nonlinear equation for $w_{k'}^{(0)}$. Thus the stabilization is self-consistent. The lefthand side of (61) is approximately $\delta n_e^2/n_o^2$. The criterion of modulation instability for $k_o \approx k_*$ gives $W^{(0)}/n_oT_e > \mu$, and at the threshold (61) gives $\delta n_e/n_o \approx \mu$ ,which means that the modulation disturbances have velocities of the order of the sound velocity.

4. The spectra of the inertia region were found in different parts of $k$ -space [25] .We mention here only that in the vicinity of the energy-containing region the condensation acts as a source which transfers the energy in a single step to higher $k$. The spectra found in this region are

$$W_\kappa^{1D} \propto \int I_\kappa^{(in)} d\omega \propto \frac{1}{\kappa^2}\ ; W_\kappa^{3D} \propto \int 4\pi\kappa^2 I_\kappa^{(in)} d\omega \propto \text{Const} \qquad (62)$$

For higher $\kappa$ the scattering processes of type (4) become important. They are essentially modified by modulation disturbances. The scattering on electrons in a single step transfers the energy of Langmuir waves from the energy-containing region to $\kappa$ which are even close to $\kappa_d$. This creates a source of Langmuir waves with the power of the source being proportional to $1/\kappa^3$. The scattering on ions is the same as in the case of the absence of condensation. The balance of the two processes of scattering gives the spectrum

$$W_\kappa^{1D} \propto \frac{1}{\kappa^2}\ ; \ W_\kappa^{3D} \propto \frac{1}{\kappa^2} \qquad (63)$$

5. The influence of the modulation disturbances on the distribution of the low-velocity particles is determined in the first approximation by the energy-containing region. We find [12] in the 1D case:

$$\frac{\partial \Phi}{\partial t} = -\frac{4\pi e^2}{m_e^2}\int \frac{w_{\kappa'}^{(0)} \mathrm{Jm}\, D_{\kappa'}^{(0)}}{|1+D_{\kappa'}^{(0)}|^2}\left(\frac{1}{v}\frac{\partial \Phi}{\partial v} + \frac{\partial^2 \Phi}{\partial v^2}\right) d\kappa' \qquad (64)$$

This equation is valid for $v \gg \omega'/\kappa'$. The stationary solution of it, $\Phi = \text{const}\, \ln v_*/v$, supports the possibility mentioned above that in the strongly turbulent state, $T^{ef}$ could be essentially different from $T_e$.

In conclusion we may emphasize that the generality of the statistical approach and the possibilities opened by it to describe the strong Langmuir turbulence in the 3D case are of interest for experiments and also illustrate the necessity of the future development of the statistical theory. We feel that this will be the high-road to the proper correspondence between the experimental data and theory in this very rapidly developing field of plasma physics.

## REFERENCES

[1] Tsytovich, V.N. and Liperovsky V.A. "On the interactions of lasers with plasmas".The Proc.of the 7-th Int.Conf. on Phen. in Ion.Gases.Belgrade, (1965).
[2] Forslund D.W., Kindel J.M. and Lindman, E.L. Phys. Fl. 18, 1002 (1975)
[3] Rosenbluth M.N. Proc.5-th Eur.Conf.on Contr.Fusion and Plasma.Phys.Grenoble(1972).
[4] Makhan'kov,V.G., Tsytovich,V.N.,Physica Scripta 7,234 (1973).

[5] Ginsburg,V.L., Tsytovich V.N. JETP 65, 1818 (1973).
[6] Gailitis A.A., Tsytovich V.N.JETP 46, 1726(1964).
[7] Ginsburg V.L., Tsytovich V.N.Izv.Vusov"Radiophysics" 18,173(1975).
[8] Bloembergen N. Nonlinear Optics New York (1965).
[9] Falk L.R., Tsytovich V.N.Physica Scripta 11, 285(1975).
[10] Silin V.P. JETP 40, 1679 (1965).
[11] Nishikawa K., J.Phys. Soc. Japan 24,916, 1152 (1968).
[12] Khakimov F.Kh. and Tsytovich V.N., JETP 64,1261(1973).
[13] Zeldovich Ya.B., Levich E.V. JETP 35 , 2323 (1968).
[14] Tsytovich V.N. "Theory of turbulent plasma" Atomizdat (1971) and Plenum press(1976).
[15] Vedenov A.A. and Rudakov L.I.Dokl.Ak.Nauk.USSR (Sov.Phys.Doklady) 159,767 (1964).
[16] Gailitis A., Candidate Thesis,Lebedev Phys.Inst.(1964); Izv. Latv.Ac.Sc.USSR Series-phys.and Eng.Sc.4,13(1965).
[17] Zacharov V.E. JETP 62,1745 (1972).
[18] Tsytovich V.N. JETP 57,141 (1969).
[19] Kruer W.L. et al prepr.UCRL 74947(1974),Thomson J.J. et al. Phys.Rev.Lett.31,918 (1973).
[20] Litvak A.G. et al. JETP Lett. 19, 23 (1974).
[21] Gorev V.V. et al. Izv. Vuzov "Radiophysics" (1976).
[22] Tsytovich V.N.,Proc.of XII-th Int.Conf. on Phen. in Ion.Gases,part 2.Inv.papers Eindhoven,Netherland, North-Holland publ. corp. p.141 (1975).
[23] Rudakov L.I.Doklady Ak.Nauk USSR(Sov.Phys.Doklady)207, 821 (1972).
[24] Litvak A.G. , Fraiman G.M.JETP 69, N 4 (1975).
[25] Khakimov F.Kh. and Tsytovich V.N. JETP 67, 312 (1976).
[26] Veryaev A.A., Tsytovich V.N.Plasma Phys.(Sov.Un.)2 N5 (1976) to be published
[27] Khakimov F.Kh., Tsytovich V.N.Plasma Phys.(Sov.Un.) 2, N 6(1976) to be published
[28] Lebedev A.A. , Tsytovich V.N.Physica scripta 11,266(1975).
[29] Degtyarev L.M.et al. JETP 67,533 (1974).
[30] Gorev A.A.and Kingsep A.S.,JETP 66,2048(1974).
[31] Nishikawa K. et al. "Comments on plasma physics and controlled fusion research." 2, N 5, 63 (1975).
[32] Degtyarev L.M., et al. Plasma Physics(Sov.Un.) 2, N3 (1976).
[33] Degtyarev L.M. and Zakharov V.E. JETP Lett.20,№6(1974).
[34] Khakimov F.Kh., Tsytovich V.N. preprint .Lebedev Phys. Inst. N 103 (1975) N 43 (1976).

# MACROSCOPIC CONSEQUENCES OF THE LANGMUIR WAVES COLLAPSE

A.A. Galeev, R.Z. Sagdeev, V.D. Shapiro, V.I. Shevchenko

Institute of Space Research

Moscow, USSR

1. Since the dominant nonlinear process in the weak Langmuir turbulence theory results in wave energy flux towards the long wave length region of phase space where there is no energy sink we are faced with the problem of plasmon condensate formation [1]. For the case when the Langmuir turbulence is pumped by strong electromagnetic waves there were attempts to explain the Langmuir turbulence spectrum [2] involving into consideration the energy transfer due to the oscillating two stream instability. But the coupling approach used in these considerations really fails in the case of so strong parametric interaction.

As was proposed in [3] the selfmodulational instability of plasmon condensate can be an effective mechanism to produce the short wavelength oscillations and therefore to remove all the problems with plasmon condensate. The rigorous theory providing the wave energy flux towards the short scales and subsequent wave absorption by electrons [4] is based on Zacharov's idea about the Langmuir wave collapse [5].

The mechanism of the wave collapse is the following. In the place of large plasmon energy density the electric field pressure forces plasma out and results in the creation of a plasma cavity. The plasma cavity in its turn serves as a potential well for plasmons and leads to an additional concentration of wave energy. A very simple argument gives us a hint whether this collapse is possible or not. These arguments are based on the use of the following expression for the wavelength of plasmons trapped in the density cavern

$$k^2\lambda_D^2 \simeq \frac{\delta n}{n_0} \tag{1}$$

and of the conservation of plasmon number in cavern:

$$\int d^s\underset{\sim}{r}|\underset{\sim}{E}|^2/4\pi\omega_p = \text{const} \tag{2}$$

Here $\omega_p$ is the plasma frequency and $\lambda_D$ is the Debye length.

According to Eq. (2) the electric field pressure increases with a decrease of the cavern volume as $|\underset{\sim}{E}|^2 \sim l^{-s}$ where s is the dimensionality of the cavern and l is its size. At the same time to force plasma out of cavern we need to create a pressure greater than the plasma pressure $\delta n\, T \sim l^{-2}$. Therefore in one-dimensional case under the given initial plasmon energy concentration the narrowing of the cavern stops when the pressure balance will be reached. In the two-dimensional case the collapse can not be stopped if the initial wave concentration is sufficiently strong. And finally in the three-dimensional case the plasmon collapse seems unavoidable.

Energy transfer towards the short wavelength region can be represented as the wavelength reduction of plasmons trapped in collapsing caverns. Therefore the speed of the energy transfer (or flux in the phase space) is defined by the rate of cavern collapse.

2. Equations describing the modulational instability of Langmuir waves and their collapse were obtained in [5] by averaging the hydrodynamical equations over the fast time

$$\operatorname{div}\left(i\,\frac{\partial \underset{\sim}{E}}{\partial t} + \frac{3}{2}\,\omega_p\lambda_D^2\,\nabla\,\operatorname{div}\underset{\sim}{E} - \frac{\omega_p}{2n_0}\,\delta n\,\underset{\sim}{E}\right) = 0 \tag{3}$$

$$\left[\frac{\partial^2}{\partial t^2} - (T_e/M)\Delta\right]\delta n = \frac{1}{16\pi M}\,\Delta|\underset{\sim}{E}|^2 \tag{4}$$

where $\underset{\sim}{E}(\vec{r},t)$ is the complex amplitude of the Langmuir waves electric field

$$\underset{\sim}{E}_e = \frac{1}{2}\,\underset{\sim}{E}(t,\underset{\sim}{r})e^{-i\omega_p t} + \text{c.c.} \tag{5}$$

Equations (3)-(4) have selfsimilar solutions describing the collapsing caverns with plasmons trapped in them. If the speed of the cavern collapse essentially exceeds the sound speed ("supersonic" collapse) then taking into account that according to Eq. (1) the size of the cavern decreases as $l \sim 1/\sqrt{\delta n}$ we obtain from Eq. (4) the explosive dependence of the wave amplitude on time $|E|^2 \sim (t_0-t)^{-2}$. The rate of collapse is defined then by the conservations of plasmon energy within the cavern

$$l(t) \sim \delta n^{-1/2} \sim (t_0-t)^{2/s} \tag{6}$$

The complete selfsimilar for the supersonic collapse was found in [4,6] in the form

$$\underline{E} \sim \frac{1}{t_0-t} \exp\left[i\lambda \int \frac{dt}{(t_0-t)^{4/s}}\right] \underline{\psi}\left(\frac{\underline{r}}{(t_0-t)^{2/s}}\right) \tag{7}$$

$$\delta n \sim \frac{1}{(t_0-t)^{4/s}} \phi\left(\frac{\underline{r}}{(t_0-t)^{2/s}}\right)$$

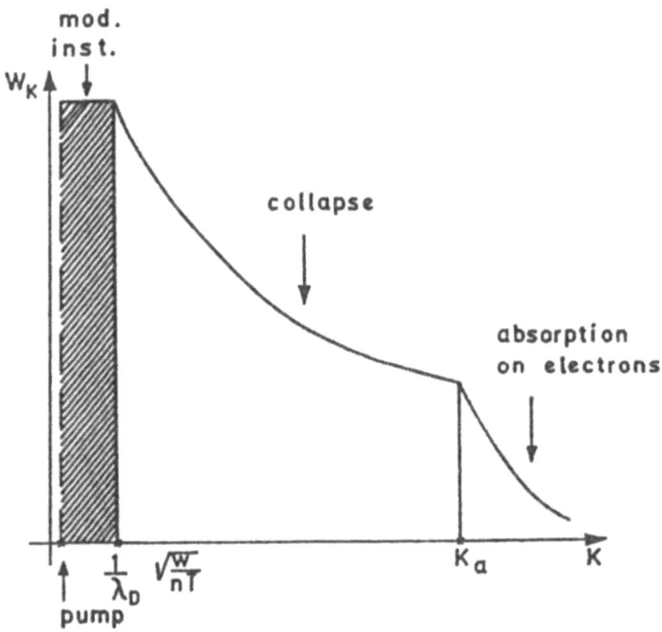


Fig. 1. Spectrum of strong Langmuir turbulence

The spectrum of isotropic plasmon turbulence in the wave number range between the source of plasmons in the long wavelength region and the sink of plasmons in the short scales due to electron Landau damping (see Fig. 1) can be found assuming the flux of the wave energy to be constant within this interval

$$\frac{W_k dk}{dt(k)} \simeq \text{const}$$

where $W_k dk$ is the plasmon energy within the wave number range $(k, k+dk)$ and $dt(k)$ is the time of cavern size change from the $l(t) \simeq 1/k$ to $l(t+dt) \sim 1/(k+dk)$. Using Eq. (6) we rewrite this equation in the form [4,6]:

$$W_k \sim \frac{dt(k)}{dk} \sim \frac{1}{k^{1+s/2}} \tag{8}$$

As was shown in [3,7] the resonant absorption of plasmons by electrons is the dominant dissipational mechanism of plasmons energy in the short scales. Since the main contribution to the wave absorption comes from the wave resonance with high energy tail of electron distribution ($v \gg v_T$) the parameter $k\lambda_D$ is always small for the described Langmuir wave turbulence model. Under this condition the nonlinearity of high and low frequency oscillations and resulted both electron and ion trajectories overlapping is unessential and the energy given to the ion species during a collapse is always small as $k_{max}^2\lambda_D^2$ in comparison with the wave energy.

The turbulence spectrum and high energy tail of the electron distribution can be found from the equation for the wave energy transfer including the Landau damping:

$$\frac{\partial}{\partial t} W_k = -\frac{\partial}{\partial k}\left(W_k \frac{dk}{dt(k)}\right) + \frac{\pi\omega_p^3}{k^3}\, \mathbf{k}\cdot\frac{\partial f}{\partial \mathbf{v}}\Big|_{\mathbf{k}\mathbf{v}=\omega_p} W_k \tag{9}$$

and the quasilinear equation for electrons

$$\frac{\partial f}{\partial t} = \frac{\pi e^2}{m^2}\int dk\, W_k k^{-2}\, \mathbf{k}\cdot\frac{\partial}{\partial \mathbf{v}}\,\delta\left(\omega_p - \mathbf{k}\mathbf{v}\right)\mathbf{k}\cdot\frac{\partial}{\partial \mathbf{v}} f \tag{10}$$

These equations have solutions corresponding to a constant particle flux towards the tail of the distribution [4]:

$$W_k \sim k^{-(3+s/2)}, \quad f(v)v^{s-1} \sim v^{-(1+s/2)} \tag{11}$$

Under the constant pump energy rate the maximum particle velocity in the tail grows as $v_{max}(t) \sim t^{1/(2+s)}$.

3. In order to be able to quantitatively describe the macroscopic consequences of a plasmon collapse we should find the rate of energy pumped into plasma caverns due to modulational instability. We investigate this process in the case of a uniform pump wave with a frequency equal to the plasma frequency. Since the space average value of the electric field in this case is $\overline{\underline{E}} = \underline{E}_0$ = const we can rewrite Eq. (3) in the form

$$\frac{d}{dt} \int |\underline{E}|^2 d\underline{r} = \frac{i\omega_p}{2n_0} \underline{E}_0 \int \delta n \, \underline{E} \, dr + c.c. \tag{12}$$

The wave energy dissipation in the isolated cavern can not be monotonic since the phase oscillations of the electric field amplitude in Eq. (7) lead to the change of dissipation sign and restrict the increase of the plasmons number. For example in the one-dimensional subsonic case Eq:s (3) and (4) have solution in the form of standing soliton with the electric field:

$$\frac{E}{E_0} = \mathcal{E} \, \frac{\exp(i \int \lambda_0 d\tau)}{ch(\zeta/\Delta)}$$

where

$$\lambda_0 = \frac{1}{\Delta^2} = \frac{\mathcal{E}^2}{4}, \quad \tau = \omega_p t \, \frac{E_0^2}{16\pi n_0 T}, \quad \mathcal{E} = \frac{z}{\lambda_D} \sqrt{\frac{E_0^2}{24\pi n_0 T}}$$

Then from Eq. (12) we find that sufficiently small pump wave $E_0^2/16\pi n_0 T \ll m/M$ without changing the form of soliton lead to the following change of its amplitude.

$$\frac{d\mathcal{E}}{d\tau} = \frac{1}{2} \mathcal{E}^2 \sin\left[\frac{1}{4} \int_0^\tau \mathcal{E}^2 d\tau\right] \tag{13}$$

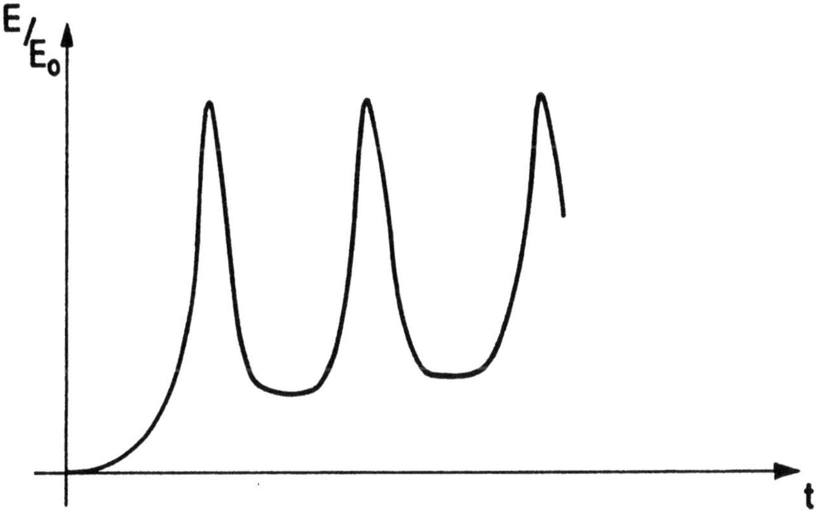


Fig. 2. Energy pumping in isolated soliton

The initial increase of the amplitude has the explosion character $E \sim 1/(\tau_{*}-\tau)$ but at the point where

$$\int_0^{\tau} E^2 d\tau = 4\pi$$

(this corresponds to $E^2 = 26$) the increase is saturated and the wave amplitude starts to oscillate (Fig. 2). Computer simulations of the one-dimensional soliton dynamics [8] at the initial phase of soliton growth had led the authors to uncorrect conclusion about the possibility of an unlimited energy pump into an isolated solitor Later on the computer study [9] was continued up to long time perioc and it was shown that the initial increase of the amplitude transforms to amplitude oscillations in accordance with Eq. (13). Soliton-like solutions of Eq:s (3) and (4) are possible also in the supersonic one-dimensional case. The law of the amplitude change of this soliton qualitatively the same as in the subsonic case; initially explosive growth of the amplitude according to the law: $E \sim 1/\tau_{*}-\tau$, $\delta n \sim 1/(\tau_{*}-\tau)^2$, then the growth is saturated and the amplitude starts to oscillate. But as it may be concluded from the computer simulations [9] the maximum value of amplitude is essentially larger in the supersonic case: $E_{max}/E_0 = 6$ for $E_0^2/16\pi n_0 T = 1.5 \cdot 10^{-3}$ and $E_{max}/E_0 = 17$ for $E_0^2/16\pi n_0 T = 1.5 \cdot 10^{-1}$. It is necessa to note that the numerical factor for the wave amplitude growth in the one-dimensional supersonic case is so large that for sufficient large initial wave amplitude ($E_0^2/16\pi n_0 T > 10^{-2}$) the wavelength of maximally grown wave becomes of the order of the Debye length and therefore they can be absorbed by electrons (the one-dimensional forced collapse).

It should also be noted that the found deep amplitude oscillations are characteristic only for the one-dimensional case. In the three-dimensional case the number of plasmons in the cavern is constant within an accuracy of the order $\sim(\tau_*-\tau)^{2/3}$ as follows from Eqs. (7) and (13). Therefore the pump is switching off gradually from the cavern during its collapse.

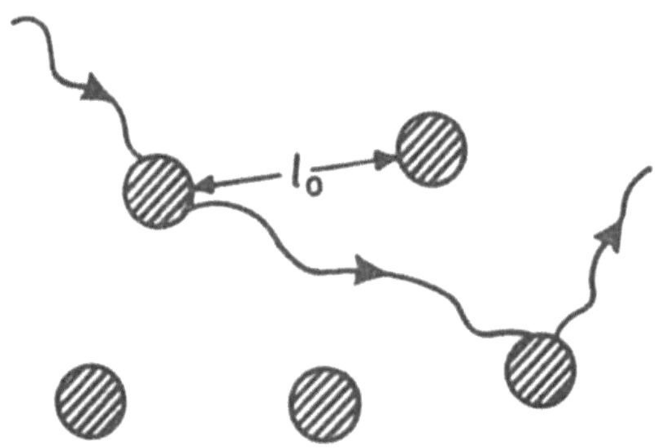


Fig. 3. Scattering of long wave-length plasmons on density fluctuations in caverns: $l_0 \sim \lambda_D\sqrt{nT/W}$.

In three-dimensional case the dissipation of the pump energy is related to the contineous creation of the long wave-length plasmons scattered on density fluctuations in the collapsing caverns (see Fig. 3). Then the correlation function of the plasmons can be represented in a form

$$<\exp i\left[\phi(\underline{r}) - \phi(\underline{r}')\right]> = \exp\left(-\frac{|\underline{r}-\underline{r}'|}{l_0}\right)$$

where $l_0$ is the distance between the caverns. The effective collision frequency for the plasmon scattering on density fluctuations can be defined as

$$\nu_{eff} = \frac{v_g}{l_0} \quad , \quad \text{where } v_g = \lambda_D^2 \, \omega_p / l_0$$

is the group velocity for plasmons. Therefore

$$\nu_{eff} \sim \omega_p \frac{\delta n}{n_0}$$

with the help of Eq. (1). The rate of energy dissipation into plasmon turbulence is defined by this collision frequency:

$$\frac{d}{dt} \int |\underline{E}|^2 d\underline{r} = \frac{\omega_p^2}{4n_0^2} E_0^2 \int d\underline{r} \; \delta n^2 \times$$

$$\times \int^t dt' < \exp\left[ i\phi(\underline{r}) - i\phi(\underline{r}') \right] > \approx \nu_{eff} E_0^2 \qquad (14)$$

Dissipation takes a form of plasmon cavern creation with a characteristic size

$$l_0 \sim \lambda_p \sqrt{\frac{16\pi n_0 T}{<E^2>}}$$

Therefore $\nu_{eff}$ can be represented as:

$$\nu_{eff} = \alpha \, \omega_p \frac{<E^2>}{16\pi n_0 T} \qquad (15)$$

where $\alpha$ is a numerical factor. The computer simulation of the energy pumping into plasmon turbulence supports this formula and gives the numerical factor $\alpha = 0.3 - 0.4$ [9].

For the steady state turbulence the energy pumping by the pump wave is balanced by the energy flux to the short scale due to caverns collapse:

$$\nu_{eff} \, E_0^2 = \gamma_M <E^2>$$

where

$$\gamma_M \simeq \omega_p \sqrt{\frac{m}{M}} \frac{< E^2 >}{16\pi n_0 T}$$

is the growth rate of the modulational instability defining the rate of cavern creation. From this equation we obtain:

$$\nu_{eff} \approx \alpha \omega_p \frac{<E^2>}{16\pi n_0 T} \approx \frac{M\alpha^3}{m} \omega_p \left(\frac{E_0^2}{16\pi n_0 T}\right)^2 \tag{16}$$

4. Thus the steady state of the strong Langmuir turbulence is reached, when the energy extracted from the pump wave is transformed to the short wave length in cavern and then absorbed by electrons. One essential feature of this turbulence - collapsing caverns are the source of the intensive ion-sound waves: when cavern collapses up to the small scales for which absorption is essential the balance between the plasma pressure and the wave pressure will be destroyed which will cause the ion-sound waves radiation from cavern.

Balancing the ion-sound waves creation rate in a course of collapse by the damping rate due to electron Landau damping we find that the level of ion-sound density fluctuations is quite high:

$$\sum_k \frac{|\delta n_k|^2}{n_0^2} = 8\pi \sqrt{\frac{M}{m k \lambda_D}} \left(\frac{<E^2>}{16\pi n_0 T}\right)^{9/4} \tag{17}$$

Intensive short wave length ion-sound turbulence leads to the stabilization of collapse. This stabilization is a result of the direct conversion of long wave length Langmuir wave into the short one on the short wave length ion-sound pulsations, since this process supresses cavern creation.

The characteristic damping rate for this process

$$\gamma_{conv} = \frac{1}{18} \sum_k \frac{|\delta n_k|^2}{n_0^2} \frac{1}{k^4 \lambda_D^4} \Gamma_k$$

($\Gamma_k$-decrement for Landau damping of short wave-length plasmons). The ion-sound density fluctuations cannot exceed the level for which $\gamma_{conv} = \gamma_M$ (for larger levels collapse is stabilized and the source of ion-sound turbulence is lost).

These arguments should be taken into account for Langmuir wave energy above some value

$$\frac{<E^2>}{8\pi n_0 T} > \frac{1}{(8\pi)^{2/3}} \left(\frac{81m}{M}\right)^{1/3} k_{max}^2 \lambda_D^2 \tag{18}$$

defined by the condition $\gamma_{conv} \gtrsim \gamma_M$.

Assuming that the balance between the creation of long wave length Langmuir turbulence due to modulational instability and electron Landau damping in the short wave length region is reached, we obtain

$$\gamma_M W = \Gamma_k W'$$

where

$$W' = W\left(\frac{k_0}{k}\right)^{3/2}$$

is the wave energy in the absorption region.

Then using the threshold condition we find the level of short wave length ion density fluctuations [10]

$$\sum_k \frac{|\delta n_k|^2}{n_0^2} \simeq g(k_{max}\lambda_D)^{5/2}\left(\frac{<E^2>}{16\pi n_0 T}\right)^{3/4} \qquad (19)$$

It is easy to show that now conversion is the dominant mechanism for the creation of short wave length plasmons.

Conversion on ion-sound pulsations also leads to the appearance of the additional term in effective collision frequency which defines the speed of pump wave energy dissipation: $\nu_{eff} \to \nu_{eff} + 2\gamma_{conv}$ but for the super-sound case $W/r_0T >> m/M$ this term is neglibly small.

5. One of the most interesting applications of the plasmons collapse is the relaxation of nonrelativistic and relativistic electron beams entering the cold plasma. The familiar quasilinear theory of beam relaxation fails if the energy of the plasma waves excited by the beam becomes sufficiently large

$$W > n_0T\left(V_T^2/V_b^2\right) \qquad (20)$$

where $V_b$ is the beam velocity.

Actually the importance of the nonlinear effects creating the shorter scale turbulence for the beam relaxation was shown in [11]. But the quantitative analytical theory of steady state turbulence based as in [11] on the exploration of the strong parametric interaction is impossible due to the failure of the mode coupling approach.

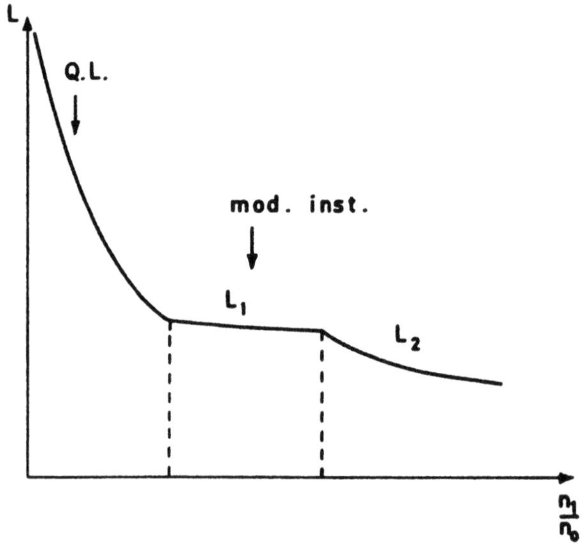


Fig. 4. Beam relaxation length as function of beam density.

The aim of this report is to find the beam relaxation lengtn increase due to wave growth saturation caused by supersonic collapse of excited Langmuir waves. Combining expression (20) and the results of quasilinear treatment of the beam relaxation in a plasma we find that this treatment fails for beam desities:

$$n_1 > 20\, n_0 \left(\frac{T}{mV_b^2}\right)^3 \qquad (21)$$

$n_0$-plasma density. For such beam density the energy of this wave is frozen on the threshold level (20) due to fast dissipation of plasma waves created by modulational instability. In this case the relaxation length does not depend on beam density (relaxation length as function of beam density is shown in Fig. 4):

$$L_1 \simeq \frac{1}{5} \frac{v_b}{\omega_p} \left(\frac{mV_b^2}{T}\right) \qquad (22)$$

This regime streches up to the densities

$$\frac{n_b}{n_0} = \frac{M\alpha^3}{m} \left(\frac{T}{mV_b^2}\right)^2 \qquad (23)$$

For larger densities the energy of plasma waves may exceed threshold level (20) and the saturation level of the beam excited long wave length oscillations is obtained from the equation of balance between the wave growth due to beam instability and wave dissipation due to creation of cavern:

$$\gamma_b E_0^2 = \nu_{eff} E_0^2 \tag{24}$$

For the nonrelativistic beam instability the growth rate is

$$\gamma_b = \omega_p \frac{n_b}{n_0} \frac{V_b^2}{(\Delta V)^2}$$

where $\Delta V$ is the beam velocity spread. Using this equation and the expression for the effective collision frequency through the pump amplitude (see Eq. (16)) we easily find the beam relaxation length:

$$L_2 \simeq \frac{V_b}{4\omega_p} \frac{mV_b^2}{T} \left(\frac{n_0}{n_b} \frac{M\alpha^3}{m}\right)^{1/2} \tag{25}$$

Relaxation of the relativistic beam in plasma has been investigated in quasilinear approximation [12] and weak turbulence approximation [13]. In the last case the induced scattering of plasmons on ions has been considered as the main mechanism for stabilization of beam-plasma instability. Using the level of plasma wave energy obtained in [13] it is easy to show that this approximation fails at beam densities:

$$\frac{n_b}{n_0} = \frac{1}{4\pi} \frac{E}{mc^2} \frac{m}{M} \Delta\theta \left(\frac{T_i}{T_e}\right)^2 \tag{26}$$

where $\Delta\theta$ is the angular spread of the beam particles velocity and $E$ is the beam particles energy. It is assumed that

$$\Delta\theta > \sqrt{\frac{m}{M} \frac{T_i}{T_e} \frac{mc^2}{T_e}}$$

At the larger densities the dissipation of plasma waves caused by modulational instability becomes to be essential and that leads to the slowing down of the relaxation process. For beam densities near threshold value (26) the relaxation length is equal [14]:

$$L_1 \simeq \frac{c}{\omega_p}\left(\frac{E}{T_e}\right)^2 \tag{27}$$

For densities sufficiently exceeding this threshold

$$\frac{n_b}{n_0} \gtrsim \left(\frac{M\alpha^3}{m}\frac{T_e}{mc^2}\frac{T_e}{T_i}\right)^{3/2}\left(\frac{m}{M}\frac{T_e}{mc^2}\frac{T_i^3}{T_e^3}\right)^{1/2}\Delta\theta^2 \tag{28}$$

the relaxation length is changing with beam density in accordance with the relation

$$L_2 \simeq \frac{c}{\omega_p}\left(\frac{E}{mc^2}\right)^{5/2}\frac{mc^2}{T_e}\sqrt{\frac{n_0}{n_b}\frac{M\alpha^3}{m}}\,\Delta\theta^3 \tag{29}$$

It is important to note that both in nonrelativistic and relativistic cases the wave growth saturation due to modulational instability causes the very large increase of relaxation length in comparison with that found in the framework of weak turbulence theory.

## REFERENCES


1. Galeev, A.A., and R.Z. Sagdeev, Nuclear Fusion, 13, 603, 1973.
2. Thomson, J.J., R.J. Faehl, W.L. Kruer, Phys. Lett., 31, 918, 1973.
3. Galeev, A.A., V.N. Oraevskii, R.Z. Sagdeev, Zh.Eksper Teor. Pis'ma Red., 16, 194, 1972
4. Galeev, A.A., R.Z. Sagdeev, Yu.S. Sigov, V.D. Shapiro, V.I. Fizika, Plasmi 1, 11, 1975 (in Russian).
5. Zacharov, V.E., Zh. Eksper. Teor. Fiz., 62, 1745, 1972.
6. Degtyarov, L.M., V.E. Zacharov, Zh. Eksper. Teor. Fiz. Pis'ma Red. 21, 1975.
7. Rosenbluth, M.N., R.Z. Sagdeev, Comments on plasma physics and thermonuclear research, 1, 10, 1973.

8. Morales, G.J., Y.C. Lee, R.B. White, Phys. Rev. Lett. 32, 457 1974.

9. Lyubarskii, M.I., I.P. Panchenko, V.D. Shapiro, V.I. Shevchenko, Zh. Technikal Fiz. Pis'ma Red, in press.

10. Galeev, A.A., R.Z. Sagdeev, V.D. Shapiro, V.I. Shevchenko, Zh. Eksper. Teor. Fiz. Pis'ma Red, 23, 1976.

11. Thode, L.E., R.N. Sudan, Phys. Rev. Lett., 30, 732, 1973; K. Papadopoulos, Phys. Fluids 18, 1769, 1975.

12. Fainberg, Ya.B., V.D. Shapiro, V.I. Shevchenko, Zh. Eksper. Teor. Fiz. 57, 966, 1969; L.I. Rudakov, ibid., 59, 2091, 1970, B. Breizman, D.D. Ryutov, ibid., 60, 408, 1971.

13. Breizman, B.N., D.D. Ryutov, P.L. Chebotaev, Zh. Eksper.Teor. Fiz., 62, 1409, 1972.

14. Breizman, B.N., D.D. Ryutov, Nuclear Fusion, 14, 873, 1974.

# LANGMUIR SOLITONS AND ION-SOUND

D. ter Haar

Magdalen College

Oxford, England

I want to report here on some work from our group at Oxford, mainly by John Gibbons, Steve Thornhill, and Martin Wardrop, details of which can be found in a forthcoming paper (Gibbons, Thornhill, Wardrop and ter Haar 1976). We have in Oxford for some time been interested in various manifestations of plasma effects in astrophysics. Due to the enormous amounts of energy available, it is likely that many astrophysical objects, such as pulsar magnetospheres, accretion discs, galactic nuclei, quasars, and supernova remnants may contain turbulent plasmas. One way to approach this problem was the introduction by Tsytovich and Chikhachev of plasma turbulent reactors (for a review of their theory see Norman and ter Haar 1975), but it now appears that these systems - at least in the simplified form in which they have been considered so far - are not adequate. It was therefore felt advisable to see whether one might find a theory of plasma turbulence which could easily be built into astrophysical theory. The work by Kingsep, Rudakov, and Sudan (1973) suggested that perhaps a strongly Langmuir-turbulent plasma might be described as a gas of Langmuir solitons. This was the reason for a programme for studying the properties of Langmuir solitons and I am presenting some of the results of this programme. I should perhaps add that we are no longer very optimistic about the soliton approach to strong turbulence.

I shall restrict my discussion mainly to the one-dimensional case and only occasionally refer to what happens in the three-dimensional case. The basic equations are those given by Zakharov (1972):

$$iE_t + E_{xx} = nE, \tag{1}$$

$$n_{tt} - n_{xx} = (|E|^2)_{xx}, \tag{2}$$

where E is the electrical field, n the deviation of the plasma ion density from its equilibrium density, and subscripts indicate partial differentiation. We are using dimensionless quantities (for the units used see Thornhill and ter Haar 1975 or Gibbons, Thornhill, Wardrop, and ter Haar 1976) in which velocities are measured in such units that the ion-sound velocity is unity.

It is convenient to introduce the hydrodynamic flux potential u through the equation

$$u_t = n + |E|^2, \tag{3}$$

so that (2) can be written in the form

$$n_t = u_{xx}. \tag{4}$$

As E is a complex quantity, we must treat E and $E^*$ as independent variables, and take u, E, and $E^*$ as our field variables. Equations (1) and (4) can be derived from the Lagrangian density

$$\mathcal{L} = \frac{1}{2} i\,(E^*E_t - E\,E^*_t) - |E_x|^2 - u_t|E|^2 + \frac{1}{2}\,|E^2|^2 + \frac{1}{2}(u_t^2 - u_x^2). \tag{5}$$

As the Langrangian,

$$L = \int \mathcal{L} dx, \tag{6}$$

is translationally invariant, does not contain the time explicitly, and is invariant under the transformation $E \to E\,e^{i\phi}$, the total linear momentum P, the total energy H, and the the total plasmon number N are conserved. These quantities are given by the equations

$$P = \tfrac{1}{2} \int_{-\infty}^{+\infty} [i(EE^*_x - E^*E_x) + 2n\mathcal{V}]dx, \tag{7}$$

$$H = \int_{-\infty}^{+\infty} [|E_x|^2 + n\,|E|^2 + \tfrac{1}{2}\,n^2 + \tfrac{1}{2}\,\mathcal{V}^2]dx, \tag{8}$$

$$N = \int_{-\infty}^{+\infty} |E|^2\,dx, \tag{9}$$

where we have introduced the hydrodynamic flux $\mathcal{V}$ through the equation

$$\mathcal{V} = -\,u_x; \tag{10}$$

$\mathcal{V}$ satisfies the equation of continuity

$$n_t + \mathcal{V}_x = 0. \tag{11}$$

A similar analysis can be given for the three-dimensional case, when apart from the linear momentum, the energy, and the phasmon number also the angular momentum is conserved.

Equations (1) and (2) are satisfied by the well-known two-parameter family of Langmuir-soliton solutions:

$$E(m,v) = \frac{m}{[2(1-v^2)]^{\frac{1}{2}}} \operatorname{sech} \frac{m(x-vt)}{2(1-v^2)} \exp i[\tfrac{1}{2}vx-(\tfrac{1}{4}v^2 - \frac{m^2}{4(1-v^2)^2})t], \tag{12}$$

$$n = - \frac{|E|^2}{1-v^2} . \tag{13}$$

The solution (12) and (13) gives the following values for N, P, and H:

$$N = 2m, \tag{14}$$

$$P = mv + \frac{2m^3 v}{3(1-v^2)^3}, \tag{15}$$

$$H = \tfrac{1}{2} mv^2 + \frac{m^3(5v^2-1)}{6(1-v^2)^3}. \tag{16}$$

It is interesting to note that, if we take the limit as $m \to o$ so that (12) becomes

$$E \to \frac{m}{[2(1-v^2)]^{\frac{1}{2}}} e^{ikx-i\omega t}, \quad k = \tfrac{1}{2}v, \; \omega = \tfrac{1}{4} v^2 = k^2, \tag{17}$$

equations (14) to (16) gives us simply the plasmon N, the momentum Nk, and the energy $N\omega$ (= $Nk^2$) of the linear wave (17). As, however, solitons are essentially non-linear entities, we see that the last terms on the right-hand sides of P and H usually dominate in cases of interest, e.g., when $1-v^2 << 1$.

For an isolated ion-sound train travelling in the positive x-direction the density $n_+$ and hydrodynamic flux $\mathcal{V}_+$ (the index "+" indicating the direction of propagation), which are both functions of the argument x-t, satisfy the relation

$$n_+ = \mathcal{V}_+, \tag{18}$$

while for an isolated ion-sound train travelling in the negative x-direction we have

$$n_- = -V_- . \tag{19}$$

If there is a number of such sound trains and as long as those moving in the positive x-direction are well separated from those moving in the negative x-direction and from any solitons or other disturbances, these trains make the following contributions to the integrals of motion:

$$N = 0, \tag{20}$$

$$P = p_+ - p_-, \tag{21}$$

$$H = p_+ + p_-, \tag{22}$$

where

$$p_\pm = \int_{-\infty}^{+\infty} n_\pm^2 \, dx, \tag{23}$$

and we note that $p_+$ and $p_-$ are positive definite quantities.

We now assume that both at t=o and as $t \to +\infty$ the state of the system governed by equations (1) and (2) can be described as a collection of well-separated solitons and ion-sound trains. We then have for both the initial and final state

$$N = 2 \sum_i m_i, \tag{24}$$

$$P = \sum_i \left[ m_i v_i + \frac{2m_i^3 v_i}{3(1-v_i^2)^3} \right] + p_+ - p_-, \tag{25}$$

$$H = \sum_i \left[ \tfrac{1}{2} m_i v_i^2 + \frac{m_i^3 (5v_i^2 - 1)}{6(1-v_i^2)^3} \right] + p_+ + p_-. \tag{26}$$

We shall now use a graphical method to discuss what kind of reactions are allowed by the fact that N, P, and H are conserved quantities, and we shall compare our results with those obtained in numerical experiments. From equation (24) it follows immediately that the process $nL \to S$, that is, a reaction in which n solitons (L) decay into ion-sound (S), is forbidden.

We now plot H as a function of P for solitons of fixed mass m, but varying velocity v (Fig.1). In Fig.1 we have also indicated by dashed lines the corresponding curve for ion-sound trains. From equations (15) and (16) we find the following properties of the H-P curves.

The curves for different m become, as $|v| \to 1$, infinitely

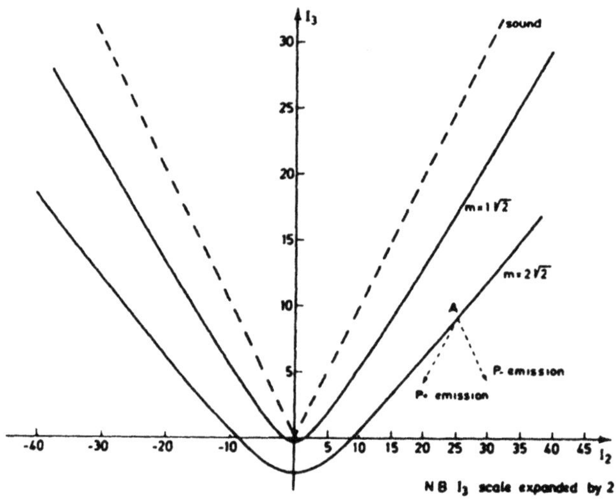


Fig.1. P-H-plots for single-soliton solutions (full-drawn lines) and for ion-sound trains (dashed lines). Note that the scales along the P- and H-axes are different ($I_2 = P/\sqrt{2}$, $I_3 = H/\sqrt{2}$).

separated from each other and from the sound lines, and curves for different m neither cross nor touch. Emission of ion-sound moving in the positive x-direction ($p_+$-emission) is described by a translation of the representative point in the P-H-plane in the direction indicated in Fig.1.

I shall not discuss here the justification for assuming the final state of the system to be one of separated solitons and ion-sound trains, except for noting that computer experiments seem to indicate that these products, if not the only ones, are at least the most important ones. For a more detailed discussion I refer to a forthcoming paper (Gibbons, Thornhill, Wardrop, and ter Haar (1976).

One sees easily that the processes

$$nL \rightleftarrows S \tag{27}$$

and

$$L \rightleftarrows L + S \tag{28}$$

are forbidden as they violate N-conservation.

Consider next the process

$$L + S \rightleftarrows L' + S'. \tag{29}$$

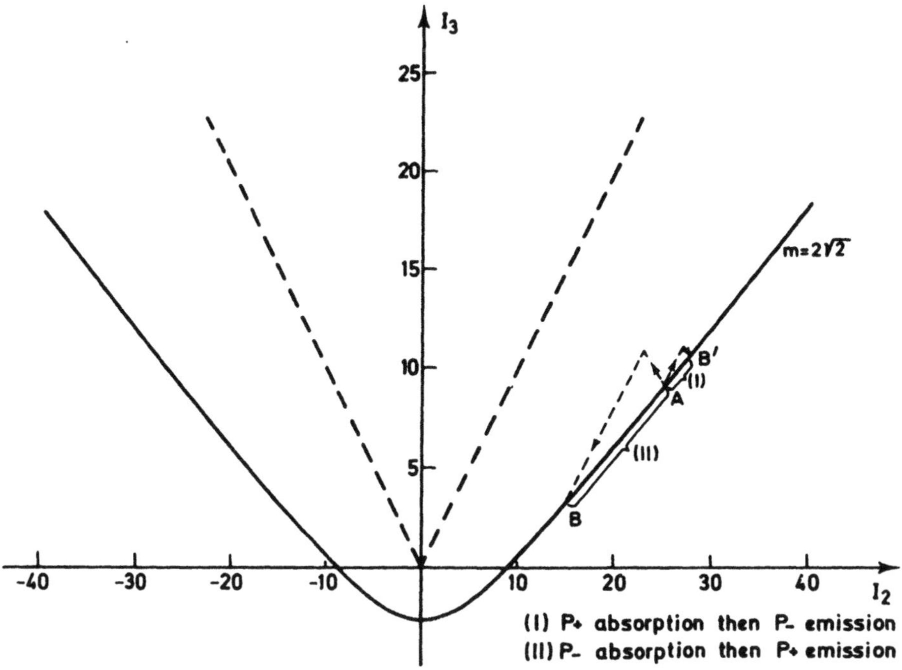


Fig.2. P,H-plot illustrating the process $L + S \rightarrow L' + S'$ ($I_2 = P/\sqrt{2}$, $I_3 = H/\sqrt{2}$).

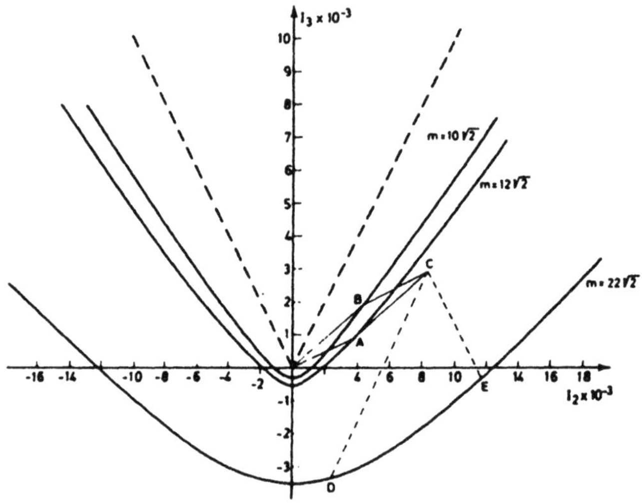


Fig.3. P,H-plot illustrating the process $L_1 + L_2 \to L_3 + S$ ($I_2 = P/\sqrt{2}$, $I_3 = H/\sqrt{2}$).

From N-conservation it follows that the representative points A and B of the initial and final states must lie on the same single-soliton curve (see Fig.2). Process (i) in Fig.2 corresponds to a deceleration of the soliton and process (ii) to an acceleration. A comparison of these two processes shows that a soliton moving through a small-amplitude random background of ion-sound will on average be decelerated.

Let us now discuss the processes

$$L_1 + L_2 \rightleftarrows L_3 + S, \tag{30}$$

or in general

$$nL \leftrightarrows L + S, \tag{31}$$

and

$$L_3 \rightleftarrows L_1 + L_2 + S, \tag{32}$$

or in general

$$L \rightleftarrows nL + S. \tag{33}$$

Consider Fig.3. Let the solitons $L_1$ and $L_2$ correspond to the points A and B. From the properties of H and P it follows that the point C, which is determined by the vector addition of the P,H points for the two solitons, lies above the curve corresponding to the mass $m_3$ ($=m_1 + m_2$). Hence it follows not only that the processes (32) and (33), but also that the processes

$$L \rightleftarrows nL \tag{34}$$

are forbidden. It follows thus also that a soliton on its own cannot break up.

The processes (30) and (31) are allowed. If we consider again Fig.3 we see that if $L_1$ and $L_2$ correspond to the points A and B, $L_3$ in process (30) will correspond to a point of the arc DE (note that the representative point must lie on that curve because of N-conservation).

I emphasize that we can only show that certain processes are allowed by the conservation laws - not that they will actually occur. The computer experiments of Abdulloev, Bogolyubskij, and Makhan'kov (1975) and of Degtyarev, Makhan'kov and Rudakov (1975) indicated that Langmuir solitons fuse for certain ranges of their parameters, but pass through one another, with perhaps slight changes - for other parameter values. I shall briefly and very qualitatively discuss this problem for the case of the "interaction" of two solitons of the same mass m and the same, but opposite velocities, $\pm$ v.

From equations (1) and (2) one can derive an expression for the total sound generated $p_{gen}$ for which we find

$$p_{gen} \lesssim \frac{2m^3}{v(1-v^2)^2} \; . \tag{35}$$

On the other hand, for m >> 1 we find from the energy conservation law that

$$p_{gen} \sim \frac{4}{3} \; m^3 \; \frac{5v^2-1}{(1-v^2)^3} + \frac{16}{3} \; m^3, \tag{36}$$

and on inspection we see that fusion cannot take place unless v is smaller than some critical value of the order of 0.6, in reasonable agreement with the computer result of Abdulloev, Bogolyubskij, and Makhan'kov (1975); the sound-emission mechanism may well be the governing factor for soliton fusion.

From our analysis of the corresponding equations for the three-dimensional case we conclude that, unfortunately, the consequences of the equations which are the analogues of equations (1) and (2) are not at all similar to the picture I have just presented. The nL + S state which seems to be stable in the one-dimensional case is unstable in the three-dimensional case. We also doubt whether the inclusion of a magnetic field in order to stabilize the solitons will, in fact, succeed in its aim. All approaches which take non-linear terms in one mode and produce soliton solutions are open to the criticism of ignor-ing mode-mode interactions which should become important at the high amplitudes involved. We feel that for a three-dimensional plasma one may well have to look for a different approach to describe strong turbulence, perhaps based on the products of collapsing groups of plasmons.

## References

Kh.D. Abdulloev, I.L. Bogolyubskij, and V.G. Makhan'kov 1975 Nucl. Fusion 15, 21.

L.M. Degtyarev, V.G. Makhan'kov, and L.I. Rudakov 1975 Sov.Phys. JETP 40, 264.

J. Gibbons, S.G. Thornhill, M.J. Wardrop, and D. ter Haar 1976 Oxford preprint 36/76. Submitted to J. Plasma Phys.

A.S. Kingsep, L.I. Rudakov, and R.N. Sudan 1973 Phys. Rev. Lett. 31, 1482.

C.A. Norman and D. ter Haar 1975 Phys. Repts. 17, No.6.

S.G. Thornhill and D. ter Haar 1975 Oxford preprint 80/75.

V.E. Zakharov 1972 Sov. Phys. JETP 35, 908.

# LANGMUIR SOLITONS AND RESONANCE ABSORPTION IN LASER-IRRADIATED PLASMAS

Hsing-Hen Chen and Chuan-Sheng Liu

Department of Physics and Astronomy
University of Maryland
College Park, Maryland 20742

## ABSTRACT

Nonlinear Langmuir wave in inhomogeneous plasmas is studied analytically in the model of nonlinear Schrödinger equation. Exact solutions in the form of multisolitons are obtained. The theory is applied to the linearly-transformed Langmuir waves at resonance with a driving field.

For a laser beam of frequency $\omega_0$ obliquely incident onto a nonuniform plasma slab, there is a component of the electric field along the density gradient driving the density oscillation. Because of Budden tunneling, the residual electric field beyond the cutoff can drive the plasma wave at resonance where $\omega_p(x) = \omega_0$. [1] This transformation of the electromagnetic wave into electrostatic waves which are subsequently absorbed by the plasma particles constitutes one of the most important absorption mechanisms of the laser radiation. Experiments on laser-pellet irradiation and computer simulations[2] have indicated that the major absorption process is consistent with the resonance absorption.

The structures of the resonance - the widths and heights of the critical surface affect the absorption as well as the characteristics of the reflection and the second harmonic generation.[3]

In the usual linear theory, a fixed density profile is assumed and the modification of the density profile by the large-amplitude plasma wave generated at resonance is neglected. In reality, even for a relatively weak pump wave, the resonantly-driven plasma wave can significantly modify the density profile near the critical density region by its ponderomotive force. Because of the localized structure of the field near the resonance, the ponderomotive pressure tends to drive the plasma out of the resonance region, thus depleting the local plasma density. This cavity in density in turn traps more plasmons leading to the modulational instability.[4] In a homogeneous plasma, it has been shown that the modulational instability evolves into multisolitons[5] in one-dimension when the nonlinear condensation is balanced by the wave diffraction or leads to spatial collapse for three-dimensional Langmuir waves.[6] In an inhomogeneous plasma, the localized field naturally arises at the resonance and one would expect that the solitons will be formed there if the nonlinear effect is taken into account. They will subsequently propagate away from the resonance. It is of interest to inquire how does the inhomogeneity affect the structure and propagation of solitons. In particular, one may ask whether multisolitons exist in inhomogeneous media. In this paper, we studied analytically the soliton formations at the resonance point by solving exactly the nonlinear Langmuir wave equation in an inhomogeneous plasma. The solitons, unlike those in homogeneous media, are found to be accelerated by the density gradient with rapid phase variations. The threshold for the soliton formations at the resonance with a driving field $E_d$ is given. Some of the consequences of these solitons to the absorption and scattering of laser light are also discussed.

For simplicity, consider an expanding, inhomogeneous plasma slab with flow velocity u(x) and density n(x). The profiles of density and flow velocity are supposed to be constant in time, i.e. the steady-state hydrodynamic equations

$$\frac{\partial n}{\partial \tau} = - \frac{\partial (nu)}{\partial x} = 0 \tag{1}$$

$$nM \frac{\partial u}{\partial \tau} = - \frac{nM}{2} \frac{\partial u^2}{\partial x} - \frac{\partial}{\partial x}\left(nT\right) - \frac{\omega_p^2}{2\omega_o^2} \frac{\partial}{\partial x} \frac{|E_L|^2}{8\pi} = 0 \tag{2}$$

are satisfied, where T is the temperature and the last term is due to the ponderomotive force of the laser light.

The nonlinear equations for the converted Langmuir wave $\mathcal{E}(x,y,\tau) = E(x,y,\tau)\, e^{-i\omega_p\tau + ik_y y} + c.c.$ in a nonuniform plasma near the critical density (x = 0), after the pump is switched off, is

$$i2\omega_p \frac{\partial E}{\partial \tau} + i6k_y v_e^2 \frac{\partial E}{\partial y} + 3v_e^2 \frac{\partial^2 E}{\partial x^2} - \omega_p^2 \left(\frac{x}{L} + \frac{\delta n_i}{n_o}\right) E = 0 \tag{3}$$

where $k_y$ is the y-component wave vector of the incident light because of momentum conservation, $L = (d\ln n/dx)^{-1}$ is the density scale length, $\delta n_i$ is the low frequency ion-density fluctuation due to nonlinearity, satisfying the following equation:

$$\left(\frac{\partial}{\partial\tau}-u\frac{\partial}{\partial x}\right)^2 \delta n_i - c_s^2\frac{\partial^2\delta n_i}{\partial x^2} = n_o\frac{\partial^2}{\partial x^2}\left|\frac{e^2|E|^2}{2m\omega_o^2}\right| \tag{4}$$

where $c_s = (T_e/M)^{1/2}$ is the sound speed. Near the critical surface, the group velocity of the Langmuir wave is much less than the sound speed and we have

$$\frac{\delta n_i}{n_o} = -\frac{e^2|E|^2}{2(c_s^2-u^2)m\omega_o^2} \tag{5}$$

It is of interest to note that the relative sign of the nonlinear density fluctuation and $|E|^2$ depends on whether the flow is supersonic or subsonic. We consider in the following only the case of subsonic flows for which the nonlinear Langmuir Eq. (3) becomes

$$i\frac{\partial E}{\partial t} + \frac{\partial^2 E}{\partial x^2} + (-2\alpha x + 2|E|^2)\,E = 0 \tag{6}$$

where $t = \tau - \omega_p y/3k_y v_e^2$ is expressed in units of $\omega_p^{-1}$, x in units of $(\sqrt{3}/2)\,\lambda_D$, $2\alpha = 1/L$ and E is in the unit of $4(\pi nT)^{1/2}$ $(1 - u^2/c_s^2)^{1/2}$. Remembering that the propagation of the usual linear wave packet in an inhomogeneous medium obeys the following Hamiltonian equations of motion

$$v_g \equiv \frac{dx}{dt} = \left.\frac{\partial\omega}{\partial k}\right|_x, \quad \frac{dk}{dt} = -\left.\frac{\partial\omega}{\partial x}\right|_k \tag{7}$$

For Langmuir waves in an inhomogeneous plasma with density variation $n(x) = n_o(1 + x/L)$, we find that the wave packet is accelerated with an acceleration

$$\dot{v}_g = \frac{3v_e^2}{\omega_p}\dot{k} = \frac{3v_e^2}{\omega_p}\frac{\partial\omega}{\partial x} = \frac{3v_e^2}{L} = -4\alpha \tag{8}$$

whereby the wave packet slows down while moving toward the turning point and speeds up while moving towards the underdense region.

A one-soliton solution of Eq. (6) can be easily obtained by assuming $E = A(x,t)\exp(i\phi(x,t))$ with $A,\phi$ real. The resulting equations for A and $\phi$ are

$$(A^2)_t + 2(A^2\phi_x)_x = 0$$

and

$$-[2\alpha x + (\phi_t + \phi_x^2)]A + A_{xx} + 2A^3 = 0 \tag{9}$$

Anticipating solutions having a time dependent velocity $v_g = 4\xi - 4\alpha t$ due to the inhomogeneity with $4\xi$ the initial velocity, we assume that $A(x,t)$ is a function with argument $2\eta(x+2\alpha t^2 - 4\xi t - x_0)$, where $x_0$ is the initial position of the packet. Eq. (9) then yields the solution for the phase $\phi$ and amplitude A

$$\phi = 2(\xi-\alpha t)x - 4\left(\frac{\alpha^2 t^3}{3} - \alpha\xi t^2 + (\xi^2-\eta^2)t\right) + \phi_0$$

and (10)

$$A = 2\eta \operatorname{sech} 2\eta(x+2\alpha t^2-4\xi t-x_0)$$

valid for $\partial\phi/\partial t \ll 1$. For this solution, we may define a generalized, time dependent wavenumber $k_x \equiv \partial\phi/\partial x = 2(\xi-\alpha t)$ and frequency shift $\Omega \equiv -\partial\phi/\partial t = 2\alpha x + 4[(\xi-\alpha t)^2 - \eta^2] = 2\alpha x + k_x^2 - 4\eta^2$. Then remarkably, the usual Hamiltonian equations for the wave packet propagation in a nonuniform medium remains valid for the soliton

$$\frac{dk_x}{dt} = -\left.\frac{\partial\Omega}{\partial x}\right|_{k_x}, \quad \frac{dx}{dt} = \left.\frac{\partial\Omega}{\partial k_x}\right|_x = 4(\xi-\alpha t). \tag{11}$$

The frequency shift $\Omega$ retains the meaning of a Hamiltonian even for this highly complicated nonlinear wave packet, and

$$\frac{d\Omega}{dt} = \left.\frac{\partial\Omega}{\partial k_x}\right|_x \frac{dk_x}{dt} + \left.\frac{\partial\Omega}{\partial x}\right|_{k_x} \frac{dx}{dt} = 0$$

as in the usual stationary linear medium. We note that this complicated space-time dependence of the frequency shift may account for the rich structures of the second harmonic emission and the reflected light observed in laser-pellet irradiation experiments.[3,7-10] We also note that the single soliton solution assumes a parabolic wave front in the xy-plane with its apex moving at a constant speed in the y-direction tangential to the density gradient. In the one-dimensional case, the wave front moves in the x-direction. If $\xi > 0$, it travels toward the overdense region initially, is slowed down by a deceleration $4\alpha$, until $t = \xi/\alpha$, when it reaches the turning point $x_T = x_o + 2\xi/\alpha$ and the velocity changes sign. The soliton is then reflected and accelerated down the density gradient toward the underdense region. The density modification by the solitons does not alter their propagation characteristics. The distance of possible penetration into the overdense region depends on the initial velocity $4\xi$, which, in turn, depends on the initial wave profile. For multisolitons or nonlinear stationary solutions in two space dimensions, the corresponding ion density fluctuations may have periodic structure in the y-direction as observed in computer simulations.[12,13]

In order to solve the complete time evolution problem for Eq. (6), or in particular, to find the N-soliton solutions, we need to introduce the inverse scattering transform method. This powerful method in solving nonlinear evolution equation such as Eq. (6)

has recently been studied enthusiastically in various branches of physical and mathematical research. Basically, it is a nonlinear generalization of the linear Fourier transform. In the present case, we will transform our unknown function E(x,t) into a set of scattering data [$C_k$, $\zeta_k$, $\rho(\xi)$ defined below] via the linear scattering problem[5,14] (eigenvalue problem).

$$\begin{aligned} V_{1x} + i\zeta V_1 &= E(x,t)V_2 \\ V_{2x} - i\zeta V_2 &= -E^*(x,t)V_1 \end{aligned} \tag{12}$$

where $V_1, V_2$ are components of a vector wave function $v = \begin{pmatrix} V_1 \\ V_2 \end{pmatrix}$, $\zeta$ is the eigenvalue, E(x,t) serves as the potential function of the scattering problem and is assumed to approach zero as $|x|\to\infty$. The time variable t plays only the role of a parameter in the direct scattering problem (12). To determine the time dependence of the wave function v we assume

$$\begin{aligned} V_{1t} &= AV_1 + BV_2 \\ V_{2t} &= CV_1 - AV_2. \end{aligned} \tag{13}$$

with $A = 2i(i\zeta)^2 + i|E|^2 - i\alpha x$, $B = -2iE(i\zeta) + iE_x$, $C = 2iE^*(i\zeta) + iE_x^*$ and $\zeta_t = -\alpha$. Then the compatibility condition for Eqa. (12) and (13) i.e., differentiating Eq. (12) with respect to x and Eq. (13) with respect to t and demanding $V_{xt} = V_{tx}$, yields precisely our basic Eq. (6). We note that the major difference introduced by the inhomogeneity is that the eigenvalues $\zeta$ are no longer constants but depend linearly on time. By allowing the time-varying eigenvalue, we have therefore greatly enlarged the set of exactly solvable nonlinear time evolution equations.

The procedure for the inverse scattering method is the following:[5,14] (i) For a given initial profile E(x,0), find the eigenvalues for both the continuum $\xi$, and bound states (discrete spectrum) $\zeta_k = \xi_k + i\eta_k$. Then solve the scattering problem for the incident wave function $\exp(i\,\xi\, x)$ at $x \to -\infty$ to find the reflection coefficient $\rho(\xi)$ and a set of constant coefficients $C_k$ for the normalization of bound state wave functions. These constitute the "initial scattering data". (ii) Solve Eq. (13) to find the time evolution of the eigenfunctions and eigenvalues and therefore the time-dependence of the scattering data. Since these data do not depend on x and also $E(x,t)\to 0$ as $|x|\to\infty$, we can easily figure out the time dependence of the scattering data by going into the limit $|x|\to\infty$. These are

$$\rho(\xi,t) = \rho(\xi,0)\exp\{-\frac{4i}{3\alpha}[(\xi-\alpha t)^3-\xi^3]\} \tag{14}$$

$$C_k(\zeta_k,t) = C_k(0)\ \exp\{-\frac{4i}{3\alpha}[(\zeta_k-\alpha t)-\zeta_k^3]\} \tag{15}$$

(iii) From these scattering data at time t, reconstruct the "potential" E(x,t) by the inverse scattering method. In this way, nonlinear evolution equation Eq. (6) is solved for the time-evolution of the initial profile E(x,0). The solution is given by [5,14]

$$E(x,t) = -2K_1(x,x;t) \tag{16}$$

obtained by solving the Gelfand-Levitan-Marchenko equations:

$$\begin{aligned} K_1(x,y) &= F^*(x+y) + \int_x^\infty K_2^*(x,z)F^*(z+y)dz \\ K_2(x,y) &= -\int_x^\infty K_1(x,z)F(z+y)dz \end{aligned} \tag{17}$$

where

$$F(x,t) = \sum_k^N C_k(t)\, e^{i\zeta_k(t)x} + \frac{1}{2\pi}\int_{-\infty}^{\infty} \rho(\xi,t)e^{i\xi(t)x}d\xi \tag{18}$$

is constructed from the scattering data at time t and $d/dx\, K_2(x,x;t) = 1/2\, |E(x,t)|^2$. Because these are linear integral equations, solutions can be more easily obtained.

We discuss the following solutions of Eq. (6), i.e. $\rho(\xi) = 0$, N = finite. For clarity we can write Eq. (18) in an abbreviated form,
$\sum_k^N \bar{C}_k e^{i\lambda_k x}$ with $\bar{C}_k \equiv C_k e^{-4i[(\zeta_k-\alpha t)^3 - \zeta_k^3]/3\alpha}$ and $\lambda_k \equiv \zeta_k - \alpha t$. The GLM equation can be solved by linearly expanding

$$K_{\substack{1\\2}}(x,y) = \sum_k^N K_{\substack{1k\\2}}(x)e^{-i\lambda_k^* y} \tag{19}$$

to obtain

$$\begin{aligned} K_{2k}^*(x)e^{i\lambda_k x} &= -i\sum_{m=1}^N \frac{\bar{C}_k c^{2i\lambda_k x}}{\lambda_k-\lambda_m^*} K_{1m}(x)e^{-i\lambda_m^* x} \\ K_{1k}(x)e^{-i\lambda_k^* x} &= \bar{C}_k^* e^{-i\lambda_k^* x} + i\sum_{m=1}^N \frac{\bar{C}_k e^{2i\lambda_k x}}{\lambda_m - \lambda_k^*} K_{2m}(x)e^{i\lambda_m x} \end{aligned} \tag{20}$$

These algebraic equations can be solved by the method of determinant to obtain the N-soliton solutions. The one-soliton solutions are especially simple. Setting N=1 in Eq. (12), we get

$$E(x,t) = -2K_1(x,x) = -\frac{2\bar{c}^* \, e^{-2i\lambda^* x}}{1 + \frac{c\bar{c}^*}{4\eta^2} e^{-4\eta x}}$$

$$= 2\eta \operatorname{sech} 2\eta (x+2\alpha t^2 - 4\xi t - x_0) \tag{21}$$

$$\times \exp i\left\{2(\xi-\alpha t)x - 4\left(\frac{\alpha^2 t^3}{3} - \alpha\xi t^2 + (\xi^2-\eta^2)t\right) + \phi_0\right\}$$

It agrees with our previous formula for one-soliton Eq. (10). We notice that the real part of the eigenvalue $\xi$ is responsible for the initial speed of the soliton, while the imaginary part of the eigenvalue $\eta$ is responsible for the height and width of the soliton. In case of N-solitons, the solutions are similar to those found in a homogeneous plasma, differing only in that they are now accelerated. When far apart from each other, they are essentially superposition of one-soliton solutions given by Eq. (10). Upon collision with each other, they emerge with their identities unaltered except for a minor shift in position. Sometimes, all eigenvalues or a subset of eigenvalues have identical real parts, corresponding to solitons moving together with the same initial velocity. Because they experience the same acceleration, they remain together forming a bound state called breathers.[14] A breather containing many solitons will in general have a symmetrical irregular shape oscillating in time periodically. Under perturbations to the equation, breather will break up and release all its solitons. We may now state the following threshold conditions for soliton formation.

As the linearly converted Langmuir wave reaches a critical amplitude, solitons are generated and propagate out of the resonance as shown in Eq. (10). The threshold for the soliton formation in terms of the initial profile E(x,o) can be shown to be[14]

$$\int |E(x,o)|\,dx \geq 0.904 \,, \tag{22}$$

unchanged by the presence of the inhomogeneity.

If the intensity of the incident radiation much exceeds the threshold, then multisoliton formation is expected. For the initial profile of the form $E(x,o) = E \operatorname{sech}[x/\Delta]\, e^{ikx}$, it can be shown that, for $N \leq E\Delta < N+1$, there will be exactly N solitons generated. They travel together with the same initial speed 2k and therefore form a breather. The fraction of the initial energy that does not go over to the solitons is $(1-N/E)^2$, which is negligibly small when $N \gg 1$. For more general initial profiles, WKB approximation incorporated with the Bohr quantization rule gives the following condition for

N-soliton formation:

$$\pi(N + \frac{3}{2}) \geq \int |E| dx \geq \pi(N + \frac{1}{2}) \tag{23}$$

This is a good approximation for sufficiently smooth profile with N >> 1.

In order to find the threshold condition in terms of the driving field $E_d$ and to look into the details of soliton formation with the driving field in action, we have to solve the nonlinear Schrödinger equation with an external source.

$$iE_t + E_{xx} + (-2\alpha x + 2|E|^2)E = E_d \tag{24}$$

Since E(x,o) = 0 initially, we may neglect both the diffraction term and the nonlinear term in the early stage of wave conversion. The simplified equation

$$iE_t - 2\alpha x E = E_d \tag{25}$$

is easily solved.

$$E = -\frac{E_d}{2\alpha x}(1 - e^{-2i\alpha x t}) \tag{26}$$

This solution describes a resonance centered around the point x = 0. A consequence of this solution is that the quantity

$$\int |E| dx = \frac{\pi |E_d|}{\alpha}$$

does not depend on time. If we combine it with Eq. (22), we then find the threshold driving field amplitude for soliton formation at resonance.

$$|E_d| \geqslant 0.904\ \alpha/\pi \tag{27}$$

We notice that this threshold condition for soliton formation at resonance has been observed before in a numerical work by Morales and Lee.[15] Our result agrees well with their calculations. If the driving field is below the threshold, the modulational instability is not effective, the Langmuir wave converted at the resonance point will spread out along the density gradient into an Airy-like structure. Once the threshold amplitude is exceeded, the solitons are generated immediately at the resonance. They eventually move away from the resonance and down the density gradient due to the acceleration that is characteristic to solitons in nonuniform media.

Since the resonance region is concentrated on the point x = 0, once solitons have moved out of the resonant region, the effect of driving field on them will be negligibly small. Following Galeev et al,[17] we may calculate the time dependence of a driven soliton in a nonuniform plasma. From Eq. (24), we can derive the following equations

$$\frac{d}{dt}\int |E|^2\, dx = iE_d \int (E-E^*)\, dx \tag{28}$$

and

$$\frac{d}{dt}\int i(E\, E_x^* - E^* E_x)dx = 4\alpha\int |E|^2 dx \tag{29}$$

Substituting the soliton solution Eq. (10) into them, we get

$$\frac{d\eta}{dt} = -\frac{\pi E_d}{2}\,\text{sech}\,\frac{\pi(\xi-\alpha t)}{2\eta}\,\sin\left[-\frac{8}{3}\alpha^2 t^3 - 8\xi\alpha t^2 + 4(\xi^2+\eta^2)t + \phi_o\right] \tag{30}$$

and

$$\xi = \alpha t + (\eta_o - \alpha\int_o^t \eta\, dt)/\eta$$

It is easily seen that, at early times, the parameters change very fast in time and soon afterward, due to the sech factor, $d\eta/dt \to 0$ and therefore $\eta$, $\xi \to$ constants, the soliton approaches a steady state. It is not affected by the driving field any longer. This behavior is completely different from the case of uniform plasma. Driven solitons in uniform plasma do not saturate by convective motions.

To determine the size of solitons generated from the resonance, we can make a simple estimate. If the pump strength is not too much above the threshold, we can assume that solitons are generated one by one. Choosing our initial profile approximated by Eq. (26), we find that the quantity $\int |E|^2 dx$ increases in time linearly, i.e.

$$\int |E|^2 dx = \pi |E_d|^2 t/\alpha \tag{31}$$

On the other hand, $\int |E|^2 dx = 4\eta$ for a soliton. Therefore, we may say that the soliton formed from the initial profile Eq. (26) will have its height $2\eta$ increasing linearly with time. The growth rate depends linearly on $|E_d|^2$ in agreement with those observed in experiment.[16] The soliton saturates when it moves out of the resonance region by the acceleration ($-4\alpha$) from the density gradient. Equating the width of the soliton $1/2\eta$ with the distance it travelled in a time $\tau$ needed for it to grow to this size, we have

$$\frac{1}{2\eta} = \frac{2\alpha}{\pi |E_d|^2 \tau} = 2\alpha\tau^2 \tag{32}$$

$$\therefore \quad \tau = (\pi|E_d|^2)^{-1/3} \quad \text{and} \quad 2\eta = (\pi|E_d|^2)^{2/3}/2\alpha$$

If we substitute into it the parameters from the UCLA experiment,[16] i.e. $|E_d|^2 \sim 10^{-5}$, $2\alpha \sim 10^{-2.5}$ we then have $1/2\eta \sim 10\ \lambda_D$. It agrees very well with the experiment. The enhancement of electric field at the resonance can also be obtained as

$$\frac{2\eta}{|E_d|} = \frac{(\pi^2|E_d|)^{1/3}}{2\alpha}$$

It is proportional to the density scale length and increases weakly with the driving field.

These solitons generated at resonance point result in the non-stationary modifications of density profile around the resonance region. It is expected to have major effects on many nonlinear processes occurred there.

## REFERENCES

1. V. L. Ginzburg, The Propagation of Electromagnetic Waves in Plasmas, Second Edition, Pergamon Press, (1970).

2. W. L. Kruer et al, Proceeding of Nobel Symposium (1976).

3. R. Sigel, K. Eidmann, H. C. Pant and P. Sachsenmaier, Phys. Rev. Lett. (1976) (to be published). K. Eidmann and R. Sigel, in Laser and Interaction and Related Plasma Phenomena Vol 3, Ed. by H. J. Schwartz and H. Hora, (Plenum, 1974), p. 667.

4. R. Sagdeev and A. A. Galeev, Nonlinear Plasma Theory (Benjamin 1968).

5. V. Zakharov and A. Shabat, Soviet Phys. JETP 34, 62 (1972).

6. V. Zakharov, Soviet Phys. JETP 35, 908 (1972).

7. J. L. Bobin, M. Decroisette, B. Meyer, and Y. Vitel, Phys. Rev. Lett. 30, 594 (1973).

8. C. Yamanaka, T. Yamanaka, T. Sasaki, J. Mizui and H. B. Kang, Phys. Rev. Lett. 32, 1038 (1974).

9. L. M. Goldman, J. Soures and M. J. Lubin, Phys. Rev. Lett. 31, 1184 (1973).

10. P. Lee, D. V. Gievanelli, R. P. Godwin and G. H. McCall, App. Phys. Lett. 24, 406 (1974).

11. J. Satsuma and N. Yajima, Suppl. Prog. Theo. Phys. Japan 55, 284 (1974).

12. E. J. Valeo and K. G. Estabrook, Phys. Rev. Lett. 34, 1008 (1975).

13. D. Biskamp and H. Welter, in Plasma Physics and Controlled Nuclear Fusion, Vol. II, p. 507 [Int. Atomic Energy Commission, Vienna (1975)].

14. M. Ablowitz, A. Newell, D. Kaup and H. Segur, Study App. Math. 53, 249 (1974).

15. G. Morales and Y. C. Lee, Phys. Rev. Lett. 33, 1016 (1974).

16. H. Kim, R. Stenzel and A. Wong, Phys. Rev. Lett. 33, 886 (1974).

17. A. Galeev, R. Sagdeev, Y. Sigov, V. Shapiro and V. Shevchenko, Sov. J. Plasma Phys. 1, 5 (1975).

# PHOTON SOLITON DUE TO INDUCED COMPTON SCATTERING

Carlos Montes

Laboratoire de Physique de la Matière Condensée*

Parc Valrose, 06034 Nice Cedex
and
Observatoire de Nice, B.P. 252, 06007 Nice Cedex

The induced (nonlinear) Compton interaction between a narrow incoherent photon spectrum - excited to high intensity in a population inversion medium (maser or laser) - and the surrounding plasma, gives rise to a strongly modulated kinetics: the narrow amplified line of width $\delta\nu$ exhibits large amplitude relaxation oscillations, generating at each oscillation a photon soliton at the Compton-Doppler distance $\overline{\Delta\nu_0} >> \delta\nu$ on the red side of the line, which shifts rapidly towards lower frequencies at constant speed $\sigma$ and constant amplitude, its logarithmic spectral intensity log $N(\nu,t)$ keeping the form:

$$\log[N(\nu,t)/N_0] = [\log(N_m/N_0)]\exp\{-[(\nu_s-\sigma t)-\nu]^2/2(\overline{\Delta\nu_0})^2\}$$

where $N_0$ is the noise and $N_m$ the maximum amplitude generated by the source at $\nu_s = \nu_0 - \overline{\Delta\nu_D}$, $\nu_0$ being the pumping line frequency. Moreover this photon soliton is the basic element composing any high intensity electromagnetic spectrum in interaction with the plasma electrons via the nonlinear Compton scattering. A narrow spectrum ($\delta\nu \leq \overline{\Delta\nu_D}$) transforms into one photon soliton and a broad spectrum ($\delta\nu > \overline{\Delta\nu_D}$) breaks into a set of such solitons.

## I. INTRODUCTION

It is exciting how a narrow photon spectrum, amplified to a high intensity level in an active physical system like a maser or laser, and interacting nonlinearly with matter (electrons and/or protons) may exhibit such a morphogenetic kinetics. In a preceding paper [1] the author has analysed the kinetics of an unstable plasmon spectrum (in the wave-number k space) interacting nonlinearly (nonlinear Landau effect) with the plasma particles. The nonlinear wave-particle interaction saturates the unstable spectrum generating a satellite near $k = 0$, and a regime of high amplitude relaxation oscillations sets in between the unstable spectrum and the satellite. It is tempting to transpose this mechanism to the nonlinear induced Compton scattering of photons by particles in a population inversion medium. The induced or stimulated Compton scattering has recently been treated in the literature [2]-[8], especially in connection with the study of powerful compact astronomical radio-sources and with laser fusion laboratory experiments. The author has also proposed this mechanism to interpret the strong variable of intensity observed in interstellar maser spectra [9]. The purpose of the present paper is to show that the nonlinear induced Compton kinetics of an unstable photon spectrum (line) not only presents strong amplitude relaxation oscillations like the plasmon kinetics [1], but also generates a Doppler satellite on the red side of the line which then behaves as an electromagnetic soliton in frequency space, moving downward on the frequency axis. This high amplitude soliton has a very simple form and satisfies a linear propagation equation at constant velocity $\sigma$, $\sigma$ being proportional to the maximum amplitude generated by the pumping source. This new soliton, which is different in nature from those encountered in strong electron-plasma-wave turbulence [10]-[12], may be rather important in laser fusion experiments or in the interpretation of quasar, pulsar and interstellar maser spectra.

Moreover we show that this photon soliton is the asymptotic solution satisfying the general nonlinear integrodifferential kinetic equation for the induced Compton scattering (which is of the Boltzmann type [2],[3]) for any initial form of the electromagnetic spectrum. In fact its spectral form shapes the form of the transition probability, which for a Maxwellian electron distribution function reads $K(\nu,\nu') \propto (\nu-\nu')\exp\{-[(\nu-\nu')^2/2(\Delta\nu_D)^2(1-\cos\alpha)]\}$ $\Delta\nu_D$ being the Doppler width and $\alpha$ measuring the scattering angle.

Previous works have dealt with simplified approaches of the problem [2]-[8]. For a broad photon spectrum ($\delta\nu >> \Delta\nu_D\sqrt{1-\cos\alpha}$) the kinetic equation was approximated by the Kompaneets [2] differential equation which solution shows the steepening of the red side of the spectrum while shifting its maximum towards lower frequencies. For a narrow photon spectrum ($\delta\nu << \Delta\nu_D\sqrt{1-\cos\alpha}$) the Fokker-Planck differential approach is not possible and the approximation was made on the transition probability kernel of the integral equation, namely $K(\nu-\nu') \propto (\nu-\nu')$. Within this "local approximation" and starting with particular narrow initial conditions it was shown different red shift evolutions for the spectra with no universal character [5].

Here, considering the general kinetic equation, we show that a narrow electromagnetic spectrum ($\delta\nu \leq \Delta\nu_D\sqrt{1-\cos\alpha}$) in the presence of a noise spectrum - which ensures the non-local coupling of the nonlinear Compton interaction - transforms with time into the photon soliton of universal form whatever be the initial spectral shape; whereas a broad electromagnetic spectrum ($\delta\nu > \Delta\nu_D\sqrt{1-\cos\alpha}$) breaks into a set of such solitons showing that the photon soliton is the basic element composing the high intensity incoherent photon spectrum in interaction with the electron gaz via the Compton collisions.

## II. KINETIC EQUATION

The general kinetic equation for the photon occupation number $n(\nu,\Theta,\phi,t) = c^2 I(\nu,\Theta,\phi,t)/h\nu^3$, where $I(\nu,\Theta,\phi,t)$ is the spectral intensity in a given solid angle $d\Omega = d\phi \sin\Theta \, d\Theta$, which governs the nonlinear Compton interaction $[n(\nu,\Theta,\phi,t) \gg 1]$ of a narrow unstable spectrum of width $\delta\nu \ll \Delta\nu_D\sqrt{1-\cos\alpha}) = [2k_BT_e(1-\cos\alpha)/m_ec^2]^{1/2}\nu$ with free Maxwellian electrons $[\nu_p = (n_e e^2/\pi m_e)^{1/2} \ll \Delta\nu_D]$, has the integro-differential form [5],[6]:

$$\frac{\partial n(\nu,\Theta,\phi,t)}{\partial t} = \gamma(\nu,\Theta,\phi)n(\nu,\Theta,\phi,t) - \frac{3}{16\pi}\sigma_T n_e c n(\nu,\Theta,\phi,t) \times$$

$$\times \int d\nu' K(\nu',\nu,\alpha)n(\nu',\Theta',\phi',t)(1+\cos^2\alpha)\sin\Theta' d\Theta' d\phi' \tag{1}$$

where: $\gamma(\nu,\Theta,\phi)$ is the unsaturated amplification coefficient of the active medium (laser or maser), $\sigma_T = 8\pi/3(e^2/m_ec^2)^2$ the Thomson scattering cross section, $n_e$ and $m_e$ the electron density and mass, $\alpha$ the photon scattering angle defined by

$$\cos\alpha = \cos\Theta \cos\Theta' + \sin\Theta \sin\Theta' \cos(\phi-\phi')$$

and where the transition probability kernel

$$K(\nu',\nu,\alpha) = A(\nu',\nu,\alpha) - A(\nu,\nu',\alpha) \tag{2}$$

describes the frequency shift of the photons in stimulated scattering by free electrons. $A(\nu,\nu',\alpha)$ is the spontaneous transition probability which describes the broadening of the radiation by the Compton collisions of photons on Maxwellian electrons and is given by

$$A(\nu,\nu',\alpha) = \left[\frac{m_e c^2}{4\pi k_B T_e \nu\nu'(1-\cos\alpha)}\right] \exp\left\{-\frac{m_e c^2(\nu-\nu')^2}{4k_B T_e \nu\nu'(1-\cos\alpha)}\right\}$$

$$= \frac{1}{\sqrt{2\pi}\,\Delta\nu_D(1-\cos\alpha)^{1/2}} \exp\left\{-\frac{(\nu-\nu')^2}{2\Delta\nu_D^2(1-\cos\alpha)}\right\} \tag{3}$$

if we neglect the quantum corrections to the Doppler width $\Delta\nu_D = [2k_eT_e/(m_ec^2)\nu\nu']^{1/2}$. The symmetric expression (3) would render Eq. (2) identically zero, thus expanding (2) with respect to the quantum correction to the Doppler shift of the frequency $\Delta\nu_1/\nu' = h\nu'(1-\cos\alpha)/m_ec^2$, and assuming $(\nu\nu')^{1/2} = \nu_0$ in $\Delta\nu_D$ we obtain

$$K(\nu',\nu,\alpha) = \frac{\partial A(\nu',\nu,\alpha)}{\partial\nu} 2\Delta\nu_1 = \frac{2h\nu'^2(\nu-\nu')}{\sqrt{2\pi}\, m_e c^2(\Delta\nu_D)^3(1-\cos\alpha)^{1/2}}$$

$$\exp\left\{-\frac{(\nu-\nu')^2}{2\Delta\nu_D^2(1-\cos\alpha)}\right\} \tag{4}$$

which ensures the conservation of the total number of photons in equation (1) in the absence of the source term ($\gamma = 0$).

For isotropic radiation, the angular integration is easy to perform since $\cos\alpha = \cos\Theta'$. It is also easy when the inversion population medium acts uniformly in two opposite directions inside an active (maser or laser) system of angular aperture $2\Theta_0$. In this case the spectral width satisfies $\delta\nu \ll \Delta\nu_D(1-\cos\alpha)^{1/2} \simeq 2\Delta\nu_D$, denoting the fact that induced Compton interaction is accompanied by the backscattering of the radiation beam, since deflection through an angle $\alpha = \pi$ has maximum probability.

Therefore performing integration over angles [5],[6],[9] we are led to study the kinetics in the frequency space alone. For an unsaturated population inversion line which presents a Doppler profile centered at $\nu = \nu_0$ the kinetic equation is the following:

$$\frac{\partial N(\nu,t)}{\partial t} = N(\nu,t)\left[\gamma_m \exp\left\{-\left(\frac{\nu-\nu_0}{\delta\nu}\right)^2\right\} + W\int d\nu'\left\{\frac{\partial}{\partial\nu} f(\nu-\nu')\right\} N(\nu',t)\right] \tag{5}$$

where the photon occupation number is defined by

$$N(\nu,t) = \frac{\nu^2}{\nu_0^2}\, n(\nu,t) \,, \tag{6}$$

the maximum growth rate by

$$\gamma_m = \frac{\Delta n c^3 B_{12} h\nu}{\sqrt{\pi}\,\delta\nu} = \frac{\Delta n c^3 A_{12}}{8\pi^{3/2}\nu^2\delta\nu} \tag{7}$$

($A_{12}$ and $B_{12}$ being the Einstein coefficients for the two level maser or laser transition and $\Delta n$ the inverted particle density), the transition probability constant by

$$W = \frac{3\Theta_0^2}{\sqrt{2\pi}}\, \frac{hc\sigma_T n_e}{m_e c^2} \left(\frac{\nu_0}{\Delta\nu_D}\right) \nu_0 \tag{8}$$

and the transition probability function describing the finite transfer between $\nu$ and $\nu'$ by

$$\frac{\partial}{\partial\nu} f(\nu-\nu') = - \frac{(\nu-\nu')}{2(\Delta\nu_D)^2} \exp\left\{ -\left(\frac{\nu-\nu'}{2\Delta\nu_D}\right)^2 \right\} \tag{9}$$

## III. CHARACTERISTIC TIMES

Starting from a white electromagnetic noise function $N(\nu,0) = N_0$ the unstable line at $\nu = \nu_0$ is amplified during the linear stage. The spectrum grows exponentially and after a time $\gamma_m t \gg 1$ we can write:

$$N(\nu,t) = N_0 \exp\left\{\gamma_m t \exp\left[-\left(\frac{\nu-\nu_0}{\delta\nu}\right)^2\right]\right\}$$
$$\simeq N_0 \exp\left\{\gamma_m t\left[1 - \left(\frac{\nu-\nu_0}{\delta\nu}\right)^2\right]\right\} = N_0 a \exp(\gamma_m t)\left[\frac{\exp\left\{-\left(\frac{\nu-\nu_0}{a}\right)^2\right\}}{a}\right] \tag{10}$$

where $a^2 = (\delta\nu)^2/\gamma_m t$ tends to zero as $\gamma_m t$ increases and we are allowed to substitute the function in brackets in the last expression (10) by $\sqrt{\pi}\,\delta(\nu-\nu_0)$. Introducing this limit into the integral of equation (5) and performing the integraion over $\nu'$, we obtain

$$\frac{\partial N(\nu,t)}{\partial t} = N(\nu,t)\left[\gamma_m \exp\left\{-\left(\frac{\nu-\nu_0}{\delta\nu}\right)^2\right\} - \frac{W\sqrt{\pi}\,\delta\nu\, N_0 e^{\gamma_m t}(\nu-\nu_0) e^{\left(\frac{\nu-\nu_0}{2\Delta\nu_D}\right)^2}}{2(\Delta\nu_D)^2(\gamma_m t)^{1/2}}\right] \tag{11}$$

Equation (11) shows how the nonlinear growth rate is itself an exponential function of time and dominates the linear one after a time $t = t_{int}$. It has a maximum on the red side of $\nu_0$ at $\nu_s = \nu_0 - \sqrt{2}\,\Delta\nu_D$ and a minimum on the blue side at $\nu_s = \nu_0 + \sqrt{2}\,\Delta\nu_D$. An order of magnitude estimate for time $t_{int}$ can be obtained by comparing the maxima of the linear and nonlinear growth rates in equation (11), thus

$$\gamma_m \simeq W\sqrt{\frac{2\pi}{e}}\left(\frac{\delta\nu}{\Delta\nu_D}\right)\frac{N_0 \exp(\gamma_m t_{int})}{(\gamma_m t_{int})^{1/2}} \tag{12}$$

After time $t_{int}$ a satellite spectrum is generated at $\nu_s$ very rapidly [$\exp\{\exp(\gamma_m t)\}$] and as its amplitude becomes comparable with the line amplitude the $\delta(\nu-\nu_0)$ function approximation for the spectrum is no longer valid. The weight of the spectrum is transferred to the satellite, and the line, which lies on the blue side of it, must relax. Introducing expression (12) into equation (5) we can iterate once more to find the characteristic time $t = t_{sat}$ at which the line begins to relax and the satellite saturates, but a more simple approach can give us a good order of magnitude estimate for $t_{sat}$. Indeed, let us compare the logarithmic growth of the spectrum in (11): after time $t_{int}$ the satellite grows as $\exp\{\gamma_m t\}$ starting with the rate $\gamma_m$ at $t_{int}$, and from nearly the initial white noise. while the line amplitude $N(\nu_0, t_{int})$ continues to grow as $\gamma_m t$. Therefore we can determine $t_{sat} = t_{int} + \Delta t$ by putting $\log N(\nu_s, t_{sat}) = \log\{\delta\nu/(\sqrt{2}\,\Delta\nu_D) N(\nu_0, t_{sat})\}$ yielding

$$\gamma_m \Delta t \simeq \log\left[\gamma_m t_{int} + \frac{1}{2}\log(\gamma_m t_{int}) + \log\left(\frac{\delta\nu}{\sqrt{2}\,\Delta\nu_D}\right)\right] \tag{13}$$

Numerical example: We have carried out computations based on the astrophysical $H_2O$ maser line at $\nu_0 = 22{,}235$ GHz detected in the W49 interstellar source, which presents strong and rapid intensity variations [9]. The dimensionless parameters are: $\delta\nu/\nu_0 = 1{,}6 \cdot 10^{-5}$,

$\Delta\nu_D/\nu_0 = 4 \cdot 10^{-4}$, $N_0 = 10^2$ and $W(\Delta\nu_D/\nu_0)^{-2}\gamma_m^{-1} = 10^{-6}$. The numerical solution of equation (5) is shown in figs. 1-5 in which $\log_{10}N(\nu,s)$ is plotted against $(\nu_0-\nu)/\Delta\nu_D$ for different values of the time parameter $s = \gamma_m t$. Fig. 6 shows $\log_{10}N(\nu_0)$ versus $s = \gamma_m t$. In fig. 1 we describe the nonlinear interaction kinetics by showing the spectra at three characteristic times. The nonlinear interaction starts at time $s_1 = s_{int} = \gamma_m t_{int} = 29{,}04$ when the line intensity attains $\log_{10}N(\nu_0,s_{int}) = 14{,}61$. Then in a short time interval ($\Delta s = 3{,}55$) the line reaches its maximum $\log_{10}N(\nu_0,s_2) = 16{,}14$ and begins to relax, while the satellite intensity, amplified to $\log_{10}N(\nu_s,s_2) = 12{,}78$, still grows until $s = s_3$. As can be seen in fig. 1 the line saturation happens at time $s_2 = s_{sat,line} = 32{,}59$ before the satellite's saturation at time $s_3 = s_{sat,sat} = 32{,}85$. At $s = s_3$ the satellite saturates at the intensity $\log_{10}(\nu_s,s_3) = 14{,}62$ while the line has relaxed to nearly the same level $\log_{10}N(\nu_0,s_3) = 14{,}40$. The interaction time $s_{int} = \gamma_m t_{int}$ and the saturation time $s_{sat} = \gamma_m t_{sat}$ agree to better than 10% with the respective approximate expressions (12) and (13).

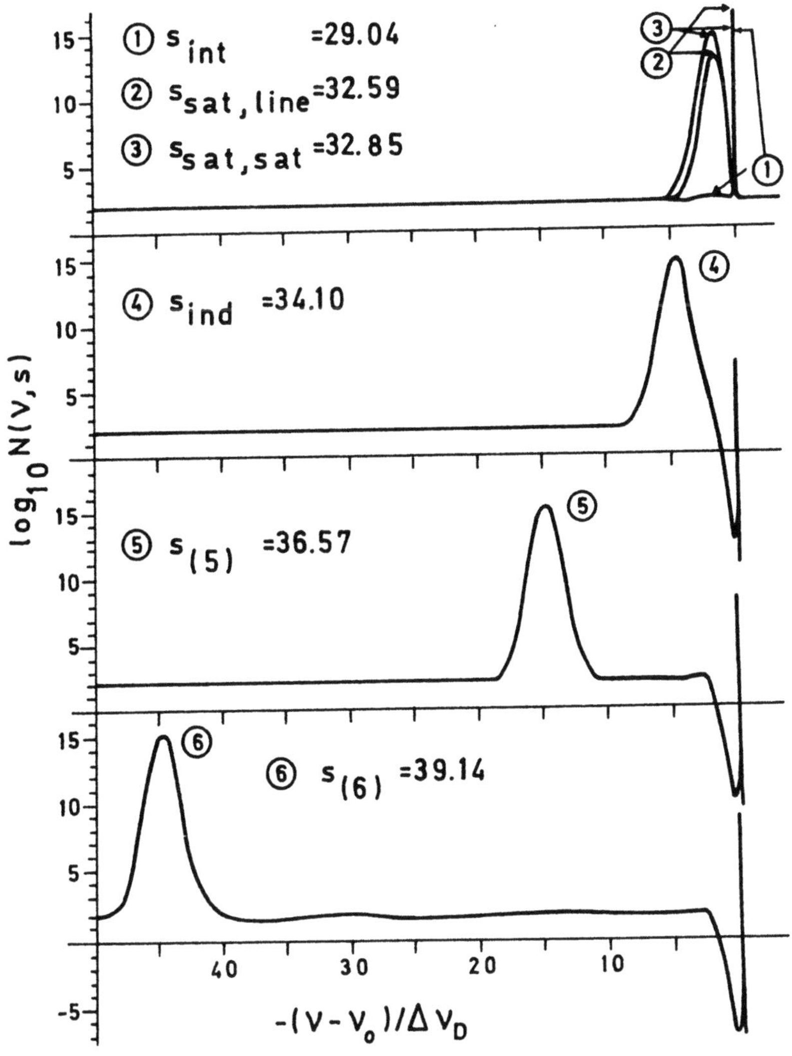


Figs. 1-4: Spectral evolution with time $s = \gamma_m t$ of a narrow maser or laser line. The unstable line saturates when the Doppler satellite generated on the red side, via the induced Compton scattering, dominates the nonlinear kinetics ($s_{sat,line}$). Then the satellite saturates too ($s_{sat,sat}$) and behaves as a photon soliton moving downwards on the frequency axis while the line relaxes and grows again after time $s_{ind}$.

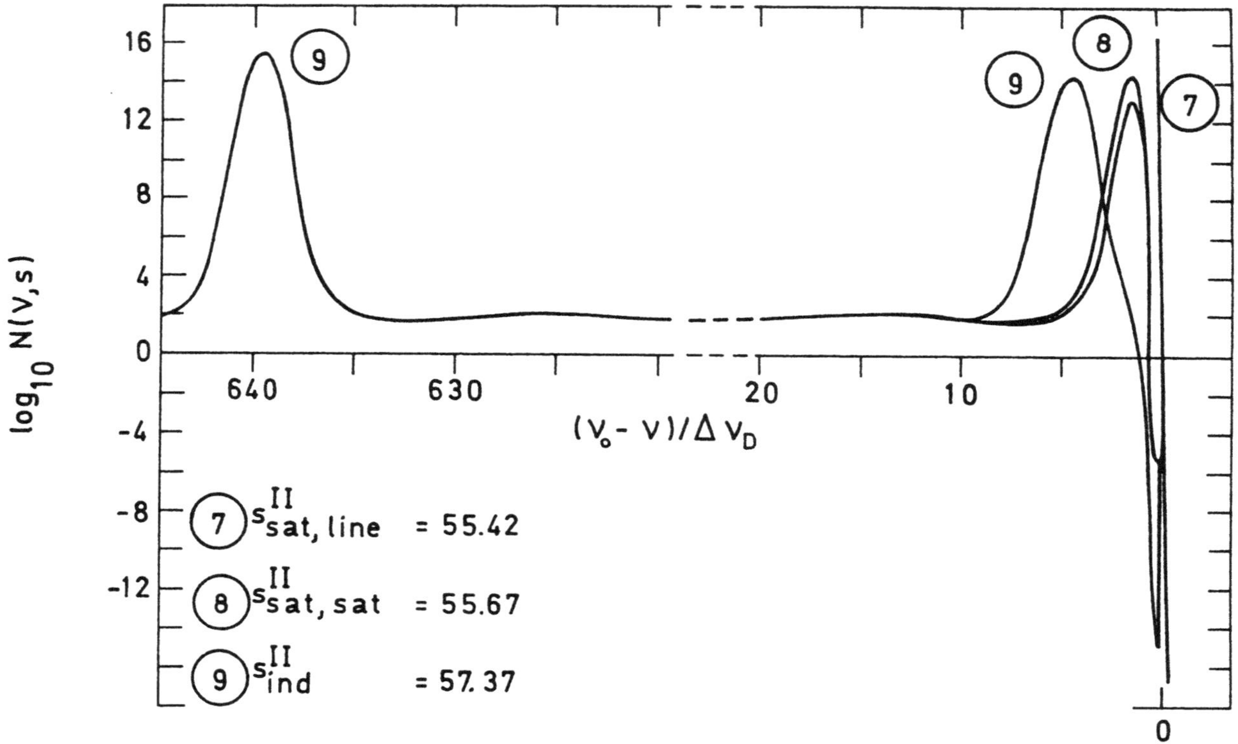


Fig. 5: Second relaxation oscillation of the source line generating a second Doppler satellite which gives rise to the second photon soliton. At time $s^{II}_{ind}$ the first soliton is at a frequency interval $\Delta\nu/\nu_0 = \sigma_s(s_{(9)} - s_{(4)}) \approx 640\Delta\nu_D\,\nu_0$.

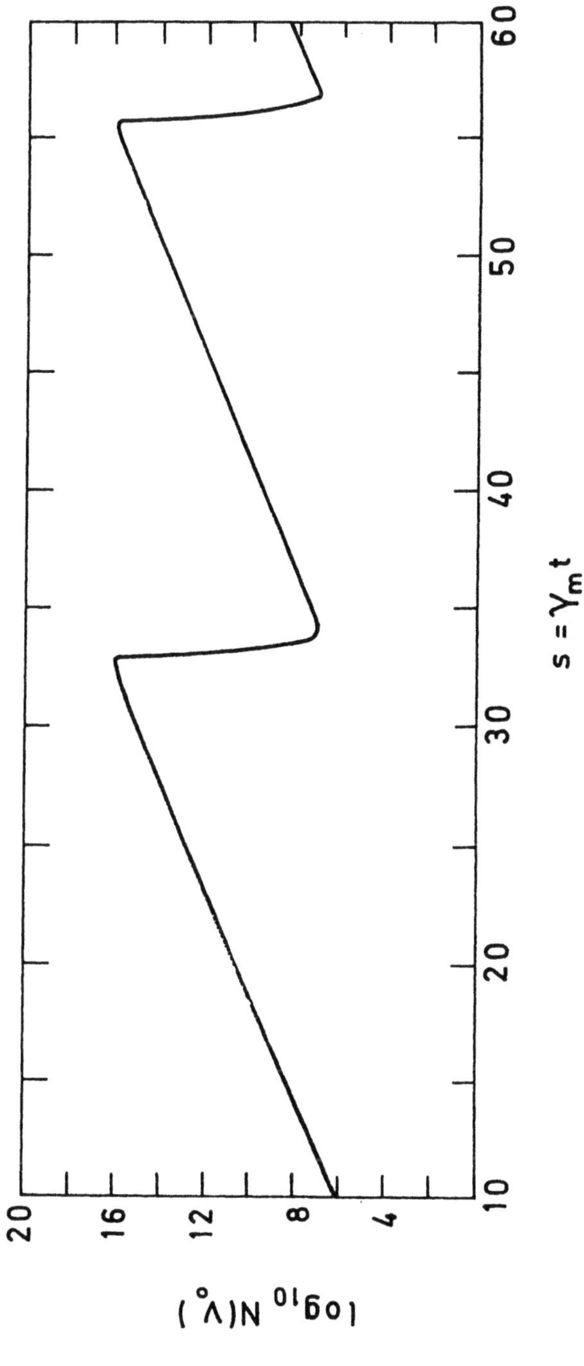


Fig. 6: Relaxation oscillations of the unstable line amplitude. Plot of the maximum intensity $\log_{10}N(\nu_0)$ versus time $s = \gamma_m t$.

## IV. THE PHOTON SOLITON

When the line relaxes and the Doppler satellite saturates, a new stage for the satellite kinetics begins which is mainly governed by the nonlinear induced Compton term in equation (5). The antisymmetric transition probability which enhances the red side of the spectrum while damping the blue, has the effect of transferring the satellite as a whole downward through the frequency spectrum. At a time $s_{ind} = \gamma_m t_{ind}$ the satellite is sufficiently far from the line - at a distance greater than the transition probability width $2\Delta\nu_D$ - so that correlation no longer exists between the line and the satellite, and the unstable line again grows exponentially as shown in fig. 6.

In our numerical example $S_{ind} - S_{sat,line} = 1{,}59$ and $S_{ind} - S_{sat,sat} = 1{,}25$. The numerical treatment shows a transient stage ($S_{ind} < S < S_{sol} \simeq 40$) during which the satellite shifts to the red with a slowly increasing rate and amplitude (figs. 2,3 and 4), after which not only the shift-velocity but also the form and amplitude of the spectrum remain constant, the logarithmic spectral intensity $Z(\nu,t)$ maintaining the form

$$Z(\nu,t) - Z_0 = (Z_m - Z_0)\exp\left\{-\left[\frac{(\nu_s-\sigma t)-\nu}{2\Delta\nu_D}\right]^2\right\} \tag{14}$$

where $Z_0 = \log N_0$ is the noise spectrum that is necessary for the nonlinear evolution, and $Z_m = \log N(\nu_s, s_{sat})$ is the maximum satellite amplitude generated by the pumping line. This extremely simple form allows us to call the satellite a soliton is frequency space or <u>photon soliton</u>. It is interesting to obtain the equation of motion governing (14) directly from the kinetic equation (5) in the absence of the linear source term and in the asymptotic limit $(Z_m - Z_0) \gg 1$.

Let us perform an expansion like (10) in expression (14)

$$\begin{aligned} N(\nu,t) &= N_0\exp(Z - Z_0) = N_0\exp\left\{(Z_m - Z_0)\exp\left[-\left(\frac{\nu_s-\sigma t-\nu}{2\Delta\nu_D}\right)^2\right]\right\} \\ &\simeq N_0\exp(Z_m - Z_0)\exp\left[-\frac{[\nu_m(t)-\nu]^2}{4(\Delta\nu_D)^2/(Z_m-Z_0)}\right] \\ &\to N_0\,\frac{2\sqrt{\pi}\,\Delta\nu_D}{(Z_m-Z_0)^{1/2}}\exp(Z_m - Z_0)\delta[\nu_m(t)-\nu] \end{aligned} \tag{15}$$

where $\nu_m(t) = \nu_s - \sigma t$.

The intense and narrow form of the satellite line - its width is smaller by the factor $(Z_m - Z_0)^{-1/2}$ than the transition probability $2\Delta\nu_D$ - allow us to perform an asymptotic analytical treatment of equation (5). Introducing the limit expression (15) into the integral of equation (5) and integrating over $\nu'$ we obtain

$$\frac{1}{N_0}\frac{\partial N(\nu,t)}{\partial t} = \frac{N(\nu,t)}{N_0}\frac{2\sqrt{\pi}\,\Delta\nu_D W N_0}{(Z_m - Z_0)^{1/2}} \exp(Z_m - Z_0)\frac{\partial}{\partial\nu}\exp\left\{-\left[\frac{\nu_m(t)-\nu}{2\Delta\nu_D}\right]^2\right\} \tag{16}$$

This may be written as an equation of linear uniform motion in frequency space

$$\frac{\partial}{\partial t}[Z(\nu,t)-Z_0] = \sigma\frac{\partial}{\partial\nu}[Z(\nu,t)-Z_0] \tag{17}$$

where the soliton profile $(Z - Z_0)$, or $\log(N/N_0)$, is given by (14) and where the soliton's velocity $\sigma$ is

$$\sigma = 3\sqrt{2}\,\theta_0^2\,\frac{h\,c\,\sigma_T n_e}{m_e c^2}\,\nu_0^2\,\frac{N_{max}}{(Z_m - Z_0)^{3/2}} \tag{18}$$

which differs by only a factor $(Z_m - Z_0)^{-3/2}$ from the red shift speed obtained for a broad photon spectrum [7],[8].

In the numerical application of the astrophysical example we have treated, the explicit expression for $\sigma$, given by (18), agrees to better than 1% after a time s = 39 ($\sigma_s \equiv \sigma/\gamma_m\nu_0 = 1{,}096\times10^{-2}$). We therefore conclude that by this time the satellite's kinetics will have become soliton.

During its downward motion in the frequency axis the logarithmic spectral profile of the photon soliton is that of the electron distribution function of the medium, with the electron momentum replaced by the photon frequency shift due to Compton collisions. In our case the electron distribution function is Maxwellian and gives rise to the solution (14). We note that the kinetics of the soliton is very sensitive to the noise spectrum. Indeed, the soliton propagates uniformly if the noise spectrum has a constant level. Of course, if the noise level is extremely low ($N_0 \ll 1$) the non local Doppler satellite can not build up. Moreover, if absorption occurs for $\nu < \nu_0$ then the satellite-soliton will be damped out before reaching the lowest frequencies.

For the case we are considering in which absorption is neglected a second soliton is generated at a time $s_{int}^{(2)} = 55{,}42$ and acquires an amplitude: $\log_{10} N(\nu_{sat}^{(2)}) = 14{,}52$ which is almost the same as the first soliton's. While the evolution of the second soliton is similar to that of the first its spectrum in the red neighbourhood of the line is more strongly damped (fig. 5). This is not surprising since it is the result of the reduction of the induced damping exerted by the first satellite in this region. Within the frame of the nonlinear kinetics and in the absence of spontaneous source terms the lowered spectrum domain is weakly time dependent and will not exhibit a frequency shift. We can ask ourselves: what will be the asymptotic state in time after the generation of many solitons and whether a stationary state can be reached? But physically this question is not realistic since the photons kinetics described is not self consistent. We have assued a constant Maxwellian distribution function for the electrons and we have ignored their heating. In fact the electron heating rate is simply given by $h\sigma$. The problem of photon condensation [8] or conversion into plasma waves is beyond the scope of this article.

General asymptotic solution: The photon soliton is in fact the general asymptotic solution of the nonlinear integrodifferential equation (5) in the absence of the linear source term. We have perform computations for both narrow ($\delta\nu \ll \Delta\nu_D$) and broad ($\delta\nu \gg \Delta\nu_D$) initial electromagnetic spectra and we are lead to the important result that the photon soliton is the basic element composing any initial high intensity photon spectrum in interaction with the Maxwellian electron plasma via the induced Compton collisions.

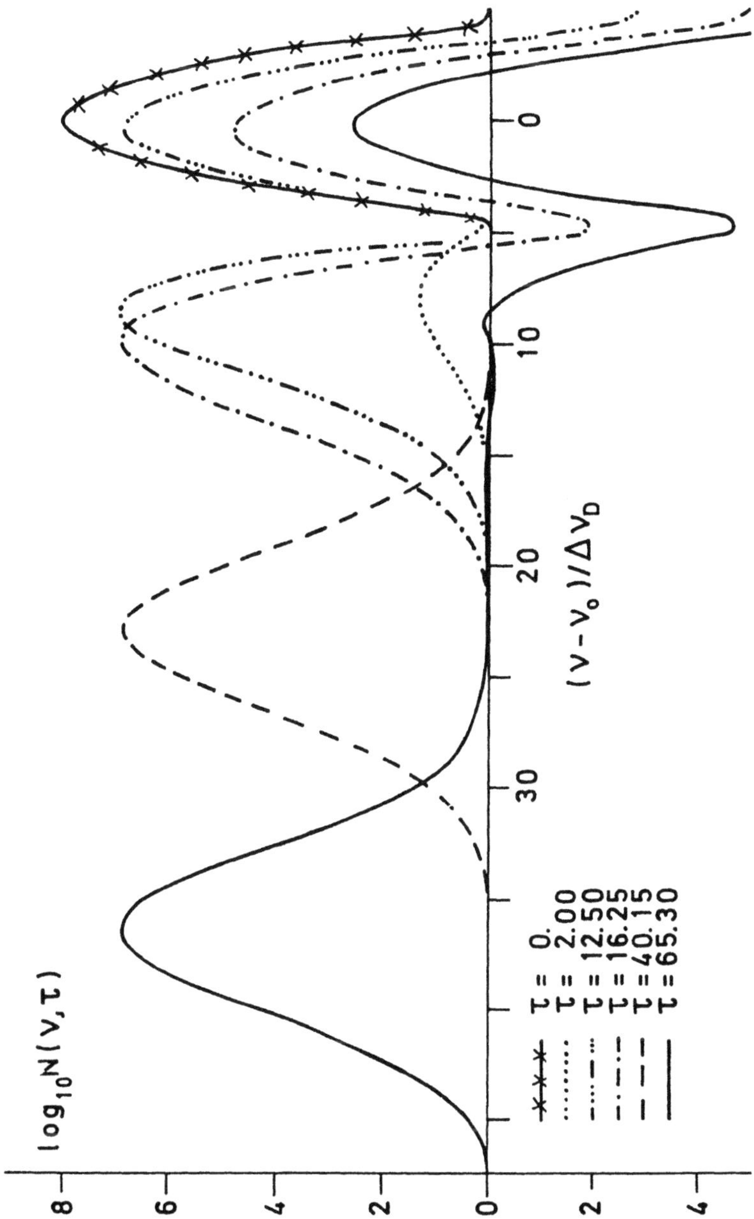


Fig. 7: Induced Compton kinetics of a narrow electromagnetic spectrum ($\delta\nu << \Delta\nu_D\sqrt{1-\cos\alpha}$): The initial spectrum transforms into one photon soliton.

Indeed, a narrow electromagnetic spectrum of Gaussian profile $N(\nu,0) = N_m \exp\{-[(\nu-\nu_0)/\delta\nu]^2\}$ and high intensity (we have taken $N_m = 10^8$ and $\delta\nu/\Delta\nu_D = .04$), in the presence of a low noise spectrum ($N_0 = 1$), transforms with time into the photon soliton (14), as we can see on fig. 7, where the time parameter is

$$\tau = t/[W(\Delta\nu_D/\nu_0)^{-2}]$$

At the first time, its maximum begins to shift towards lower frequencies disymmetrizing the spectrum. Then it converts rapidly with time into the Doppler satellite on the red side by the non-local nonlinear interaction with the noise spectrum. When the main part of the spectral intensity has been transferred to the satellite, the satellite itself moves towards lower frequencies and behaves as the photon soliton (14) with the same asymptotic form. The spectral evolution is therefore similar to that described before for the narrow unstable line after time $t_{int}$. However, it is interesting to remark that the evolution is completely different to that obtained by Zel'dovich et al. [5] for the case of a narrow initial Gaussian profile within the $\nu-\nu'$ approximation for $K(\nu,\nu')$ and in the absence of any noise level other than the Gaussian wings. The amplitude of these are so small at a distance $\sqrt{2}\,\Delta\nu_D \gg \delta\nu$ from the abscissa $\nu_0$ of the maximum, that prevents the non-local coupling between these regions of the spectrum, where the nonlinear transition probability is maximum or minimum, and that region near $\nu_0$ where the spectral intensity is strong (main spectrum). So, they describe a particular situation of local coupling [$\nu-\nu' \sim \delta\nu \ll \Delta\nu_D$ and the exponential function in (9) is replaced by unity] near the main spectrum and obtain that the Gaussian spectrum shifts towards lower frequencies without distortion, the $\nu-\nu'$ approximation being sufficient. This must not be confounded with our photon soliton which universal form (i.e. independent of the initial spectral shape) is due to the satellite construction on the noise and must be described by taking into account the complete transition probability kernel (9). A simple comparison between the nonlinear characteristic growth rate at $\nu = \nu_0 - \sqrt{2}\,\Delta\nu_D$ (satellisation) and at $\nu = \nu_0 - \delta\nu/\sqrt{2}$ (shift of the main spectrum) in equation (11), regardless the linear growth rate, gives as threshold for the soliton formation $N_0 > \exp(-\sqrt{2}\,\Delta\nu_D/\delta\nu)$.

## V. KINETICS OF A BROAD SPECTRUM

The induced Compton scattering of a broad electromagnetic spectrum ($\delta\nu \gg \Delta\nu_D$) has been described up to now by the Kompaneets differential equation [2]-[7]

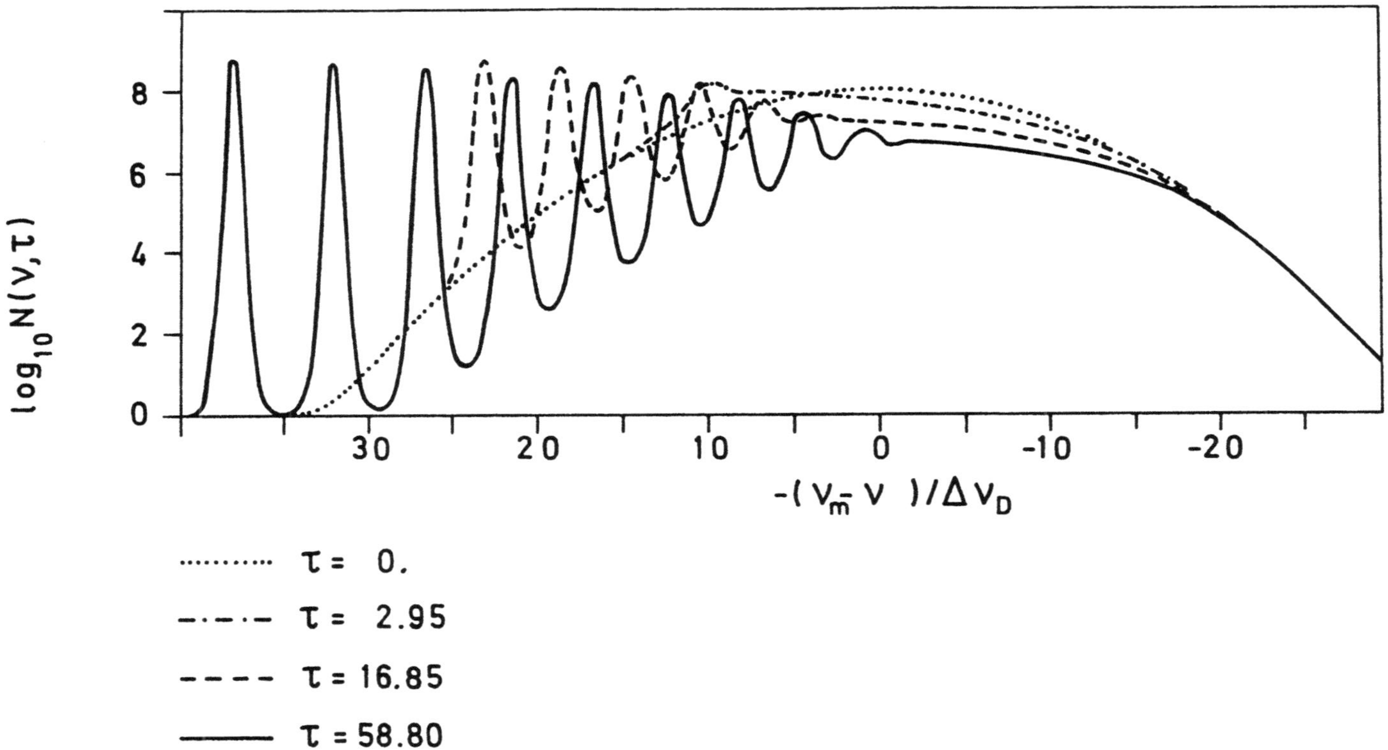


Fig. 8: Induced Compton kinetics of a broad electromagnetic spectrum ($\delta\nu \gg \Delta\nu_D\sqrt{1-\cos\alpha}$): The initial spectrum breaks into a set of photon solitons during its evolution with time. Plot of $\log_{10}N(\nu,\tau)$ versus $(\nu_m-\nu)/\Delta\nu_D$ for different values of the time parameter $\tau = t/[W(\Delta\nu_D/\nu_O)^2]$.

$$\frac{\partial n(\nu,t)}{\partial t} = \frac{\sigma_T n_e h}{m_e c} \frac{1}{\nu^2} \frac{\partial}{\partial \nu} \nu^4 \left\{ n(\nu,t)[n(\nu,t)+1] + \frac{kT_e}{h} \frac{\partial n(\nu,t)}{\partial \nu} \right\} \tag{19}$$

i.e. within an "hydrodynamical" approximation by a Fokker Planck type equation of the general nonlinear integrodifferential kinetic euqation, which in the absence of any linear term ($N(\nu,t) = = \nu^2/(\nu_0)n(\nu,t) >> 1$ reads

$$\frac{\partial N(\nu,t)}{\partial t} = N(\nu,t)W \int d\nu' \left\{ \frac{\partial}{\partial \nu} f(\nu-\nu') \right\} N(\nu',t) \tag{20}$$

with the above defined quantities.

For $n(\nu,t) >> 1$ the nonlinear term in eq. (19) tends to distort the spectrum by shifting its maximum towards lower frequencies with a speed proportional to $n(\nu,t)$, steepening its red side. The presence of the linear diffusion term in (19) prevents the solution to be multivalued, Reinisch [7] has studied in detail the triangular evolution which follows and which is described by a Burger's type equation. For the strong nonlinear case without diffusion, Peyraud [3] has made a previous attempt in order to prevent the multivalued evolution by adding, heuristically, a dispersive term of the Korteweg de Vries type $\propto \partial^3 n(\nu,t)/\partial\nu^3$ to the nonlinear Kompaneets equation (19). He explains so qualitatively a high intensity laser experiment [13] which shows a breaking of the initial profile into secondary ones or "solitons" following the KdV terminology.

In the light of the general result obtained for a narrow spectrum, we have computed directly the integrodifferential equation (20) for a broad initial electromagnetic spectrum with a Gaussian profile

$$N(\nu,0) = N_m \exp\left\{ -\left[\frac{\nu-\nu_m}{\delta\nu}\right]^2 \right\}$$

We have taken the following values of the parameters

$$N_m = 10^8 \;;\quad N_0 = 1 \;;\quad \frac{\delta\nu}{\nu_0} = 2\times10^{-4} \;;\quad \frac{2\Delta\nu_D}{\nu_0} = 1{,}5\times10^{-3}$$

The spectral evolution of $\log_{10}N(\nu,\tau)$ is plotted on figure 8 for different values of the characteristic time $\tau = t/[W(\Delta\nu_D/\nu_0)^2]$. We can see, at the first time, the whole distortion of the spectrum and the shift of its maximum towards lower frequencies. Then, the spectrum sinks behind the maximum generating another maximum. In the following steps the spectrum is completely breaked into secondary ones which move towards lower frequencies approaching more and more the photon soliton form (14) when they leave the initial spectrum. The number of photon solitons generated is proportional to the total number of photons of the initial spectrum as is common to other types of nonlinear differential equations [10]. In their motion downward on the frequency axis they are ordered by their amplitudes, or what is the same by their speeds, therefore increasing the distances between them. This spectral evolution of the general kinetic equation may explain satisfactorily the laser experiment [13] without adding any heuristic term to the approximate Kompaneets equation.

## VI. CONCLUSION

The novel contribution of the present paper has been to describe the induced Compton kinetics governed by the general nonlinear integrodifferential equation for any kind of initial conditions, either narrow or broad, and to obtain the general asymptotic form of the solution which is one photon soliton for a narrow ($\delta\nu < \Delta\nu_D$) initial spectrum or a set of such solitons for a broad ($\delta\nu > \Delta\nu_D$ one. In the case of a broad spectrum and even in the absence of linear diffusion terms the computation of the exact nonlinear kinetic equation doesn't lead to multivalued solutions. It has the advantage of showing the fine structure of the high intensity electromagnetic spectrum in interaction with the electrons via the Compton collisions. We expect, the last numerical results show it, that for the case when a nearly strong diffusion takes place the finite structure is smeared out and the evolution is of the triangular type.

## VII. ACKNOWLEDGMENTS

The author is indebted to Dr. G. Reinisch and Dr. M. Henon for basic suggestions. He also thanks Dr. J. Peyraud for stimulating discussions.

## REFERENCES

1. Montes, C., J. Plasma Phys., 11, (1), 141 (1974).
2. Kompaneets, A.S., Zh. Eksp. Teor. Fiz., 31, 876 (1956) [Sov. Phys. JETP 4, 730 (1957)]
3. Peyraud, J., J. de Phys. (Paris), 29, 88, 806, 872 (1968)

4. Zel'dovich, Ya.B. and E.V. Levich, Zh. Eksp. Teor. Fiz., 55, 2423 (1968) [Sov. Phys. JETP 28, 1287 (1969)].

5. Zel'dovich, Ya.B., E.V. Levich and R.A. Syunyaev, Zh. Eksp. Teor. Fiz., 62, 1392 (1972) [Sov. Phys. JETP 35, 733 (1972)]

6. Galeev, A.A. and R.A. Syunyaev, Zh. Eksp. Teor. Fiz., 63, 1266 (1972) [Sov. Phys. JETP, 36, 669 (1973)]

7. Reinisch, R., Physica 81C, 339 (1976).

8. Coste, J. and J. Peyraud, Phys. Rev. A, 12, 2144 (1975).

9. Montes, C., Ap. J. (to be published).

10. Zakharov, V.E., Zh. Eksp. Teor. Fiz., 62, 1475 (1972) [Sov. Phys. JETP, 35, 908 (1972)]

11. Kingsep, A.S., L.I. Rudakov and R.N. Sudan, Phys. Rev. Lett., 31, 1482 (1972)

12. Valeo, E.J. and W.L. Kruer, Phys. Rev. Lett., 33, 750 (1974).

13. Decroisette, M., J. Peyraud and G. Piar, Phys. Rev. A 5, 1391 (1972).

* Laboratoire Associé au C.N.R.S. n° 190

# UTILIZATION OF THE FORMALISM OF LAGRANGIAN VARIABLES FOR THE INVESTIGATION OF SOME NONLINEAR PLASMA PHYSICS PROBLEMS

L. M. Kovriznych

P. N. Lebedev Physics Institute
of the USSR Academy of Sciences
Moscow, USSR

## I. Introduction

A number of plasma physics phenomena may be sufficiently well described in the frame of a two-fluid hydrodynamic model. In particular for the collisionless plasma such an approximation can give a quite satisfactory description of the structure and character of propagation of waves when their energy exceeds the particle thermal energy density (cold plasma approximation), or in the case of a nonisothermal plasma ($T_e \quad T_i$), if we are interested in the phenomena which are not affected essentially by the fine structure of the particle distribution function.

On the other hand, it is well knwon that hydrodynamic equations may be written in two equivalent forms: in Eulerian variables (the usual form of hydrodynamic equations) when density and velocity are functions of time t and of the spatial coordinate x, and in Lagrangian variables when a fluid element displacement is an unknown function of time t and of the initial position $x_0$.

The utilization of Lagrangian variables has in our opinion a number of advantages especially in applications to one-dimensional nonlinear problems in inhomogeneous plasmas. This is due to the fact that Lagrangian variables permit us to reduce two unknown functions to one and in some cases to reduce the nonlinear equations in partial derivatives to ordinary equations.

A shortcoming of the Lagrangian approach accounting

for its comparatively rare utilization (especially for the analytical treatment of problems) is the difficulty of the conversion from Lagrangian variables t, $x_0$ to the ordinary variables t, x. However, with the present broad utilization of computers this difficulty happens to be not so essential and the advantages of Lagrangian variables become more important than their shortcomings, especially in solving problems numerically.

In this paper, which is more or less a review, some nonlinear problems of plasma physics are considered using the one-dimensional equations of motion in Lagrangian variables.

## 2. Equations of Motion in Lagrangian Variables

Consider the simplest one-dimensional model of collisionless two-fluid hydrodynamics with constant electron temperature and zero ion temperature:

$$\begin{aligned} &\dot{v}_j + v_j v_j' = \frac{e_j}{m_j}(E+E_0) - v_{Tj}^2 \frac{n_j'}{n_j}; \quad v_{Tj}^2 = T_j/m_j \\ &\dot{n}_j + (n_j v_j)' = 0; \quad E' = 4\pi \sum_j n_j e_j \end{aligned} \tag{1}$$

Here $n_j$, $v_j$, $e_j$, $m_j$, $T_j$ are the density, velocity, charge, mass and temperature of the electrons ( $j = e$ ) and ions ( $j = i$ ), respectively, and $E_0$, $E$ the external and plasma fields, which will be considered as potential ones.

For conversion to Lagrangian variables we introduce as usual functions $\sigma_j(x_0,t)$ as displacements of a fluid element from its initial position $x_0$ at the time moment $t$.

Densities and velocities are related to the function $\sigma_j$ as follows:

$$n_j(x,t)\Big/_{x=x_0+\sigma_j} = \frac{n_0(x_0)}{1+\sigma_j'}; \quad v_j(x,t)\Big/_{x=x_0+\sigma_j} = \dot{\sigma}_j \tag{2}$$

where the prime is the derivative with respect to the initial coordinate $x_0$, the dot means the time derivative ( at $x_0 = const$) and $n_0(x_0)$ is the initial density value, which is assumed to be the same for electrons and ions.

After some trivial transformations we obtain from the system (I) the following system of equations for the function[I] $\sigma_j$:

$$\ddot{\sigma}_e = \Gamma_e - \Gamma_o(x_o+\sigma_e,t) - v_T^2 \frac{(n_o N_e)'}{n_o}; \quad N_e = \frac{1}{1+\sigma_e'} \tag{3}$$

$$\ddot{\sigma}_i = \delta\left[\Gamma_i + \Gamma_o(x_o+\sigma_i,t)\right] \tag{4}$$

$$\Gamma_j(x_o,t) = -\int_{x_o}^{x_o+\lambda_j} \omega_o^2(x)\,dx \quad ; \quad (T_e = T,\ T_i = 0) \tag{5}$$

where $\delta = m_e/m_i$, $\omega_o^2(x) = 4\pi e^2 n_o(x)/m_e$, $\Gamma_o(x,t) = \frac{|e_e| E_o(x,t)}{m_e}$ and the functions $\lambda_j(x_o,t)$ are to be defined from the following relations:

$$\sigma_e(x_o,t) = \lambda_e + \sigma_i(x_o+\lambda_e,t) \tag{6}$$

$$\sigma_i(x_o,t) = \lambda_i + \sigma_e(x_o+\lambda_i,t) \tag{7}$$

Note that the electric field in the Eulerian variables is related to the functions $\sigma_j$ in the following way:

$$\Gamma_j(x_o,t) = \frac{e_j}{m_e} E(x,t)\Big/_{x = x_o+\sigma_j} \tag{8}$$

So far we have used no approximations and the system (3)-(7) is as correct as the system (I).

However for problems where the quasineutrality condition is kept well enough, i.e. the electron displacement $\xi = \sigma_e - \sigma_i$ relative to ions is small as compared with the characteristic length of inhomogeneities, one may simplify the relations (6)-(7) and obtain expressions of the power series

$$\lambda_e = \sum_{n=1}^{\infty} \frac{\xi^n}{n!}\left(N_i \frac{\partial}{\partial x_o}\right)^{n-1} N_i \quad ; \quad N_i = \frac{1}{1+\sigma_i'} \tag{9}$$

$$\lambda_i = \sum_{n=1}^{\infty} \frac{(-\xi)^n}{n!}\left(N_e \frac{\partial}{\partial x_o}\right)^{n-1} N_e \quad ; \quad N_e = \frac{1}{1+\sigma_e'} = \frac{N_i}{1+\xi' N_i} \tag{10}$$

---

[I] Note that to take into account rare electron collisions one has to replace $\ddot{\sigma}_j$ in (3),(4) by $\ddot{\sigma}_j + \frac{m_e}{m_j}\nu_o \dot{\lambda}_j$, where $\nu_o$ is the electron ion collision frequency for the initial density.

where $\zeta = \sigma_e(x_0 t) - \sigma_i(x_0, t)$

Note that it follows from (5),(9),(10) that the functions $\Gamma_j$ are quite simply related to the density values $n_j$ in Eulerian variables:

$$\Gamma_e = -\frac{4\pi e^2}{m_e}\int_{x_0+\sigma_i}^{x_0+\sigma_e} n_i(x,t)dx \ ; \ \Gamma_i = -\frac{4\pi e^2}{m_e}\int_{x_0+\sigma_e}^{x_0+\sigma_i} n_e(x,t)dx \qquad (11)$$

Substracting (4) from (3) and neglecting small corrections of the order of $\delta = \frac{m_e}{m_i} \ll 1$ we obtain

$$\ddot{\zeta} = \Gamma_e - \Gamma_0(x_0+\sigma_i+\zeta, t) - \frac{v_T^2}{n_0}\left(\frac{n_0 N_i}{1+\zeta' N_i}\right)' ; \quad \ddot{\sigma}_i = \delta \Gamma_i \qquad (12)$$

For sufficiently small $\zeta$ when one can omit all but the two first terms of the expansion we have [2]

$$\lambda_e = N_i \zeta + \frac{\zeta^2}{2} N_i N_i' \ ; \ \lambda_i = -\lambda_e + \frac{1}{2}(\zeta^2 N_i^2)' \qquad (13)$$

Note also that in the homogeneous initial distribution case

$$\Gamma_j = -\omega_0^2(x_0)\lambda_j \qquad (14)$$

## 3. Nonlinear Potential Waves in Homogeneous Plasma

Let us suppose that the initial plasma density is homogeneous, the typical spatial scales are sufficiently great to use the expressions (9),(10) for the values $\lambda_j$, and an external field oscillates in time with frequency $\omega$ of the order of the electron plasma frequency $\omega_0$. Then due to the smallness of the ratio $\delta = m_e/m_i$

---

2) It is interesting to note that the expansion (9) does not contain $\zeta'$. That permits us to treat problems with strong electron nonlinearity in the case of fixed ion motion (or when $\delta$ can be regarded to be equal to 0)

the motion may be presented as a superposition of fast oscillations with frequency of the order of $\omega$ and of slow variations with a time scaling determined by ion inertia.

Consequently assuming $\xi = \tilde{\xi} + \bar{\xi}$, where $\tilde{\xi}$ is a fast and $\bar{\xi}$ a slow function of time (the bar means the averaging over the "fast" time)[3] and keeping only linear terms in the equation for $\tilde{\xi}$

$$\ddot{\tilde{\xi}} = -\omega_0^2 N_i \tilde{\xi} + v_T^2 (\tilde{\xi}' N_i^2)' - \Gamma_0 (x_0 + \sigma_i + \bar{\xi}, t) \tag{15}$$

$$\ddot{\sigma}_i = -\delta \left\{ (\overline{\Gamma_0' \tilde{\xi}}) + v_T^2 N_i' + \frac{\omega_0^2}{2} (\overline{\tilde{\xi}^2} N_i^2)' \right\} \tag{16}$$

$$\bar{\xi} = -\frac{\overline{\tilde{\xi}^2} N_i'}{2} - \frac{1}{\omega_0^2 N_i} \left\{ (\overline{\Gamma_0' \tilde{\xi}}) + v_T^2 N_i' \right\}$$
$$\Gamma_0' = \frac{\partial \Gamma_0 (x,t)}{\partial x} \Big|_{x = x_0 + \sigma_i + \bar{\xi}} \tag{17}$$

These equations will be the subject of our following consideration.

a) <u>Modulation instability of the Langmuir wave of finite amplitude</u>

Suppose the external field is absent ($\Gamma_0 = 0$) and in the equilibrium state $N_i = 1$, and $\xi = \xi_0 \exp i[\omega_k t - k x_0]$ where $\xi_0 = const$, $\omega_k^2 = \omega_0^2 + k^2 v_T^2$. Assuming as usual that the density perturbation is $n = N_i - 1 = A \exp i[\Omega t - q x_0] + c.c.$, we obtain for low-frequency oscillations the following dispersion equation

$$\frac{\Omega^2}{q^2 v_s^2} = 1 + W \left\{ 1 + \frac{2\omega_0 \omega_k^* - \Omega^2}{4[\Omega - \omega_{k+q}^* + \omega_k^*][\Omega + \omega_{k-q}^* - \omega_k^*]} \right\} \tag{18}$$

where $v_s^2 = \delta v_T^2$, $\omega_k^* = k^2 v_T^2 / 2\omega_0$ ; $W = \omega_0^2 |\xi_0|^2 / 2 v_T^2 = \frac{E^2}{8\pi n_0 T}$

The modulation instability has recently been studied in detail and apparently the most complete investigation is given in [I] . However the above-mentioned dispersion equation is distinguished from that obtained

---

3) This is valid for $k\tilde{\xi} \ll 1$ and $k r_d \ll 1$ where $k$ is a typical wave number, and $r_d = v_T/\omega_0$ is the Debye electron radius.

in [I] by two additional terms (by unity in the brackets and by the second term $-\Omega^2$ in the numerator), which become essential at values of $\delta W > 1$ leading to the instability increment saturation with increasing $W$. The analysis of equation (18) is straightforward and here we shall restrict ourselves only to a brief resumé. Depending on the relation between $q r_d$ and $W$ the maximum increment $\gamma$ behaves as follows[4)] (for $(\kappa r_d)^2 > \delta$):

$$
\begin{aligned}
&\text{for } W < 4\delta^{1/2} q r_d && \gamma^2 = \delta^{1/2}\omega_0 \frac{W q r_d}{4} \\
&\text{for } 4\delta^{3/2} q r_d < \delta W < 4(q r_d)^4; && \gamma^3 = \omega_0^3 \frac{\delta W (q r_d)^2}{4} \\
&\text{for } 4(q r_d)^4 < \delta W < 4/g && \gamma^2 = \omega_0^2 \frac{\sqrt{\delta W}(q r_d)^2}{2} \\
&\text{for } 4/g < \delta W && \gamma^2 = \omega_0^2 \frac{(q r_d)^2}{3}
\end{aligned}
\tag{19}
$$

In particular, it follows from (19) that the increment increases with growing $W$, its maximum value being limited by the value $\gamma_m = \omega_0 q r_d/\sqrt{3}$

b) Nonlinear electroacoustic waves

For $v_T = 0$ equation (18) gives:

$$\Omega^2 = \frac{3}{8} V_0^2 q^2 \quad ; \quad V_0^2 = \delta \omega_0^2 \xi_0^2 \tag{20}$$

It follows from this that in a cold plasma in the presence of Langmuir oscillations a propagation of a low-frequency electroacoustic wave is possible. Equations (15)-(17) allow us to trace the nonlinear evolution of these waves. Supposing $V_T = 0$, $\Gamma_0 = 0$ and taking into account that $N_i$ is a slow function of time, we find

$$
\begin{aligned}
&\tilde{\xi} = \frac{\xi_0}{N_i^{1/4}} \exp i\left\{\omega_0 \int^t N_i^{1/2} dt - \kappa x_0\right\} \\
&\ddot{\sigma}_i = -\frac{1}{4}\left(V_0^2 N_i^{3/2}\right)' \; ; \quad V_0^2 = \delta\omega_0^2 |\xi_0|^2
\end{aligned}
\tag{21}
$$

---

4) For $(q r_d)^2 < \delta$ instability is possible only for $W > (q r_d)^2$

Bearing in mind that $\sigma_i' = N_i^{-1} - 1$ the last equation may be rewritten in the form

$$\ddot{(N_i^{-1})} + \frac{1}{4}\left(V_0^2 N_i^{3/2}\right)'' = 0 \qquad (22)$$

However, it is easy to see that in the case of homogeneous Langmuir oscillations when $V_0^2 = const$ one may rewrite equation (22) in Eulerian variables in the form

$$\dot{v}_i + v_i v_i' = -\frac{p'}{n_i}\,;\quad \dot{n}_i + (n_i v_i)' = 0\,;\quad p = \frac{V_0^2}{4} n_i^{3/2} \qquad (23)$$

These equations are known to have a solution in the form of a simple Riemann wave whose front becomes steeper in due time up to the moment $t = t^*$ when $\partial v_i/\partial x$ diverges ("wave breaking"). Actually, it is not difficult to make sure that the solution (28) representing the travelling wave is[5]

$$n_i = \left[1 \pm \frac{1}{\sqrt{6}}\frac{v_i}{V_0}\right]^4;\quad v_i = f\left(x - \left[\frac{5}{4}v_i \pm \sqrt{\frac{3}{8}}V_0\right]t\right) \qquad (24)$$

where $f(x)$ is an arbitrary function.

Denoting the maximum velocity by $v_m$ and typical wave number by $q = f'/f$, we find from (24) the "breaking time" $t^* \approx 1/qv_m = 4/5f'$

Thus, although the expansions used to derive (21) fail directly before the "breaking" one may suppose that the arbitrary initial perturbation should lead to the kind of a shock wave profile whose width is of the order of the oscillation amplitude $\xi_0$ (the only length-dimension parameter).[6]

---

5) Analogous results by means of rather complex expansions of the equations in Eulerian variables were obtained in [2].

6) It is worthy of note that since at the moment of breaking the density remains finite it is not associated with the selfcrossing of ion trajectories and can not lead to multistream flux formation.

On the other hand, due to the dependence of the high-frequency oscillations on the coordinate even for $\kappa\xi_0 < 1$ the self-crossing of electron trajectories and multistream flux formation may occur if the ion density modulation is deep enough. Then equations (15)-(17) and consequently (21) certainly fail. Taking into account that $N_i \simeq n_i \simeq 1 + v/V_0$ and $\partial v/\partial x \simeq \partial v/\partial t \cdot u_0^{-1}$ it is easy to show that the linearization with respect to $\xi$ is valid if the amplitude of the ion density oscillations satisfies the condition

$$|N_i - 1| \ll \sqrt{\delta} \quad \text{or} \quad \frac{v_m}{V_0} \ll \sqrt{\delta} \tag{25}$$

c) Solitary waves in the presence of an external electric field

Let us suppose now that the external electric field has the form: $\Gamma_0(x,t) = \omega_0 V_E \exp i[\omega t - \kappa x]$, where $V_E$ is a real constant and let us look for a solution of the system (15)-(17) where the amplitude and the density have the form of solitary waves.

Assuming that $\xi = a_0 Z(\zeta) \exp i[\omega t - \kappa x_0]$, where $a_0$ is a constant equal to the amplitude at infinity, and $Z(\zeta)$ is a real function of $\zeta = x_0 - Vt$, with $Z(\infty) = 1$, that the ion density perturbation is $n = N_i - 1 = n(\zeta) \ll 1$, with $n(\infty) = 0$, and linearizing equations (15)-(16) with respect to $n$, we find the following equation for $Z(\zeta)$:

$$r_d^2 Z'' + (\Delta - n) Z = \Delta$$
$$n = \frac{\alpha(1 - Z^2)}{2(1 + \alpha Z^2)} \; ; \quad a_0 = V_E/\omega_0 \Delta \tag{26}$$

where $\alpha = \delta V_E^2 / 2\Delta^2 (V_s^2 - V^2)$, $\Delta = (\omega^2 - \kappa^2 v_T^2)/\omega_0^2 - 1$, $V = \frac{\kappa v_T^2}{\omega}$ and the prime designates the derivative with respect to $\zeta$.

In deriving this equation we have supposed that $\Gamma_0(x_0 + \sigma_i + \xi, t) \simeq \Gamma_0(x_0, t)$. It may be shown that that is valid only for $\alpha, \Delta \ll 1$ or at $\kappa = 0$. Otherwise, it is necessary to take into account the slow change of the high-frequency wave phase, i.e. to consider $Z(\zeta)$ as a complex value. So in what follows we limit ourselves to small detuning values $\Delta \ll 1$ and to sufficiently weak external fields $\delta V_E^2 \ll 2\Delta^2 / V_s^2 - V^2$[7]

[7] The case of arbitrary $\Delta$ and $\alpha$ at $\kappa = 0$, which corresponds to solitons at rest, may be easily considered on the basis of equation (26), with the soliton-type solutions existing near the frequencies satisfying $\Delta \simeq V_E^2 / 2\Delta^2 V_T^2 + V_E^2$

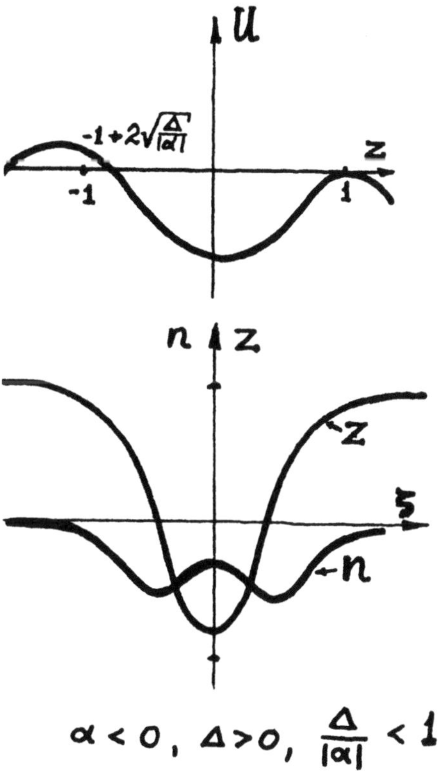


Fig. 1a

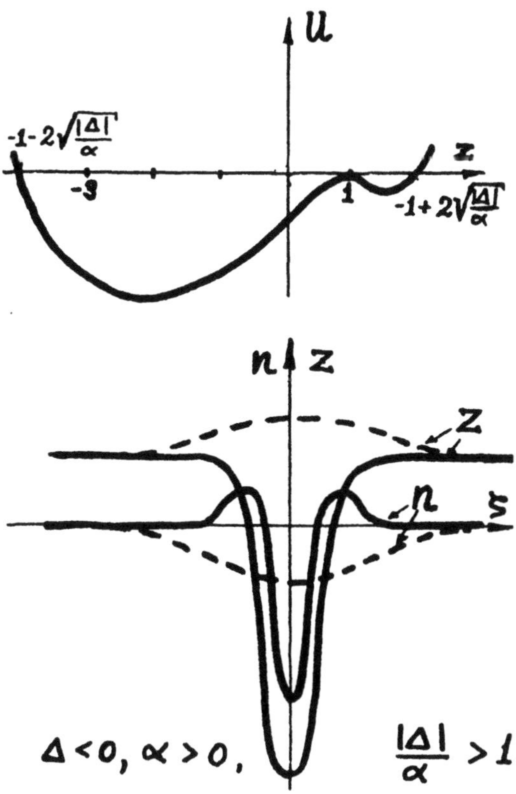


Fig. 1b

In this case it is convenient to represent equation (26) in the form

$$r_d^2 Z'' = -\frac{\partial U}{\partial Z}; \quad U(Z) = \frac{(1-Z)^2}{2}\left[\Delta + \frac{\alpha}{4}(1+Z)^2\right] \quad (27)$$

From the expression for the potential $U(Z)$ it is easy to see that soliton-type solutions are possible only for $\alpha+\Delta \leq 0$ and $\alpha \cdot \Delta \leq 0$.

Fig. 1a presents the potential form $U(Z)$, the dimensionless amplitude $Z(\zeta)$, and the density $n(\zeta)$ as a function of the coordinate $\zeta$ for the supersonic soliton case with $V^2 > V_s^2$ (Fig.1a) and the subsonic soliton case $V^2 < V_s^2$ (Fig.1b). When the external field is $\Gamma_0 = 0$ only supersonic solitons exist[8]. In this case the oscillation frequency is determined from the condition $\Delta = 0$ , the amplitude $a_0$ and wave number $K$ are arbitrary, and the solutions have the form

$$Z = \pm th\, q\zeta \; ; \; n = \frac{\alpha}{2ch^2 q\zeta} \; ; \; q^2 r_d^2 = -\frac{\alpha}{4} \; ; \; \alpha = \frac{8\omega_0^2 a_0^2}{2(V_s^2 - V^2)} \quad (28)$$

For $\Delta > 0$ and $\alpha < -\Delta$ the amplitude $Z$ varies within the limits $1 - 2(\Delta/|\alpha|)^{1/2} \leq Z \leq 1$, and the density oscillation amplitude $n_m$ and characteristic wave number $q$ defining the soliton width go to zero at $\Delta \to |\alpha|$ (Fig.1a).

For negative values $\Delta$ and subsonic velocities $V^2 < V_s^2$ (Fig.1b) two solutions are possible. For one of them the oscillation amplitude always exceeds unity and lies in the interval $[1, 2\sqrt{|\Delta|/\alpha} - 1]$ , with $n_m$ and $q$ as in the previous case tending to zero at $|\Delta| \to \alpha$ (the dashed curves in Fig.1b). The second solution (the solid curves in Fig.1b) does not reduce to the trivial one at $|\Delta| \to \alpha$ . The amplitude always passes through zero and changes from $-1-2\sqrt{|\Delta|/\alpha}$ to $1$ .

For sufficiently weak external fields when $\alpha \ll |\Delta|$, it is easy to obtain an analytical expression for both solutions. It has the form

$$Z = 1 \pm \frac{A}{ch\, q\zeta} \; ; \; q^2 r_d^2 = -\Delta \; ; \; A = 2\sqrt{\frac{-\Delta}{\alpha}} \; ; \; n = \frac{\alpha}{2}(1 - Z^2) \quad (29)$$

It is interesting to note that in this limiting case, both the ion density amplitude $n_m$ in the centre of the soliton and the maximum electron oscillation amplitude $\xi_m$ do not depend on the external electric field and are equal to

$$n_m = 2\Delta \; ; \; \xi_m = r_d \left[ -\left(1 - \frac{V^2}{V_s^2}\right) 2\Delta \right]^{1/2} \quad (30)$$

---

8) To avoid misunderstanding note that we consider solutions for which the high-frequency oscillation amplitude does not tend to zero as $\zeta \to \infty$ . Otherwise, as is known, only subsonic solitons exist [3].

**By analogy to the transverse wave case [4] one may suppose that by the action of the external field on a bounded plasma whose density is above the critical value, perturbations propagating from the boundary will take on the form of solitary waves considered above. The typical time of formation of such a wave will be of the order of $t \simeq \frac{V_T \Delta}{V_s \omega_0 \sqrt{\delta}}$.**

## 4. Electric Field Structure in a Plasma Resonance Region

Recently interest has been aroused in the problem of the intense radiation-inhomogeneous plasma interaction. There is good reason, both theoretical and experimental, to believe that the plasma resonance region, i.e., the plasma region where the external field frequency appears to be close to the local value of the plasma frequency $\omega \simeq \omega_0(x)$, is responsible for many phenomena. As is known [5], the electromagnetic field may achieve rather large values in this region, and this considerably affects the motion of the plasma as a whole as well as the motion of individual particles strongly interacting with plasma waves.

The singularity of the electric field at the point $\omega^2 = \omega_0^2(x)$ is related to the fact that for the cold collisionless plasma the dielectric constant $\varepsilon(\omega) = 1 - \omega_0^2(x)/\omega^2$ may be equal to zero. In this case due to continuity of the inductance $\varepsilon(\omega)E$ the electric field in the plasma $E = E_0/\varepsilon$ diverges at this point. However, even in the linear theory, taking account of small but finite absorption effects or of the spatial dispersion (thermal motion) removes this singularity. Nevertheless, for a sufficiently low temperature and rare plasma when the above-mentioned effects are small enough, the linear theory leads to such large field values that one cannot neglect nonlinear effects. Taking account of these nonlinear effects leads, as we will show in what follows, to the limitation of the field in the resonance region at a far lower level.

This section[9)] is devoted to the brief consideration

---

**9) The results described in this section were obtained by the author in collaboration with S.V.Bulanov and A.S.Sakharov [6,7]. The analogous problems were discussed also in [8-10].**

of some of these effects. Here we confine ourselves for simplicity to the case when the inhomogeneous plasma is in the external homogeneous field $E_0 \sin\omega t$ changing harmonically in time and directed parallel to the inhomogeneity gradient. Then the equations describing the plasma behaviour will have the form (see (12) )

$$\ddot{\xi} - \Gamma_e + \frac{v_T^2}{n_0}\left(\frac{n_0 N_i}{1+\xi' N_i}\right)' = -\Gamma_0 \sin\omega t \; ; \quad \ddot{\sigma}_i = \delta \Gamma_y \tag{31}$$

moreover, according to (11)

$$\begin{aligned} \Gamma_e &= -\sum_{n=1}^{\infty} \frac{(\xi)^n}{n!}\left(N_i \frac{\partial}{\partial x_0}\right)^{n-1} \omega_0^2(x_0) N_i \; ; \quad N_i = \frac{1}{1+\sigma_i'} \\ \Gamma_i &= -\sum_{n=1}^{\infty} \frac{(-\xi)^n}{n!}\left(N_e \frac{\partial}{\partial x_0}\right)^{n-1} \omega_0^2(x_0) N_e \; ; \quad N_e = \frac{N_i}{1+\xi' N_i} \end{aligned} \tag{32}$$

a) Linear approximation

Let us consider the equation for the electron component and denote the typical size of the electric field inhomogeneity by $\Delta X$, and the ratio of the electron oscillation amplitude $r_0 = e_e E_0/m_e\omega^2$ in the external field to the amplitude $\xi_m$ in the resonance region by $S = E_0/E_m$

If the nonperturbed density in the resonance region is represented by the expression $\omega_0^2(x) = \omega^2[1-(-x/L)^\kappa]$ then according to the linear theory[7,11,12] $\Delta x = S^{1/\kappa} L$ and the value $S$ for ions at rest is equal to

$$S = \max\left\{ S_\nu = \frac{\nu_0}{\omega} \; , \; S_T = \left(\frac{r_d}{L}\right)^{\frac{2\kappa}{\kappa+2}} \right\} \tag{33}$$

which corresponds to the resonance saturation either due to the collision absorption or due to the energy transport from the resonance region by plasma waves with the group velocity $v_g = \kappa v_T^2/\omega_0 \simeq v_T^2/\Delta x\,\omega_0$. Here the time to reach the maximum value of the field will be of the order of $t_0 = \pi/S\omega_0$

One more linear mechanism of the field restriction can be pointed out which is associated with the ion motion. Actually, if $\dot{N}_i/N_i \ll \omega$ then the linearized equation (31) can be solved in the adiabatic approximation

$$\xi(x_0,t) = -\frac{\Gamma_0}{\sqrt{\omega_p(t)}} \int_0^t dt' \frac{\sin\omega t'}{\sqrt{\omega_p(t')}} \sin\int_{t'}^{t} \omega_p(t'')dt'' \tag{34}$$

where $\omega_p^2(t) = \omega_0^2(x_0)\, N_i(x_0,t)$ is the instantaneous value of

Langmuir frequency. It follows from this, in particular, that the maximum value of the field is determined by the rate of the ion density change and corresponds to the value[10)]

$$S = S_t = \sqrt{\frac{\dot{\omega}_p}{\omega_p^2}} \simeq \sqrt{\frac{\dot{N}_i}{\omega}} \qquad (35)$$

b) Nonlinear effects

In the case when the high-frequency oscillation amplitude is much less than a typical size of the ion density inhomogeneity it is sufficient to keep only three terms of the expansion in the expression for $\Gamma_e$ and equation (31) takes the form

$$\ddot{\xi} + \omega_p^2 \xi + \frac{\xi^3}{3!} N_i (N_i \omega_p^{2\prime})' - \frac{v_T^2 N_i \xi''}{(1+\xi' N_i)^2} = -\Gamma_0 \sin\omega t - v_T^2 \frac{(n_0 N_i)'}{n_0(1+\xi' N_i)} - N_i \omega_p^{2\prime} \frac{\xi^2}{2}; \qquad (36)$$

One can show that the two last terms on the right-hand side of (36) do not affect directly the field structure in the resonance region and determine the average (in time) displacement of electrons, which in its turn causes the ion density change. Consequently in the case of given ion motion (and, in particular, for ions at rest) these terms may be ignored.

It should be noted that as the expression (32) for $\Gamma_e$ does not contain the spatial derivatives $\xi'$, then for $\xi \omega_p^{2\prime}/\omega_p^2 \sim \xi/L \ll 1$ and at sufficiently low temperatures equation (36) will be weakly nonlinear up to values of $\xi' N_i$ close to -1, corresponding to the moment of plasma wave "breaking" associated with the selfcrossing of electron trajectories[11)]. At the same time hydrodynamic equations in Eulerian variables become strongly nonlinear long before the "breaking" moment and a much stronger requirement $\xi' N_i \ll 1$ is the condition for their nonlinearity.

Thus equation (36) even when linearized with respect to $\xi$ permits us to trace the nonlinear field evolution up to the moment of time directly preceding the breaking.

---

[10)] A more detailed analysis of this problem is given in [7].

[11)] Note that since $\sigma_e' = \sigma_i' + \xi'$ the selfcrossing of electron trajectories always precedes the selfcrossing of ion trajectories.

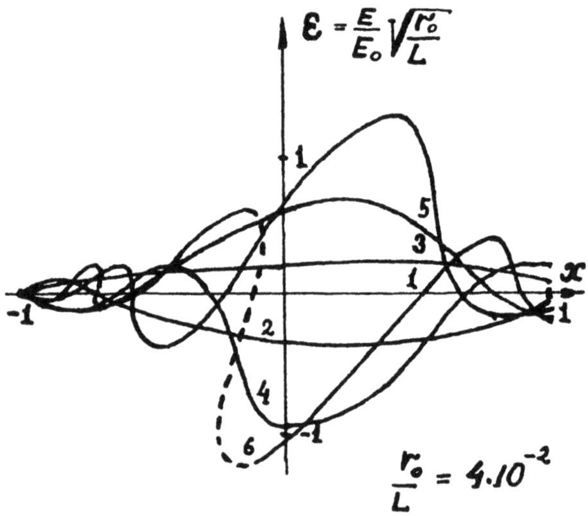


Fig. 2

The curves in Fig. 2 illustrate the dependence of the dimensionless electric field $\mathcal{E} = E\sqrt{r_0}/E_0\sqrt{L}$ represented in Eulerian variables versus the dimensionless coordinate $x = x_0/L$ for time moments $\omega t/\pi = 1, 2, 3, 4, 5, 6$. The dependence is obtained from an analytical solution of the linearized equation (36) in the case of ions at rest $(N_i \equiv 1)$ for a linear profile of the initial density $\omega_0^2 = \omega^2(1 + X/L)$, zero electron temperature $v_T^2 = 0$ and ratio $r_0/L = 4 \cdot 10^{-2}$. The transition from Lagrangian coordinates to the Eulerian coordinates was carried out with a computer. One may show here that the nonlinear terms in (36) practically do not affect the field structure.

Coming back to the analysis of equations (36) and assuming for simplicity $N_i = 1$, note that since at $\xi' = -1$ the electron density tends to infinity ( for $1 + \xi' < 0$ equations (36) have no physical meaning as they lead to a negative density) the maximum oscillation amplitude $\xi_m$ is of the order of the typical size of its inhomogeneity $\Delta X$. If the first term of the expansion $\omega_0^2(X)$ in the vicinity of the resonance point $X = 0$ begins with $X^k$ $(k = 1, 2, 3 \ldots)$ i.e. $\omega_0^2(x) \simeq \omega_0^{2(k)} X^k / k!$, $\omega_0^{2(k)} = \partial^k \omega_0^2 / \partial x$ then as follows from (36) $\Delta X = S^{1/k} \sqrt{k!\, \omega_p^2 / \omega_p^{2(k)}} \simeq S^{1/k} L$. Supposing $\xi_m = r_0/S$ and taking into account that $\xi_m \simeq \Delta X$ we find that the electric field amplitude in any case cannot exceed the value $E_0/S_N$, where

$$S_N \simeq (r_0/L)^{\frac{k}{1+k}} \tag{37}$$

It is interesting to note that since the electron oscillation amplitude proves to be of the order of the typical inhomogeneity size then passing through the plasma resonance region the electrons, in principle, may increase their energy by a value of the order of $W_e = m_e r_o^2 \omega^2 / 2 S_N^2 \simeq eE_o L / S_N^{\frac{\kappa-1}{\kappa}}$. Besides, the ponderomotive forces associated with the inhomogeneity of the high-frequency electric field $F_p \simeq \frac{\partial}{\partial x}(e^2 E^2 / m_e \omega^2)$ seem to be so great that they lead to a practically complete removal of electrons from the field localization region. Indeed, the average quasistatic field $\bar{E}$ arising due to the charge separation and compensating the pondermotive forces should be of the order of $\bar{E} \approx F_p / e \simeq m_e \omega^2 \xi_m^2 / e \Delta x$. Thus the relative change of the average electron density $\Delta n / n \simeq \bar{E} / 4\pi e n \Delta x \simeq (\xi_m / \Delta x)^2 \simeq 1$ proves to be of the order of one. Under the action of the field $\bar{E}$ ions begin to shift, and this in turn leads to the ion density change. If for the typical field saturation time $t_o = \pi / \omega S$ the density profile change does not greatly affect the field structure in the resonance, then some ions may also be accelerated up to the energies of the order of $e\bar{E}\Delta x \simeq W_e$

Elementary estimations show that if the temperature is sufficiently low so that the inequality

$$S_N \gg S_T \quad \text{or} \quad \frac{r_d}{L} \ll \left( \frac{r_o}{L} \right)^{\frac{2+\kappa}{2+2\kappa}} \tag{38}$$

is fulfilled, then the last term on the left-hand side of equation (36), which describes the spatial dispersion, may be neglected up to the moments directly preceding the appearance of multibranch solutions ("breaking"). Then equation (36) takes the form of the equation for a forced anharmonic oscillator. The analysis of such equations is well known [13]. The stationary amplitude $\xi_o$ of the forced oscillations is to be determined from the equation

$$\xi_o \left[ \frac{\omega_o^{2(\kappa)}}{\omega^2} \frac{x_o^\kappa}{\kappa!} + \frac{\omega_o^{2\,\prime\prime}(x_o)}{8\omega^2} \xi_o^2 \right] = r_o \tag{39}$$

generally yielding three values for the amplitude $\xi_o$. However, since the solution of the initial problem is unique, the value of the stationary amplitude $\xi_m = r_o / S$ can be estimated assuming in (39) $x_o = \Delta x = S^{1/\kappa} L$. Thus the amplitude limitation associated with the presence of the anharmonicity (due to the density inhomogeneity) determines one more value $S_a$. In each specific case the value $S_a$ is readily determined from equation (39); however, by order of magnitude

$$S_a = \begin{cases} (r_0/L)^{2/3} & \text{for } \kappa = 1 \\ (r_0/L)^{\frac{\kappa}{\kappa+1}} \simeq S_N & \text{for } \kappa \geq 2 \end{cases} \qquad (40)$$

Therefore, in the case when the ion density derivative $n_0'(x)$ in the resonance region is not close to zero ($\kappa = 1$), the influence of anharmonicity turns out to be negligible. Otherwise ($\kappa \geqslant 2$) it should be taken into account, in particular, at the last stage of the process, when $\xi'$ becomes of the order of one.

Summerizing one may say that the maximum field amplitude in resonance is equal to $E_0/S$, the typical size of the field inhomogeneity $\Delta x = L S^{\frac{1}{2}}$, and the saturation time $t_0 = \pi/\omega S$ where

$$S = S_e = max \{ S_\nu , S_T , S_t , S_N \} \qquad (41)$$

and the values $S_\nu, S_T, S_t, S_N$ are determined by relations (33),(35),(37).

As has already been noted for $S_N > S_T$ the last term on the left-hand side of (36), taking into account the kinetic pressure and preventing the wave from breaking, begins to play a significant role only at the stage when oscillations have an essentially nonlinear character (i.e. for $\xi'$ close to one). Physically, this is associated with the fact that in the nonlinear regime the "group velocity" of the plasma waves[12] depends on their amplitude, increasing with it. This in turn leads to a faster energy transport from the plasma resonance region.

By way of example Fig.3 shows the initial electric field $\mathcal{E} = E\sqrt{r_0}/E_0\sqrt{L}$ versus $x = X/L$ (Fig.3a) and the value of the field at the time moment $\tau = \frac{\omega t}{\pi} = 3$ (Fig.3b) for the case of plasma at rest ($N_i = 1$), a linear density profile and parameters $S_N = (r_0/L)^{1/2} = 0.05$, $S_T = (r_d/L)^{4/3} = 0.04$. The cold plasma solution for $\tau = \frac{\omega t}{\pi} = 12$ is taken as the initial distribution.

It follows from comparison of Figs.3a and 3b that at the nonlinear stage the energy transport from the resonance region is much faster than follows from the linear theory. Indeed, according to the linear theory the perturbation should cover the distance $L/2$ propa-

---

[12] More precisely, the velocity of the energy being carried by plasma waves.

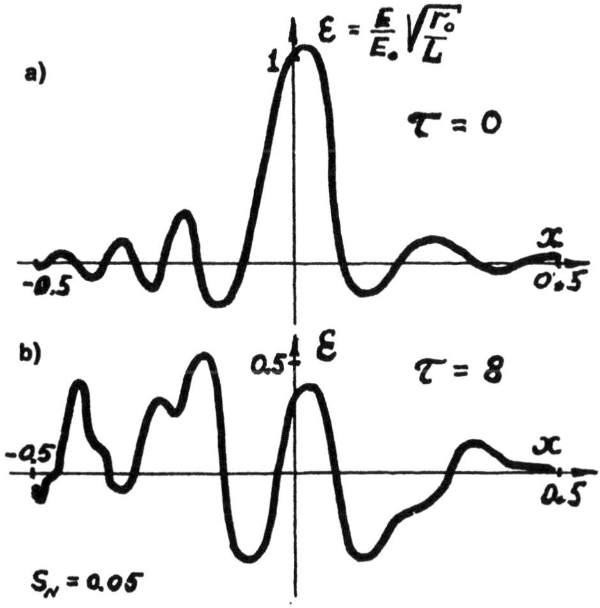


$S_N = 0.05$
$S_T = 0.04$

Fig. 3

gating with the "linear group velocity" $v_g = v_T^2/\omega\Delta x$ for a time of the order of $\tau = \omega t/\pi \simeq L\omega^2\Delta x/2\pi v_T^2 \simeq 10^2$ while it follows from Fig.3b that it spreads much faster covering a distance of the order of $L/2$ in time $\omega t/\pi = 8$.

For $T_e = 0$ the equation (36) describes the plasma behaviour quite well up to the moment of "breaking", corresponding to the divergence of the electron density $N_e$ and to the appearance of multiple solutions. After the "breaking" the basic equations (3)-(5), and more precisely, the equation (5) for the selfconsistent field $\Gamma_j$ , fail. For the sake of analysis of the further field evolution and, especially, of the particle dynamics it is necessary to solve a kinetic equation which in the case of cold plasma actually coincides with equations (3), (4) (or (31') ), with the expressions for the selfconsistent field $\Gamma_j$ being the sum of individual particle fields.

The results of a numerical solution of the "cold" kinetic equation, illustrating the field and particle behaviour at "breaking" are given in Fig.4 a,b,c,d. A plasma has been simulated by several hundreds of particles, the density profile was chosen as a linear one, and the parameter $S_N = \sqrt{r_0/L}$ was supposed to be equal to 0,1. In each part of Fig.4 the "particle" distribution is shown in the phase plane $x, v$ in dimensionless variables $x = X/L$

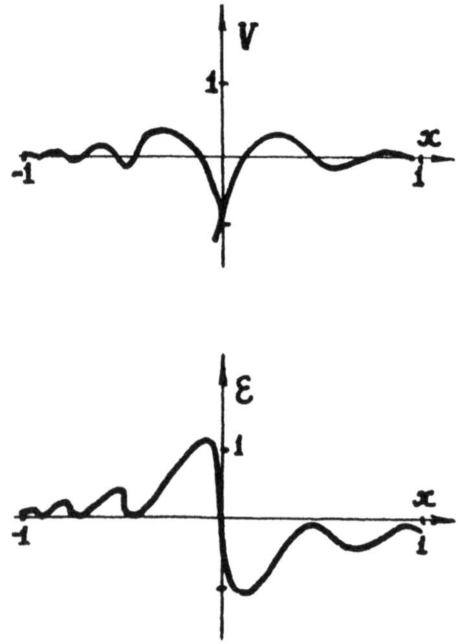


Fig. 4a $\tau = 10.5$

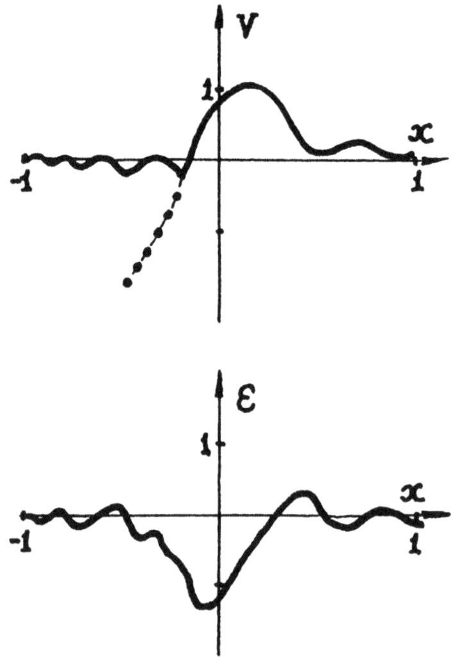


Fig. 4b $\tau = 11$

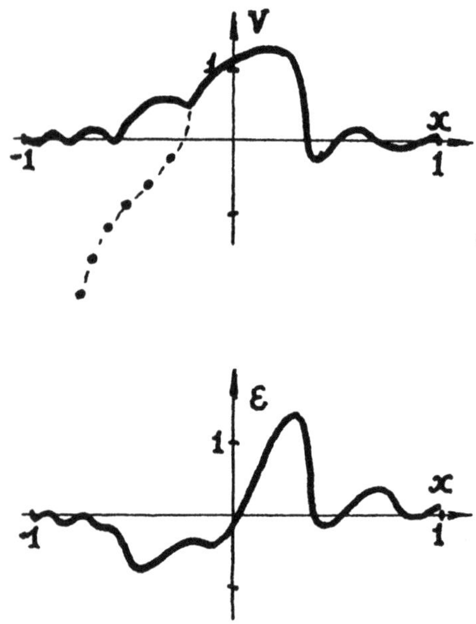


Fig. 4c $\tau = 11.5$

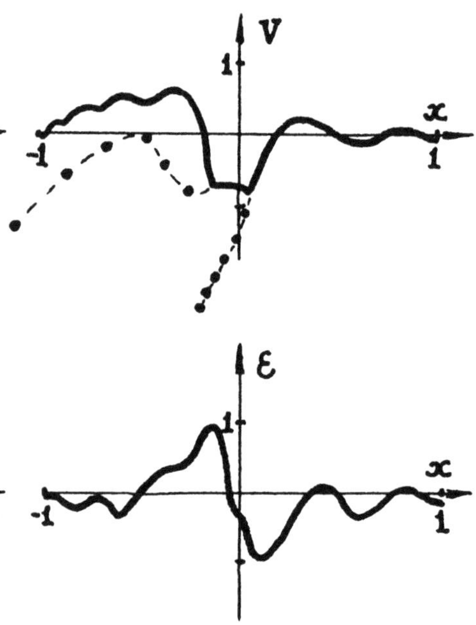


Fig. 4d $\tau = 12.5$

$V = vS_N/\omega r_c$ [13]; also shown is the dependence of a dimensionless electric field $\mathcal{E} = \frac{E S_N}{E_0}$ on $x$ for various time moments $\tau = \omega t/\pi = 10.5$; $11$, $11.5$; $12.5$.

The above figures show that at the "breaking" moment $\tau \simeq S_N^{-1} = 10$. The representing line has a sharp peak at the point $x = 0$ (Fig.4a), then some particles move from its apex to negative $x$ (the two-stream flow occurs, Fig.4b). These particles then separate from the main group and move with a velocity of the order of $\omega r_0/S_N \simeq (eE_0L/m_e)^{1/2}$ to the plasma boundary and, finally, cross it (Fig.4c). After one period the general features of the picture repeats (Fig.4d) and a new portion of accelerated particles leaves the plasma.

As estimations show, the total number of electrons accelerated up to the energies of the order of $W_e = eE_0L$ during one period of oscillation amounts to $S_N^{3/2}$ of the total number of particles in a layer of width $L$ and the energy carried by them is approximately $\sqrt{S_N}$ of the whole field energy contained in the plasma resonance region. These energy losses correspond to the effective collision frequency $\nu_{eff} \simeq \omega S_N^{3/2}$.

Since these losses are quite large then after several periods the oscillation amplitude should decrease so that the particle acceleration process stops. Then the oscillation amplitude begins to increase again until the accelerated particles appear again and the process repeats once more. For a time of the order of $1/\omega S_N$ about $S_N^{-1/2}$ ejections should take place on the average so that the average effective collision frequency would be of the order of $\nu_{eff} \simeq \omega S_N$.

In conclusion, let us point out the conditions under which the ion motion may be neglected and their density may be considered as constant in time (i.e. $N_i = 1$). Using, for example, equation (16) we find that the order of the magnitude of the relative change of the ion density for time $t$ is

$$\Delta N_i \simeq \frac{\delta}{4} \frac{S_N^4 S^{-4}}{(\omega t)^{-2} + 8 S_N^4 S_i^{-3} S^{-2}} \tag{41}$$

where $S_N = (r_0/L)^{1/2}$, $S_i = (r_0/r_d)^{2/3}$, $S = r_0/\xi$

---

[13] In the cold plasma case the particle position in the phase plane is obviously represented by points of a line intersecting the abscissa at the initial moment of time.

The corrections associated with taking account of the ion density change will apparently become important for $\Delta N_i \gtrsim \delta_e$ so for $\Delta N_i \ll \delta_e$ one may ignore them. Thus from (41) it follows that for $\delta_i \ll \delta_e$, where $\delta_e$ is determined according to (41), the ion motion does not affect the field structure in the plasma resonance region. If $\delta_i >> \delta_e$ then up to the value $t \approx \pi/\omega\delta$, until the ion density has no time to change, the field amplitude increases up to $E = E_0/\delta$ where $\delta$ is determined from the equation

$$4\delta^3 = \frac{\delta\, \delta_N^4}{\delta^4 + \delta\, \delta_N^4 \delta_i^{-3}} \qquad (42)$$

Then it begins to fall and reaches its "stationary" value determined by the ion density. Here $\Delta x \simeq L\delta_i$, $E = E_0/\delta_i$.

The author is grateful to S.V.Bulanov and A.S.Sakharov for many useful discussions.

## Bibliography


1. В.Е.Захаров. ЖЭТФ 62, 1745 (1972)
2. А.А.Иванов,М.Г.Никулин. ЖЭТФ 65, 168, (1973)
3. Л.И.Рудаков. ДАН СССР, 207, 821, (1972)
4. В.И.Карпман. "Нелинейные волны в диспергирующих средах". "Наука", Москва 1973г.
5. В.Л.Гинзбург. "Распространение электромагнитных волн в плазме", "Наука", Москва, 1967г.
6. С.В.Буланов, Л.М.Коврижных, Физика плазмы 2,105, (1976)
7. С.В.Буланов,Л.М.Коврижных,А.С.Сахаров."Ленгмюровские колебания конечной амплитуды в области плазменного резонанса" Препринт ФИАН 1976г.
8. P.Koch, J.Albritton,Phys.Rev.Lett. 32,1420,(1974)
9. G.J.Morales,Y.C.Lee,Phys.Rev.Lett. 33,1016,(1974)
10. J.Albritton,P.Koch,Phys.Fluids,18,1136, (1975)
11. Н.Г.Денисов, ЖЭТФ, 31,609, (1956)
12. И.Г.Кондратьев, М.А.Миллер, Изв.высших учебных заведений - Радиофизика, 6, 885, (1968)
13. Н.Н.Боголюбов,Ю.А.Митропольский,"Асимптотические методы в теории нелинейных колебаний","Наука", Москва 1974г.

# THEORY OF MAGNETIC-FIELD TURBULENCE IN DISK PLASMAS AND ITS APPLICATION TO THE GALAXY AND ACCRETION MODEL OF COMPACT X-RAY BINARIES

Setsuo Ichimaru

Department of Physics, University of Tokyo

Bunkyo-ku, Tokyo, Japan

## I. INTRODUCTION

It has been recognized that the differential rotation and radial flow of plasmas in a disk geometry act to generate and amplify the magnetic-field fluctuations in it. The presence of such a magnetohydrodynamic turbulence plays an essential part in the physical processes involved in various astrophysical objects, such as the Galaxy and the accretion model of compact X-ray sources (Prendergast and Burbidge 1968, Shakura and Sunyaev 1973, Eardley and Lightman 1975, and many others). In the former case the theory should be relevant directly to the question on the origin of the Galactic magnetic field; the spectral distribution of magnetic-field fluctuations may then be correlated with the observational data such as the energy dependence of the anisotropy and the total path length of the cosmic rays. In the latter case the flux of angular momentum carried away by the stress tensor of the magnetic field enables the matter to flow toward the accreting star ; the static and dynamic properties of such an accretion disk are vitally controlled by the rate of such an angular-momentum transfer in the plasma. In the cases of accretion onto a magnetic neutron star, the disk's inner boundary is determined by the pressure balance between the stellar magnetic field and the accreting plasma (the Alfvén surface). The eventual fall of the plasma onto the stellar surface must be accounted for in terms of the theory of plasma diffusion across the Alfvén surface; the basic processes involved are the anomalous electric resistivity in the boundary domain caused by the presence of magnetic-field fluctuations in the disk (Ichimaru, to be published).

In this paper we thus develop a first-principle theory of magnetohydrodynamic turbulence appropriate to the plasmas in such a disk geometry. Turbulence is generated mainly by the differential rotation of the plasmas; it decays through current dissipation due to anomalous magnetic viscosity. Based on the microscopic formulation of transport processes (Ichimaru 1975b) we derive a set of equations describing space-time evolution of the magnitudes and the spectral distributions of both the diagonal and off-diagonal elements of the magnetic - field fluctuation tensor. Its stationary solution yields explicit expressions for turbulence in terms of the macroscopic physical parameters of the system, such as the disk thickness and the rate of differential rotation. In Section III, the results are applied to the Galaxy, providing an essential account of the observed energy density of the magnetic fields in it. The calculated spectral distribution of the magnetic-field fluctuations is tentatively compared with cosmic-ray observations.

In Section IV, we apply the turbulence theory to analysis of accretion disks around compact objects such as a black hole. The diagonal elements of the stress tensor associated with the fluctuating magnetic field act to heat the accreting plasma; the off-diagonal elements produce outward transport of angular momentum, enabling the plasma to fall radially toward the compact star. Combining these calculations with the standard hydrostatic pressure-balance equation and conservation laws, we determine the disk thickness, the radial and azimuthal flow velocities, the density and temperatures of the plasma as functions of the radial distances from the accreting star; thermal stabilities of such solutions are investigated. It is shown that two physically distinct states exist in the middle part of the accretion disk surrounding a black hole; a transition from one state to the other takes place when the physical conditions of the accreting plasma near the disk outer boundary cross over the critical conditions separating the two states. We present a definite numerical model of Cygnus X-1 accounting for observed features of its transitions, such as magnitudes and spectral distributions of the X-ray luminosity, transition times, and variabilities.

## II. GENERATION AND DISSIPATION OF MAGNETIC-FIELD FLUCTUATIONS IN DISK PLASMAS

### 2.1 Formulation of Turbulence Problem in Disk

Consider an axisymmetric disk configutation of a plasma with an effective thickness $2H$. We employ the cylindrical system of coordinates $(r,\phi,z)$ with the z axis chosen in the direction perpendicular to the disk. Along the z direction across the disk, the plasma is assumed to be uniform with the number density n, the temperature T (in energy units), and the flow velocity,

$$\vec{v} = -u\hat{r} + v\hat{\phi} , \tag{1}$$

where $\hat{r}$ and $\hat{\phi}$ are the unit vectors in the r and $\phi$ directions; u and v are functions of r alone.

The induction equation of the magnetic field reads

$$\frac{\partial \vec{B}}{\partial t} = \vec{\nabla} \times (\vec{v} \times \vec{B}) + \frac{\eta c^2}{4\pi} \nabla^2 \vec{B} \tag{2}$$

where $\eta$ is the electric resistivity operator. We expand the magnetic field as

$$\begin{aligned} \vec{B} &= \sum_k \vec{B}_k \exp[i(\vec{k}\cdot\vec{r}-\omega t)] \\ &= \sum_k (B_k^r \hat{r} + B_k^\phi \hat{\phi} + B_k^z \hat{z}) \exp[i(\vec{k}\cdot\vec{r}-\omega t)] \ . \end{aligned} \tag{3}$$

The expansion is carried out with the periodic boundary conditions appropriate to a cube of volume $(2H)^3$, so that the domain of the $\vec{k}$ summation is restricted to $k \geq \pi/H$. The spectral amplitudes, $B_k^r$, $B_k^\phi$, and $B_k^z$, are slowly varying functions of r and t; the scales of those r and t variations are greater than $2\pi/k$ and $2\pi/|\omega|$.

We substitute (3) into (2) to obtain

$$\begin{aligned} \frac{\partial \vec{B}_k}{\partial t} - i\omega \vec{B}_k &= \vec{\nabla} \times (\vec{v} \times \vec{B}_k) + i\vec{k} \times (\vec{v} \times \vec{B}_k) \\ &\quad - \frac{\eta_k c^2 k^2}{4\pi} \vec{B}_k \ . \end{aligned} \tag{4}$$

In this equation we have retained only the lowest-order contribution with respect to $|\partial \vec{B}_k/\partial r|/|k\vec{B}_k|$ arising from the last term of (2); inclusion of the next-order contribution would slightly modify the dispersion relation to be obtained presently. The quantity $\eta_k$ denotes the electric resistivity for the current fluctuation with wave vector $\vec{k}$.

The frequencies of fluctuations are determined from the equation,

$$\omega \vec{B}_k + \vec{k} \times (\vec{v} \times \vec{B}_k) = 0 \ , \tag{5}$$

which reveals two transverse modes,

$$\omega = \vec{k}\cdot\vec{v} \qquad \text{with} \qquad \vec{k}\cdot\vec{B}_k = 0 \ . \tag{6}$$

The remaining solution, $\omega = 0$ with $\vec{v} \times \vec{B}_k = 0$, is devoid of coupling between the velocity field and the magnetic field; its amplitude simply decays through current dissipation. We thus consider only the two transverse modes of (6) in the subsequent analysis. We note that the frequencies of those modes vanish in the frame comoving with the plasma; their spectrum is broadened around $\omega = 0$ mainly through imaginary frequency arising from the magnetic viscosity, i.e., the last term of (4). This situation is therefore quite analogous to the cases of ordinary hydrodynamic turbulence obeying the Navier-Stokes equation.

With the aid of (5), we obtain from (4) the equations describing space-time evolution of the spectral amplitudes of the magnetic-field fluctuations:

$$\frac{\partial B_k^r}{\partial t} = - \frac{\eta_k c^2 k^2}{4\pi} B_k^r \tag{7}$$

$$\frac{\partial B_k^\phi}{\partial t} = \frac{\partial}{\partial r} (B_k^\phi u + B_k^r v) - \frac{\eta_k c^2 k^2}{4\pi} B_k^\phi \tag{8}$$

$$\frac{\partial B_k^z}{\partial t} = \frac{1}{r} \frac{\partial}{\partial r} (B_k^z ur) - \frac{\eta_k c^2 k^2}{4\pi} B_k^z \quad . \tag{9}$$

The first terms on the right-hand sides of (8) and (9) represent compressional effects on the magnetic field arising from the radial flow of the plasma. The second term on the right-hand side of (8) describes creation rate of $B_k^\phi$ out of $B_k^r$ owing to the differential rotation in the plasma. This may be clearly seen as we note

$$\frac{\partial}{\partial r} (B_k^r v) = B_k^r (\frac{\partial v}{\partial r} - \frac{v}{r}) = B_k^r \, r \frac{\partial}{\partial r} (\frac{v}{r}) \tag{10}$$

stemming from $\vec{\nabla} \cdot \vec{B} = 0$. The last terms of (7)∿(9) describe decay of field fluctuations arising from the finite electric resistivity or the magnetic viscosity of the plasma.

To obtain a description of the turbulence spectrum, we multiply (7)∿(9) by $B_{-k}^r$ etc. and carry out a statistical average denoted by $< >$. Space-time evolution of the diagonal and off-diagonal elements of the spectral tensor of the magnetic-field fluctuations is thus described by the equations,

$$\frac{\partial}{\partial t} <\vec{B}_{-k} \cdot \vec{B}_k> = u \frac{\partial}{\partial r} \left[ <B_{-k}^\phi B_k^\phi> + <B_{-k}^z B_k^z> \right] + 2 \frac{\partial u}{\partial r} \left[ <B_{-k}^\phi B_k^\phi> + <B_{-k}^z B_k^z> \right]$$

$$+ 2\,\frac{u}{r}\,\langle B^z_{-k} B^z_k\rangle + 2r\,\frac{\partial\Omega}{\partial r}\,\langle B^r_{-k} B^\phi_k\rangle - \frac{\eta_k c^2 k^2}{2\pi}\,\langle \vec{B}_{-k}\cdot\vec{B}_k\rangle\ , \qquad (11)$$

$$\frac{\partial}{\partial t}\,\langle B^r_{-k} B^\phi_k\rangle = \frac{\partial}{\partial r}\left[u\langle B^r_{-k} B^\phi_k\rangle\right] + \frac{u}{r}\,\langle B^r_{-k} B^\phi_k\rangle$$

$$+ r\,\frac{\partial\Omega}{\partial r}\,\langle B^r_{-k} B^r_k\rangle - \frac{\eta_k c^2 k^2}{2\pi}\,\langle B^r_{-k} B^\phi_k\rangle\ , \qquad (12)$$

where $\Omega \equiv v/r$ is the angular velocity. The mean square value $\langle\delta B^2\rangle$ of the fluctuations and the r-$\phi$ element $\chi/4\pi$ of the stress tensor are then calculated as

$$\langle\delta B^2\rangle = \sum_k \langle \vec{B}_{-k}\cdot\vec{B}_k\rangle \qquad (13)$$

$$\chi = \sum_k \langle B^r_{-k} B^\phi_k\rangle\ . \qquad (14)$$

## 2.2 Anomalous Magnetic Viscosity

To complete the formulation, we must establish an explicit way to calculate the electric resistivity $\eta_k$ for the current fluctuation or the magnetic viscosity $\eta_k c^2/4\pi$ for the magnetic-field fluctuation with wavenumber k. Generally the dc electric resistivity of an electromagnetically turbulent plasma may be given by

$$\eta = -\,\frac{1}{e^2 n^2 V}\sum_k \int d\omega\,\frac{k_\parallel}{\omega}\,\overleftrightarrow{\kappa}\,(\vec{k},\omega):\langle\overleftrightarrow{EE}^*(\vec{k},\omega)\rangle\ .$$

Here $\overleftrightarrow{\kappa}(\vec{k},\omega)$ refers to the ac conductivity tensor of the particles of the particular species which carries the dc current $en\vec{V}$ under consideration; e and m are the electric charge and the mass of such a particle; $k_\parallel$ is the component of the wave vector in the direction of the drift velocity $\vec{V}$ ; the spectral tensor of the electric-field fluctuations is defined as

$$\langle\overleftrightarrow{EE}^*(\vec{k},\omega)\rangle = \frac{1}{2\pi}\int d\vec{r}\int_{-\infty}^{\infty} dt\,\langle\vec{E}(\vec{r},\,t)\vec{E}(0,0)\rangle\,\exp[-i(\vec{k}\cdot\vec{r} - \omega t)]\ . \qquad (15)$$

In the presence of magnetohydrodynamic turbulence, the major mechanism for the resistivity is provided by scattering of the current by the low-frequency fluctuating magnetic fields; the result is (Ichimaru 1975b)

$$\eta = \frac{1}{3nmc^2}\left(\frac{\pi m}{2T}\right)^{1/2} \sum_k \frac{1}{k} \langle \vec{B}_{-k} \cdot \vec{B}_k \rangle . \tag{16}$$

In the treatment of the turbulent magnetic viscosity for the magnetic-field fluctuation with wavenumber k, we regard it essential to note a hierarchy among the time scales involved in the fluctuations with different wavelengths. We thus calculate the effective resistivity $\eta_k$ as a summation of all the contributions of scattering acts produced by those fluctuations with wavenumbers greater than k. These fluctuations vary faster in space and time than the fluctuation with wavenumber k; the interactions may thus be looked upon as producing impulsive, collisional effects (e.g., Tajima, Ichimaru, and Nakano 1974). Those fluctuations with wavenumbers smaller than k vary slowly in space and time; their effects may thus be regarded as causing adiabatic deformation and diffusion of the fluctuation under consideration, resulting in isotropization of the turbulence spectrum. This is the only hypothesis involved in the present theory; a hypothesis similar to this is implicit in the Kolmogorov-Heisenberg theory of turbulent viscosity (Kolmogorov 1941, Heisenberg 1948; see also Nakanò 1972).

Adopting the hypothesis stated above, we express

$$\eta_k = \frac{1}{3nmc^2}\left(\frac{\pi m}{2T}\right)^{1/2} \sum_{q(>k)} \frac{1}{q} \langle \vec{B}_{-q} \cdot \vec{B}_q \rangle . \tag{17}$$

Substituting this expression into (11) and (12), we find that

$$\langle \vec{B}_{-k} \cdot \vec{B}_k \rangle = \left(\frac{\pi}{H}\right)^3 \frac{\langle \delta B^2 \rangle}{4Hk^4} \tag{18}$$

$$\langle B^r_{-k} B^\phi_k \rangle = \left(\frac{\pi}{H}\right)^3 \frac{\chi}{4Hk^4} \tag{19}$$

constitute consistent solutions for the k spectrum. The amplitudes now satisfy the differential equations,

$$\frac{\partial}{\partial t} \langle \delta B^2 \rangle = \frac{2}{3} u \frac{\partial}{\partial r} \langle \delta B^2 \rangle + \frac{4}{3} \frac{\partial u}{\partial r} \langle \delta B^2 \rangle + \frac{2}{3} \frac{u}{r} \langle \delta B^2 \rangle$$

$$+ 2r \frac{\partial \Omega}{\partial r} \chi - 2\nu \frac{\langle \delta B^2 \rangle^2}{8\pi nT} , \tag{20}$$

$$\frac{\partial}{\partial t}\chi = \frac{1}{r}\frac{\partial}{\partial r}(\chi u r) + \frac{1}{3} r \frac{\partial \Omega}{\partial r} \langle \delta B^2 \rangle - 2\nu \frac{\langle \delta B^2 \rangle \chi}{8\pi n T} , \qquad (21)$$

where

$$\nu \equiv \frac{\pi}{3H}\left(\frac{\pi T}{2m}\right)^{1/2} . \qquad (22)$$

In the derivation of (20) and (21), we have assumed isotropy in the diagonal elements of the spectral tensor, so that

$$\langle B^r_{-k} B^r_k \rangle = \langle B^\phi_{-k} B^\phi_k \rangle = \langle B^z_{-k} B^z_k \rangle = \frac{1}{3} \langle \vec{B}_{-k} \cdot \vec{B}_k \rangle . \qquad (23)$$

Barring the cases that $\langle B^r_{-k} B^r_k \rangle = 0$, a deviation from this assumption would require no essential corrections in the subsequent arguments.

The wavenumber-dependent magnetic viscosity is calculated from (17) as

$$\nu_M = \frac{\nu}{k^2} \frac{\langle \delta B^2 \rangle}{8\pi n T} . \qquad (24)$$

As k increases, $\nu_M$ decreases. We thus find an upper bound in k space for the applicability of the turbulence theory; the upper bound stems from comparison between the anomalous magnetic viscosity (24) and the ordinary magnetic viscosity, $\nu_{M0} = \eta_c c^2/4\pi$, arising from the Coulomb collisions between ions and electrons, that is, the standard Coulomb resistivity $\eta_c$.

It is important at this moment to examine the effects of renormalization brought about by the presence of turbulence. We remark that the calculation leading to (16) was based on the use of a nonturbulent (unrenormalized) evaluation of the ac conductivity of the plasma. Enhanced collision rates in turbulence act to reduce the effective lifetimes of the fluctuations, resulting in broadening of the spectral width. Such a spectral broadening or the finiteness of the correlation times then introduces an additional dissipation mechanism of the fluctuating fields in the plasma (e.g., Ichimaru 1975a). The effects of renormalization are therefore such as to increase the magnitude of the effective resistivity (17), which in turn acts to decrease the spectral intensity of turbulence.

In the light of (4) and (24), we estimate the increase $\Delta\omega$ in the spectral width brought about by turbulence as

$$\Delta\omega = \nu_M k^2 = \nu \frac{\langle\delta B^2\rangle}{8\pi n T} . \tag{25}$$

In judging the effectiveness of renormalization, we compare this quantity with the typical frequency, $k(T/m)^{1/2}$, of the single-particle excitations; the unrenormalized ac conductivity has been obtained from consideration of these excitations alone. We thus find that the effects of renormalization become most important in the limit of long wavelengths. With the aid of (22) we evaluate the ratio between $\Delta\omega$ and $k(T/m)^{1/2}$ at the maximum wavelength, i.e., $k = \pi/H$:

$$\left.\frac{\Delta\omega}{k(T/m)^{1/2}}\right|_{k=\pi/H} = \frac{1}{3}\left(\frac{\pi}{2}\right)^{1/2} \frac{\langle\delta B^2\rangle}{8\pi n T} . \tag{26}$$

From this comparison we may conclude that the effects of renormalization remain small over the entire domain of relevant wavelengths unless the magnetic energy density substantially exceeds its equipartition value around nT.

### 2.3 Stationary Solution and Turbulent Viscosity

In many practical cases including the accretion disk around a compact star and the Galaxy, we may assume $|u| \ll |v|$ and $|\partial u/\partial r| \ll |\partial v/\partial r|$. In these circumstances the stationary solution to (20) and (21) is obtained from the last two terms as

$$\frac{\langle\delta B^2\rangle}{8\pi n T} = \frac{1}{\sqrt{6}\,\nu} \left| r \frac{\partial\Omega}{\partial r} \right| , \tag{27}$$

$$\frac{\chi}{4\pi n T} = \frac{1}{3\nu}\, r \frac{\partial\Omega}{\partial r} . \tag{28}$$

Outward transport of angular momentum takes place when $\partial\Omega/\partial r < 0$. Both the diagonal and off-diagonal elements of the stress tensor associated with the magnetic-field fluctuations are produced by the differential rotation of the plasmas.

It is instructive to consider the magnitude of turbulent viscosity $\nu_T$ predicted in the present theory. Since the turbulent viscosity may be defined through $|\chi/4\pi| = \nu_T\,(r/2)\,|\partial\Omega/\partial r|$, we find from (28)

$$\nu_T = \left(\frac{2}{\pi}\right)^{3/2} nm \left(\frac{T}{m}\right)^{1/2} H . \tag{29}$$

This result may be interpreted in accord with the concept of a mixing-length theory; (29) indicates that momenta in the plasma are effectively mixed over a length comparable to H. This enhancement of the mixing length is a consequence of the $k^{-4}$ spectrum obtained in (18) and (19); the bulk of turbulence is contained in the long-wavelength domain, $k \gtrsim \pi/H$, so that mixing takes place over large distances.

## III. MAGNETIC FIELD IN THE GALAXY

### 3.1 Origin of Galactic Magnetic Field

The theory of magnetohydrodynamic turbulence described in the preceding section is equally applicable to the current fluctuations produced by ions and electrons. Dependence on the particle species enters in the formalism through the relaxation rate $\nu$ defined in (22). Since one can ordinarily assume $T_e/m_e > T_i/m_i$, the decay rate is larger for the electrons than for the ions. Hence one finds that the level of turbulence remains smaller for the electron-current fluctuations than for the ion-current fluctuations [cf. Eq. (30) below]. In this section we thus consider the intensity of the magnetic-field fluctuations produced by ions and apply the result to estimate the magnetic energy density in the Galaxy.

We first substitute (22) in (27) to obtain

$$\frac{\langle \delta B^2 \rangle}{8\pi} = \frac{1}{\pi} \left(\frac{3}{\pi}\right)^{1/2} nm \left(\frac{T}{m}\right)^{1/2} H \left| r \frac{\partial \Omega}{\partial r} \right| . \tag{30}$$

The Galactic plasma consists mostly of protons (and electrons) with $n \simeq 1\ \mathrm{cm}^{-3}$ and $T \simeq 10^4$ K. We use the values of the parameters (Allen 1973): $2H = 2$ kpc and $|r\partial\Omega/\partial r| \simeq 25$ km/sec kpc near the Sun. Equation (30) is then computed as

$$\frac{\langle \delta B^2 \rangle}{8\pi} \simeq 0.74\ \mathrm{eV/cm}^3 \quad .$$

Observationally (e.g., Morrison 1961), the magnetic energy density in the Galaxy is said to be $\sim 1\ \mathrm{eV/cm}^3$. We thus find that the magnetohydrodynamic turbulence generated and maintained by the differential rotation of the plasma essentially accounts for the strength of the magnetic field in the Galaxy.

The rate at which the field fluctuations are generated by the differential rotation may be estimated from (20) and (21). Since $\langle \delta B^2 \rangle \simeq \chi$ [cf. Eqs. (27) and (28)], we find the characteristic time $\simeq |r\partial\Omega/\partial r|^{-1}$, which takes on a value $\sim 4 \times 10^7$ yr in the vicinity of the Sun. When the field strength is given by its stationary value (30), its decay rate through anomalous magnetic viscosity is naturally balanced by the generation rate. If the level of field fluctuations is greater than (30), the decay rate would be faster than the generation rate computed above. It thus appears that the "fossil" theory of the Galactic magnetic field encounters a difficulty when the anomalous magnetic viscosity is taken into consideration.

### 3.2 Comparison with Cosmic-Ray Observations

The present theory predicts not only the total intensity but also the spectral distribution of the magnetic-field fluctuations. Introducing the spectral index $\alpha$ via $\langle \vec{B}_{-k} \cdot \vec{B}_k \rangle \sim k^{-\alpha}$, we find $\alpha = 4$ from (18). This spectrum is valid in the wavelength domain down to $\lambda_{min} \simeq 10^{-6}$ pc, where the anomalous viscosity (24) and the Coulomb viscosity become comparable. Inclusion of the spectral broadening effect arising from turbulence, as discussed in Section 2.2, would act to decrease the magnitude of $\alpha$ slightly.

In principle, the spectral properties of the magnetic field can be correlated with the observational data, such as the energy dependence of the anisotropy and the total path length of the cosmic rays. In this section we describe an attempt to investigate the structure of magnetic-field fluctuations in the Galaxy through such cosmic-ray observations. An investigation along these lines has been carried out recently by Somogyi (1975) with somewhat different conclusions.

The energy range of cosmic-ray particles to be considered is typically $10^{12}$ eV $\lesssim \varepsilon \lesssim 10^{15}$ eV; according to the relation to be derived shortly, such particles couple with the electromagnetic fluctuations with wavelengths, $10^{-3}$ pc $\sim$ 1 pc, much shorter than $2H \simeq 2$ kpc or the thickness of the disk. The bulk of field fluctuations contained in the long-wavelength domain near $k = \pi/H$ thus provides an average background of magnetic field for the behavior of the relativistic cosmic-ray particles.

Anisotropy, $\delta = (I_{\parallel} - I_{\perp})/(I_{\parallel} + I_{\perp})$, in the cosmic-ray flux distribution I between the directions parallel and perpendicular to such an average magnetic field is governed by the coefficient $D_{\mu}$ of pitch-angle diffusion. The pitch-angle diffusion stems from scattering by fluctuating electromagnetic fields; for transverse fluctuations propagating along the magnetic lines of force, we have (Jokipii 1968; Ichimaru, to be published)

$$D_{\mu} = \sum_{k} \int d\omega \frac{\pi e^2}{p^2} \langle |E^2|(\vec{k},\omega) \rangle^{T} (1 - \mu^2)\left(\mu - \frac{kv}{\omega}\right)^2 \delta(\Omega_c + kv\mu - \omega) \quad . \tag{31}$$

Here p and v are the momentum and velocity of the particle; $\mu$ is its direction cosine with respect to the average magnetic field; $\Omega_c = ec[\langle\delta B^2\rangle]^{1/2}/\varepsilon$ is the cyclotron frequency, and $\langle|E^2|(\vec{k},\omega)\rangle^T$ is the transverse part of the spectral tensor (15). For a relativistic particle ($v \simeq c$) scattered by low-frequency fluctuations under present consideration, (31) reduces to

$$D_\mu = \frac{e^2}{2\pi c p^2} \frac{1-\mu^2}{|\mu|} k^2 \left.\langle \vec{B}_{-k} \cdot \vec{B}_k \rangle\right|_{k = |\Omega_c/\mu c|} \tag{32}$$

The relation, $k = |\Omega_c/\mu c|$, comes from the $\delta$ function in (31), representing the resonant contribution between particle and fluctuation; it establishes correspondence between particle energies and wavelengths of fluctuations.

The anisotropy is determined in terms of the coefficient of pitch-angle diffusion as $\delta D_\mu X \sim 1$, where X is the effective life of cosmic rays. When the effective size of confinement is L, we have a relation $X \sim L^2/D$, where D is the coefficient of spatial diffusion along the magnetic lines of force; since $D_\mu$ gives also a measure of effective collision frequency, we have $D \sim c^2/D_\mu$.

Energy dependence of the quantities such as $\delta$ and X can be deduced from observational data; we thus write $\delta \sim \varepsilon^\xi$, $X \sim \varepsilon^{-\zeta}$, and $L \sim \varepsilon^{-\theta}$. Consideration of the pitch-angle diffusion and spatial diffusion with the aid of (32) then yields the relations,

$$-\zeta + (\alpha - 4) + \xi = 0, \tag{33}$$

$$-\zeta - (\alpha - 4) + 2\theta = 0 \quad . \tag{34}$$

Over the energy range, $10^{12}$ eV $\lesssim \varepsilon \lesssim 10^{15}$ eV, there is an indication that $0.3 \lesssim \xi \lesssim 0.5$ (Gombosi et al. 1975); in the energy interval, $10^9$ eV $\lesssim \varepsilon \lesssim 5 \times 10^{10}$ eV, it was reported that $0.1 \lesssim \zeta \lesssim 0.3$ (Ramaty, Balasubrahmanyan, and Ormes 1973). Assuming that the latter values of $\zeta$ apply also to the energy interval in the former observations, we may deduce $\alpha \simeq 3.8$ from (33). Alternatively an additional assumption that L is independent of $\varepsilon$, i.e., $\theta = 0$, would likewise yield $\alpha \simeq 3.8$ from (33) and (34). Although these results indicate a reasonable agreement with the theoretical prediction, $\alpha = 4$, a substantial degree of uncertainty seems to be involved both in the interpretation of the observed data and in the arguments connecting the observed results and the spectral index; more work would be needed before one can draw any definite conclusion on the spectral properties of the magnetic field.

## IV. ACCRETION MODEL OF COMPACT X-RAY BINARIES

### 4.1 Basic Parameters of Accretion Disk near Its Outer Boundary

Gas accretion onto a compact star which constitutes a part of a close binary system has been studied as a standard model of compact X-ray sources such as Cyg X-1 (e.g., Shakura and Sunyaev 1973, Novikov and Thorne 1973). Various plasma phenomena are involved in the accretion disks so formed around compact stars. In this section we apply the results obtained in Section II to investigate static and dynamic properties of such an accretion disk around a black hole; we thereby construct a theoretical model explaining observational features of Cyg X-1.

We begin with the outer boundary of the accretion disk, where the gas stream originating from the Roche-lobe overflow may splash. The azimuthal flow velocity of gas within such a boundary is approximately Keplerian, $v_K = (GM/r)^{1/2}$. (G: the gravitational constant; M: mass of the accreting star) The half thickness H of the disk is determined from the condition of hydrostatic pressure balance in the z direction. Taking account of the magnetic-pressure contribution (27), we have

$$H = 1.18\ r(2T/m)^{1/2}/v_k \qquad (35)$$

where m is the mass of an ion.

The infall velocity u is determined from the conservation law of angular momentum,

$$nmrv_K u = w_{r\phi} r + \dot{J}/4\pi rH \quad , \qquad (36)$$

where $w_{r\phi}$ is the r-$\phi$ element of the stress tensor, and $\dot{J}$ is the rate at which angular momentum is deposited in the accreting star. The contribution of the $\dot{J}$ term in (36) becomes significant only in the vicinity of the star; we shall henceforth neglect this term. With the aid of the explicit calculation (28) and the relation $w_{r\phi} = -\chi/4\pi$, we obtain

$$u = 0.636\ T/mv_K \quad . \qquad (37)$$

It is instructive to compare the present theory with a standard "$\alpha$ model" assumed in a conventional treatment of the disk problem (e.g., Shakura and Sunyaev 1973); in the latter, one assumes that $w_{r\phi}$ is proportional to the total pressure P, i.e., $w_{r\phi} = \alpha P$. The results of the magnetohydrodynamic-turbulence theory described in the foregoing sections are in fact compatible with the $\alpha$ model in the outer domain of the disk; there, we obtain $\alpha = 0.23$. In the inner domain, however, the $\alpha$ model is not justified.

The plasma density n is determined from the continuity equation, $\dot{M} = 4\pi rHnmu$, where $\dot{M}$ is the mass accretion rate; with the aid of (35) and (37) we have

$$n = 7.50 \times 10^{-2} \frac{\dot{M} v_K^2}{mr^2 (T/m)^{3/2}} . \tag{38}$$

Basic parameters of the disk plasmas, such as the density, the flow velocities, and the disk thickness, can thus be determined once the mass accretion rate and the temperature are known at a given radial distance.

### 4.2 Bimodal Behavior

In addition to the pressure-balance condition and conservation laws of matter and angular momentum, we need to examine energy budget in the gas. Viscous dissipation takes place at a rate proportional to a product between shear stress $w_{r\phi}$ and shear $\sim (GM/r^3)^{1/2}$, or what is the same thing, the last term on the right-hand side of (20). This energy input is processed in two ways: radiative loss and gas heating. When the effective opacity $\tau_{eff}$ is greater than one, the rate of radiative loss may be determined by radiative diffusion in an optically thick medium; when $\tau_{eff} < 1$, bremsstrahlung is the principal mechanism. Assuming the electron scattering opacity $\tau_{es}$ is greater than the average free-free opacity $\bar{\tau}_{ff}$, we calculate the effective opacity with the aid of (35) and (38) as

$$\tau_{eff} = 3.2 \times 10^{-7} \dot{M}_{17}^{3/2} M_* T_6^{-7/2} r_{11}^{-3} ,$$

where $M_* = M/M_\odot$, $\dot{M}_{17} = \dot{M}/(10^{17}$ gr/sec), $T_6 = T/(10^6$K), and $r_{11} = r/(10^{11}$cm); when $\tau_{es} < \bar{\tau}_{ff}$, $\tau_{eff} = \bar{\tau}_{ff}$. In the vicinity of the disk outer boundary, we ordinarily have $\tau_{eff} \ll 1$.

Suppose that the rate of radiative loss exceeds the rate of viscous dissipation near the outer boundary of the disk. The gas temperature T then decreases as it falls toward the accreting star; $\tau_{eff}$ further increases, resulting in more efficient radiative losses. After passing a transient region, the gas evolves into an optically thick (geometrically thin) disk. If we begin with opposite boundary conditions (viscous dissipation > radiative loss), the situations will be reversed and the gas will end up with an optically thin (geometrically thick) disk, i.e., an onset of thermal runaway (e.g., Pringle, Rees, and Pacholczyk 1973). Those two cases are schematically illustrated in Fig. 1.

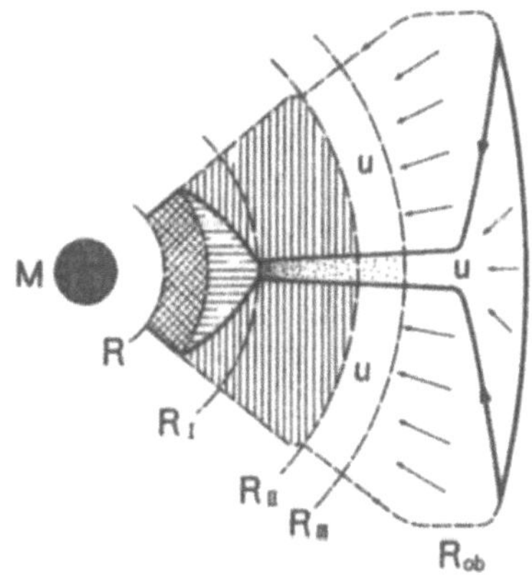


Fig. 1. Schematic diagram (cross-sectional view) of accretion disk around a black hole. Solid lines depict a disk state corresponding to high state of Cyg X-1; dashed lines, low state.

Whether the accreting gas evolves into an optically thick disk or an optically thin disk, therefore, depends on comparison between the rates of viscous dissipation and radiative loss at the disk outer boundary. To carry out a quantitative comparison, we use the result of the explicit formulation of shear stress developed in Section II. The rate of viscous dissipation per unit volume is then obtained as

$$\dot{\mathcal{E}}_V = \frac{nT}{3\nu}\left(r\,\frac{\partial\Omega}{\partial r}\right)^2 = 0.954\ nT\,\frac{v_K}{r} \quad . \tag{39}$$

The rate of radiative energy loss per unit volume in an optically thin medium is given by

$$\dot{\mathcal{E}}_R = (1.6\times 10^{-27}\ \mathrm{erg/cm^3\ sec}\ [\mathrm{cm^3}]^2[\mathrm{K}]^{-1/2})\ n^2T^{1/2} \quad . \tag{40}$$

We thus find the condition for the thermal runaway at r as

$$\dot{M} < (3.8\times 10^{17}\ \mathrm{gr/sec})\ M_*^{-1/2}\ T_7^2\ r_{11}^{3/2} \quad . \tag{41}$$

The nature of bimodal behavior is thus clear. The physical conditions of the accreting matter (e.g., $\dot{M}$ and T) at the disk outer boundary may fluctuate. However, as long as (41) is satisfied, the disk evolves into an optically thin state; if not satisfied, it evolves into an optically thick state.

It is important to examine thermal stability of those two disk states. The state of an optically thick disk is determined through balance between viscous dissipation and radiative loss. Taking account of (38), we have $\dot{\mathcal{E}}_V \sim T^{-1/2}$ as a function of the temperature. In an optically thick medium, $\dot{\mathcal{E}}_R \sim T^{9/2}$ or $T^{19/2}$, depending on the relative magnitude between $\tau_{es}$ and $\bar{\tau}_{ff}$. We thus find that an optically thick disk state is stable against a perturbation of T. The state of an optically thin disk is determined through balance between viscous dissipation and gas heating. The rate of gas heating per unit volume is given by

$$\dot{\mathcal{E}}_G = 2nu \frac{\partial}{\partial r} c_p T \sim T^{1/2} \quad , \tag{42}$$

where $c_p$ is the constant-pressure specific heat per particle. Clearly we see that the optically thin disk state is also a thermally stable one.

### 4.3 Inner Domain

Let us follow the accretion disk evolved into an optically thick state. In the middle domain of an optically thick disk, the temperature increases to keep up with increase of viscous dissipation,

$$T = (1.7 \times 10^6 \text{ K})\dot{M}_{17}^{2/5} M_*^{3/10} r_8^{-9/10} \quad . \tag{43}$$

At a radius $R_I$, the radiation pressure ($\sim T^4$) becomes equal to the gas pressure. Inside $R_I$, the disk thickness, determined from the hydrostatic pressure balance condition, bulges up rapidly ($\sim r^{-21/2}$), accompanied by sharp decrease in density ($\sim r^{12}$). The effective opacity thus decreases quickly ($\sim r^{51/4}$), which in turn reduces the efficiency of radiative losses and acts to raise the gas temperature. An onset of thermal runaway takes place at

$$R_I = (1.0 \times 10^7 \text{ cm})\dot{M}_{17}^{16/21} M_*^{1/3} \quad . \tag{44}$$

For $r < R_I$, the disk goes over to an optically thin state (see Fig. 1). The solution of inner disk ($r < R_I$) as described above in this paragraph differs markedly from the corresponding one presented by Shakura and Sunyaev (1973). We remark that the latter solution is thermally unstable. Such an instability develops in the direction to heat the gas, resulting in loosening of coupling between electrons and ions; and hence between radiation field and ions. Our solution has been obtained by taking these effects into account.

It is thus important to note the following: Even though the disk has evolved into one of the two physically distinct states in the middle domain depending on the criterion (41) at its outer boundary, it develops into a single (optically thin, geometrically thick) state inside the radius $R_I$ irrespective of the criterion (41).

The optically thick middle disk ($R_I < r < R_{III}$) emits soft X rays corresponding to temperatures (43), where $R_{III}$ is the radius at which $\tau_{es}$ and $\bar{\tau}_{ff}$ become equal (e.g., Novikov and Thorne 1973); its total luminosity is

$$L_S = (1.9 \times 10^{36} \text{ erg/sec}) \dot{M}_{17}^{5/21} M_*^{2/3} . \tag{45}$$

To investigate the properties of the optically thin inner disk, we follow the disk evolved into an optically thin state (see Fig. 1); temperatures in this state are substantially higher than those in the optically thick state. As T increases, the energy equipartition time between ions and electrons (e.g., Spitzer 1962) increases. As matter falls inward, the mass density $\rho$ and hence $\tau_{es}$ increase. Combined effect of increase in T and $\tau_{es}$ is to enhance the efficiency of Comptonization of low-energy photons. Electrons lose energy through such Comptonization; photon energies and the total luminosity are boosted. In the domain where those energy transfer processes become important, ion temperatures are substantially greater than electron temperatures (Eardley, Lightman and Shapiro 1975).

The radius $R_{II}$ at which such a two-temperature situation begins to develop is given by

$$R_{II} = (2.1 \times 10^9 \text{ cm}) M_* \left( \frac{\dot{M}_{17}}{\mathcal{L}_{37}} \frac{\ell n\Lambda}{15} \right)^{-1/2} , \tag{46}$$

where $\mathcal{L}$ is the luminosity of Compton unsaturated photons, and $\ell n\Lambda$ is the Coulomb logarithm (e.g., Spitzer 1962). The electron temperatures inside $R_{II}$ are

$$T_e = (7.3 \times 10^8 \text{ K}) M_*^{1/5} \left( \frac{\dot{M}_{17}}{\mathcal{L}_{37}} \frac{\ell n\Lambda}{15} \right)^{2/5} r_7^{-1/5} \tag{47}$$

The inner boundary R of the disk may be determined by a radius at which relativistic corrections start to become important (Novikov and Thorne 1973); we take $R = 6GM/c^2$. The total luminosity from the inner domain ($R \lesssim r < R_I$ or $R_{II}$) is

$$L_H = (1.6 \times 10^{34}\ \mathrm{erg/sec}) A(\rho, T_e) \dot{M}_{17}^2 M_*^{-1} \left( \frac{\dot{M}_{17}}{\mathcal{L}_{37}} \frac{\ell n \Lambda}{15} \right)^{1/5} . \tag{48}$$

where $A(\rho,T_e)$ represents the enhancement factor over bremsstrahlung due to Comptonization around $r = R$; we may approximate $\mathcal{L} \simeq L_H$ when Comptonization is not saturated.

The factor $A_\infty$ for saturated Comptonization ($\tau_{es} = \infty$) has been obtained by Kompaneets (1956) as

$$A_\infty = (3/4)[\ell n(4/\gamma x)]^2 , \tag{49}$$

where $\ell n\ \gamma = 0.577\cdots$, and x is a solution of $\beta x^{-2} \ell n(4/\gamma x) = 1$ with $\beta = (7.1 \times 10^9 [\mathrm{cm}^3][\mathrm{K}]^{9/2}) n T_e^{-9/2}$. For the case of a finite $\tau_{es}$, we then find an interporation formula (Ichimaru, to be published)

$$A = \{\exp[y(\tau_{es}^2\ \ell n\ t - 1) - 1\}/[\tau_{es}^2\ \ell n\ t - 1] , \tag{50}$$

where $y = \ell n(A_\infty \tau_{es}^2\ \ell n\ t)/\tau_{es}^2\ \ell n\ t$, and $t = 1 + (4T_e/m_e c^2)$ .

4.4 Application to Cygnus X-1.

Observations of Cyg X-1 (Agrawal et al. 1972, Tananbaum et al. 1972, Baity et al. 1973, Heise et al. 1975, Holt et al. 1975, Sanford et al. 1975) appear to indicate that its time-averaged X-ray spectrum usually assumes one of two states. In a "low state", the bulk of the luminosity, of the order of a few times $10^{37}$ erg/sec, is contained in the high-energy part ($E \gtrsim 10$ keV) of the spectrum, with a continuous tail extending into its low-energy part ($E \lesssim 10$ keV). In a "high state", the luminosity in the high-energy part appears to remain virtually unchanged, while the luminosity in the low-energy part increases substantially to a magnitude comparable to that in the high-energy part. [See Thorne and Price (1975) for a resume of the spectral variations.] Transitions from high state to low state were observed in April 1971 and in May 1975, both of which required a time of the order of a week (Tananbaum et al. 1972, Sanford et al. 1975). A transition from low state to high state, observed in April 1975, required a time $\lesssim$ 1 day (Holt et al. 1975). Time variabilities of the spectrum with inclucion of millisecond bursts were noted (Oda et al. 1971, Rappaport et al. 1971, Schreier et al. 1971, Rothschild et al. 1974, Weisskopf et al. 1975).

A number of theories, based on a "standard" accretion disk

model surrounding a black hole, have been advanced recently to account for those Cyg X-1 observations. Thorne and Price (1975) considered Comptonized photons arising from optically thin inner region and thermal photons emitted from optically thick middle disk; Eardley, Lightman, and Shapiro (1975) showed possible existence of two-temperature inner disk which reproduced the observed hard X-ray spectrum; in an unpublished report Thorne noted possibility of hot-gas cloud lying in the outer regions of the Roche lobe. While those theories should correspond to physical reality in Cyg X-1 and their basic features are retained in the present theory as well, one important question has remained unanswered: How does a transition from one state to the other take place? In the light of the analyses described above it would be natural to assume that the origin of the transition is the bimodal behavior of an accretion disk.

In applying the foregoing results to interpretation of Cyg X-1 observations, we tentatively identify the high state with a state in which (41) is not satisfied. In the middle domain an optically thick disk is realized, emitting soft X-rays with luminosity (45). The low state is identified with a state in which (41) is satisfied; in the middle domain the disk is optically thin, emitting lesser amount of radiation. In either of those two states, a permanent hard X-ray luminosity is provided by (48).

For numerical comparison, we adopt the following set of parameters derivable from orbital analysis (Bolton 1975; Thorne, unpublished): $M_* = 13$, the mass of the primary = 21 $M_\odot$, the binary separation $a = 3.3 \times 10^{12}$ cm, the inclination angle = 26°. The radius $R_{ob}$ of the disk outer boundary may then be estimated from an equation, $(GMR_{ob})^{1/2} \simeq a^2/(\text{period of the binary})$; we find $R_{ob} \simeq 3 \times 10^{11}$cm, about a quarter of an estimated Roche-lobe radius, $1.1 \times 10^{12}$ cm. At $R_{ob}$, a temperature corresponding to the gravitational free-fall velocity is $4.7 \times 10^7$ K, while the optically thin disk solution predicts $T \simeq 2.1 \times 10^7$ K at the same radius. Physical processes near the outer boundary such as shock heating would bring the gas temperatures close to the values cited above; we adopt $T_{ob} = 2 \times 10^7$ K. The critical condition (41) then is $\dot{M} < 2.2 \times 10^{18}$ gr/sec. For transitions to take place, an average value of $\dot{M}$ should be in the vicinity of its critical value; hence $\dot{M}_{17} = 22$($\dot{M} = 3.5 \times 10^{-8}$ $M_\odot$/yr). Correspondingly we compute $R = 1.2 \times 10^7$ cm, $R_I = 2.4 \times 10^8$ cm, $R_{II} = 6 \times 10^9$ cm, and $R_{III} = 3 \times 10^{10}$ cm. The hard X-ray luminosity (48) is $4.0 \times 10^{37}$ erg/sec. The soft X-ray luminosity (45) in the high state is $2.2 \times 10^{37}$ erg/sec.

Approximate features of the spectral distribution predicted with those parameters are depicted in Fig. 2. The spectral distribution arises mainly due to spatial variation of radiation characteristics. Although $T_e$ in (47) depends rather weakly on r, average energies of photons exhibit considerable r dependence because of different rates of Comptonization. In Fig. 2, the

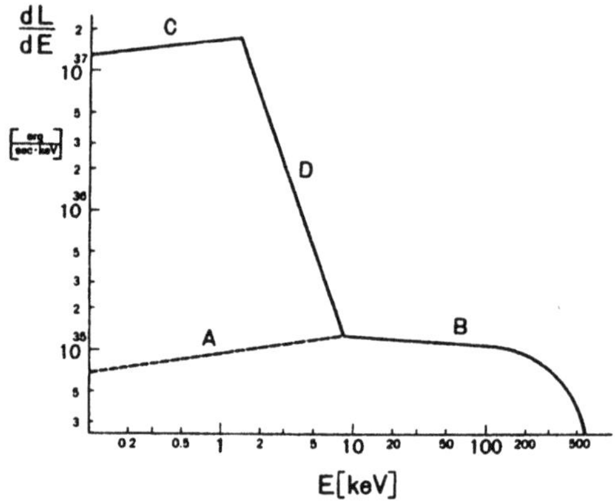


Fig. 2. Approximate shape of X-ray spectrum predicted in high state (solid lines) and low state (AB) of accretion disk around a black hole with $M = 13\ M_\odot$ and $\dot{M} = 3.5 \times 10^{-8}\ M_\odot/yr$. The spectral index $\alpha$ (i.e., $dL/dE \propto E^{-\alpha}$) is $\sim 0.1$ for B; $\alpha \simeq 2.7$ for D.

spectrum B stems from those photons emitted from the cross-hatched domain in Fig. 1; photons around the high-energy edge are mainly those emitted from the vicinity of R. The turn-over at $E \simeq 8$ keV has been assessed from the condition $\tau_{es}^2\ (4T_e/m_ec^2) \simeq 1$, which takes place at $r \simeq 5 \times 10^7$ cm. The spectrum A represents the contribution from the vertically hatched domain in Fig. 1.

The spectrum C originates from the optically thick middle disk (the dotted region in Fig. 1). Again the spatial variation is the major cause of the spectral distribution. Luminosity around the high-energy edge ($E \simeq 1.4$ keV) comes from those photons emitted from the vicinity of $R_I$. The spectrum D stems from the transitional domain (horizontally hatched in Fig. 1) between $R_I$ and the fully developed inner domain.

The characteristic time, $t_{up}$, for a transition from low state to high state is estimated to be a time in which to build up an optically thick middle domain after the condition (41) is crossed over:

$$t_{up} = (r/u)_{\text{thick middle}} = (3.2 \times 10^4\ \text{sec})\dot{M}_{17}^{-2/5} M_*^{1/5} r_9^{7/5} \tag{51}$$

where u is the radial inflow velocity. Choosing $r_9 = 2$ close to luminous region near $R_I$, we obtain $t_{up} \simeq 0.5$ days.

When T in the outer boundary exceeds its critical value, hot matter starts to accrete with velocities much faster than cool matter in the existing disk; the hot matter abrades the cool disk. The time, $t_{down}$, for a transition from high state to low state is thus estimated from the rate of hot gas fed into the disk and the ratio between specific heat per unit area in the hot and cool disks; we find

$$t_{down} = (1.8 \times 10^5\ sec)\dot{M}_{17}^{3/5} M_*^{7/10} r_{10}^{-3/5} (T_7 r_{11})_{ob}^{1/2} , \quad (52)$$

where $r_{10}$ refers to a radius at which transient behavior arising for the outer boundary conditions has been quenched; we may estimate $r_{10} = 3.8 \times 10^7 M_*/T_{ob}$. Use of our model parameters then yields $t_{down} \simeq 1.5$ weeks, in close agreement with observations.

The fast time variabilities may be explained in our model as arising from the magnetohydrodynamic turbulence in the inner disk. The calculated level of magnetic-field fluctuations is rather high, i.e., $\langle \delta B^2 \rangle / 8\pi$ (gas pressure) $\simeq 0.59$. The bulk of those fluctuations is contained in the long-wavelength domain comparable to the disk thickness; their periods range from a few milliseconds to a few tenths of a millisecond near R. Similarly, in the vicinity of $R_I$, a hydrodynamic time scale across the disk is 0.3 sec; a time scale for photon diffusion is 0.5 sec.

We remark that within the framework of an accretion model the observed features of spectral transitions support a black hole, rather than a neutron star, as the central object of Cyg X-1. The orbital data (Bolton 1975) and the millisecond variabilities cannot distinguish between a model with a black hole as the unseen secondary of HDE 226868 and a model which has a neutron star as a companion of an unseen secondary normal star. Suppose that a neutron star with a radius $R_*$ is involved. Kinetic energy associated with the Keplerian velocity of accreting matter must be released on or near the stellar surface; this energy must subsequently be radiated away from the star, which may resemble a black body. To radiate at a luminosity L, the radiation temperature is $\sim (10^7\ K)[L/(10^{37}\ erg/sec)]^{1/4}[R_*/(10^6\ cm)]^{-1/2}$. For accretion onto a neutron star one thus expects a permanent soft X-ray emission to be observed, contrary to the observed features in Cyg X-1 transitions. On the other hand, as we have shown, a model based on accretion into black hole is consistent with observed transitions in the low-energy part of the X-ray spectrum.

## ACKNOWLEDGMENTS

I wish to express my gratitude to Professor David Pines for his hospitality at the University of Illinois and for important suggestions and stimulating discussions during the course of this work. I should like to thank Professor Kip S. Thorne for his unpublished report which has given me much insight and stimulation on the Cyg X-1 problems. This work was supported in part by the Japan Society for the Promotion of Science and the National Science Foundation under the Japan-U.S. Cooperative Science Program.

## REFERENCES

Agrawal, P. C., Gokhale, G. S., Iyengar, V. S., Kunte, P. K., Manchanda, R. K., and Sreekantan, B. V. 1972, Ap. and Space Sci., 18, 408.

Allen, C. W. 1973, Astorphysical Quantities, 3rd ed. (Athlone Press, London), Sec. 134.

Baity, W. A., Ulmer, M. P., Wheaton, W. A., and Peterson, L. E. 1973, Nature Phys. Sci., 245, 90.

Bolton, C. T. 1975, Ap. J., 200, 269.

Eardley, D. M., Lightman, A. P.1975, Ap. J., 200, 189.

Eardley, D. M., Lightman, A. P., and Shapiro, S. L. 1975, Ap. J. (Letters), 199, L153.

Gombosi, T., Kota, J., Somogyi, A. J., Varga, A., Betev, B., Katsarsky, L., Kavlakov, S., and Khirov, I. 1975, Nature, 255, 687.

Heise, J., Brinkman, A. C., Schrijver, J., Mewe, R., den Boggende, A., Gronenschild, E., Parsignault, D., Grindlay, J., Schnopper, H., Schreier, E., and Gursky, H. 1975, Nature, 256, 107.

Heisenberg, W. 1948, Z. Physik, 124, 628.

Holt, S. S., Boldt, E. A., Kaluzienski, L. J., and Serlemitsos, P. J. 1975, Nature, 256, 108.

Ichimaru, S. 1975a, J. Phys. Soc. Japan, 39, 1373.

Ichimaru, S. 1975b, Ap. J., 202, 528.

Jokipii, J. R. 1968, Ap. J., 152, 997.

Kolmogorov, A. N. 1941, Compt. Rend. Acad. Sci. USSR, 30, 301.

Kompaneets, A. S. 1956, Zh. Eksp. Teor. Fiz., 31, 876 (English transl. in Soviet Phys. JETP, 4, 730 [1957]).

Morrison, P. 1961, Handbuch der Phys., 4611, 1.

Nakano, T. 1972, Ann. Phys. (N. Y. ), 73, 326.

Novikov, I. D., and Thorne, K. S. 1973, in Black Holes, ed. C. DeWitt and B. Dewitt (New York: Gordon and Breach).

Oda, M., Gorenstein, P., Gursky, H., Kellogg, E., Schreier, E., Tananbaum, H., and Giacconi, R. 1971, Ap. J. (Letters). 166, L1.

Prendergast, K. H. and Burbidge, G. R. 1968, Ap. J. (Letters), 151, L83.

Pringle, J. E., Rees, M. J., and Pacholczyk, A. G. 1973, Astr. and Ap., 29, 179.

Ramaty, R., Balasubrahmanyan, V. K., and Ormes, J. F. 1973 Science, 180, 731.

Rappaport, S., Doxsey, R., and Zauman, W. 1971. Ap. J. (Letters),

168, L43.
Rothschild, R. E., Boldt, E. A., Holt, S. S., and Serlemitsos, P. J. 1974, Ap. J. (Letters), 189, L13.
Sanford, P. W., Ives, J. C., Bell Burnell, S. J., Mason, K. O., and Murdin, P. 1975, Nature, 256, 109.
Schrier, E., Gursky, H., Kellogg, E., Tananbaum, H., and Giacconi, R. 1971, Ap. J. (Letters), 170, L21.
Shakura, N. I. and Sunyaev, R. A. 1973, Astr. and Ap., 24, 337.
Somogyi, A. J. 1975, Nature, 255, 689.
Spitzer, L., Jr. 1962, Physics of Fully Ionized Gases (New York: Wiley).
Tajima, T., Ichimaru, S., and Nakano, T. 1974, J. Plasma Phys., 12, 381.
Tananbaum, H., Gursky, H., Kellogg, E., Giacconi, R., and Jones, C. 1972, Ap. J. (Letters), 177, L5.
Thorne, K. S. and Price, R. H. 1975, Ap. J. (Letters), 195, L101.
Weisskopf, M. C., Kahn, S. M., and Sutherland, P. G. 1975, Ap. J. (Letters), 199, L147.

PLASMA HEATING BY HIGH-CURRENT RELATIVISTIC ELECTRON BEAMS

D.D. Ryutov

Institute of Nuclear Physics

Novosibirsk, 630090, USSR

## 1. INTRODUCTION

The relaxation of a beam is due to the scattering of beam electrons on the small-scale oscillations excited by the beam itself; the subsequent dissipation of energy of these oscillations results in plasma heating. The dispersion properties and the growth rates of oscillations are defined by the plasma density, the temperature, the magnetic field intensity, the beam density, and the beam angular spread, etc. Below we will consider the case when the magnetic field is small ($\omega_{He} << \omega_{pe}$) and the most unstable waves are the Langmuir oscillations.

At a sufficiently large beam angular spread $\Delta\theta$ (practically at $\Delta\theta \gtrsim 15$-$20^{o}$, cf. Ref. [1]) the instability is of kinetic type. The beam excites a broad spectrum of oscillations with random phases, so that a statistical approach can be used for the description of the system. At not too high beam-to-plasma density ratios $n_b/n$, the energy density of the Langmuir oscillations, U, is also small, and one can neglect the nonlinear corrections to the dispersion relation. In other words, at $\Delta\theta > 15^{o}$ and at small values of $n_b/n$ one can use the weak turbulence theory when determining the oscillation spectrum excited by the beam.

Within the framework of this theory, the equation which determines the spectral energy density of Langmuir oscillations $W_{\vec{k}}$ is:

$$\frac{\partial W_{\vec{k}}}{\partial t} = 2\gamma_{\vec{k}} W_{\vec{k}} + \hat{A} W_{\vec{k}} \qquad (1)$$

where $\gamma_{\vec{k}}$ in the growth rate of the beam instability and the term $\hat{A}W_{\vec{k}}$ describes the interaction between the oscillations ($\hat{A}$ is a nonlinear operator).

When writing down equation (1), we considered the plasma as homogeneous (effects of inhomogeneity are discussed in Refs. [1-3]). Besides that, we neglected the convection of oscillations at the group velocity. The last assumption is usually valid for Langmuir oscillations because of their small group velocity. The situation when the group velocity is significant (but the nonlinear interaction term $\hat{A}W_{\vec{c}}$ is small) has been treated in Ref. [4].

In $\vec{k}$-space, the linear growth rate of the beam instability $\gamma_{\vec{k}}$ is essentially different from zero in the vicinity of the plane $k_z = \omega_{pe}/c$ (c is the speed of light, and the z-axis is directed along the beam axis). The characteristic value of $\gamma_{\vec{k}}$ for oscillations with $k_\perp \lesssim \omega_{pe}/c$ is (see Ref. [1]:

$$\gamma_b \sim \omega_{pe} \frac{n_b}{n} \frac{mc^2}{\varepsilon} \frac{1}{\Delta\theta^2} \tag{2}$$

where $\varepsilon(> mc^2)$ is the energy of the beam electrons. At $k_\perp >> \omega_{pe}/c$ the growth rate decreases as $k_\perp^{-2}$.

As was already mentioned, equation (1) is valid only at sufficiently small values of $n_b$. At growing beam densities, the energy density of the waves also grows, and at some point the weak turbulence theory breaks down. Since such a situation is rather often encountered in the experiments, we will below also consider the effects of strong turbulence.

## 2. BEAM RELAXATION IN AN ISOTHERMAL PLASMA IN A WEAK MAGNETIC FIELD

In this section we shall consider a plasma with $T_e = T_i = T$. As we shall show below, the parameter which characterizes the effect of the magnetic field on the relaxation process is $\beta = 8\pi n T_e/H^2$: at $\beta >> 1$ the role of the magnetic field is more or less insignificant and at $\beta \lesssim 1$ it determines some very essential features of the relaxation process. The condition $T_e = T_i = T$, $\beta >> 1$ will be satisfied in the experiments on dense plasma heating that are planned at present [5,6].

As has been shown previously (see Refs. [7,8] and the review [1])the most important nonlinear process in the range of temperatures interesting for thermonuclear applications ($T \lesssim mc^2(m/10M)^{2/5}$, i.e.

at $T \lesssim 20$ keV for Hydrogen plasma), is the induced scattering of Langmuir oscillations on ions. This process results in a spectral transfer of Langmuir oscillations into the long wave-length part of the spectrum. As a result, the phase velocity of Langmuir oscillations becomes higher than the speed of light and the oscillations cannot any more interact with the beam and, consequently, their amplitude becomes saturated. This is the stabilizing effect of scattering on ions.

For the very soft condition (see [1])

$$T > m^2c^2/10M \tag{3}$$

(i.e. at $T \gtrsim 30$ eV for Hydrogen plasma) the change $\Delta k$ of the modulus of the wave vector in a separate scattering event is small as compared with the wave vector of the unstable oscillations $k \sim \omega_{pe}/c$, and the spectral transfer is of a "differential" form (cf.[9]). Under these conditions the characteristic time of spectral transfer of oscillations via the most unstable region of wave numbers of width $\sim \omega_{pe}/c$ appears to be of the order of (cf. [9]):

$$\tau \sim \frac{10}{\omega_{pe}} \frac{nT}{U} \frac{M}{m} \frac{T}{mc^2} \tag{4}$$

In the stationary state the condition

$$\tau\gamma_b \sim 1 \tag{5}$$

must be satisfied. This condition allows one to estimate the energy density of the oscillations in the region $k > \omega_{pe}/c$ (we shall call this region "resonant" - just to show that the Cherenkov resonance condition which determines the possibility of interaction of the oscillations with the relativistic beam, is satisfied in this region). The energy density of the resonant oscillations appears to be

$$U \sim 10\, nT \frac{M}{m} \frac{T}{\varepsilon} \frac{n_b}{n\Delta\theta^2} \tag{6}$$

Substituting this estimate into the quasi-linear equation for the relativistic electrons, one can find the length L at which the beam has lost half of its initial energy (relaxation length):

$$L \simeq \frac{1}{20} \frac{c}{\omega_{pe}} \frac{n}{n_b} \frac{m}{M} \left(\frac{mc^2}{T}\right)^2 \left(\frac{\varepsilon}{mc^2}\right)^3 \tag{7}$$

The characteristic feature of the interaction of the beam electrons with the three-dimensional spectrum of Langmuir oscillations is that the energy losses of the beam are accompanied by an increase of the beam angular spread. One can show that the relative energy losses $\Delta\varepsilon/\varepsilon$ are connected with the arising angular spread $\Delta\theta$ through the relationship [7,8,10]: $\Delta\varepsilon/\varepsilon \sim \Delta\theta$. This means that the most essential fraction of the beam energy is released when the beam is already strongly scattered, $\Delta\theta \sim 1$. For this reason we shall make all the subsequent estimates by assuming that $\Delta\theta \sim 1$.

Formula (7) was obtained by Breizman and Ryutov and published (cf. [7,8]).Obviously, it correctly reflects the basic features of the considered phenomena. For a more precise description of the relaxation process one not only needs to estimate the energy density of the oscillations, but one should also know the details of the spectrum. According to equation (1), the stationary spectrum should be determined from the condition

$$W_{\vec{k}} \left( \gamma_{\vec{k}} + \frac{1}{W_{\vec{k}}} \hat{A} W_{\vec{k}} \right) = 0$$

For the induced scattering by ions, $W_{\vec{k}}^{-1} \cdot \hat{A}$ is a linear integral operator and consequently, the problem of determining of the stationary spectrum is reduced to solving a Fredholm integral equation of the 1-st kind. As is well known, such equations have, generally speaking, only singular solutions, that is, the stationary spectrum must be singular. In explicit form, this singular solution has been found by Breizman et al [1]. An estimate of the relaxation length based on this solution coincides with (7). In principle, it is also possible to find the exact formula for L. However, it is not obvious that the initial spectrum of oscillations will evolve to the singular stationary solution. It is quite possible that this stationary state will never be reached (or will be reached in a too long time).

As was already mentioned, the oscillations excited by the beam are transferred to the nonresonant region $k < \omega_{pe}/c$. In this region the same scattering mechanism works, resulting in a further decrease of k to the values $k \lesssim \Delta k \sim (1/r_D)(m/M)^{1/2}$. The time of this process is of the order of (4). If there is no sink of Langmuir oscillations on their way from $k \sim \omega_{pe}/c$ to $k \ll \omega_{pe}/c$, then at $k \lesssim \Delta k$ a region with a very high concentration of Langmuir oscillations is formed ("Langmuir condensate"), and all the energy released by the beam comes finally to this "condensate".

It is obvious that pair collisions cannot prevent the formation of the condensate. Indeed, the condition of absorption of Langmuir oscillations within the time of spectral transfer ($\nu_{ei}\tau > 1$) contradicts the condition of excitation of the beam instability ($\gamma_b > \nu_{ei}$). This can be seen from equation (5). However, if the instability is not too strong, $\gamma_b - \nu_{ei} \lesssim \nu_{ei}$, then one can be sure that at least half of the energy released by the beam will go to plasma heating and only half of it will reach the condensate. The estimate (7) of the relaxation length can be presented at $\gamma_b - \nu_{ei} \sim \nu_{ei}$ as: $L \sim (c/20\ \nu_{ei})(m/M)(\varepsilon/T)^2$.For the case of a sufficiently dense plasma this estimate gives values of the relaxation length which are not too large. For example, at $n = 10^8\ cm^{-3}$, $T = 10^4$ eV, and $\varepsilon = 10^6$ eV, L appears to be about 2 m.

In the dense plasma the formation of a condensate can be strongly influenced even by a very weak magnetic field. The role of the magnetic field consists in changing the dispersion law for Langmuir oscillations. That dispersion relation at $\omega_{He} \ll \omega_{pe}$ has the form:

$$\frac{\omega-\omega_{pe}}{\omega_{pe}} = \frac{3}{2}\frac{k^2 v_{Te}^2}{\omega_{pe}^2} + \frac{1}{2}\frac{\omega_{He}^2}{\omega_{pe}^2}\frac{k_1^2}{k^2}\left(1 - \frac{\omega_{pe}^2}{k^2 c^2}\right) \qquad (8)$$

If the condition $\beta \gg 1$ is satisfied, then the last term in equation (8) is much smaller than the first one for the resonant oscillations ($k \sim \omega_{pe}/c$), that is, the magnetic field does not strongly affect the induced scattering of resonant oscillations and consequently the time of the spectral transfer through the resonant region remains of the order of (4). On the other hand, at $k \lesssim \omega_{pe}/c\beta^{1/4}$ the last term in equality (8) becomes much greater than the first one and strongly influences the time of spectral transfer from $k \sim \omega_{pe}/c\beta^{1/4}$ to $k \sim \Delta k \sim (1/r_D)\sqrt{m/M}$ (we imply that the quite nonrestrictive condition $\beta \ll (MT/m^2c^2)^2$ is satisfied, that is, $\omega_{pe}/c\beta^{1/4} \gg \Delta k$). As was shown in Ref. [12], this time significantly grows as compared to the estimate (4). If it becomes greater than $\nu_{ei}$, then the oscillations are absorbed by pair collisions and the condensate is not formed.

The sink for nonresonant oscillations can also be provided by the process of merging of two Langmuir waves into an electromagnetic wave at twice the plasma frequency. This process can be quite essential if the condition $T/mc^2 > \sqrt{m/10M}$ (see Ref. [8] is satisfied. If the plasma density grows more than 4 times in all directions from the relaxation region, then this radiation turns out to be trapped and is effectively absorbed near the reflection points due to effects of spatial dispersion. On the other hand, if the radiation is not trapped, then it is lost from the system and the plasma heating becomes less effective (but the system becomes an effective source of microwave radiation).

One should note that all the considered mechanisms of dissipation of Langmuir oscillations results almost entirely in electron heating; ion heating is inefficient. The only reliable source of ion heating is collisional energy exchange between electrons and ions; however, in a dense plasma this source is rather effective.

## 3. THE EFFECTS OF STRONG TURBULENCE

At growing beam densities the oscillation energy also grows and, finally, the question of the applicability of the weak turbulence theory arises. The violation of this theory manifests itself in a change of the dispersion properties of the plasma. In particular, in the case under consideration a new - and unstable - branch of oscillations arises. This phenomenon was discovered by Vedenov and Rudakov [13]. The reason for the excitation of this instability is that the pressure of Langmuir oscillations is negative. The condition of modulational instability has the form:

$$\frac{U}{nT} > (k\, r_D)^2 \tag{9}$$

where k is the characteristic wave number of the Langmuir oscillations.

Let us first consider the case where the mechanisms mentioned in the previous sections can not prevent condensate formation. In this case one should obviously identify U and k entering into inequality (10) as the energy density of the Langmuir condensate (which we will denote by $U_*$) and the characteristic wave number of the condensate (which will be denoted by $k_*$), respectively.

The question of plasma heating is equivalent to the question of condensate dissipation. If all the energy released by the beam is accumulated in the condensate region, it is clear that in the dense plasma the dissipation can be provided just by binary collisions. In this case one can estimate $U_*$ from the condition

$$U_* \nu_{ei} = U_{res} \gamma_b \tag{10}$$

where $U_{res}$ is the energy density of the resonant Langmuir oscillations.

One can see from equation (10), that $\gamma_b \gg \nu_{ei}$ automatically gives $U_* \gg U_{res}$, so that the weak turbulence theory is violated at an early stage. The modulational instability then develops, and the plasma becomes inhomogeneous. The fastest growing fluctuations are those with the scale close to the characteristic wave-length of the condensate $k_*^{-1}$. When the fluctuation level reaches the value $\Delta n/n \sim (k_* r_D)^2$, the inhomogeneities begin to influence the spectrum of the Langmuir condensate which results in a broadening of the spectrum in the direction of large k. Together with changes in the Langmuir wave spectrum the scale of the modulational fluctuations also changes (remaining always close to the characteristic wavelength of the condensate). Finally, the system will reach the stable state in which $k_*$ corresponds to marginal stability,

$$\frac{U_*}{nT} \sim (k_* r_D)^2 \tag{11}$$

and the fluctuation level is determined by the equality:

$$\frac{\Delta n}{n} \sim (k_* r_D)^2 \tag{12}$$

The characteristic time of the development of the acoustic oscillations under typical conditions is very small as compared to all other time scales of the problem, so that the state determined by conditions (11) and (12) is reached very quickly. The constancy of $U_*$ is provided by the collisional dissipation (see Eq. (10)).

The existence of long-wavelength ($k_* << \omega_{pe}/c$) low-frequency density fluctuations results in angular scattering of the resonant oscillations. At sufficiently high values of $\Delta n$ the scattering frequency $\nu_{scatt}$ which is equal to $\omega_{pe}(\Delta n^2/n^2)(k_*/k^3 r_D^2)$ could become larger than $\gamma_b$ in the most essential region ($k \sim \omega_{pe}/c$), that is, the spectrum in this region might get almost isotropic. In this case the beam instability would be completely suppressed [14].

Indeed, for the isotropic oscillation spectrum the power generated by the beam per unit volume is

$$Q = \int d^3\vec{k}\, W(k)\gamma_b(\vec{k}) = \int k^2 dk\, W_k(k) \int do\, \gamma_b(\vec{k})$$

where do represents integration over solid angles in $\vec{k}$-space. By noting that

$$\gamma_b(\vec{k}) = 2\pi^2 e^2 \omega_{pe} \int d^3\vec{p}\, \frac{\vec{k}}{k^2} \frac{\partial f}{\partial \vec{p}}\, \delta(\omega_{pe} - \vec{k}\vec{v})$$

one can find (see [14]):

$$\int do\, \gamma_b(\vec{k}) = -\frac{4\pi^3 e^2 \omega_{pe}^2}{k^3 c^2} \times$$

$$\times \left\{ \int_{v > \frac{\omega_{pe}}{k}} \frac{d^3\vec{p}}{v^3} \frac{f}{p} + \int_{v = \frac{\omega_{pe}}{k}} d^3\vec{p}\, f \left(1 + \frac{p^2}{m^2 c^2}\right) \right\} < 0$$

that is $Q < 0$. Consequently, for the isotropic spectra the absorption of oscillations by the beam itself always overbalances the beam pumping, and the instability disappears. In fact, this means that the fluctuation level $\sqrt{\Delta n^2}/n$ will always be close to its marginal value corresponding to $\nu_{scatt} \sim \gamma_b$, and the system will stay close to the border of instability ($Q \simeq 0$). The decrease in effective growth rate arising from this effect results in an increasing relaxation length and in smaller heating efficiency. However, in a dense plasma with frequent collisions this undesirable effect manifests itself only at high beam densities (for more details see Ref. [14]).

Up to now, we considered a dense plasma where the dissipation of the Langmuir condensate by pair collisions was more or less effective. From now on, we will proceed with the case of a collisionless plasma. In this plasma, after the beam has been introduced, the energy density of the condensate will grow with time linearly: $U_* = \gamma_b U_{res} t$. The modulational instability will proceed, and the plasma will become inhomogeneous. The characteristic wave number of the "condensate" will grow according to the relationship:

$$(k_* r_D)^2 \sim \sqrt{\Delta n^2}/n \sim U_*/n T \propto t$$

Finally, with growing $k_*$, absorption of Langmuir oscillations by thermal electrons will be allowed and a new channel for dissipation will be opened.

What should be noted here is that this channel of dissipation will be opened already at a very small fluctuation level $\sqrt{\Delta n^2}/n$. Indeed, even for a small $\Delta n^2$, high negative spikes will be formed somewhere in space due to occasional phasing of fluctuations, and dissipation of oscillations captured inside these spikes will start. Note that the formation of the spikes is a completely kinematical effect of phasing of low frequency fluctuations. To obtain these spikes one does not need any dynamical effect like plasmon attraction. Such effects can of course increase the probability of spike formation, but one can see that even without these dynamical effects one can obtain quite a sufficient amount of spikes due to occasional phasing. For instance, if one assumes that the distribution of low frequency fluctuations follows the Normal Gaussian Law, one immediately obtains the following spectrum for the Langmuir oscillations of the condensate:

$$W_{*\vec{k}} = \frac{U_*}{\pi^{3/2} k_*^3} \exp\left(- \frac{k^2}{k_*^2}\right)$$

The high k tail of this distribution ($k \gg k_*$) is a manifestation of the kinematic formation of the spikes. By means of this formula, one can easily calculate the energy dissipation by the thermal electrons:

$$Q = \omega_{pe}\sqrt{\frac{\pi}{8}} \int d^3\vec{k} \frac{W_{*\vec{k}}}{(kr_D)^3} \exp\left(- \frac{1}{2k^2 r_D^2}\right) \simeq$$

$$\simeq 2^{3/4} \sqrt{\frac{\pi}{3}} \frac{hT}{\sqrt{k_* r_D}} \exp\left(- \frac{\sqrt{2}}{k_* r_D}\right)$$

From the balance condition

$$Q \sim \gamma_b U_{res}$$

one can new easily estimate $k_*$:

$$k_* \sim 1/(r_D \Lambda)$$

where $\Lambda \simeq \ln[(\omega_{pe}/\gamma_b)(nT/U_{res})]$ in a weakly varying logarithmic factor which is usually of the order of 12-16. The fluctuation level, which is needed, appears to be relatively small:

$$\sqrt{\overline{\Delta n^2}}/n^2 \sim 1/\Lambda^2 \sim (10^{-2} - 3 \cdot 10^{-3})$$

The oscillations are mainly absorbed by suprathermal electrons with energies of the order of $T\Lambda$, that is beam energy is mainly deposited in the high-energy tail of the electron distribution. In principle, the fast particles can heat the main body of the distribution by binary collision. The problem whether they will not leave the plasma earlier depends on the details of the confinement system.

It is quite probable that, because of the tendency of plasmons to group together, the probability of spike formation will be higher than that estimated above. However, the obtained estimate

$$\sqrt{\overline{\Delta n^2}/n^2} \sim (10^{-2} - 3\cdot 10^{-3})$$

seems to be rather insensitive to some more refined properties of Langmuir turbulence which are discussed in References [16-18].

When considering the collisionless dissipation of the Langmuir condensate one should as before keep in mind the possibility that the low-frequency oscillations will affect the pumping efficiency. However, because of lack of space we will not discuss this effect here in detail.

Now we proceed to strongly nonlinear effects for the case where a condensate is not formed at all, and the modulational instability is excited by the resonant oscillations themselves. Since $k \sim \omega_{pe}/c$ for these oscillations, the critical $U_{res}$ value found by means of the estimate (9) appears to be

$$nT \frac{T}{mc^2} \qquad (13)$$

It is reached at $\gamma_b/\omega_{pe} \sim m/M$. The corresponding relaxation length is (see Refs. [1,7])

$$L_{min} \simeq \frac{c}{\omega_{pe}} \left(\frac{\varepsilon}{T}\right)^2 \qquad (14)$$

(the subscript "min" indicates that this is the minimum relaxation length which can be attained within the framework of weak turbulence theory).

Since the e-folding time of the modulational instability is usually much shorter than the beam pulse length, the value of U at $\gamma_b/\omega_{pe} > m/M$ must be "frozen" at the level (13). This is provided by the development of low-frequency inhomogeneites of the plasma density with a characteristic scale length of the order of $c/\omega_{pe}$, which arise due to the modulational instability. At the condition (3), the phase velocity of these inhomogeneities is smaller than the group velocity of the resonant Langmuir oscillations and the interaction of resonant oscillations with them is almost elastic and leads to an isotropization of the Langmuir wave spectrum, resulting in the required reduction of the pumping. Because of this reduction, the efficiency of the heating decreases at $\gamma_b/\omega_{pe} > m/M$.

The case of plasma heating at $\gamma_b/\omega_{pe} > T/mc^2$, where even the dispersion of the Langmuir oscillations completely changes remains still unclear (at $\gamma_b/\omega_{pe} > T/mc^2$ the contribution of the beam to the dispersion relation overbalances the usual thermal correction $(kr_D)^2$).

## IV. PLASMA HEATING AT MODERATE MAGNETIC FIELDS ($\beta<1$)

At $\beta<1$, the magnetic correction to the dispersion relation (8) becomes more important than the thermal one even in the resonant region ($k \sim \omega_{pe}/c$). The estimation for the time $\tau$ of spectral transfer through the resonant part of k-space is now also different:

$$\tau \sim \frac{10}{\omega_{pe}\beta^2} \frac{nT}{U} \frac{M}{m} \frac{T}{mc^2}$$

Determining U from condition (5) one can obtain the following estimation for the relaxation length:

$$L \simeq \frac{\beta^2}{20} \frac{c}{\omega_{pe}} \frac{n}{n_b} \frac{m}{M} \left(\frac{mc^2}{T}\right)^2 \left(\frac{\varepsilon}{mc^2}\right)^3$$

One can see that L becomes $\beta^2$ times smaller as compared to the estimation by means of equation (7).

The condition for the validity of the weak turbulence approach also changes significantly. In particular, for the case where a condensate is not formed this condition is (see Refs. [1,7]):

$$\frac{U}{nT} < \frac{T}{mc^2\beta} \simeq \frac{\omega_{He}^2}{\omega_{pe}^2}$$

(we assume that $\omega_{He} \ll \omega_{pe}$). Consequently, instead of (14) one obtains:

$$L_{min} \sim \frac{c}{\omega_{pe}} \left(\frac{\varepsilon}{T}\right)^2 \beta$$

This estimation corresponds to $\gamma_b/\omega_{pe} \sim (m/M)\beta$. At $\gamma_b/\omega_{pe} > (m/M)\beta$ the efficiency of the heating will decrease.

## V. NONISOTHERMAL PLASMA ($T_e >> T_i$)

The main distinction between this case and those already considered is connected with the fact that at $T_e >> T_i$ a new branch of oscillations (ion sound oscillations) appears in the plasma. Under condition (3) there is coupling with ion acoustic waves by decay processes. The set of equations describing these processes is as follows:

$$\dot{n}_{\vec{q}} = \int P(\vec{k},\vec{q}) N_{\vec{k}} N_{\vec{k}-\vec{q}} d\vec{k} + $$

$$+ n_{\vec{q}} \int P(\vec{k},\vec{q})(N_{\vec{k}} - N_{\vec{k}-\vec{q}}) d\vec{k} - \gamma_s(\vec{q}) n_{\vec{q}} \tag{15}$$

$$\dot{N}_{\vec{k}} = N_{\vec{k}} \int d\vec{q} \left[ P(\vec{k}+\vec{q},\vec{q}) N_{\vec{k}+\vec{q}} - P(\vec{k},\vec{q}) N_{\vec{k}} \right] +$$

$$+ \int d_{\vec{q}} n_{\vec{q}} \left[ P(\vec{k}+\vec{q},\vec{q})(N_{\vec{k}+\vec{q}} - N_{\vec{k}}) - P(\vec{k},\vec{q})(N_{\vec{k}} - N_{\vec{k}-\vec{q}}) \right] +$$

$$+ N_{\vec{k}} \gamma_b(\vec{k}) \tag{16}$$

where $n_{\vec{q}}$ and $N_{\vec{k}}$ are the occupation numbers for ion acoustic and Langmuir waves, respectively, $\gamma_s(\vec{q})$ is the linear damping rate of the ion acoustic waves, and $P(\vec{k},\vec{q})$ is the matrix element. One needs to find the stationary solution of these equations. Let us start with equation (15). The second integral term in this equation describes nonlinear damping (or growth) of ion acoustic waves due to their interaction with Langmuir waves. It is well known that, for the case of isotropic distribution of Langmuir waves, ($N_{\vec{k}} \equiv N(k)$) it always corresponds to damping. However, for the distributions which are formed in the considered problem, this term is positive for some $\vec{q}$. The order of magnitude of it is:

$$\int P(\vec{k},\vec{q})(N_{\vec{k}} - N_{\vec{k}-\vec{q}}) d\vec{k} \sim \omega_{pe} \frac{U_{res}}{nT} \frac{mc^2}{T} \frac{s}{c}$$

where s is the ion acoustic velocity, and q is assumed to be of the order of $\omega_{pe}/c$. At $T_e >> T_i$, the damping rate of ion acoustic waves with $q \sim \omega_{pe}/c$ is $\gamma_s \sim \omega_{pe} \sqrt{m/M}\,(s/c)$. One then observes that at

$$\frac{U_{res}}{nT} < \frac{T}{mc^2} \sqrt{\frac{m}{M}} \tag{17}$$

i.e. at a very low beam density, the last term in eq. (15) is larger than the second one and ion acoustic waves are stable. Their level for this case can be found by equating the first and the last terms in eq. (15). Substituting this estimated value for $n_{\vec{q}}$ into eq. (16), one finds that the term containing $n_{\vec{q}}$ is negligible as compared to the first term. This means in fact that, if condition (17) is satisfied, the equation for $N_{\vec{k}}$ becomes of the same type as for the case of induced scattering by ions, and one recovers the same results as those described in Section II.

The situation becomes completely different when condition (17) is not satisfied. Then, at the first moments, the level of the ion acoustic oscillations grows exponentially and finally the term containing $n_{\vec{q}}$ becomes the most essential one in equation (16). This in turn results in strong changes of the spectrum of the Langmuir oscillations and in a suppression of the instability of the ion waves. Up to now no reliable estimations of the relaxation length have been done for this regime.

One should also note that for the case where $T_e >> T_i$ the ion acoustic turbulence can be excited by the instability of the return current. The interference of this instability with the beam instability was discussed in Ref. [1].

## VI. ACKNOWLEDGMENTS

The author is indebted to Professor L. Stenflo for valuable comments.

## REFERENCES

1. Breizman, B.N., D.D. Ryutov, Nucl. Fusion, 14, 933 (1974).

2. Breizman, B.N., D.D. Ryutov, Sov. Phys.-JETP Lett., 11, 421 (1970).

3. Breizman, B.N., D.D. Ryutov, O.P. Sobolev, Fizika Plasmi, 2, 86 (1976) - in Russian.

4. Breizman, B.N., D.D. Ryutov, Sov. Phys. - JETP, 33, 220 (1971).

5. Ryutov, D.D., Uspekhi Fiz. Nauk, 116, 341 (1975) - in Russian.

6. Benford, J., Talk presented at the Int. Topical Conf. on Relativistic Electron Beams, Albuquerque, N.-M., Nov. 1975.

7. Altyntsev, A.T., B.N. Breizman, A.G. Es'kov, O.A. Zolotovskii, V.I. Koroteev, R.Kh. Kurtmullaev, V.L. Masalov, D.D. Ryutov, V.N. Semenov, International Conference on the Physics of Plasma, paper CN-28/E-20, Madison, Wis., 1971.

8. Breizman, B.N., D.D. Ryutov, P.Z. Chebotaev, Sov. Phys. - JETP, 35, 741 (1972).

9. Tsytovich, V.N., Nonlinear Effects in Plasma, Plenum, New York London, 1970.

10. Breizman, B.N., D.D. Ryutov, Proc. of the Conf. on Electrostatic and Electromagnetic Plasma Confinement and Phenomenology of Relativistic Electron Beams, N.-Y. Ac. Sci., N.-Y., 1974, p. 458.

11. Breizman, B.N., V.E. Zakharov, S.L. Musher, Sov. Phys. - JETP, 37, 658 (1974).

12. Breizman, B.N., D.D. Ryutov, Sov. Phys. - JETP Lett., 21, 192 (1975).

13. Vedenov, A.A., L.I. Rudakov, Sov. Phys. - Doklady, 9, 1073 (1965)

14. Nishikawa, K., D.D. Ryutov, J. Phys. Soc. Japan, to be published.

15. Vedenov, A.A., D.D. Ryutov, Review of Plasma Physics, v.6 (M.A. Leontovich ed., Consultants Bureau, 1975).

16. Zakharov, V.E., Sov. Phys. - JETP, 35, 908 (1972).

17. Tsytovich, V.N., These Proceedings.

18. Galeev, A.A., R.Z. Sagdeev, V.D. Shapiro, V.I. Shevchenko. These proceedings.

THE COLD-GAS BLANKET PROBLEM

FOR TURBULENT PLASMAS

F. Engelmann and A. Nocentini*)

Association Euratom-FOM
FOM-Instituut voor Plasmafysica
Rijnhuizen, Jutphaas, The Netherlands

The cold-gas blanket, at least as far as its function as refuelling device is concerned, is a consistent concept only if there exist appropriate steady states of the plasma on a time scale larger than the particle confinement time. The implications of this condition are discussed in general, and the collisional transport model for toroidal geometry is considered as an illustrative example. The importance of a more detailed knowledge about the actual anomalous transport properties for treating the cold-gas blanket problem realistically, is displayed.

## 1. THE COLD-GAS BLANKET CONCEPT

Using a cold-gas blanket [1,2,3,4,5,6] to separate a hot plasma core from the vessel wall in a closed confinement system is under consideration for two main reasons. The first is that the presence of a blanket of cold gas and plasma may be the simplest means of reducing the impurity content of a magnetically confined plasma. In fact, if there is an outside layer having temperatures not higher than some tens of eV and a width larger than the mean free path with respect to charge exchange of hot neutrals that might be created in the hot core, bombardment of the wall by highly energetic particles and consequently release of impurities from the wall would be strongly reduced. Furthermore, it is thinkable that in such a layer the transport properties of highly

---

*) On leave of absence from Istituto di Meccanica, University of Trieste, Trieste, Italy.

charged ions are such that no appreciable penetration into the plasma of impurities, which nevertheless might be created at the wall, is possible. The second reason is that a cold blanket may serve to refuel the hot core of a reactor plasma through a flux of deuterium and tritium ions flowing from the blanket region into the core. Of course, to make this a useful device, one has to require that at the same time the α-particles created in fusion events can flow from the core to the blanket instead of accumulating in the plasma, which would lead to an increasing "fusion-generated" impurity content.

The property of the cold blanket to screen plasma and wall from each other is important also for pulsed systems in which the discharge time is shorter than the particle confinement time $\tau_p$. Recent experiments on Alcator [7] and Pulsator [8] have indicated that the presence of a cold plasma layer, impenetrable for neutrals, is indeed beneficial to reduce the impurity content of the plasma. In pulsed operation the generation of such a layer is, however, only a "technical" problem: one has to reach plasma densities n such that

$$na \gtrsim 10^{15}\ \mathrm{cm}^{-2} \quad [4] \quad , \tag{1}$$

where a measures the linear dimension of the plasma, and possibly manipulate the profiles such that a reasonably dense and wide cool outside layer appears. In the experiments cited this is achieved by introducing cold gas into the discharge vessel after having heated the plasma.

On the other hand, using a cold blanket as a refuelling device is possible only if such a blanket naturally appears under steady-state conditions. In fact, refuelling of a reactor plasma has to be done on a time scale

$$\tau_R = \left(\frac{\tau_E}{\tau_p} \cdot \frac{U_\alpha}{6\kappa T}\right) \tau_p \quad , \tag{2}$$

where $\tau_E$ is the energy confinement time of the plasma, T its temperature, and $U_\alpha$ the energy of fusion α-particles at generation. Relation (2) is a consequence of the steady-state energy balance of an ignited reactor,

$$\tau_H = \tau_E \tag{3}$$

with

$$\tau_H \equiv \frac{3n\kappa T}{U_\alpha(\dot{n}/2)}$$

$$= \frac{6n\kappa T}{U_\alpha(n/\tau_R)} \tag{4}$$

the typical nuclear heating time. This implies that $\tau_R$ tends to be large compared with $\tau_p$, in particular in turbulent plasmas, so that refuelling is a problem only for pulse durations larger than $\tau_p$, that is, when steady-state conditions are approached.

We shall concentrate here on this latter problem, displaying what kind of information is needed to say something about the appearance and the properties of a cold blanket on a time scale larger than $\tau_p$. After having described the general framework, we shall discuss, as an example, some results obtained in the case of collisional transport in toroidal geometry, as this model is the only one worked out in sufficient detail to give a non-trivial answer.

## 2. THE STEADY STATE

It is clear from the preceding that the question of the presence of a cold blanket on a time scale larger than $\tau_p$ is equivalent with asking which shape the steady-state profiles of the plasma temperature and density have in the discharge considered. All necessary information is therefore contained in the local (steady-state) balances of particles and energy, supplemented by appropriate boundary conditions at the plasma edge.

Let us concentrate, in particular, on the particle balances. Because of relation (2) there is no essential difference between a burning and a non-burning plasma, as the sink term appearing due to the presence of fusion reactions tends to be a small correction to the ion particle balance. Therefore, disregarding, for simplicity, possible recycling processes in the outermost plasma layer into which neutrals penetrate, fulfilling the particle balances implies

$$\Gamma_j = 0 \tag{5}$$

with $\Gamma_j$ the radial fluxes of the different particle species j present. In the absence of impurity ions, conditions (5) reduce to just one relation, which for simply charged plasma ions reads

$$\Gamma_e \equiv \Gamma_i = 0 \quad , \tag{6}$$

where e and i refer to electrons and ions, respectively. Relation (6) can be taken as a condition for the plasma density profile. In the more general case of a plasma with impurities, relations (5) yield also conditions for the profiles of the impurity concentrations.

The essential point, following readily from this discussion, is now that the shapes of the particle density profiles depend sensitively on the form of the $\Gamma_j$ which, in turn, are determined by the details of the plasma transport properties. Considering, for instance, a pure plasma with equal electron and ion temperatures ($T_e = T_i = T$), the general structure expected for the radial (r) flux of plasma particles is

$$\Gamma = L_n \frac{1}{n}\frac{dn}{dr} + L_T \frac{1}{T}\frac{dT}{dr} + L_E E_{\|} \quad , \tag{7}$$

where $E_{\|}$ is the component of the external electric field parallel to the confining magnetic field and $L_n$, $L_T$, and $L_E$ are coefficients depending on the plasma and discharge parameters. The latter contain the information required to make condition (6) for the plasma density n(r) explicit. Of course, in order to obtain both n and T as functions of r this condition must be complemented by the energy balance.

## 3. COLLISIONAL TRANSPORT IN TOROIDAL GEOMETRY

Let us consider as an instructive example a plasma in toroidal geometry, which is subject to collisional transport. For simplicity, we disregard the Ware effect, that is, we take $L_E = 0$ throughout. This is justifiable for large poloidal β. As long as the ration $L_T/L_n$ is constant, which is true in each transport regime separately [9,10], equation (6) can then readily be integrated to yield relations between the plasma density and temperature for these different regimes, their general form being

$$n\, T^{(L_T/L_n)} = \text{const} \quad . \tag{8}$$

From the known expressions for the particle flux [9,10,11] one has

$$\frac{L_T}{L_n} = -0.28 \tag{9a}$$

for the banana regime,

$$\frac{L_T}{L_n} = 1.46 \tag{9b}$$

for the plateau regime,

$$\frac{L_T}{L_n} = 0.58 \tag{9c}$$

for the Pfirsch-Schlüter regime, and

$$\frac{L_T}{L_n} = 1 \tag{9d}$$

when the collisionality of the plasma is so high as to fulfill the condition

$$\frac{\nu_i}{\omega_{bi}} \gg \left(\frac{m_i}{m_e}\right)^{\frac{1}{4}} \tag{10}$$

with $\nu_i$ the collision frequency of the ions, $\omega_{bi} = (\kappa T/m_i)^{\frac{1}{2}}/qR$ the frequency of the poloidal motion of a typical ion, $m_i$ and $m_e$ the ion and electron mass, respectively, q the safety factor, and R the major radius of the torus.

To show that a steady state can consistently exist, the energy balance must also be solved, which requires a numerical approach. Moreover, the actual radial dependence of the temperature and density profiles is determined by this balance. Of course, the temperature profile must result essentially bell-shaped, a situation for which equations (9) indeed imply the presence of a cold dense blanket of relatively low pressure (cf. Fig. 1). On the one hand, this is due to the fact that the plasma density decreases with temperature in the banana regime, which in a quiescent thermonuclear plasma would be present over a wide range of temperatures. On the other hand, in the colder outside regimes the density increases again with further decreasing temperature. This circumstance leads, in particular, to a situation where Bremsstrahlung losses are mostly concentrated in the hot plasma core, so that satisfying the steady-state energy balance should not imply difficulties.

It may be added that the collisional transport model, for profiles fulfilling relation (8) with $L_T/L_n$ according to (9), predicts the possibility of accumulation of impurities and $\alpha$-particles in the outside layer, yielding a clean plasma core. Hence, within the framework of this model (disregarding consistently the problem of stability of the profiles found), a cold blanket having the required properties turns out to exist. The width of the cold blanket may, in particular, be adapted to the necessary dimension by manipulating its impurity content to ensure the appropriate cooling.

## 4. CONCLUSIONS

For the realistic case of a turbulent plasma, the preceding considerations imply, that much more detailed knowledge about the relevant transport properties, and, in particular about the Nernst effect, i.e., the particle flux driven by the temperature gradient, are required to be able to give an answer to the question whether or not a cold blanket surrounding a hot plasma is possible and what its properties may be. These transport properties must be known all over the plasma, not only in the colder outside region. A pre-requisite for this would, of course, be to know which type of modes is responsible for the transport properties in different plasma regimes in practical cases. This information is not available at the moment for hot plasmas confined in toroidal geometry. As far as instabilities driven by gradients are important, a self-consistent picture must be constructed, taking into account the coupling between the shape of the profiles and the dynamics of the modes. To obtain a guide line, more experimental results on the modes excited and on the time evaluation of the density profiles over times at least comparable with the particle confinement time would be essential.

This work was performed as part of the research programme of the association agreement of Euratom and the "Stichting voor Fundamenteel Onderzoek der Materie" (FOM) with financial support from the "Nederlandse Organisatie voor Zuiver-Wetenschappelijk Onderzoek" (ZWO) and Euratom.

References

[1] H. Alfvén and E. Smårs, Nature 188, 801 (1960).
[2] H. Alfvén et al., Nucl. Fusion, Suppl., Part I, 33 (1962).
[3] B. Lehnert, Nucl. Fusion 8, 173 (1968).
[4] B. Lehnert, Nucl. Fusion 13, 781 (1973).
[5] B. Lehnert, Proc. 3rd Int. Symp. on Toroidal Plasma Confinement, Garching 1973, Paper C1-I.
[6] B. Brandt and C.M. Braams, Nucl. Fusion, Spec. Suppl., 385 (1974).

[7] R.R. Parker, private communication (contribution to the Meeting on High-Density Tokamak Results and Implications on JET, Culham, 26 and 27 February 1976).
[8] O. Klüber, Nucl Fusion 15, 1194 (1975).
[9] F.L. Hinton and R.D. Hazeltine, Rev. Mod. Phys. 48, 239 (1976).
[10] F. Engelmann and A. Nocentini, submitted to Nucl. Fusion.
[11] F.L. Hinton and M.N. Rosenbluth, Phys. Fluids 16, 836 (1973).

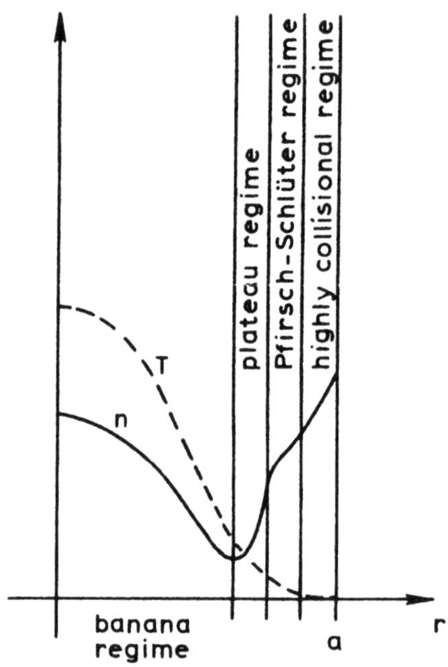


Fig. 1

Schematic plot of the radial dependence of the steady-state plasma density n(r) for a bell-shaped temperature profile T(r), if plasma particle transport is collisional.

# CROSS-FIELD ANOMALOUS RESISTIVITY ASSOCIATED WITH THE LOWER-HYBRID-DRIFT INSTABILITY IN STRONGLY INHOMOGENEOUS PLASMAS

Ronald C. Davidson

Department of Physics and Astronomy
University of Maryland, College Park, Maryland 20742 USA
and
Naval Research Laboratory, Washington, D. C. 20375 USA

ABSTRACT

The cross-field anomalous resistivity associated with the lower-hybrid-drift instability is investigated. Particular emphasis is placed on the low-drift-velocity regime characterized by $r_{Li}/2\Delta L_n \lesssim 1$, where $r_{Li}=v_i/\omega_{ci}$ is the ion Larmor radius, and $\Delta L_n=-(\partial \ell nN/\partial x)^{-1}$ is the length scale for density inhomogeneity. The predicted value of anomalous resistivity (for $T_e \ll T_i$) is $\eta_{AN}=4\pi(\pi/2)^{1/2}\Omega_{LH}^{-1}\mathcal{E}_F/NT_i$, where $\Omega_{LH}=\omega_{pi}/(1+\omega_{pe}^2/\omega_{ce}^2)^{1/2}$ is the lower hybrid frequency, and $\mathcal{E}_F=\langle\delta \underset{\sim}{E}^2\rangle/8\pi$ is the energy density of the unstable field fluctuations.

## 1. INTRODUCTION

Anomalous transport processes associated with microinstabilities influence diffusion and heat conductivity in steady-state experiments [1], and play an important role in implosion heating and in determining the shape of the imploding magnetic field and density profiles in rapidly pulsed high-density pinch experiments [2]. For example, in linear theta-pinch geometry, a relative azimuthal drift between electrons and ions,

$$V_d = V_{e\theta} - V_{i\theta} = \frac{c}{4\pi eN}\frac{\partial B_z}{\partial r},$$

is associated with the radial gradient in axial magnetic field $B_z\hat{\underset{\sim}{e}}_z$. Depending on the size of the drift velocity $V_d$, various cross-field instabilities (e.g., Buneman, ion acoustic, lower-hybrid-drift and modified-two-stream instabilities) can develop. Since $V_d$ varies as a function of position, different instabilities can dominate at different radial locations. As a general remark, the instabilities generate turbulent field fluctuations, and the particles in turn scatter from the waves, i.e., interact nonlinearly with the field fluctuations. The net effect of these interactions is to reduce the available free energy that drives the instability, which is typically manifest through a decrease in $V_d$, and a turbulent heating of electrons and ions. The reduction in $V_d$ can be thought of in terms of an anomalously large resistivity [3] produced by the instability. Of course, the reduction in $V_d$ is also accompanied by a decrease in $\partial B_z/\partial r$ (an increase in sheath thickness), indicating the important effect that microinstabilities can have on the sheath structure.

The purpose of this paper is to give an overview of the anomalous transport properties, particularly the anomalous resistivity, associated with the lower-hybrid-drift instability [4-6]. The lower-hybrid-drift instability is a high-frequency ($\omega_{ci} << |\omega + i\gamma| << \omega_{ce}$) drift-wave instability driven by cross-field currents in the presence of inhomogeneities in density N(x) and magnetic field $B_z(x)$ [Fig. 1]. Moreover the characteristic growth rate $\gamma$ and perpendicular wavenumber $k_y$ satisfy (assuming flute perturbations with $\underset{\sim}{k}\cdot\underset{\sim}{B}=0$)

$$\gamma \underset{\sim}{\sim} \Omega_{LH} \quad , \quad k_y^2 r_{Le}^2 \underset{\sim}{\sim} 1 \ ,$$

for $T_e \underset{\sim}{\sim} T_i$ and $V_E/v_i \underset{\sim}{\sim} 1$. Here $V_E = -cE_x^0/B_z$ is the cross-field $\underset{\sim}{E}\times\underset{\sim}{B}$ velocity, $v_i = (2T_i/m_i)^{1/2}$ is the ion thermal speed, $r_{Le} = v_e/\omega_{ce}$ is the electron Larmor radius, and $\Omega_{LH} = \omega_{pi}/(1+\omega_{pe}^2/\omega_{ce}^2)^{1/2}$ is the lower-hybrid frequency. Unlike the ion acoustic instability, the lower-hybrid-drift instability can operate when $T_i \gtrsim T_e$. Moreover, recent theoretical studies [4,7,8] indicate that the lower-hybrid-drift instability (and the associated anomalous transport) persist in the low-drift-velocity regime, where

$$V_E \lesssim v_i \ .$$

The significance of this result is evident. Assuming perturbations about a quasisteady state characterized by $\partial T_i/\partial x = 0$, $V_{yi}=0$ and $NeE_x^0 = T_i \partial N/\partial x$ (electrostatically confined ions), we note that

$$\frac{V_E}{v_i} = \frac{r_{Li}}{2\Delta L_n} \ ,$$

where $\Delta L_n = -(\partial \ell n N/\partial x)^{-1}$ is the length scale for density variation. Expressing $r_{Li} = \beta_i^{1/2} c/\omega_{pi}$, where $\beta_i = 8\pi N T_i/B_z^2$ is the local ion beta, the inequality $V_E/v_i \lesssim 1$ becomes

$$\Delta L_n \gtrsim \frac{1}{2} \beta_i^{1/2} \frac{c}{\omega_{pi}} \ ,$$

or equivalently,

$$\Delta L_B \gtrsim \frac{\beta_i^{1/2}}{\beta_i + \beta_e} \frac{c}{\omega_{pi}} \ ,$$

where $\Delta L_B=(\partial \ell n B_z/\partial x)^{-1} \simeq 2(\beta_e+\beta_i)^{-1}\Delta L_n$ is the characteristic length scale for magnetic field variation. Therefore, for plasma betas of order unity, the persistence of instability and anomalous transport for $V_E/v_i \lesssim 1$ indicates that the sheath region continues to exhibit instability and long-time transport (broadening) even when the sheath thickness exceeds $c/\omega_{pi}$.

There are several experimental motivations for these studies. We note that linear theta-pinch experiments [9-12] during the late stages of implosion often exhibit considerably more diffuse magnetic field profiles than predicted by numerical studies [2,13] that model anomalous transport processes only for $V_E>v_i$. Moreover, recent experimental studies [9] of the post-implosion resistivity in the Maryland 40 kV theta pinch give measured values of $\eta$ about fifty times the classical [14] cross-field value. The above experimental results are strongly suggestive that substantial anomalous resistivity persists in the low-drift-velocity regime $V_E \lesssim v_i$. In addition, the implementation of the staging concept as a means of achieving wall stabilization has also provided strong motivation for understanding microstability behavior in theta-pinch experiments with $V_E \lesssim v_i$. In particular, the turbulence associated with the existence of microinstabilities in implosion sheaths manifests itself, macroscopially, as an anomalous resistivity which strongly affects the way in which magnetic fields interpenetrate the plasma. Since to scale up staging from current theta-pinch experiments to reactor-like devices will probably require knowledge about the scaling of anomalous transport coefficients for $V_E \lesssim v_i$, it is desirable to understand the microinstability transport properties as well as possible in this regime.

Moreover, the question of sheath stability and long-time transport also arises in other high-beta fusion concepts such as Tormac [15] and imploding liners [16]. The implementation of implosion heating techniques as a means of forming high-beta Tokamak plasmas also provides strong motivation for understanding long-time transport. In this regard, we emphasize that the present analysis does not include the important influence of magnetic shear [17] on the lower-hybrid-drift instability.

The organization of this paper is the following. In Sec. 2, we discuss the unperturbed plasma configuration and basic assumptions [Sec. 2.A], the linear dispersion relation for the lower-hybrid-drift instability [Sec. 2.B], and linear stability properties in the low-drift-velocity regime with $V_E<v_i$ [Sec. 2.C]. In Sec. 3, a general formalism is developed for describing anomalous transport in spatially nonuniform plasmas, including derivation of the macroscopic transport equations [Sec. 3.A], and calculation of the rate of anomalous momentum transfer [Sec. 3.B]. Finally, in Sec. 4, we make use of the formalism described in Sec. 3 to determine the

anomalous resistivity associated with the lower-hybrid-drift instability [Sec. 4.A]. We also compare the relative sizes of classical and anomalous resistivity for theta-pinch parameters of fusion interest [Sec. 4.B].

## 2. THE LOWER-HYBRID-DRIFT INSTABILITY

For future reference, we summarize here several important properties of the lower-hybrid-drift instability [4-6], including the unperturbed plasma configuration and basic assumptions [Sec. 2.A], the linear dispersion relation [Sec. 2.B], and linear stability properties in the low-drift-velocity regime with $V_E < v_i$ [Sec. 2.C].

### A. Plasma Configuration and Basic Assumptions

The plasma configuration and slab geometry employed in the present analysis are illustrated in Fig. 1. The electrons drift across the magnetic field $B_z\hat{e}_z$ with mean fluid velocity ($m_e \to 0$)

$$V_d = V_{ye} = V_E + V_{de} , \tag{1}$$

where $V_E = -cE_x^0/B_z$ is the $E \times B$ drift velocity, $V_{de} = -(T_e/m_e\omega_{ce})(\partial/\partial x)\ell nN$ is the electron diamagnetic drift velocity ($\partial T_e/\partial x = 0$ is assumed), $\omega_{cj} = |e_j|B_z/m_jc$ is the cyclotron frequency, $T_e$ is the electron temperature, and $N(x) = N_e(x) \simeq N_i(x)$ is the density. Moreover, force balance (in the x-direction) on an ion fluid element can be expressed as

$$V_{yi} = V_E + V_{di} + \omega_{ci}^{-1} \frac{D}{Dt} V_{xi} , \tag{2}$$

where $V_{di} = (T_i/m_i\omega_{ci})\partial\ell nN/\partial x$ is the ion diamagnetic drift velocity ($\partial T_i/\partial x = 0$) is assumed). For zero ion current in the y-direction ($V_{yi} = 0$), Eq. (2) reduces to

$$0 = V_E + V_{di} + \omega_{ci}^{-1} \frac{D}{Dt} V_{xi} , \tag{3}$$

which relates $V_E$ to $V_{di}$ and the local acceleration of the fluid element. If $DV_{xi}/Dt = 0$, then Eq. (3) reduces to

$$V_E = - V_{di} , \tag{4}$$

or equivalently, $NeE_x^0 = T_i\nabla N$, corresponding to electrostatically confined ions.

We consider wave perturbations with

$$k_y^2 \gg k_x^2 \gg (\partial\ell nN/\partial x)^2 \quad , \quad (\partial\ell nB_z/\partial x)^2 \tag{5}$$

and employ the local approximation throughout the linear stability

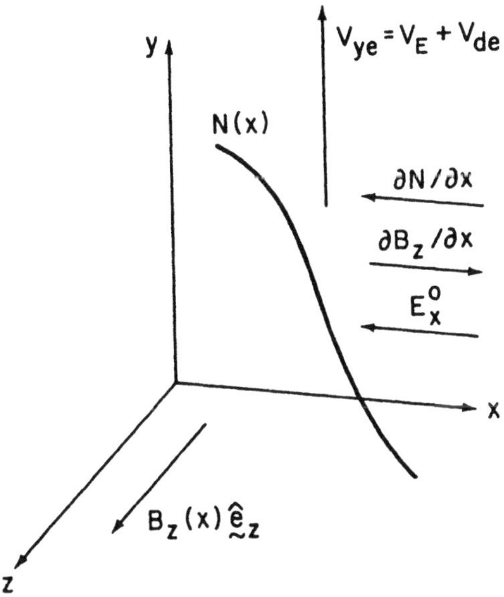


Fig. 1. Plasma Configuration and Slab Geometry

analysis. To elucidate the basic physics and simplify the analysis, the following assumptions are also made:

(a) The ion and electron temperatures satisfy $T_i \gtrsim T_e$ and the ion acoustic instability is not operative in the parameter range under consideration.

(b) Flute perturbations with $\underset{\sim}{k} \cdot \underset{\sim}{B} = 0$ are considered. The primary motivation for this assumption is the parameter study by Gladd [5] of the modified-two-stream and lower-hybrid-drift instabilities, which shows that maximum growth occurs for $k_z = 0$, when $T_i \gtrsim T_e$.

(c) The present analysis neglects transverse electromagnetic wave perturbations, which is valid if $\beta = 8\pi N(T_e + T_i)/B_z^2$ is sufficiently small. The inclusion of electromagnetic corrections [18] to the electrostatic dispersion relation discussed below gives modifications to the growth rate of order $\beta$, but does not change the qualitative features of the instability when $T_i > T_e$.

## B. Linear Dispersion Relation

The lower-hybrid-drift instability is characterized by strongly magnetized electrons and unmagnetized ions, with perturbation frequencies and wavenumbers satisfying

$$\omega_{ci} \ll |\omega + i\gamma| \ll \omega_{ce} \;, \qquad k_\perp^2 r_{Li}^2 \gg 1 \;, \quad k_\perp^2 r_{Le}^2 \lesssim 1 \;, \tag{6}$$

where $r_{Lj} = (2T_j/m_j\omega_{cj}^2)^{1/2}$ is the thermal Larmor radius, and $\omega + i\gamma$ is

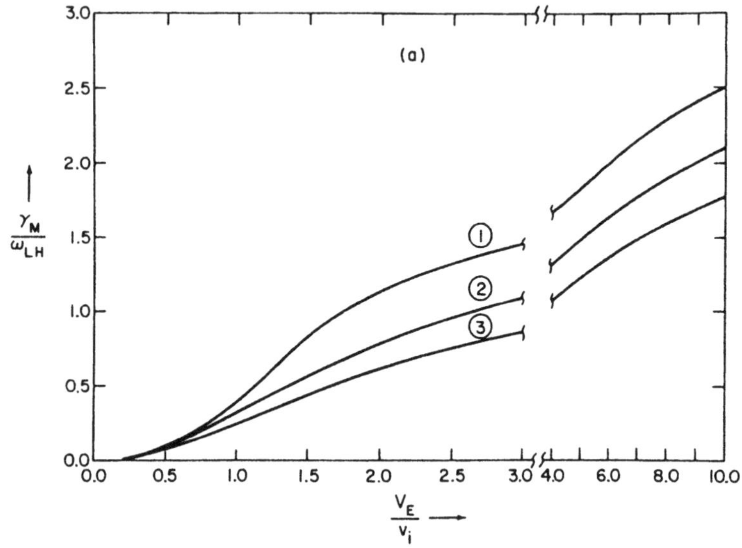


Fig. 2. Plot of $\gamma_M/\omega_{LH}$ Versus $V_E/v_i$ [Eq. (7)]

the complex eigenfrequency. For Maxwellian electrons drifting through a stationary Maxwellian ion background ($V_{yi}$=0), the local dispersion relation for electrostatic perturbations with $\underset{\sim}{k}=k_y\hat{\underset{\sim}{e}}_y$ can be expressed as [4-6]

$$D(k_y,\omega+i\gamma) = 1 + \chi_i(k_y,\omega+i\gamma) + \chi_e(k_y,\omega+i\gamma)=0 \ , \tag{7}$$

where

$$\chi_i = \frac{2\omega_{pi}^2}{k_y^2 v_i^2}\,[1+\xi_i Z(\xi_i)] \ , \tag{8}$$

$$\chi_e = \frac{\omega_{pe}^2}{\omega_{ce}^2}\,\frac{1-I_0(b)\exp(-b)}{b} + \frac{2\omega_{pe}^2}{k_y^2 v_e^2}\,\frac{k_y V_\Delta(k_y)}{\omega-k_y V_E+i\gamma} \ . \tag{9}$$

In Eqs. (8) and (9), $\omega_{pj}^2=4\pi Ne_j^2/m_j$, $v_j^2=2T_j/m_j$, $b=k_y^2(T_e/m_e\omega_{ce}^2)$, $I_0(b)$ is the modified Bessel function of the first kind of order zero, $V_E=-cE_x^0/B_z$ is the $\underset{\sim}{E}\times\underset{\sim}{B}$ drift velocity, $V_\Delta(k_y)$ is defined locally by

$$V_\Delta(ky) = -\frac{T_e}{m_e\omega_{ce}}\,I_0(b)\exp(-b)\left\{\frac{1}{N}\frac{\partial N}{\partial x} - \frac{1}{B_z}\frac{\partial B_z}{\partial x}\left|1-b\left(1-\frac{I_1(b)}{I_0(b)}\right)\right|\right\}, \tag{10}$$

and $Z(\xi_i)$ is the plasma dispersion function with $\xi_i=(\omega+i\gamma)/k_y v_i$. In Eq. (10), $\partial \ln B_z/\partial x$ is related to other quantities by the $\underset{\sim}{\nabla}\times\underset{\sim}{B}$ Maxwell equation, i.e.,

$$\frac{1}{B_z}\frac{\partial}{\partial x}B_z = -\frac{1}{2}\beta_e\frac{\partial}{\partial x}\ell nN + \frac{\omega_{pe}^2}{c^2\omega_{ce}}V_E , \quad (11)$$

where $\beta_e=8\pi NT_e/B_z^2$.

It is important to note in Eqs. (8) and (9) that the customary assumption of small electron Larmor radius ($b<<1$) has not been made, since the wavenumber that maximizes the growth rate for $T_i/T_e \gtrsim 1$ generally satisfies [4]

$$k_M r_{Le} \approx 1.4\sqrt{\frac{T_e}{T_i}} . \quad (12)$$

It is also apparent from Eqs. (7)-(10) that the electron $\mathbf{E}\times\mathbf{B}$ drift motion and the inhomogeneities in N and $B_z$ provide the free energy to drive the instability. As an historical note, we note that the lower-hybrid-drift instability (for the case $\partial N/\partial x\neq 0$, $\partial B_z/\partial x=0=\partial T_e/\partial x$) was first discussed for electrically nonneutral plasmas [19,20].

Equation (4) has been solved numerically to determine the values of $k_y$, $\gamma$ and $\omega$ corresponding to maximum growth rate [4]. Typical results are summarized in Fig. 2, where the maximum growth rate $\gamma_M$ (maximized with respect to $k_y$) is plotted versus the normalized drift velocity $V_E/v_i$, for the choice of local equilibrium parameters: $\partial T_i/\partial x=0$, $\omega_{pe}^2/\omega_{ce}^2=100$, $m_e/m_i=1/1836$, and (1) $T_e/T_i=2$, $\beta_e=0.4$, (2) $T_e/T_i=1$, $\beta_e=0.2$, and (3) $T_e/T_i=0.2$, $\beta_e=0.04$ [4]. For simplicity, $DV_{xi}/Dt=0$ has been assumed in Fig. 2, which corresponds to the quasiequilibrium constraint $V_E=-V_{di}$ [Eq. (4)]. For $\partial T_i/\partial x=0$, it follows that

$$\frac{V_E}{v_i} = -\frac{V_{di}}{v_i} = \frac{r_{Li}}{2\Delta L_n} , \quad (13)$$

where $\Delta L_n \equiv -(\partial \ell nN/\partial x)^{-1}$ is the length scale for density variation. Therefore, the abscissa $V_E/v_i$ in Fig. 2 is equivalent to $r_{Li}/2\Delta L_n$.

Several features are noteworthy in Fig. 2. First, the maximum growth rate $\gamma_M$, measured in units of $\omega_{LH}=(\omega_{ce}\omega_{ci})^{1/2}$, is substantial, even in the low-drift-velocity regime, $V_E/v_i=r_{Li}/2\Delta L_n \lesssim 1$. For example, when $T_e/T_i=1$ and $V_E/v_i=1$, then $\gamma_M \simeq 0.36\ \omega_{LH}$. Moreover, for fixed $V_E/v_i$, the maximum growth rate is an increasing function of $T_e$ [compare curves (1), (2) and (3)], since the electron diamagnetic current is correspondingly larger.

### C. Low-Drift-Velocity Regime ($V_E<v_i$)

For inhomogeneity length scale satisfying $\Delta L_n \gtrsim r_{Li}$, it is of particular interest to examine the linear dispersion equation (7) in the low-drift-velocity regime characterized by $V_E<v_i$. To illustrate

the essential features of the instability in this regime, we assume (for simplicity) that $V_E \ll v_i$ and $T_e \ll T_i$. In this case, $b = k_y^2 r_{Le}^2/2 \ll 1$ and $Z(\xi_i) \simeq i\sqrt{\pi} k_y/|k_y|$, so that Eq. (7) can be approximated by [4]

$$D(k_y,\omega+i\gamma) = 1 + \frac{2\omega_{pi}^2}{k_y^2 v_i^2}\left(1 + \frac{\omega+i\gamma}{|k_y| v_i}\, i\sqrt{\pi}\right) + \frac{\omega_{pe}^2}{\omega_{ce}^2}$$
$$- \frac{2\omega_{pe}^2}{k_y^2 v_e^2}\,\frac{(k_y T_e/m_e\omega_{ce})}{\omega - k_y V_E + i\gamma}\left\{\frac{1}{N}\frac{\partial N}{\partial x} - \frac{1}{B_z}\frac{\partial B_z}{\partial x}\right\} = 0 . \quad (14)$$

Assuming $|\gamma| \ll |\omega|$ and $|\partial \ell n B_z/\partial x| \ll |\partial \ell n N/\partial x|$, we solve Eq. (14) for $\omega$ and $\gamma = -D_i/(\partial D_r/\partial \omega)$. This gives

$$\omega - k_y V_E = -\frac{2\omega_{pi}^2}{k_y^2 v_i^2} k_y V_E \left(1 + \frac{\omega_{pe}^2}{\omega_{ce}^2} + \frac{2\omega_{pi}^2}{k_y^2 v_i^2}\right)^{-1} ,$$
$$\gamma = \sqrt{\pi}\,\frac{(\omega - k_y V_E)^2}{|k_y| v_i}\,\frac{\omega}{k_y V_E} , \quad (15)$$

where $V_E = -V_{di} = -(T_i/m_i\omega_{ci})\,\partial \ell n N/\partial x$. Equation (15) gives instability for phase velocities in the range $0 < \omega/k_y < V_E$, with maximum growth rate ($\gamma_M$) and phase velocity at maximum growth ($\omega_M/k_M$) given by [4]

$$\gamma_M = \frac{\sqrt{2\pi}}{8}\left(\frac{V_E}{v_i}\right)^2 \Omega_{LH} ,$$
$$\frac{\omega_M}{k_M} = \frac{V_E}{2} , \quad (16)$$

where $k_M^2 = (2\omega_{pi}^2/v_i^2)(1+\omega_{pe}^2/\omega_{ce}^2)^{-1}$, and

$$\Omega_{LH} = \frac{\omega_{pi}}{(1+\omega_{pe}^2/\omega_{ce}^2)^{1/2}} \quad (17)$$

is the generalized lower-hybrid frequency. [Note that $\Omega_{LH} \to \omega_{LH} = (\omega_{ce}\omega_{ci})^{1/2}$ for $\omega_{pe}^2/\omega_{ce}^2 \gg 1$.] Although derived for $V_E \ll v_i$, Eq. (16) gives a reasonably good estimate of maximum growth rate $\gamma_M$ even when the drift velocity is comparable with the ion thermal velocity ($V_E \lesssim v_i$). Note also that $\gamma_M$ can be a substantial fraction of $\Omega_{LH}$, e.g., $\gamma_M = 0.08\,\Omega_{LH}$ when $V_E/v_i = 0.5$. In this regard, it is important to keep in mind that the range of validity of Eqs. (14)-(16) is limited by the unmagnetized ion assumption, $|\omega_M + i\gamma_M| \gg \omega_{ci}$.

## 3. ANOMALOUS TRANSPORT THEORY

In this section we discuss a general formalism for describing anomalous transport in spatially nonuniform plasmas. The macroscopic transport equations are derived in Sec. 3.A, and in Sec. 3.B the expression for $R_j^{AN}$, the rate of anomalous momentum transfer, is simplified for subsequent application to the lower-hybrid-drift instability [Sec. 4].

## A. Transport Equations

A complete description of collective interactions in a collisionless plasma is given by the nonlinear Vlasov-Maxwell equations,

$$\left\{\frac{\partial}{\partial t} + \mathbf{v} \cdot \frac{\partial}{\partial \mathbf{x}} + \frac{e_j}{m_j}\left(\mathbf{E} + \frac{\mathbf{v}\times\mathbf{B}}{c}\right) \cdot \frac{\partial}{\partial \mathbf{v}}\right\} f_j = 0 \ , \tag{18}$$

$$\nabla\times\mathbf{E} = -\frac{1}{c}\frac{\partial}{\partial t}\mathbf{B} \ , \tag{19}$$

$$\nabla\times\mathbf{B} = \frac{4\pi}{c}\sum_j e_j \int d\mathbf{v}\, \mathbf{v}\, f_j + \frac{4\pi}{c}\mathbf{J}_{ext} + \frac{1}{c}\frac{\partial}{\partial t}\mathbf{E} \ , \tag{20}$$

$$\nabla\cdot\mathbf{E} = \sum_j 4\pi e_j \int d\mathbf{v}\, f_j \quad , \quad \nabla\cdot\mathbf{B} = 0 \ , \tag{21}$$

where $f_j(\mathbf{x},\mathbf{v},t)$ is the density of particles in the phase space $(\mathbf{x},\mathbf{v})$, $\mathbf{E}(\mathbf{x},t)$ is the electric field, $\mathbf{B}(\mathbf{x},t)$ is the magnetic field, $e_j$ and $m_j$ are the charge and mass, respectively, of a particle of species j, and $\mathbf{J}_{ext}$ is the current density associated with current sources external to the plasma. We now divide each of $f_j$, $\mathbf{E}$ and $\mathbf{B}$, into a (slowly varying) average part plus an (instability produced) fluctuation, i.e.,

$$\begin{aligned} f_j(\mathbf{x},\mathbf{v},t) &= \langle f_j\rangle(\mathbf{x},\mathbf{v},t) + \delta f_j(\mathbf{x},\mathbf{v},t) \ , \\ \mathbf{E}(\mathbf{x},t) &= \langle \mathbf{E}\rangle(\mathbf{x},t) + \delta\mathbf{E}(\mathbf{x},t) \ , \\ \mathbf{B}(\mathbf{x},t) &= \langle \mathbf{B}\rangle(\mathbf{x},t) + \delta\mathbf{B}(\mathbf{x},t) \ , \end{aligned} \tag{22}$$

where < > denotes ensemble average. Moreover, the length scale for spatial variation of fluctuating quantities is assumed short in comparison with the length scale for spatial variation of average quantities, e.g., $|\partial \ln \delta f_j/\partial\mathbf{x}| \gg |\partial \ln\langle f_j\rangle/\partial\mathbf{x}|$. For present purposes, we also assume electrostatic perturbations with $\delta\mathbf{B}\simeq 0$ and $\delta\mathbf{E}\simeq-\nabla\delta\phi$. Substituting Eq. (22) into Eqs. (18)-(21) and taking the average, we find [21]

$$\left\{\frac{\partial}{\partial t} + \mathbf{v} \cdot \frac{\partial}{\partial \mathbf{x}} + \frac{e_j}{m_j}\left(\langle\mathbf{E}\rangle + \frac{\mathbf{v}\times\langle\mathbf{B}\rangle}{c}\right) \cdot \frac{\partial}{\partial \mathbf{v}}\right\} F_j = \left(\frac{\partial F_j}{\partial t}\right)_{AN} \ , \tag{23}$$

$$\nabla \times \langle\mathbf{E}\rangle = -\frac{1}{c}\frac{\partial}{\partial t}\langle\mathbf{B}\rangle \ , \tag{24}$$

$$\nabla \times \langle \underset{\sim}{B} \rangle = \frac{4\pi}{c} \sum_j e_j \int d\underset{\sim}{v}\, \underset{\sim}{v}\, F_j + \frac{4\pi}{c} \underset{\sim}{J}_{ext} + \frac{1}{c}\frac{\partial}{\partial t} \langle \underset{\sim}{E} \rangle , \tag{25}$$

$$\nabla \cdot \langle \underset{\sim}{E} \rangle = \sum_j 4\pi e_j \int d\underset{\sim}{v}\, F_j , \tag{26}$$

$$\nabla \cdot \langle \underset{\sim}{B} \rangle = 0 , \tag{27}$$

where $F_j(\underset{\sim}{x},\underset{\sim}{v},t) \equiv \langle f_j \rangle(\underset{\sim}{x},\underset{\sim}{v},t)$. In obtaining Eqs. (23)-(27), use has been made of $\langle \delta f_j \rangle = 0 = \langle \delta \underset{\sim}{E} \rangle$, and $(\partial F_j/\partial t)_{AN}$ is defined by

$$\left(\frac{\partial F_j}{\partial t}\right)_{AN} = - \frac{e_j}{m_j} \langle \delta \underset{\sim}{E}(\underset{\sim}{x},t) \cdot \frac{\partial}{\partial \underset{\sim}{v}} \delta f_j(\underset{\sim}{x},\underset{\sim}{v},t) \rangle . \tag{28}$$

Note that $(\partial F_j/\partial t)_{AN}$ describes the response of the average distribution function $F_j$ to the turbulent field fluctuations. Subtracting Eq. (23) from Eq. (18) gives

$$\left\{\frac{\partial}{\partial t} + \underset{\sim}{v} \cdot \frac{\partial}{\partial \underset{\sim}{x}} + \frac{e_j}{m_j}\left(\langle \underset{\sim}{E} \rangle + \frac{\underset{\sim}{v} \times \langle \underset{\sim}{B} \rangle}{c}\right) \cdot \frac{\partial}{\partial \underset{\sim}{v}}\right\} \delta f_j + \frac{e_j}{m_j} \delta \underset{\sim}{E} \cdot \frac{\partial}{\partial \underset{\sim}{v}} F_j$$
$$= - \frac{e_j}{m_j}\left(\delta \underset{\sim}{E} \cdot \frac{\partial}{\partial \underset{\sim}{v}} \delta f_j - \langle \delta \underset{\sim}{E} \cdot \frac{\partial}{\partial \underset{\sim}{v}} \delta f_j \rangle\right) , \tag{29}$$

where $\delta \underset{\sim}{E} = -\nabla \delta \phi$ is determined self-consistently from

$$\nabla \cdot \delta \underset{\sim}{E} = \sum_j 4\pi e_j \int d\underset{\sim}{v}\, \delta f_j . \tag{30}$$

Equations (23)-(27) provide a complete description of the evolution of average properties of the system, as well as the fluctuations $\delta f_j$ and $\delta \underset{\sim}{E}$. For purposes, here, however, we make use of a simplified version of Eq. (29), which is equivalent to neglecting the influence of nonlinear wave-particle and wave-wave interactions. In particular, note that the right-hand side of Eq. (29) is quadratic in the fluctuation amplitude. Therefore, in a lowest-order quasilinear sense, $\delta f_j$ evolves according to

$$\left\{\frac{\partial}{\partial t} + \underset{\sim}{v} \cdot \frac{\partial}{\partial \underset{\sim}{x}} + \frac{e_j}{m_j}\left(\langle \underset{\sim}{E} \rangle + \frac{\underset{\sim}{v} \times \langle \underset{\sim}{B} \rangle}{c}\right) \cdot \frac{\partial}{\partial \underset{\sim}{v}}\right\} \delta f_j + \frac{e_j}{m_j} \delta \underset{\sim}{E} \cdot \frac{\partial}{\partial \underset{\sim}{v}} F_j = 0 , \tag{31}$$

which is identical to the linearized Vlasov equation for electrostatic perturbations about the slowly varying reference state described by $F_j$, $\langle \underset{\sim}{E} \rangle$ and $\langle \underset{\sim}{B} \rangle$.

Although desirable, a determination of the detailed evolution of $F_j(\underset{\sim}{x},\underset{\sim}{v},t)$ from Eqs. (23)-(28) is generally not tractable, analytically or numerically. Of considerable practical interest, however, is the temporal and spatial development of various average macroscopic properties, such as density, mean velocity, etc. To this end, it is instructive to make use of Eq. (23) as a moment generating equation which determines the evolution of average macroscopic properties, including the effects of anomalous transport. We define the average density $N_j(x,t)$, particle flux $N_j(\underset{\sim}{x},t)\underset{\sim}{V}_j(\underset{\sim}{x},t)$, and stress tensor $\underset{\approx}{\Pi}_j(\underset{\sim}{x},t)$ according to

$$N_j = \int d\underset{\sim}{v}\, F_j \quad , \quad N_j\underset{\sim}{V}_j = \int d\underset{\sim}{v}\, \underset{\sim}{v}\, F_j \ , \qquad (32)$$
$$\underset{\approx}{\Pi}_j = m_j \int d\underset{\sim}{v}(\underset{\sim}{v}-\underset{\sim}{V}_j)(\underset{\sim}{v}-\underset{\sim}{V}_j)F_j \ .$$

Taking the appropriate velocity moments of Eq. (23), we find, after some straightforward algebra,

$$\frac{\partial}{\partial t} N_j + \nabla \cdot (N_j\underset{\sim}{V}_j) = 0 \ , \qquad (33)$$

$$\frac{\partial}{\partial t} (N_j m_j \underset{\sim}{V}_j) + \underset{\sim}{\nabla} \cdot (N_j m_j \underset{\sim}{V}_j\underset{\sim}{V}_j + \underset{\approx}{\Pi}_j) = N_j e_j \left( <\underset{\sim}{E}> + \frac{\underset{\sim}{V}_j \times <\underset{\sim}{B}>}{c} \right) + \underset{\sim}{R}_j^{AN} \ , \qquad (34)$$

$$\frac{\partial}{\partial t} \underset{\approx}{\Pi}_j + \nabla \cdot (\underset{\sim}{V}_j\underset{\approx}{\Pi}_j) + \underset{\approx}{\Pi}_j \cdot (\nabla\underset{\sim}{V}_j) + (\nabla\underset{\sim}{V}_j)^T \cdot \underset{\approx}{\Pi}_j$$
$$= \frac{e_j}{m_j c} (\underset{\approx}{\Pi}_j \times <\underset{\sim}{B}> - <\underset{\sim}{B}> \times \underset{\approx}{\Pi}_j) + 2\underset{\approx}{Q}_j^{AN} \ , \qquad (35)$$

where $(\ )^T$ denotes dyadic transpose, and the anomalous transport terms, $\underset{\sim}{R}_j^T$ and $\underset{\approx}{Q}_j^T$, are defined by

$$\underset{\sim}{R}_j^{AN} = e_j <\delta N_j \delta\underset{\sim}{E}> \ , \qquad (36)$$

$$\underset{\approx}{Q}_j^{AN} = \frac{1}{2} e_j <[\delta(N_j\underset{\sim}{V}_j) - \underset{\sim}{V}_j \delta N_j]\delta\underset{\sim}{E}$$
$$+ \delta\underset{\sim}{E}[\delta(N_j\underset{\sim}{V}_j) - \underset{\sim}{V}_j \delta N_j]> \ , \qquad (37)$$

with $\delta N_j \equiv \int d\underset{\sim}{v}\ \delta f_j$ and $\delta(N_j\underset{\sim}{V}_j) \equiv \int d\underset{\sim}{v}\ \underset{\sim}{v}\ \delta f_j$. In deriving Eq. (35), we have assumed negligible heat flow, and therefore approximated

$$\underset{\sim}{\nabla} \cdot [m_j \int d\underset{\sim}{v}(\underset{\sim}{v}-\underset{\sim}{V}_j)(\underset{\sim}{v}-\underset{\sim}{V}_j)(\underset{\sim}{v}-\underset{\sim}{V}_j)F_j] = 0 \ ,$$

which is a valid approximation provided the velocity-space distortions of $F_j(\underset{\sim}{x},\underset{\sim}{v},t)$ are relatively symmetric about $\underset{\sim}{v}=\underset{\sim}{V}_j$.

In derivations of the classical transport equations, which include the effects of binary collisions, the argument is made that collisions maintain $F_j$ close to a local Maxwellian, so that isotropic pressure ($\underset{\approx}{P}_j \simeq P_j\underset{\approx}{I}$) is a good approximation [22]. This leads to considerable simplification in the classical analogue of Eq. (35)

(and in its center-of-mass version). In the anomalous case, however, further simplification of the energy transport equation (35) generally requires an investigation of the degree of isotropy of the turbulence, as well as an investigation of instability mechanisms [23,24] for collective isotropization.

For our purposes here, we examine the rate of anomalous momentum transfer, $\underset{\sim}{R}_j^{AN}$, and the associated anomalous resistivity. A similar investigation of the anomalous heating rate [Eq. (37)] is given elsewhere [7].

### B. Rate of Anomalous Momentum Transfer

In this section we simplify the expression for the rate of anomalous momentum transfer, $\underset{\sim}{R}_j^{AN}$, defined in Eq. (36). Use is made of Eqs. (30) and (31) and the assumption that the turbulence is short wavelength, i.e.,

$$|\underset{\sim}{k}|^{-1} << L , \tag{38}$$

where $|\underset{\sim}{k}|^{-1}$ is the characteristic instability wavelength, and L is the characteristic length scale for spatial variation in $<\underset{\sim}{E}>$, $<\underset{\sim}{B}>$ or $F_j$. Integrating Eq. (31), $\delta f_j(\underset{\sim}{x},\underset{\sim}{v},t)$ can be expressed as

$$\delta f_j = -\frac{e_j}{m_j}\int_{-\infty}^{t} \delta\underset{\sim}{E}(\underset{\sim}{x}',t')\cdot\frac{\partial}{\partial\underset{\sim}{v}'}F_j\,dt' \tag{39}$$

where $\underset{\sim}{x}'$ and $\underset{\sim}{v}'$ are the particle trajectories in the (slowly varying) fields $<\underset{\sim}{E}>$ and $<\underset{\sim}{B}>$, with "initial" conditions $\underset{\sim}{x}'(t'=t)=\underset{\sim}{x}$ and $\underset{\sim}{v}'(t'=t)=\underset{\sim}{v}$. Fourier decomposing with respect to the fast spatial variation, the perturbed density amplitude $\delta N_j(\underset{\sim}{k})=\int d\underset{\sim}{v}\,\delta f_j(\underset{\sim}{k},\underset{\sim}{v})$ can be expressed in the form

$$\delta N_j(\underset{\sim}{k}) = \chi_j(\underset{\sim}{k},\omega+i\gamma)\,\frac{k^2}{4\pi e_j}\,\delta\phi(\underset{\sim}{k}) , \tag{40}$$

where $-i\underset{\sim}{k}\delta\phi(\underset{\sim}{k})=\delta\underset{\sim}{E}(\underset{\sim}{k})$, $\chi_j(\underset{\sim}{k},\omega+i\gamma)$ is an effective susceptibility, and $\omega+i\gamma$ is the complex oscillation frequency. Substituting Eq. (40) into Poisson's equation, $i\underset{\sim}{k}\cdot\delta\underset{\sim}{E}(\underset{\sim}{k})=\sum_j 4\pi e_j\,\delta N_j(\underset{\sim}{k})$, gives the dispersion relation (in a local sense)

$$D(\underset{\sim}{k},\omega+i\gamma) = 1 + \chi_i(\underset{\sim}{k},\omega+i\gamma) + \chi_e(\underset{\sim}{k},\omega+i\gamma) = 0 . \tag{41}$$

Since $F_j$, $<\underset{\sim}{E}>$ and $<\underset{\sim}{B}>$ are slowly varying functions of $\underset{\sim}{x}$ and t, we note that Eq. (41) is effectively an adiabatic dispersion relation which determines $\omega+i\gamma$ as a slowly varying function of $\underset{\sim}{x}$ and t.

Making use of Eq. (40), it is straightforward to show that the anomalous force term in the momentum transport equation (34) can be expressed as [7]

$$\begin{aligned} \underset{\sim}{R}_j^{AN} &= e_j\langle\delta N_j(\underset{\sim}{x},t)\delta\underset{\sim}{E}(\underset{\sim}{x},t)\rangle \\ &= 2\int d\underset{\sim}{k}\,\mathcal{E}_{\underset{\sim}{k}}\,\mathrm{Im}[\underset{\sim}{k}\chi_j(\underset{\sim}{k},\omega+i\gamma)]\ , \end{aligned} \tag{42}$$

where $\mathcal{E}_{\underset{\sim}{k}}$ is the spectral energy density associated with the electric field fluctuations, i.e., $\int d\underset{\sim}{k}\,\mathcal{E}_{\underset{\sim}{k}}=\langle|\delta\underset{\sim}{E}(\underset{\sim}{x},t)|^2/8\pi\rangle$. Since $\mathrm{Im}\ \chi_i=-\mathrm{Im}\ \chi_e$ follows from Eq. (41), we make use of Eq. (42) to obtain the momentum conservation relation

$$\underset{\sim}{R}_i^{AN} = -\underset{\sim}{R}_e^{AN}\ . \tag{43}$$

In applying Eqs. (41) and (42), it is important to make use of the conjugate symmetries

$$\begin{aligned} \omega(-\underset{\sim}{k},\underset{\sim}{x},t) &= -\omega(\underset{\sim}{k},\underset{\sim}{x},t)\ , \\ \gamma(-\underset{\sim}{k},\underset{\sim}{x},t) &= \gamma(\underset{\sim}{k},\underset{\sim}{x},t)\ , \\ \mathcal{E}_{-\underset{\sim}{k}}(\underset{\sim}{x},t) &= \mathcal{E}_{\underset{\sim}{k}}(\underset{\sim}{x},t)\ . \end{aligned} \tag{44}$$

Moreover, in the sense of a quasilinear analysis, the spectral energy density evolves according to [21]

$$\frac{\partial}{\partial t}\mathcal{E}_{\underset{\sim}{k}} = 2\gamma\mathcal{E}_{\underset{\sim}{k}} \tag{45}$$

where $\gamma$ is determined from the linear dispersion relation (41).

Thus far, no assumption has been made about the explicit form of $\chi_j$. For application to rapidly growing microinstabilities, such as the lower-hybrid-drift instability discussed in Sec. 2, we now specialize to the case of unmagnetized ions with $|\omega+i\gamma|\gg\omega_{ci}$ and $k_\perp^2 r_{Li}^2\gg1$, where $r_{Li}=(2T_i/m_i\omega_{ci}^2)^{1/2}$ is the ion Larmor radius, and $k_\perp$ is the characteristic wavenumber of the microturbulence perpendicular to $\langle B\rangle=B_z\hat{\underset{\sim}{e}}_z$. In this case, we evaluate $\delta f_i$ with $\langle\underset{\sim}{B}\rangle\to0$, and $(\partial F_i/\partial t)_{AN}$ can be expressed in the familiar quasilinear form

$$\left(\frac{\partial}{\partial t}F_i\right)_{AN} = \frac{\partial}{\partial\underset{\sim}{v}}\cdot\left[\frac{8\pi e_i^2}{m_i^2}\int d\underset{\sim}{k}\,\frac{\mathcal{E}_{\underset{\sim}{k}}\underset{\sim}{k}\underset{\sim}{k}}{\omega-\underset{\sim}{k}\cdot\underset{\sim}{v}+i\gamma}\cdot\frac{\partial F_i}{\partial\underset{\sim}{v}}\right]\ , \tag{46}$$

and $\chi_i$ reduces to

$$\chi_i(\underset{\sim}{k},\omega+i\gamma) = \frac{4\pi e_i^2}{m_i k^2}\int d\underset{\sim}{k}\,\frac{\underset{\sim}{k}\cdot\partial F_i/\partial\underset{\sim}{v}}{\omega-\underset{\sim}{k}\cdot\underset{\sim}{v}+i\gamma}\ . \tag{47}$$

In summary, Eq. (42) forms the final expression for the anomalous transport term $\underset{\sim}{R}_j^{AN}$ to be used in the macroscopic transport equation (34). Moreover, the spectral energy density $\mathcal{E}_{\underset{\sim}{k}}$ evolves according to

Eq. (45), and the complex eigenfrequency $\omega+i\gamma$ is determined from the dispersion relation (41).

## 4. ANOMALOUS RESISTIVITY ASSOCIATED WITH THE LOWER-HYBRID-DRIFT INSTABILITY

In this section, we make use of the general anomalous transport formalism discussed in Sec. 3 to investigate the anomalous cross-field resistivity $\eta_{AN}$ associated with the lower-hybrid-drift instability [Sec. 2]. An expression for $\eta_{AN}$ is obtained in Sec. 4.A, and in Sec. 4.B we compare the relative sizes of classical and anomalous resistivity for theta-pinch parameters of fusion interest.

### A. Anomalous Resistivity $\eta_{AN}$

Making use of the formalism developed in Sec. 3, we now evaluate the rate of anomalous momentum transfer [Eq. (42)] associated with lower-hybrid-drift instability. In this regard, it is convenient to define the spectral average of a function $f(\underset{\sim}{k},t)$ by

$$\langle f(\underset{\sim}{k},t)\rangle_{\underset{\sim}{k}} = \frac{\int d\underset{\sim}{k}\, f(\underset{\sim}{k},t)\mathcal{E}_{\underset{\sim}{k}}(t)}{\int d\underset{\sim}{k}\, \mathcal{E}_{\underset{\sim}{k}}(t)} , \tag{48}$$

where $\mathcal{E}_{\underset{\sim}{k}}(t)$ is the spectral energy density associated with the electric field fluctuations. It then follows from Eq. (48) and the wave kinetic equation $\partial\mathcal{E}_{\underset{\sim}{k}}/\partial t=2\gamma\mathcal{E}_{\underset{\sim}{k}}$, that the total fluctuation energy density, $\mathcal{E}_F(t)=\int d\underset{\sim}{k}\mathcal{E}_{\underset{\sim}{k}}(t)=\langle|\delta\underset{\sim}{E}(\underset{\sim}{x},t)|^2/8\pi\rangle$, evolves according to

$$\frac{\partial}{\partial t}\mathcal{E}_F = 2\langle\gamma\rangle_{\underset{\sim}{k}}\mathcal{E}_F . \tag{49}$$

To simplify notation, we carry out the stability analysis in a frame of reference where the mean ion velocity is (locally) zero [see also Sec. 2]. From Eq. (43), we note that turbulence-induced changes in $\underset{\sim}{V}_i$ satisfy $\Delta\underset{\sim}{V}_i=-(m_e/m_i)\Delta\underset{\sim}{V}_e$, so that $\underset{\sim}{V}_i\simeq 0$ and

$$\underset{\sim}{V}_d = \underset{\sim}{V}_e - \underset{\sim}{V}_i \simeq \underset{\sim}{V}_e \tag{50}$$

remain good approximations as the system evolves. [In any case, it is straightforward to extend the present analysis to circumstances where $\underset{\sim}{V}_i\neq 0$ by making the replacements $\omega\to\omega-\underset{\sim}{k}\cdot\underset{\sim}{V}_i$ and $\omega-\underset{\sim}{k}\cdot\underset{\sim}{V}_d\to\omega-\underset{\sim}{k}\cdot\underset{\sim}{V}_e$.]

It is convenient to introduce the effective collision frequency $\nu_{AN}$ for anomalous momentum exchange between the electrons and the waves along the current direction $\underset{\sim}{V}_d$. Taking the dot product of $\underset{\sim}{R}_e^{AN}$ with $\underset{\sim}{V}_d$, we define $\nu_{AN}$ by

$$\left(\frac{\partial}{\partial t} N\, m\, \underset{\sim}{V}_e\right)_{AN} \cdot \underset{\sim}{V}_d = -\nu_{AN} N\, m\, \underset{\sim}{V}_e \cdot \underset{\sim}{V}_d = \underset{\sim}{R}_e^{AN}\cdot\underset{\sim}{V}_d . \tag{51}$$

Combining Eqs. (42) and (51), we find

$$\nu_{AN} = \langle \mathrm{Im}[-\underset{\sim}{k}\cdot\underset{\sim}{V}_d\chi_e(\underset{\sim}{k},\omega+i\gamma)]\rangle_{\underset{\sim}{k}} \frac{\mathcal{E}_F}{Nm_eV_d^2/2} , \tag{52}$$

where $\langle\ \rangle_{\underset{\sim}{k}}$ is the spectral average defined in Eq. (48). Moreover, from Eq. (52), the effective anomalous resistivity $\eta_{AN}$ along the current direction $\underset{\sim}{V}_d$ can be expressed as

$$\begin{aligned}\eta_{AN} &= \frac{m_e\nu_{AN}}{N e^2}\\ &= \frac{4\pi}{\omega_{pe}^2}\langle \mathrm{Im}[-\underset{\sim}{k}\cdot\underset{\sim}{V}_d\chi_e(\underset{\sim}{k},\omega+i\gamma)]\rangle_{\underset{\sim}{k}} \frac{\mathcal{E}_F}{Nm_eV_d^2/2}\\ &= \frac{4\pi}{\omega_{pe}^2}\langle \mathrm{Im}[\underset{\sim}{k}\cdot\underset{\sim}{V}_d\chi_i(\underset{\sim}{k},\omega+i\gamma)]\rangle_{\underset{\sim}{k}} \frac{\mathcal{E}_F}{Nm_eV_d^2/2} ,\end{aligned} \tag{53}$$

where use has been made of the linear dispersion relation $\chi_e=-1-\chi_i$ [Eq. (41)]. In circumstances where it is valid to approximate the instantaneous ion distribution function $F_i$ by a local Maxwellian, we obtain $\chi_i=(2\omega_{pi}^2/\underset{\sim}{k}^2v_i^2)[1+\xi_iZ(\xi_i)]$ from Eq. (47) [see also Eq. (8)], where $Z(\xi_i)$ is the plasma dispersion function, and $\xi_i=(\omega+i\gamma)/|\underset{\sim}{k}|v_i$ for $\underset{\sim}{V}_i=0$. In this case, the expression for anomalous resistivity in Eq. (53) reduces to

$$\eta_{AN} = \frac{4\pi}{\omega_{pe}^2}\left\langle \mathrm{Im}\left(\underset{\sim}{k}\cdot\underset{\sim}{V}_d\frac{2\omega_{pi}^2}{\underset{\sim}{k}^2v_i^2}\xi_iZ(\xi_i)\right)\right\rangle_{\underset{\sim}{k}} \frac{\mathcal{E}_F}{Nm_eV_d^2/2} . \tag{54}$$

A simple estimate of the spectral average required in evaluating the transport coefficient $\eta_{AN}$ is often possible. For example, in circumstances where the spectral energy density $\mathcal{E}_{\underset{\sim}{k}}(t)$ becomes strongly peaked about the wavenumber, $\pm\underset{\sim}{k}_M(t)$, corresponding to maximum linear growth rate, then $\langle f(\underset{\sim}{k},t)\rangle_{\underset{\sim}{k}}$ and $\langle\gamma\rangle_{\underset{\sim}{k}}$ can be approximated by

$$\begin{aligned}\langle f(\underset{\sim}{k},t)\rangle_{\underset{\sim}{k}} &= \frac{1}{2}[f(\underset{\sim}{k}_M,t)+f(-\underset{\sim}{k}_M,t)] ,\\ \langle\gamma\rangle_{\underset{\sim}{k}} &= \gamma_M ,\end{aligned} \tag{55}$$

where $\gamma_M$ is the maximum growth rate satisfying the linear dispersion relation $D(\pm\underset{\sim}{k}_M,\pm\omega_M+i\gamma_M)=0$ [see Eqs. (7) and (41)].

For present purposes, we examine Eq. (54) in the low-drift-velocity ($V_E\ll v_i$) cold-electron ($T_e\ll T_i$) regime discussed in Sec. 2.C. Approximating $Z(\xi_i)\simeq i\sqrt{\pi}k_y/|k_y|$, and $V_d=V_E+V_{de}\simeq V_E=-V_{di}$ (since $T_e\ll T_i$ is assumed), we find [7,8]

$$\eta_{AN} = \frac{4\pi}{\omega_{pe}^2}\left\{\mathrm{Im}\left[k_y V_E \frac{2\omega_{pi}^2}{k_y^2 v_i^2}\frac{\omega}{k_y v_i} i\sqrt{\pi}\right]\right\}_{k_y=k_M} \frac{\mathcal{E}_F}{Nm_e V_E^2/2} , \tag{56}$$

where $\mathcal{E}_F = \int_{-\infty}^{\infty} dk_y \mathcal{E}_{k_y}(t)$, and the spectrum is assumed to be one-dimensional ($k_y^2 >> k_x^2$). Making use of the maximum growth estimates in Eq. (16), the expression for $\eta_{AN}$ in Eq. (56) reduces to

$$\eta_{AN} = 4\pi \sqrt{\frac{\pi}{2}}\, \Omega_{LH}^{-1} \frac{\mathcal{E}_F}{NT_i} \tag{57}$$

where $\Omega_{LH}$ is defined in Eq. (17). It is convenient to rewrite Eq. (57) in terms of the wave energy density $\mathcal{E}_W$ (field energy density plus oscillatory particle kinetic energy density). Making use of

$$\mathcal{E}_W = \mathcal{E}_F \left|\omega \frac{\partial D_r}{\partial\omega}\right|_{k_y=\pm k_M} , \tag{58}$$

we readily find $\mathcal{E}_W = 2(1+\omega_{pe}^2/\omega_{ce}^2)\mathcal{E}_F$. Equation (57) can then be expressed as

$$\eta_{AN} = 2\pi \sqrt{\frac{\pi}{2}} \frac{\Omega_{LH}^{-1}}{(1+\omega_{pe}^2/\omega_{ce}^2)} \frac{\mathcal{E}_W}{NT_i} . \tag{59}$$

To estimate the value of anomalous resistivity $\eta_{AN}$, we invoke a physically plausible upper bound on the wave energy $\mathcal{E}_W$, namely, the total directed kinetic energy $Nm_eV_E^2/2$, i.e.,

$$\begin{aligned}[\mathcal{E}_W]_{MAX} &\simeq Nm_e V_E^2/2 \\ &= NT_i \frac{m_e}{m_i}\left(\frac{V_E}{v_i}\right)^2 .\end{aligned} \tag{60}$$

Making use of Eqs. (13), (59) and (60), gives the resistivity bound [7,8]

$$\eta_{AN} \simeq \left(\frac{\pi}{2}\right)^{3/2} \frac{\Omega_{LH}^{-1}}{1+\omega_{pe}^2/\omega_{ce}^2} \frac{m_e}{m_i}\left(\frac{r_{Li}}{N}\frac{\partial N}{\partial x}\right)^2 \tag{61}$$

The anomalous resistivity estimate in Eq. (61) can be orders-of-magnitude larger [8] than the classical cross-field resistivity [14]. Indeed, in recent experimental studies by Commisso and Griem [9] of the post-implosion resistivity in the Maryland 40 kV theta pinch, the measured resistivity in the sheath region ($r_{Li}/\Delta L_n \sim 1/3$) is about fifty times the classical value, and in reasonably good agreement (approximately a factor of five larger) than the theoretical estimate in Eq. (61). The fact that the measured value of $\eta_{AN}$ is somewhat larger than the theoretical estimate in Eq. (61) is physically plausible, since the cold-electron assumption ($T_e << T_i$) made in deriving Eq. (61) is of marginal validity in the Maryland experiment [9]. As discussed in Sec. 2.B, the instability growth rate (and also the anomalous resistivity) increase with increasing $T_e/T_i$.

As the plasma-magnetic field interface broadens in response to the diffusion induced by the lower-hybrid-drift instability, the growth rate for this mode also drops, since the maximum growth rate $\gamma_M$ scales as $\gamma_M \sim (r_{Li}/\Delta L_n)^2$. However, as this occurs, other instabilities might be expected to grow, continuing to tap the free energy in the confined plasma. It can be shown that in the limit $\Delta L_n >> r_{Li}$ the widely studied [25] low-frequency drift waves and universal instabilities have ample time to grow and affect the late-time transport.

### B. Comparison with Classical Resistivity [8]

It is instructive to compare the relative sizes of $\eta_{an}$ [Eq. (61)] and the cross-field classical resistivity $\eta_{cl}$ [14] given by

$$\eta_{cl} = \frac{4.7\times10^{-18}}{T_e^{3/2}} \sec^{-1} , \tag{62}$$

where $T_e$ is the electron temperature in keV, and $\ell n\Lambda \simeq 13$ and $Z=1$ have been assumed. For the sake of definiteness, we assume a theta-pinch configuration with gaussian density profile

$$N = N_0 \exp\left(-\frac{x^2}{2\sigma^2}\right) , \tag{63}$$

and isothermal electrons and ions ($\partial T_j/\partial x=0$). Quasiequilibrium pressure balance gives

$$B_z = B_w[1-\hat{\beta}\exp(-x^2/2\sigma^2)]^{1/2} , \tag{64}$$

where $\hat{\beta}=8\pi N_0(T_i+T_e)/B_w^2$, $B_w$ is the magnetic field at the wall ($x=x_w$), and $N(x=x_w)\simeq0$ has been assumed. For present purposes, we also assume zero bias field on axis, with $B_z(x=0)=0$, and therefore $\hat{\beta}=1$. Moreover, for purposes of comparing $\eta_{an}$ and $\eta_{cl}$, we evaluate the resistivity at the point of <u>maximum density gradient</u>, i.e., at $x=\sigma$. Substituting $\hat{\beta}=1$ and $x=\sigma$ in Eq. (64), we find $B(x=\sigma)=0.627\ B_w$. An evaluation of $[\omega_{pe}^2/\omega_{ce}^2]_{x=\sigma}$ will also be useful. Making use of Eqs. (63) and (64), it is straightforward to show that

$$\left(\frac{\omega_{pe}^2}{\omega_{ce}^2}\right)_{x=\sigma} = \left(\frac{\hat{\beta}\ \exp(-x^2/2\sigma^2)}{1-\hat{\beta}\exp(-x^2/2\sigma^2)}\right)_{x=\sigma} \frac{m_e c^2/2}{T_i+T_e} .$$

For $\hat{\beta}=1$, $x=\sigma$, and $m_e c^2=511$ keV, we obtain

$$\left(\frac{\omega_{pe}^2}{\omega_{ce}^2}\right)_{x=\sigma} = \frac{394}{T_i+T_e} , \tag{65}$$

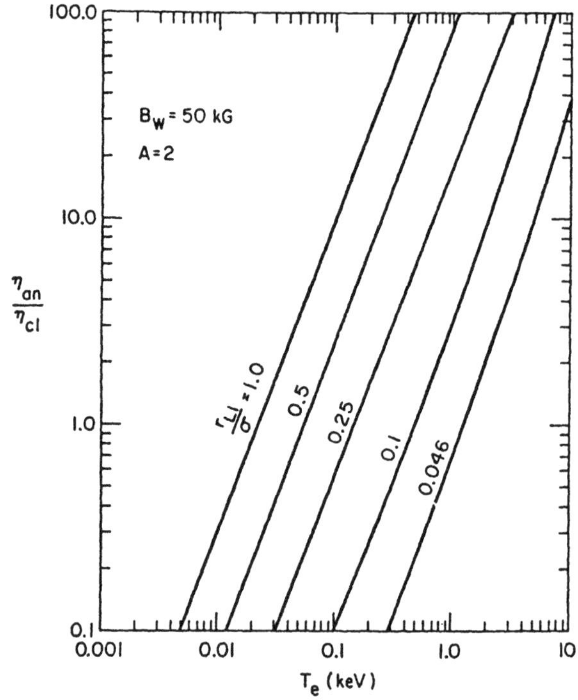


Fig. 3 Plot of $\eta_{AN}/\eta_{cl}$ versus $T_e$ [Eq. (67)]

where $T_i+T_e$ is expressed in keV. Substituting Eqs. (63)-(65) into Eq. (61) and evaluating at $x=\sigma$ gives [8]

$$\eta_{AN} \simeq 9.9\times10^{-15} \frac{(T_i+T_e)}{A^{1/2}\, B_w} \left(\frac{r_{Li}}{\sigma}\right)^2 \sec^{-1} , \tag{66}$$

where A is the ion mass number, and $B_w$ is the magnetic field at the wall in kilogauss. In obtaining Eq. (66), $[\omega_{pe}^2/\omega_{ce}^2]_{x=\sigma} >> 1$, or equivalently, $T_e+T_i << 394$ keV has been assumed.

To compare the relative size of the anomalous and classical resistivities, we form the ratio $\eta_{AN}/\eta_{cl}$. Making use of Eqs. (62) and (66), we find [8]

$$\frac{\eta_{AN}}{\eta_{cl}} \simeq 2.1\times10^{3} \frac{T_e^{3/2}(T_i+T_e)}{A^{1/2}\, B_w} \left(\frac{r_{Li}}{\sigma}\right)^2 . \tag{67}$$

In Eq. (66), $T_e$ and $T_i$ are expressed in keV, and $B_w$ in kilogauss.

Moreover, we remind the reader that $T_e \ll T_i$ has been assumed in deriving Eq. (61), and hence Eq. (67).

It is clear from the estimate in Eq. (67) that $\eta_{an} > \eta_{cl}$ unless the electrons are sufficiently cold and/or the confining field $B_w$ is sufficiently large. For example, if $T_e$=1.3 keV, $T_i$=10 keV, $B_w$= 10 kG, A=2 (deuterium) and $r_{Li}/\sigma$=0.25, then $\eta_{AN}/\eta_{cl} \simeq 10^2$ follows from Eq. (67). In Fig. 3, $\eta_{AN}/\eta_{cl}$ is plotted versus $T_e$, for A=2, $B_w$=60 kG, $T_i$=10 keV and various values of $r_{Li}/\sigma$ [8]. Note that $\eta_{AN} < \eta_{cl}$ only for extremely low values of $T_e$ and $r_{Li}/\sigma$. For example, if $r_{Li}/\sigma$=0.1, then $\eta_{AN} < \eta_{cl}$ only for $T_e \lesssim 400$ eV.

## 5. CONCLUSIONS

We have investigated the cross-field anomalous resistivity associated with the lower-hybrid-drift instability, with particular emphasis on the low-drift-velocity regime characterized by $V_E/v_i = r_{Li}/2\Delta L_n \lesssim 1$. The predicted value of anomalous resistivity (for $T_e \ll T_i$) is $\eta_{AN} = 4\pi(\pi/2)^{1/2}\Omega_{LH}^{-1}\mathcal{E}_F/NT_i$ [Eq. (57)], where $\mathcal{E}_F = \langle \delta \underset{\sim}{E}^2 \rangle /8\pi$ is the energy density of the unstable field fluctuations. Assuming that the wave energy is bounded by the directed kinetic energy $Nm_eV_E^2/2$, we obtain the simple bound on anomalous resistivity given in Eq. (61), which can be orders-of-magnitude larger than the classical value.

In conclusion, we note that the late-time resistivity measurements in the Culham 8-meter experiment [26] are often quoted as general evidence that post-implosion transport in theta pinches is classical. In this regard, we emphasize that the plasma was sufficiently cold ($\sim$150 eV) in the Culham experiment that the late-time transport was collision-dominated. Indeed, for the Culham parameters [26], it can be shown [8] that $\eta_{cl} \simeq 7\,\eta_{AN}$ in the sheath region, where $\eta_{AN}$ is the anomalous resistivity estimate in Eq. (61). For high-temperature theta-pinch parameters of fusion interest, however, the condition $\eta_{AN} \gg \eta_{cl}$ is easily satisfied [8], at least during the immediate post-implosion phase where $r_{Li}/\Delta L_n \lesssim 1$. As mentioned earlier, the measured post-implosion resistivity in the Maryland experiment [9] is about fifty times the classical value.

## 6. ACKNOWLEDGMENTS

This research was supported by the Energy Research and Development Administration, and in part by the Electric Power Research Institute Inc. under Contract #RP423-1 with Princeton Plasma Physics Laboratory. It is a pleasure to thank Professor Hans Wilhelmsson for his kind hospitality and the Nobel Foundation for sponsoring the 1976 Symposium on Nonlinear Effects in Plasmas.

## 7. REFERENCES


[1] B. B. Kadomtsev and O. P. Pogutse, Nucl. Fusion 11, 67 (1971).
[2] P. C. Liewer and N. A. Krall, Phys. Fluids 16, 1953 (1973).
[3] R. Z. Sagdeev, Proc. Symp. Appl. Math. (New York, 1965); (American Mathematical Society, Providence, 1967) Vol. 18, p. 281.
[4] R. C. Davidson and N. T. Gladd, Phys. Fluids 18, 1327 (1975).
[5] N. T. Gladd, Plasma Physics 18, 27 (1976).
[6] N. A. Krall and P. C. Liewer, Phys. Rev. A4, 2094 (1971).
[7] R. C. Davidson and N. A. Krall, "Review of Anomalous Transport in High-Temperature Plasmas", submitted for publication (1976).
[8] P. C. Liewer and R. C. Davidson, "Analytic and Numerical Studies of Sheath Broadening by the Lower-Hybrid-Drift Instability in Post-Implosion Theta Pinches", submitted for publication (1976).
[9] R. Commisso and H. R. Griem, Phys. Rev. Letters 36, 1038 (1976); R. Commisso and H. R. Griem, submitted for publication (1976).
[10] K. F. McKenna, R. Kristal, K. S. Thomas, Phys. Rev. Letters 32, 409 (1974).
[11] W. Davis and A. W. DeSilva, Phys. Fluids 16, 1740 (1973).
[12] E. L. Cantrell, W. R. Ellis, H. W. Harris, F. C. Jahoda, R. Kristal, M. D. Machalek, J. R. McConnell, W. E. Quinn, F. L. Ribe, G. A. Sawyer, F. T. Seibel and R. E. Siemon, in Plasma Physics and Controlled Nuclear Fusion Research (IAEA, Vienna, 1974), Vol. III, 13.
[13] E. Oktay, A. W. DeSilva, P. C. Liewer, Y. G. Chen, H. R. Griem, R. Hess, N. A. Krall, in Plasma Physics and Controlled Nuclear Fusion Research (IAEA, Vienna, 1974).
[14] L. Spitzer, Physics of Fully Ionized Gases (Wiley, New York, 1967).
[15] A. H. Boozer and M. A. Levine, Phys. Rev. Lett. 31, (1973) 1287.
[16] A. I. Barcilon, D. L. Book, J. P. Boris, A. L. Cooper, K. Hain, P. C. Liewer, A. E. Robson, R. A. Shanny, P. J. Turchi, and N. K. Winsor in Plasma Physics and Controlled Fusion Research (IAEA, Vienna, 1975), Vol. III, p. 365.
[17] N. A. Krall, Phys. Fluids 19, in press (1976).
[18] R. C. Davidson, N. T. Gladd, J. Huba and C. S. Wu, submitted for publication (1976).
[19] R. H. Levy, J. D. Daugherty and O. Buneman, Phys. Fluids 12, 2616 (1969).
[20] R. C. Davidson, Theory of Nonneutral Plasmas (Benjamin, New York, 1974) p. 62.
[21] R. C. Davidson, Methods in Nonlinear Plasma Theory (Academic Press, New York, 1972).
[22] S. I. Braginskii in Reviews of Plasma Physics (Consultants Bureau, New York, 1965) Vol. 1, p. 205.
[23] R. C. Davidson and J. Ogden, Phys. Fluids 18, 1045 (1975).
[24] S. Hamasaki and N. A. Krall, Phys. Fluids 16, 145 (1973).
[25] N. A. Krall, in Advances in Plasma Physics (Eds. A. Simon and W. B. Thompson, Wiley, New York 1968). Vol. 7, p. 153, and references therein.

[26] H. Bodin, J. McCartan, A. A. Newton and G. H. Wolb, in Plasma Physics and Controlled Fusion Research (IAEA, Vienna, 1969) Vol. 2, p. 533.

# NONLINEAR DYNAMICS OF JOULE HEATED TOROIDAL DISCHARGES

Michel Cotsaftis

ASSOCIATION EURATOM-CEA

Fontenay-aux-Roses, France

## 1. INTRODUCTION

A rather large and puzzling discrepancy still exists between the experimental results and the various proposed models. If ions could be almost (neo) classical, though they do not seem to switch as expected into the predicted banana regime, this is by far not the case for the electrons. For instance, numerical computations with (neo) classical MHD System produce a very strong skin effect on the electronic temperature $T_e$ taking a long time compared to the duration of the discharge to disappear, which is never observed. During the discharge also, even if one artificially eliminates the skin effect, $T_e$ does not seem to follow simple proposed laws. Moreover, the belief that the observation is the result of many antagonist effects (instabilities, impurities, neutrals) obscures again the possibility of a "theoretical" representation, as examplified by the rapidly growing numerical simulations, taking adjusted ad hoc coefficients to follow the datas, showing that classical laws are not simply verified, and limiting the theoretical arguments to pseudo qualitative ones.

Nevertheless, it is proposed here to revisit the possibility of a "theoretical" approach to the nonlinear dynamics of toroidal discharges, taking advantage that the observed datas are very smooth in space and time and thus, must in some sense originate in global phenomena. Effectively, it has now been repeatedly observed that even very complex systems are surprisingly showing very gentle and simple behavior. This rests upon the remarkable property, already shown in the other rapidly growing field of Nonlinear Partial Differential Equations, that the system will evolve along the "simplest" behavior compatible with the boundary conditions. This is especially important when, whatever complex the system can be, it possesses nice invariance properties which can be observed with adapted mathematical tools prior to any resolution, enormously simplifying this problem. This is for example the success of soliton-like behavior of solutions of some classes of nonlinear systems, which are just group invariant /1/.

Here also, instead of taking as usual a simplified system and researching a solution, as exact as possible, one will conversely take a system as complete as possible and try to show that its solution, at least to first order, is a very simple one, due to invariance properties. To this end, one will start from a general MHD System, including impurities, already introduced /2/. Its properties will be briefly reviewed first, leading to break it into different subsystems, and to keep to first order the "dominant" one which drives the dynamics. Observed datas, analyzed in next paragraph, show mainly that a discharge can be splitted into two different parts : the rising part at the beginning, where the plasma builds up and the quasi stationnary part, covering the subsequent period, where the plasma is maintained by the current. The end of the plasma will be discussed elsewhere. The main features of the behaviors during these two periods are indicated to be theoretically obtained from the analysis of the MHD System, i.e. the space and time dependence of electronic density and temperature and their relation with the current behavior. It is shown in the third paragraph that the datas exhibit a group invariance property, also shared by the dominant MHD subsystem. So, the solution will be researched in the corresponding invariant class, and the nonlinear dynamics can, to first order, be presented as an "identification" problem which, due to the group structure, is extremely simplified and can be solved. Classes of compatible transport coefficients are given, in agreement with the datas. Due to the complex structure of the discharge during this period, they are not reducible to the known classical or neoclassical ones. The quasistationary state is discussed afterwards. There again, another invariance property is shown to exist from the datas, also shared by the dominant MHD subsystem. Researching the solution as previously in the corresponding invariant class, the nonlinear dynamics also reduces to an identification problem which can likewise be solved. Various "experimental" laws are justified and compatible transport coefficients giving a good experimental fit are of the Trapped Electrons Instability type /3/. Note that static equilibria, next rapidly discussed, are not reached in this evolution scheme. In conclusion, from this approach based on invariance, a picture of the plasma discharge can be set up. Variations of observable quantities as functions of time, space and main parameters (magnetic field, size, ..) can be exhibited, in agreement with observations, then allowing for predictions and their verification. Effectively, these invariance properties from which the "simple" solutions are obtained are only first order ones, and one can expect the corresponding symmetries to be broken by higher order terms here neglected. For this reason, the discharges discussed here will be called Ideal Discharges, and it will be of interest to compare them to real ones, from the datas of the next coming generation of machines.

## 1. THE MHD SYSTEM ($\mathfrak{D}$) AND THE DOMINANT SUBSYSTEM ($\mathfrak{D}_s$)

From simplicity argument, and from the observations of the experimental datas which exhibit rather smooth variations in both space and time, one will consider as a starting hypothesis the possibility of representing the evolution of the system at the MHD level first, which is adapted for accounting of the global fluid like behavior. This does not prevent postponing for further analysis more detailed discussion on specific transport coefficients which can be modified either by an MHD branching, changing the topology of the field lines, or by a Microinstability, both affecting the net concerned fluxes. So one will take the system of usual conservations of mass, momentum and energy for each species, including the impurities, which are playing a role in the course of the discharge. This very large system, denoted ($\mathfrak{D}$) in part A of /2/, will not be repeated here. To study it, one can either keep all the terms and perform a numerical simulation to test combined effects /4/, or try to simplify it to hopefully approach analytically some of its properties. Effectively, ($\mathfrak{D}$) can be splitted into three well defined subparts :

(a) the dynamical part
(b) the mass energy part
(c) the impurity part

which are only weakly coupled. First, the observed velocities are small, so that (a) reduces to an independent static equilibrium equation fixing the magnetic structure, eventually modified when crossing a branch point of the system. This happens when the thermal current instability (/2/, part A) sets in, transforming the well behaved nested flux surfaces (branch AB') into a two domain medium, with again well behaved nested flux surfaces inside a limiting surface r (q = mo) and an outer zone with magnetic islands (branch`AB), affecting the transport coefficients such that skin effect disappears (/2/, part B). As a consequence of the two domain structure of the discharge along AB and as the time variation of the inward fluxes is long compared to the response time of the system, the temperatures can be considered as fixed to first order in (c) which becomes a closed set of equations. This nonlinear system will be discussed elsewhere. So finally, in the remaining part (b), there is a static MHD equilibrium and impurities are only entering through global quantities like $Z_{eff}$, or the loss impurity term. Furthermore, when the exchange term between ions and electrons remains modest, the electrons are dominating and drive all the dynamics in Joule heated discharges. So, to first order, the complete system ($\mathfrak{D}$) can be reduced to its dominant subpart ($\mathfrak{D}_s$), given in the cylindrical approximation, though this restriction is not necessary, by :

$$n_e\frac{\partial T_e}{\partial t}-\frac{k_1}{\rho}\frac{\partial}{\partial\rho}\rho\left\{n_e^{-\xi}T_e^{\alpha}\frac{J^{\beta}Z_{eff}^{\varpi}B^{\delta}q^{\zeta}}{a^{\sigma+\tau-2(\varepsilon-1)}}\left(\frac{q}{q}\right)^{2\varepsilon}\cdot\left(\frac{\partial n_e}{\partial\rho}\right)^{\sigma}\left(\frac{\partial T_e}{\partial\rho}\right)^{\tau}\cdot\left[n_e\frac{\partial T_e}{\partial\rho}+\frac{5}{3}T_e\frac{\partial n_e}{\partial\rho}\right]\right\}=k_2\frac{Z_{eff}J^2}{T^{\gamma}}-R \qquad (1a)$$

$$\frac{\partial J}{\partial t}=\frac{k_3}{a^2}\frac{1}{\rho}\frac{\partial}{\partial\rho}\rho\frac{\partial}{\partial\rho}Z_{eff}\frac{J}{T^{\gamma}} \qquad (1b)$$

splitting furthermore apart the mass equation for the electrons :

$$\frac{\partial}{\partial t} n_e + \operatorname{div} \mathcal{D}_e \operatorname{grad} n_e = S \tag{1c}$$

which simply determines the source term S and, assuming from previous theoretical analysis, a general monomial form, for $\mathcal{K}_e$, with : $\mathcal{K}_e = n_e \mathcal{D}_e$ One will take : $\eta = k_2 Z_{eff} T^{-\gamma}$, and : $R = k_4 T^{\theta} n_e n_I$ /5/, where $\gamma = 3/2$ in Spitzer case, and $\theta$ is a slowly increasing function of $Z_{eff}$ , The $k_j$ are numbers related to units scaling, and $\alpha, \beta, \delta, \varepsilon, \sigma, \tau, \sigma, \xi$, are numbers depending on the considered diffusion law, various forms of which will be discussed later. On the other hand, the boundary and initial conditions are : $T_e(a,t) = 0$, $2\int_0^1 \rho J(\rho,t) d\rho = I_p(t)/\pi a^2 = \bar{I}_p(t)$, the mean current density, whereas : $T_e = J = 0$ at $t = 0$, and $\frac{\partial T_e}{\partial \rho} = \frac{\partial J}{\partial \rho} = 0$ on the axis. In what follows, one will essentially try to represent the observed space and time variations of the corresponding quantities $T_e$, $n_e$ from the analysis of this system ($\mathcal{D}_p$), given by (1abc,2).

## 2. THE OBSERVED DATAS

In general, a current I (t) is programmed with a rising part and a plateau, and $n_e$ and $T_e$ profiles are measured for various times. The main observed features are represented on fig. n° 1 for the variation of $T_e$ (0,t), $n_e$ (0,t) and $I_p$ (t) vs t, and on fig. n° 2 and 3 for the profiles of $T_e$ ($\rho$,t) and $n_e$ ($\rho$,t) respectively vs (with $r = a\rho$ ) for various times.

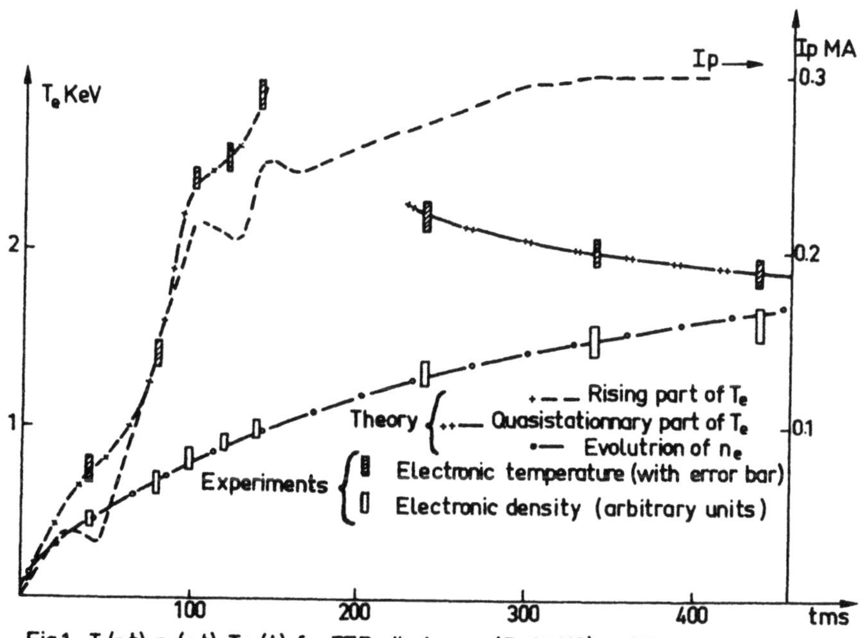


Fig.1 - $T_e$(o,t), $n_e$(o,t), Ip (t) for TFR discharge (B=50KG) and theory

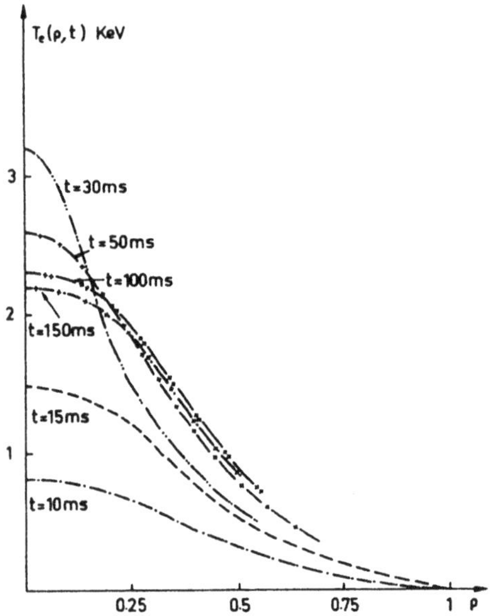


Fig.2-Electronic temperature profile $T_e(\rho,t)$ for various t in a TFR discharge

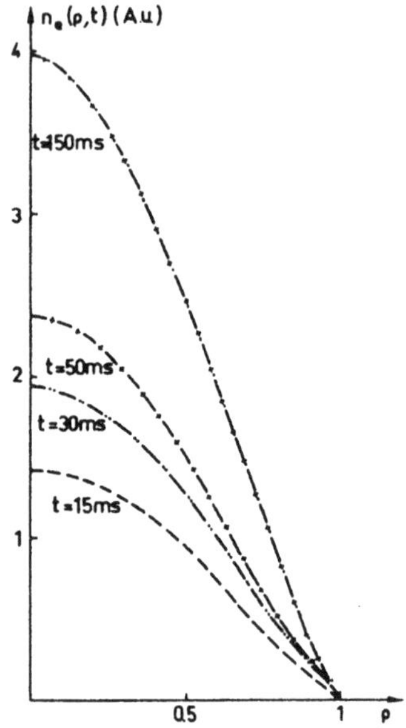


Fig.3-Electronic density profile $n_e(\rho,t)$ for various t in a TFR discharge

From these observations and many other ones, the following general behavior can be drawn. First, the plasma discharge can be separated into two very different parts, the rising up part and the quasistationnary part. The rising up part, which occurs at the beginning of the discharge, is characterized by a rapid increase of $T_e$ which is the stronger and the more peaked, the current rise is itself sharper (fig. n° 1,2). For a particular value of $T_e$ called $T_{ecrit}$, there is for high enough current a sudden change. $T_e$ (0,t) now varies very slowly, decreasing when $I_p$ is constant, and the profile of $T_e$ enlarges up to an asymptotic shape very often like $(1-\rho^2)^{2.5}$. This is the quasistationnary part. Note that, though related as discussed later, $T_{ecrit}$ does not correspond to the beginning of the current plateau. When the final current value is not large enough, like in some Ormak discharges /6/, $T_e$ keeps going on and does not know a sharp change and its behavior, as previously. These discharges will be called "collisional" for convenience. The extreme opposite and uninteresting situation is the "runaway" case, when the current rise is so fast that most of the electrons are gaining a large parallel velocity, when properly accounting for the growth of $Z_{eff}$ itself. So there is a continuous sequence of possible behaviors from collisional to runaway ones, in which the ions are less and less involved, justifying to split apart their dynamics. The situation is quite different for the density $n_e$ (fig. n°1,3). Apart possibly at the very beginning, $n_e$ grows very regularly all over the discharge, curiously like $t^{0.4-0.5}$, and its profile does not shrink nor enlarge, being of parabolic type, i.e. like $(1-\rho^2)^{0.9-1}$. So from the previous rapid description, any theoretical approach would have to solve the following problems :

(a) the laws of the temperature growth related to the current growth $I_p = I_p(t)$, and of the profile variations in the rising up and quasistationnary parts

(b) the value of $T_{ecrit}$ for the change from the rising part to the quasistationnary one

(c) the "universal" behavior of the density $n_e$, both in time and space.

This will be the main task in the following, where it will be shown that quantitative agreement on these three points can be found from ($\mathcal{D}_S$) only, justifying its distinction from the complete system ($\mathcal{D}$).

### 3. EARLY TIME GROUP BEHAVIOR AND IDEAL DISCHARGE

3.1. Consider first the rising up part of the discharge. A refined analysis of the available datas reveals the two following remarkable properties :

(a) at the same time, the $T_e$ profiles are similar for different values of $I_p$ and B, in that they can be transformed into one another by an affinity along $T_e$-axis.

(b) for a given experiment, as a function of time, there exists a relation between the value of $T_e(0,t)$ and the width h (t) of the temperature profile, measured by the coefficient of the parabolic approximation : $T_e(\rho,t) \simeq T_e(0,t).(1-h(t).\rho^2)$. Letting : $k = T_e(0,t_2)/T_e(0,t_1)$, $\mu^2 = h(t_2)/h(t_1)$, and $\lambda = t_2/t_1$, this relation takes the form :

$$\lambda = \mu^2 . k^{3/2} \tag{2}$$

These two observed properties are suggesting that the experimental discharges are enjoying, to a good accuracy, a group property. This would mean that, if one knows the law of the group, one can pass from one profile $T_e(\rho,t;\, I_p, B,a,R)$ to another one $T_e(\rho',t';\, I'_p,B',a',R')$ by a simple transformation, without solving any equation. To support that such a fundamental property could exist, one has to verify that ($\mathcal{D}_S$) is itself group invariant, knowing already that the heat equation, to which the nonlinear parabolic system ($\mathcal{D}_S$) reduces in a first approximation useful for a step by step (conveyent) approximation, is obeying a group property /7/. Likewise, neglecting the loss term compared to the Joule term in the RHS of (1a), which is justified at the beginning of the

discharge, one can show by using the classical theory of group invariance /8/ that, unless a specific relation among the exponents, the only group of transformations leaving (1ab) invariant is the similarity group ($S_0$). In other words, with the stretching transformation :
$n_e \to \nu n_e$, $T_e \to kT_e$, $J \to \ell J$, $t \to \lambda t$, $\rho \to \mu\rho$, $a \to \mu_1 a$, $R \to \mu_2 R$,
$\bar{I}_p \to m\bar{I}_p$, $B \to bB$, $Z_{eff} \to zZ_{eff}$,
the invariance under ($S_0$) means that if $T_e$ ($\rho$, t; $I_p$, B, a, R, $Z_{eff}$) is a solution of (1ab), then $kT_e$ ($\mu\rho$, $\lambda t$; $mI_p$, bB, $\mu_1 a$, $\mu_2 R$, $zZ_{eff}$) will also be a solution of this system if :

$$k = \left[\mu_1^{\gamma} \cdot z^{-\bar{\varpi}} \cdot b^{-\bar{\delta}-\bar{\xi}} \cdot m^{\bar{\zeta}+2\bar{\xi}} \cdot \mu_2^{2(2\bar{\xi}-\bar{\zeta})} \cdot \lambda^{-\bar{\varepsilon}}\right]^{\frac{1}{\bar{\alpha}-\gamma\bar{\varepsilon}}} \tag{3a}$$

$$\mu = \lambda^{1/2} \cdot k^{-\gamma/2} \cdot z^{1/2} \; ; \; \ell = m \lambda^{-1} z^{-1} \cdot k^{\gamma} \; ; \; \nu = m^2 \cdot \lambda^{-1} \cdot k^{\gamma-1} \cdot z \tag{3bcd}$$

with : $\bar{\varepsilon} = \varepsilon + \bar{\zeta} + \bar{\xi} - \frac{\sigma+\tau}{2}$, $\bar{\alpha} = \alpha + \tau + \bar{\xi}$, $\bar{\xi} = \xi - \sigma$ ; $\bar{\delta} = \delta + \beta$, $\bar{\zeta} = \zeta - \beta$, $\bar{\varpi} = \varpi + 1 + \bar{\varepsilon}$.
Further, from the group invariance, (1ab) reduces to a simple differential system when expressing it in the similarity variables /9/, manifesting that, due to this invariance, on can reconstruct the profiles at any time from a "universal" one given by the solution of the differential system, by acting the group on it. Here, the similarity variables are the functions $\phi$, G, F, H, Z(.) given by : $\rho B_\theta = t^d . \phi(u)$ ; $J = t^b . G(u)$ ; $T_e = t^a . F(u)$ ; $n_e = t^n . H(u)$ ; $Z_{eff} = t^{?} Z(u)$ ; $u = \rho t^c$, which, when reported in (1ab), give the system :

$$u G = \phi' \; ; \; k_3 u Z . (G F^{-\gamma})' = d\phi + c u^2 G \tag{4ab}$$

$$H.(aF + cuF') = k_1 u^{-1} . \left[ u^{1+2\zeta} . \frac{Z^{\varpi} . F^{\alpha} . G^{\beta} . \phi^{-\zeta} . F'^{\tau} . H'^{\sigma} . H^{-\xi}}{(2u^{-1} - uG\phi^{-1})^{2\varepsilon}} \left(HF' + \frac{5}{3} FH'\right)\right]' + k_2 G^2 Z F^{-\gamma} \tag{4c}$$

with the safety factor : $q = \rho B/RB_\theta$ where : $B_\theta = \rho^{-1} \int_0^{\rho} \rho' J(\rho', t)\, d\rho'$, and collecting the constants a, R, B in $k_1$. The quantities a, b, d, n, q, are linked to the coefficients ($\alpha, \beta \ldots \tau$) by four relations strictly equivalent to (3abcd).

3.2. Thus, from the invariance of ($\mathcal{D}_{\zeta}$) under ($S_{\delta}$), the problem of solving its evolution has been splitted into two independant parts :

(a) the "outer" time growth of the various quantities from (3abcd)
(b) the shaping of the profiles from the reduced system (4abc) with corresponding boundary conditions.

Let us rapidly indicate for (b) that eq. (4abc), with the conditions : $\phi(0) = 0$, $G(0) = G_0$, $F(0) = F_0$, $F'(0) = 0$, can be completely analysed. In particular, this system is itself invariant under the similarity group, which reduces its order. Asymptotic analysis, on an infinite interval, can also be performed, as well as expansion close to the axis, distinguishing the case where $\sigma = \tau = 0$. However, a simple way to proceed is to give : $\phi(u) = G_0 \frac{d}{du} \mathcal{V}$. Then, G, F and H are respectively given by the relations (4abc), supposing Z given aside from the impurity system. The function G(u), or $\phi$ (u), is related to the mean current by :

$$I_p(t) = t^{b-2c} \cdot \int_0^{t^c} u \cdot G(u) \cdot du \qquad (5)$$

This shows, once again, that the fixed boundary $\rho_B = 1$ does not satisfy the group property, contrarily to a "moving limiter" boundary : $\rho t^c = u_c$. Such a solution, with a regular $\phi(u)$, gives a bell shape solution for $T_e$, free of skin effect, which could be of practical interest /10/. One can expect however to be too far from the group structure as long as the part of the current flowing outside the fixed boundary (and inside the moving one) is small enough. In any case, one could numerically analyze ($\mathcal{D}_S$) directly. Unfortunately, the coefficients ($\alpha, \beta, \ldots \tau$) are yet unknown. Recall that due to the complex structure of the plasma along the branch AB, as indicated in § 2, it did not seem realistic to expect a "simple" diffusion law to rule out the behavior of the plasma there, being a mixture of different ones. Determining these coefficients from the analysis of the full solution of ($\mathcal{D}_S$), which is mathematically known as parameters identification, is hopeless. Effectively, it is required to calculate a set of numbers from the knowledge of a field T ($\rho$,t) i.e. to realize a very strong reduction, which is not possible here due to the error bars. But now, one can account of another "simple" fact, i.e. the behavior of the solution of ($\mathcal{D}_S$) itself, through the group invariance property. This is the deep reason why we insisted on the (possible) group structure of the datas and the corresponding one for ($\mathcal{D}_S$) to facilitate this reduction. Due to their importance, group invariant solutions will be called <u>ideal solutions</u> and, as already shown by the datas, real plasma discharges are close enough to them. Here now, only part (a), which gives the time growth of the various quantities from (3abcd) in terms of the ($\alpha, \beta \ldots \tau$), is involved for the identification problem. As (3bcd) are side equations giving $\mu, \ell, \nu$, this problem even reduces in the ideal case, from various experimental $I_p = I_p(t)$, ie $m = m(\lambda)$ by stretching transform, to determine ($\alpha, \ldots \tau$) such that $k = k(\lambda)$ by (3a) are best fitting the observed $T_e(0,t)$. Note that (3a) is identical to (2), for slowly experimentally varying $Z_{eff}$, providing strong argument for ideal structure. Given the experimental fit : $k \simeq m^{\theta_1} \cdot \lambda^{\theta_2} \cdot b^{\theta_3} \cdot \xi^{-\theta_4}$, one can from (3a) reexpress $\alpha+\tau, \xi-\sigma, \varepsilon-\frac{\sigma+\tau}{2}, \zeta-\beta$ and $\nu$ in terms of $\theta_j$ and two free parameters $\bar{\alpha}$ and $\bar{\delta}$. The best fit (fig. n°1) is :

$$k \simeq m^{2/3}(\lambda) \cdot \lambda^{1/3} \cdot b^{1/2} \qquad (6)$$

the variations with respect to a, R, $Z_{eff}$ being yet uneasy to appreciate. There corresponds a class of $\kappa_e$ all giving the same growth (6), to be restricted by various speculations.. For instance, if $T_e$ is parabolic close to $\rho = 0$, then : $\varepsilon = \frac{\sigma+\tau}{2}$. For $\varepsilon = \sigma = \tau = 0$, one gets : $\kappa_e \simeq n_e \cdot Z_{eff}^{-1} \cdot J^{\beta} \cdot B^{\delta} \cdot [Z_{eff}^{2/5(3\alpha+2)} \cdot T_e^{-3/5} \cdot n_e^{-4/5} \cdot q^{-8/5}]^{\delta+\beta}$. Or else, the exponents of $\kappa_e$ along the branch AB must be lying in between those of the two media, $\kappa_e^{core}$ and $\kappa_e^{Isl.}$ due to convexity properties. In any case, $T_e$ is relatively insensitive to the functional form of $\kappa_e$, which is already observable from the results of the various numerical codes, reemphasizing the importance of (6) over a detailed approach to $\kappa_e$, probably not easy to define exactly in the present case.

## 4. PLATEAU QUASI STATIONNARY BEHAVIOR

4.1. In fast enough discharges where the current rises rapidly and stops at its plateau value, the temperature on the axis grows up to a certain point and drops down slowly during all the plateau. To interpret this phenomenon, let us return back to eq. (1a) and note that $\partial T_e/\partial t$ becomes negative when the diffusion Sink overpasses the Joule source, the top value of $T_e$ (0,t) corresponding to the moment where these two terms are equal. Further note from fig. n° 2 that, at this point, the temperature gradient is the largest, suggesting an instability threshold. A valuable candidate is the trappedelectron instability /3/, many estimates of which are showing the drastic limitations it induces. Taking thus $K_e = K_e^{T.E.}$, writing that the Joule term is equal to the diffusion term somewhere in the profile, and supposing furthermore that $\frac{1}{n_e}\frac{\partial n_e}{\partial \rho}$ is fixed, because its observed variation is smaller than that of $T_e$, one finally obtains, using also the group representation of the solution from (3abc), the relation :

$$T_{ecrit}(0,t) \simeq \left[R^{3/2}.B^2.Z_{eff}^{3/4}.I_p^2.t^{-5/4}\right]^{8/31} \tag{7}$$

in compatible agreement with the datas. From (6) and (7), one could for verification program $I_p$ (t) to reach $T_{ecrit}$ at a given time. We will not insist there, but much on the possibility of a change in the diffusion regime, reorganizing the losses in the discharge.

4.2. Now, let us analyze the dynamics of the discharge during the quasistationary period, which more or less coincides also with the plateau of the current. The impurity loss term R in (1a), which now counterbalances the Joule term, cannot be neglected. Let us research as before meaningful and simple enough solutions of (1ab). Guided by the present datas (§ 2), one such class is obtained, when remarking from eq. (2), that the simplest case corresponds to J = J ($\rho$) only, i.e. is t-independent. From (1b), this is possible as long as : $Z_{eff}\ JT^{-\gamma} = E\ (t)$ which means that $T_e$ takes a form : $\varphi(t).T(\rho)$, and that the $\rho$-dependence of J and T are related by : $J(\rho) = CT^{\gamma}(\rho)$. To solve (1a), let us assume in agreement with observations that $\varphi(t) \simeq t^a$ and that $n_e = t^b.\mathcal{N}(\rho)$, and equate independently, the RHS and the LHS to 0, breaking (1a) into two equations. If one supposes also that $Z_{eff} = t^c.\mathcal{Z}(\rho)$, $n_I = t^d.\mathcal{N}_I(\rho)$, one obtains from these two equations :

$$a = -(\gamma+\theta)^{-1}.(\vartheta + d - c) \tag{8a}$$

$$(\alpha+\tau)a + (\sigma-\xi)b + \varpi c + 1 = 0 \tag{8b}$$

for the exponents a and b of the time dependence of $T_e$ ($\rho$,t) and $n_e$ ($\rho$,t) and :

$$\mathcal{Z}(\rho).T^{\gamma-\theta}(\rho) = C_1\mathcal{N}(\rho).\mathcal{N}_I(\rho) \tag{9a}$$

$$a\mathcal{N}(\rho)T(\rho) = \frac{C_2}{\rho}\frac{d}{d\rho}\rho\left\{\mathcal{Z}^{\varpi}.N^{-\xi}.T^{\alpha}.\left(\frac{dN}{d\rho}\right)^{\sigma}\left(\frac{dT}{d\rho}\right)^{\tau}.\Phi(\rho)\left[N.\frac{dT}{d\rho} + \frac{5}{3}T\frac{dN}{d\rho}\right]\right\} \tag{9b}$$

determining N, T and J as functions of $\rho$. Once again, for this class of acceptable solutions of the system $(\mathcal{D}_S)$, which we will call quasi-stationnary, because they are curiously lying in between the self similar, groups invariant, solutions discussed above and the strictly static ones to be analyzed later, it has been possible to disantangle the time problem, eq. (8ab), from the space one, eq. (9ab). So, in the same way, the identification problem already analyzed can here also be very much simplified.

Turning to the experiments, one can even simplify again by noting that $c \simeq d \simeq 0$ during this period. On the other hand, one already knows that $\chi_e^{T.E.}$ could be a valuable candidate. Then : $\varpi = -1, \sigma - \xi = -1, \alpha + \tau = 7/2$ and one obtains from (8ab), for $\gamma = \gamma_{Spit} = 3/2$ :

$$a = -(5+\theta)^{-1},\ b = (3+2\theta).(10+2\theta)^{-1} \tag{10}$$

For $\theta \simeq 1$ to 1,25, corresponding to experimental conditions, one finds the typical behaviors : $b \simeq 1/2$, $a \simeq -1/6$, which have been repeatedly observed in many cases to a good accuracy. On the other hand, supposing that : $Z \simeq N_I$, eq. (9a) gives the scaling : $T \simeq N^{\frac{1}{\gamma-\theta}}$ ; $J \simeq N^{\gamma/\gamma-\theta}$. For $\gamma = \gamma_{Spit} = 3/2$ and $\theta = 1$, one obtains the "perfect" scaling : $T \simeq N^2$, $J \simeq N^3 = NT$, sometimes observed. The first of these relations has also been observed several times indicating that in general $\theta$ varies from 1 to 1,25 depending on $Z_{eff}$. Its variation is obtained by representing the exact term R in the form of (1a). Note that correspondingly the space independent electric field E (t) would, from the previous relations, behave like : $E(t) \simeq t^{c-\gamma a}$. For c = 0, a = $-1/6$, $\gamma = \gamma_{Spit} = 3/2$, this gives : $E(t) \simeq t^{1/4}$, a small growth which is compatible with the datas, strongly depending on the self effects at the points where $dI_p/dt$ changes. Finally, the space structure of the profile is fixed by the solution of eq. (9b) written with $\chi_e = \chi_e^{T.E.}$, that is :

$$T^{5/2-\theta} + \frac{C_3}{\rho}\frac{d}{d\rho}\left[\rho^{5/2}.T^{3/2}.\left(\frac{dT}{d\rho}\right)^3\right] = 0 \tag{11}$$

with $C_3 = \frac{k_1(5+\theta)(\gamma-\theta)}{B^2.Z_{eff}\,R^{3/2}}$. The solution T of (11) has, for T (0) $\neq$ 0, a singularity on the axis. But for other reasons related to the existence of saw-tooth relaxations of the central temperature with time, one will not extend the present solution inside the domain limited by $\rho = \rho(q = 1)$ taking instead a relatively flat profile there, and keep the gradient part. Note that (11) possesses the property of invariance under the group of similarity transformations $(S_0)$. From $q = \rho B/RB_\theta$, one can as well rewrite (11) as an equation on $Q = q(a)/q(\rho)$ and using its invariance under $(S_0)$, one can show that, if q (a) is varied, the point $\rho_1 = \rho\,(q = 1)$ will vary as : $\rho(q=1) \simeq Z_{eff}^{2/5}.R^{-7/2}.q(a)^{-4/5}$, in compatible agreement with observation. When R increases for fixed a, the current channel shrinks as exhibited by the negative power of R. Other properties could be discussed as well. For slowly varying $I_p = I_p(\varepsilon t)$, where $\varepsilon = \tau_E/\tau_c \ll 1$, where $\tau_E$ is the lifetime of the energy and $\tau_c$ a characteristic time of the current, one can research a solution to $(\mathcal{D}_S)$ by expansion. For completeness, let us indicate that ideal group invariant solutions in the case where $\kappa_e = \kappa_e^{T.E.}$ gives for (3a) the laws : $k = (\frac{2}{3}^2 b^2 \mu_1^{3/2} m_e^2 \lambda^{-3/4})^{8/39}$ from which, for $I_p \simeq t^s$, one would obtain : $T_e(0,t) \simeq t^{8/39(2s-3/4)}$, in agreement with observations for some s in the interval (–0.1 ; 0.2). For s = 0, the decay is $\simeq t^{-2/13}$, very similar to (10).

Finally, let us rapidly discuss the strictly static solutions of the initial system (1ab). Such solutions however, are not likely to occur in the present scheme, unless the impurity loss term remains comparatively small with respect to the Joule term. Eliminating dependent variables, one can reduce the static system from ($\mathcal{D}_S$) to the simple form :

$$Z_{eff}^{\sigma'} \cdot \mathcal{U}^{\beta'/\gamma} \cdot \mathcal{V}^{\gamma'} \cdot \frac{d\mathcal{N}}{dx} = K_1 \cdot S_o(x) \tag{12a}$$

$$Z_{eff}^{\sigma'} \cdot \mathcal{U}^{(1+\beta')/\gamma} \cdot \mathcal{V}^{\gamma'-1} \cdot \left[\gamma \mathcal{N} \cdot \mathcal{U}^{-1} \cdot \frac{d\mathcal{U}}{dx} + \frac{5}{3}\frac{d\mathcal{N}}{dx}\right] = K_2 \tag{12b}$$

with : $\frac{d}{dx}(x\mathcal{V}) = \mathcal{U}$, where $\mathcal{U} = T^{\gamma}$, $\mathcal{V} = B_\theta/\rho$, $\mathcal{N} = n_e^{1-\xi}$, $x = \rho^2$, $K_j$ are numbers related to units, $S_0$ (x) is the source term, and with the boundary conditions : $\mathcal{U}(1) = 0$ ; $\mathcal{V}(1) = \bar{I}_p$ ; $\mathcal{N}(1) = 0$, supposing for simplicity that $\sigma = \tau = 0$. The system (12ab) is also invariant by the group ($S_0$) and its order can thus be reduced. If (12a) is supposed to relate N (x) to $S_0$ (x) unknown, (12b) becomes a first order equation which can be completely analyzed. In particular, a property is that the observed relation (10) cannot be obtained, limiting the present interest of these solutions.

## 5. DISCUSSION AND CONCLUSION

Starting from the complete MHD system ($\mathcal{D}$), it has been possible in the present analysis to reduce this very large system to its dominant subpart ($\mathcal{D}_S$), which drives to first order its dynamics, splitting apart the equilibrium and cinematics of the magnetic surfaces, the impurity cascade dynamics and the ions dynamics, to be analyzed separately.
To study the system ($\mathcal{D}_S$) which is yet a complicated nonlinear one, a specific way has been proposed. Instead of using as elsewhere the approximation or expansion techniques which could very well apply to ($\mathcal{D}_S$), due to its parabolic nature, useful for convergence, it has been remarked that ($\mathcal{D}_S$) possesses invariance property through the group of similarities ($S_0$). During the first phase of the discharge when the temperature profile concentrates and shrinks whereas the temperature rises up, the boundary effects, as verified on the datas, are not felt by the solution which follows the group invariance property. For a given value of the temperature, $T_{ecrit}$, which has been determined as the point where $\partial T_e/\partial t = 0$, i.e. the diffusion losses are balancing the Joule heat source, the behavior of the system changes and a new type of solutions, the quasistationary ones, here also partially invariant under scaling transformations, is seen to take place in agreement with observed datas, when $I_p$ (t) is a constant. In some cases, when $I_p$ is too small, $T_{ecrit}$ is never reached and the system evolves along the first phase. In the

two phases depicted above, the analysis was heavily resting upon the fact, in agreement with the datas, that the solution of ($\mathcal{D}_S$) instead of exploring its complete manifold, is staying close to first order invariant submanifolds, possessing interesting properties. It has thus been possible to completely separate the time and space behaviors, reducing the first one to a simple algebraic system, from which the problem of identification of the heat transport coefficient $K_e$ can be solved. In the first period, it has been shown that, as expected from the two domains structure of the plasma discharge, i.e. including a free internal boundary, the real $K_e$ cannot be simply reduced to elementary classical one. In the second period, on the contrary, the plasma discharge is strong enough and indication has been given that it is ruled out by its temperature gradient, from the agreement of the datas with heat conductivity produced by the trapped electron instability. The problems stated at the end of § 2 have been solved and it has been shown that agreement with the observations can be found within the scope of the dominant subsystem ($\mathcal{D}_S$) only. For this reason, and though the solutions discussed above are only approximate ones, it is interesting to call the corresponding discharges they theoretically represent, Ideal Discharges. Given a plasma current profile $I_p = I_p(t)$, and the parameters of the plasma (a,R,B), it is possible from the formulae (6), (7), (10) to follow or predict the evolution of the main quantities $T_e$, $n_e$, to be discussed in more details elsewhere. However, in the second phase, other observed phenomena are modifying the evolution described above, related to the existence of a domaine where q becomes less than 1, which is a static one, as repeatedly observed. Inside this domain, a kink mode instability can develop, and a new branching along AB occurs, where the final structure of the discharge is now a three domains one. The kink, as it will be discussed in next part, nonlinearly generates a sub-mode which induces the observed saw tooth relaxations. But the restoring time, i.e. the period, being proportional to the pressure, enlarges as t increases because $n_e$ grows like $t^{1/2}$, allowing larger displacements, until the moment when the coupling of the kink mode with the resistive m = 2 tearing mode localized around the surface where q = 2 takes place prior to the relaxation phenomenon. The two (unstable) reservoirs have enough stored energy to blow up the discharges in a compounded kink, as observed. It is important to note that, though the consequences of the saw tooth on the history of the discharge are extremely important, the dynamics is to first order ruled out by the main gradient, i.e. the body of the discharge, which justifies to study separately the saw tooth relaxations. In a similar way, the impurity cascade can be studied apart, and the global $Z_{eff}$ injected in the system ($\mathcal{D}_S$), as will be shown in another part.

Finally, there still remains to show whether the ideal discharges are stable ones or not, and how far in a functional sense, they could stay from exact solutions of ($\mathcal{D}$). For the first point, one already knows that if one perturbs (1ab) the linearized system is just the one studied in part A of /2/, showing stability against current thermal instabilities which are localized ones. For the class of nonlocal ones, one numerically knows for particular values of the exponents /10/ that the group invariant solution is stable, but a complete discussion is still lacking. As for the second point, usual expansion procedure insures that ideal discharges are first order solutions of the complete system ($\mathcal{D}$) and, as mentioned several times, higher order terms are breaking their symmetry. That the real discharges can be fitted by the ideal ones, i.e. by solutions of ($\mathcal{D}_S$) only, could be interpreted as an indication that ($\mathcal{D}_{S'}$) contains the main part of the observed information. Nevertheless, the explored domain of parameters is still a small one, and the analysis of the predicted scaling invariance on the discharges of the next coming machines will be needed to state more firmly the group properties discussed here.

## REFERENCES


1. F. Calogero, A. Degasperis : Il Nuovo Cimento Vol 32B Ser 11 n° 2 p. 201 (1976) and references therein.
2. M. Cotsaftis : Current Thermal Instability in Sheared Systems and Skin Suppression, APS Meeting S$^t$ Petersbourg Fla 5-12 Nov. 1975
3. B.B. Kadomtsev, O.P. Pogutse : JETP Vol 24 n°6 p. 1172 (1967); Rev. of Plasma Physics Ed. Leontovitch Tome V Consultant Bureau (1972)
4. D. Duchs : NRL Report n° 7340 Feb. 1972 ; J. Hogan : ORNL Report n° TM 51-53 Nov. 1975
5. E. Hinnov : PPL Report n° MATT-777 (1970)
6. M. Murakami, W.R. Wing, P.H. Edmonds : Nuclear Fusion Vol 14 p.779 (1974)
7. G.W. Bluman, J.D. Cole : Similarity Methods for Differential Equations, Springer-Verlag Applied Math. Science n° 13 (1974)
8. L.V. Ovsjannikov : Group Properties for Differential Equations, Novosibirsk (1962)
9. W.F. Ames : Nonlinear Partial Differential Equations in Engineering Academic Press New York Vol. I (1965), Vol. II (1972)
10. G. Bardotti, L. Enriques, F. Santini : Nuclear Fusion Vol.15 p.621 (1975)

# ON TEMPORAL EVOLUTION OF PLASMA DENSITY FLUCTUATIONS IN THE PRESENCE OF A STRONG ELECTROMAGNETIC WAVE

A.D.Piliya

A.F.Ioffe Physico-Technical Institute

Leningrad, USSR

Plasma heating experiments at the lower hybrid frequency have lately been carried out in many laboratories. The advantages of this frequency region are well known, as is the danger of anomalous peripheral absorption of energy due to parametric instabilities. For comparatively small amplitudes of the pump wave the non-linear mechanism of such absorption can be described in terms of the theory of weak plasma turbulence /1,2/. In this case a stationary spectrum of high-frequency noises is supposed to exist in a plasma, transferring its energy to particles mainly through collisional damping. At higher amplitude levels qualitative changes in the character of plasma turbulence occur, in particular, strong density fluctuations accompanied by distortions in the spatial distribution of high-frequency oscillations. In this paper we shall consider an instability of this type driven directly by the pump wave. We shall ignore completely the possible presence of a turbulent background, implying that the instability in question either has the fastest growth rate or may occur independently of the short-wavelength noise. Instabilities caused by pondermotive forces of highly intense waves (self-modulation, filamentation, soliton formation and collapse) have been widely discussed; however, the case of a monochromatic quasi-electrostatic pump reveals some new features which we shall first consider qualitatively.

The slow wave usually used in lower hybrid experiments has, inside a plasma volume, the refractive index $N \gg 1$ and can be well described by the WKB approximation. The wave energy propagates along trajectories which can be considered, neglecting weak regular plasma inhomogeneity, as straight lines inclined at a small angle $\alpha$ to the magnetic field. Assume now that a sporadic one-dimensional density depression has appeared in the plasma at an angle $\theta$, slightly different from $\alpha$, to the magnetic field. If the size of the density perturbation is sufficiently large, the wave propagates through it without reflections. The angle $\alpha$ increases with reduction of density so that the trajectories take the form shown in Fig.1a. Diminishing of the group velocity component along the density gradient due to variations in $\alpha$ leads to enhancement of the wave amplitude inside the depression and so provides for the instability mechanism.

With the growth of the initial perturbation the point is reached when the trajectories become perpendicular to the density gradient vector ($\alpha$ equals $\theta$) in the concentration minimum. The wave now is incapable of crossing the line of minimal density, which becomes an asymptote of the trajectories. At smaller concentrations there are two such asymptotes, as shown in Fig.1b. In a cold plasma these asymptotes would give rise to singularities in the electric field of the incident wave. In a real plasma with finite temperature the linear conversion between the incident "cold" wave and the short-wavelength plasma mode occurs instead in the vicinity of the asymptotes. The propagation of the wave now looks like this (Fig.1b). The incident wave reaches the left-hand asymptote, undergoes the conversion, propagates in the form of a plasma wave toward the right-hand asymptote, and after another conversion leaves the region of density perturbation. For sufficiently smooth density variations the mode conversion causes no reflection of the wave.

The above transformation process is, of course, the familiar case of linear wave conversion for an arbitrary angle between the density gradient and the magnetic field /3/, with the trajectory asymptotes being the lines of the oblique hybrid resonance. At these lines the component $\varepsilon_{xx}$ of the dielectric tensor vanishes; here $x$ is the coordinate on which the plasma density depends. The slow wave we consider here propagates initially at an angle very close to the resonant one so that the mode conversion readily occurs at a relatively low level of the density variations. However, the necessary condition for the linear conversion at angles different from 0 or $\pi/2$ is a quasi one-dimen-

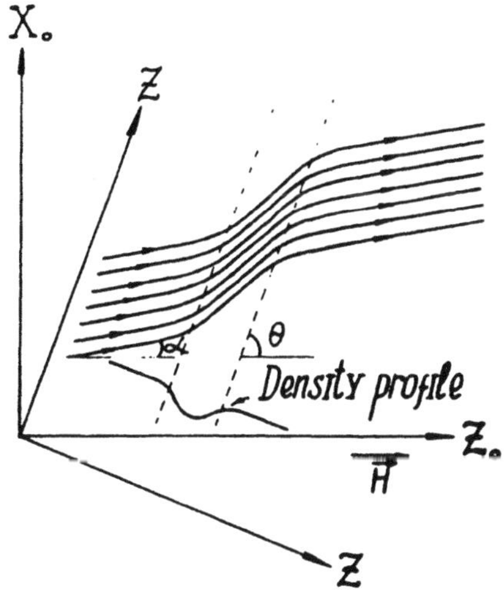


Fig. 1a.

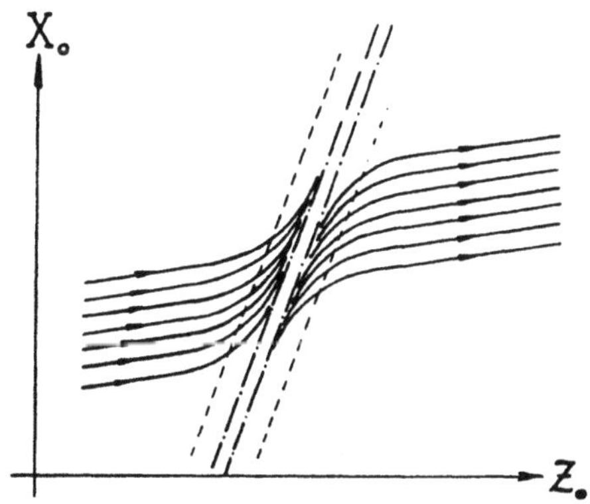


Fig. 1b.

Fig. 1. Typical wave trajectories: a) before the conversion conditions are reached; b) at the stage of the wave conversion. The plasma wave propagating between the asymptotes is not shown.

sional form of the density profile /4/. It requires that the transverse scale length of the perturbation be small as compared with its length.

The mode conversion gives a burst to the electric field so that one could expect fast and unlimited growth of the density depression after the transformation stage of the instability is reached. The plasma wave has, however, the inverse dependence of the group velocity on the density as compared with the incident mode that provides the possibility for saturation of the instability. In the plasma state saturated in this way the density depression may be, under reasonable assumtions as to the pump wave intensity, so deep that the wave undergoes strong Landau damping. Sharply localized absorption of the highly intense wave must have a strong effect on the plasma, including the production of energetic particles.

For quantative analysis of the instability just outlined we can make use of the following simplifying circumstance. Since the consideration is restricted to the case of perturbations making a finite angle with the wave trajectories, we may assume that the wave crosses the instability region in a short time interval compared with the time-scale of the density variations. (The duration of this interval is determined by the group velocity component along the density gradient, which is, taking into account the wave conversion, in some way related to the electron thermal velocity $v_{Te}$; the density variation time-scale is characterized by the velocity

$$u_E = \sqrt{\frac{m_e}{M_i}}\,\frac{eE_0}{m_e\omega_0} \ll v_{Te}$$

with $E_0$ and $\omega_0$ being the wave amplitude and frequency respectively.) In these circumstances the form of the HF-field is determined at any given moment of time by an instantaneous density profile (and boundary conditions). For the low-frequency motion we shall use in the familiar way the linearized hydrodynamic equations including the HF pressure (so-called Miller's force).

For simplicity, we shall restrict ourselves to a low-density peripheral plasma with

$$\omega_0^2 \ll \omega_{pe}^2 < \omega_0 \cdot \omega_{ce}$$

where $\omega_{pe}$ is the electron plasma frequency and $\omega_{ce}$ is the electron cyclotron frequency. Assuming, in addition, that the incident wave represents the Trivelpiece-Gould mode and that the characteristic period of low-frequency motion is small compared with $1/\omega_{ci}$ we may write a system of equations governing the wave propagation and the evolution of the density profile:

$$\mathrm{div}\left[\hat{\varepsilon}_{\alpha\beta}(\nabla\Phi)_\beta\right] = 0 \qquad (1)$$

$$\frac{\partial^2 n}{\partial t^2} - c_s^2\nabla^2 n = \frac{F}{M} \qquad (2)$$

where $\Phi e^{-i\omega t}$ is the HF-potential, $c_s = \sqrt{\frac{T_e}{M_i}}$ is the speed of sound and $F$ is the potential of the HF-pressure.

According to the above arguments, we have omitted time derivatives in the high-frequency Eq.(1). The potential $\Phi$ and the density $n$ are coupled through the dielectric tensor $\mathcal{E}_{\alpha\beta}(\omega_o, t)$, calculated for the instantaneous profile $n$. Eq.(2) includes no dissipative term and is valid only if the instability growth rate sufficiently exeeds the linear damping rate of the ion sound. It means that in case of $T_e \sim T_i$ the entire term $C_s^2 \nabla n$ must be much smaller than the right-hand side of the equation (to put it in another way, the instability growth rate must be much higher than the ion sound frequency).

In what follows we shall consider solutions of the system (1)-(2) depending on one spatial coordinate and time. In the coordinate system with the magnetic field in the $x-z$ plane and $x$ pointing in the inhomogeneity direction Eqs. (1)-(2) take the form:

$$\beta \Phi^{IV} + (\eta \Phi')' + 2i\sigma \Phi' + \mathcal{E} \Phi = 0 \tag{3}$$

$$\frac{\partial^2 n}{\partial t^2} - C_s^2 n'' = \frac{1}{M_i}\left(|\Phi'|^2\right)'' \tag{4}$$

where $\beta \simeq \frac{3 v_{Te}^2}{c^2} \frac{\omega_{pe}^2}{\omega_o^2} \cos^4\theta$, $\eta = \mathcal{E}_\perp \cdot \cos^2\theta + \mathcal{E}_\parallel \sin^2\theta$, $\sigma = n_{oz}(\mathcal{E}_\parallel - \mathcal{E}_\perp) \sin\theta \cos\theta$, $\mathcal{E} = n_{oz}(\mathcal{E}_\perp \sin^2\theta + \mathcal{E}_\parallel \cos^2\theta) + \mathcal{E}_\perp n_{oy}^2$, $n_{oz}$, $n_{oy}$ are components of the dimensionless wave vector $\vec{n}_o = \frac{c}{\omega}\vec{K}_o$ of the incident wave, $n_{oz} = n_o(\sin\alpha \cdot \cos\theta - \cos\alpha \cdot \sin\theta \cdot \cos\varphi)$, $n_{oy} = n_o \cdot \cos\alpha \cdot \sin\varphi$, $\theta$ is the angle between the $z$-axis and the magnetic field, $\alpha$ is the angle of the resonant cone in the unperturbed plasma, $tg^2\alpha = -\mathcal{E}_\perp/\mathcal{E}_\parallel$, $\varphi$ is the angle between the $XZ$ and $K_o H_o$ -planes, $\mathcal{E}_\parallel = 1 - \omega_p^2/\omega^2$, $\mathcal{E}_\perp = 1 - \frac{\omega_{pe}^2}{\omega^2 - \omega_{He}^2}$. In Eqs.(3)-(4) the dimensionless coordinate $x$ is measured in units of $c/\omega_o$ and the primes denote differentiation with respect to $x$. The first term in Eq.(3) takes into account the spatial dispersion. We shall consider nearly resonant perturbations $|\theta - \alpha| << 1$ with small relative value $\delta n/n << 1$. In this case $|\eta| << 1$ and we can regard $\eta$ as the only density-dependent coefficient in (3). The behaviour of the HF-potential $\Phi$ in response to small density variations strongly depends on the value of the angle $\varphi$. At $\varphi = 0$ and $|\theta - \alpha| << 1$ the last term in Eq.(3) is small compared with $\sigma \varphi'$ and can be neglected. For this reason, the first term responsible for the spatial dispersion becomes important when $\eta \to 0$. It just means that the linear wave conversion occurs, as has been pointed out above.

On the contrary, at large $\varphi$ the two last terms are always of importance and the solution is rather insensitive to small density variations, as a consequence of which we shall consider here only the case of $\varphi = 0$. Keeping in Eq.(3) only its three first terms and introducing instead of $\Phi$ and $n$ the new functions $\mathcal{E} = \omega\varphi'/E_o c$ (with $E_o$ being the incident wave amplitude) and $\eta$, we have

$$\beta\,\mathcal{E}''' + (\eta\,\mathcal{E})' + 2i\delta\,\mathcal{E} = 0 \tag{5}$$

$$\frac{\partial^2\eta}{\partial t^2} - c_s^2\,\eta'' = -u_E^2\left(|\mathcal{E}|^2\right)'' \qquad u_E^2 = \frac{me}{4M_i}\left(\frac{eE_o}{m\omega_o}\right)^2\cdot\sin^2\theta \tag{6}$$

Consider now the case when the dependence of $\eta$ on $x$ for all $t$ is such that the WKB approximation is valid for Eq.(5). Then

$$\mathcal{E} = A\cdot e^{i\int\kappa\,dx} \tag{7}$$

where $\kappa$ obeys the dispersion equation

$$\beta\kappa^3 - \eta\kappa + \eta_o\kappa_o = 0 \tag{8}$$

and

$$A^2 = \frac{\kappa^2\eta_o}{\kappa_o(\eta\kappa - 3\beta\kappa^3)} \tag{9}$$

In Eq.(8) we have put $2i\delta = \eta_o\kappa_o$ where

$$\eta_o \simeq \frac{\omega_p}{\omega_o}(\alpha - \theta) \tag{10}$$

is the unperturbed value of $\eta$, so that $\eta(x,t)\to\eta_o$ when $|x|\to\infty$. At $\beta = 0$, (7) is an exact solution of Eq.(5). For this reason, the WKB approximation can be used as long as the spatial scale-length of the density variations satisfies the inequality

$$\ell^2 \gg \frac{\beta}{|\eta_o|}\,\frac{c^2}{\omega_o^2}$$

It is valid, consequently, for short-wavelength perturbations as well (it is assumed that the spatial dispersion is negligible for the incident wave, so that $\beta K_0^2 \ll \eta_0$ ). We see now that the low-frequency oscillations are governed by one non-linear partial differential equation (6) where $|\mathcal{E}|^2 = A^2$ is a function of $\eta$ determined by (8) and (9). Consider for the beginning the small-amplitude waves obeying this equation. Assuming $\eta = \eta_0 + \eta_1$ , $\eta_1 \ll \eta_0$ we obtain in the linear approximation the dispersion equation for these waves:

$$\Omega^2 = \left[ c_s^2 - u_E^2 \frac{\partial A^2}{\partial \eta} \right] K^2 \tag{11}$$

The quantity $\left.\frac{\partial A^2}{\partial \eta}\right|_{\eta=\eta_0}$ is positive at $\eta_0 < 0$ and negative at $\eta_0 > 0$. As a result at $\eta_0 < 0$ and sufficiently high level of the pump the waves are driven to instability. For the instability threshold we have, taking into account that initially $A^2 = \eta_0^2/\eta^2$

$$u_{E\,thr}^2 = \frac{1}{2} c_s^2 |\eta_0| \tag{12}$$

According to (10), $u_{E\,thr} = 0$ for resonant perturbations with $\theta = \alpha$ ; as a matter of fact, however, the approximation (5) is violated when $\alpha \to \theta$ ; this case will not be considered here.

Due to the character of the dependence of $A^2$ on $\eta$ at $\eta < 0$, the monotonic growth of any initial perturbation is guaranteed at least until the value $\eta = 0$ is reached in the density minimum, i.e. the condition for the oblique hybrid resonance for a given $\theta$ . At $\eta > 0$ the plasma becomes stable with respect to small perturbations of the above form, so that the saturation at a certain level of $\eta > 0$ can be expected. To find the possible form of the limiting (for $t \to \infty$) density profile, consider the stationary solution of Eqs. (6)-(9). For $\eta(x)$ finite at $|x| \to \infty$ we have from (6):

$$-\eta + \gamma A^2(\eta) = C \tag{13}$$

where $\gamma = u_E^2/c_s^2$ and $C$ is an arbitrary constant.

The function $A^2(\eta)$ has a maximum at $\eta = 0$ and $A^2 \to 0$ as $|\eta| \to \infty$, therefore the transcendental equation (13) has three real roots (two at $\eta < 0$ and one at $\eta > 0$). The solution $\eta(x)$ is consequently a stepwise discontinuous function equal to one of the roots on different parts of the $x$ axis. Taking one of them equal to the unperturbed value $\eta_0$, we can determine the constant $C$ and then find $\eta$ and $K$ inside the depression region. At $\gamma >> |\eta_0|$, $C = \gamma A^2(\eta_0) = \gamma$ and we have the condition $A^2(\eta) = 1$ for the plasma parameters in the "well". For the wave number $K$ it gives

$$K = \frac{|\eta_0|}{3\beta K_0} = \frac{|\eta_0| c^2}{9 v_{Te}^2 n_{\parallel}^2} K_0 \qquad (14)$$

where $n_{\parallel} = \frac{k_{\parallel} c}{\omega}$ and $k_{\parallel}$ is the incident wave vector component along the magnetic field. The wave in the final stationary state is slowed down more the farther from resonance the initial density perturbation, that is, the greater $|\eta_0|$. It is seen from (14) that at $|\eta_0| \sim n_{\parallel} v_{Te}/c$, $K \cdot \cos\alpha \to c/v_{Te}$ and strong electron Landau damping occurs. According to (12), the threshold of the strong absorption is

$$u_E^2 \sim \frac{c_s^2 v_{Te}}{c} \qquad (15)$$

Solutions of Eq.(6) depending on $x - vt$ can be considered in the same way. Eq.(13) is also valid for them with $\gamma = u_E^2/c_s^2 - v^2$. Strictly speaking, the discontinuous solutions are incompatible with the WKB approximation; however, they seem to give a reasonable estimate of the instability characteristics.

The large-scale instability discussed here, although it does not have, in its linear stage, the fastest growth rate, can be readily triggered off by the intrinsic inhomogeneity of the electric field launched by a finite-sized source. An instability of this type has been observed experimentally in /6/.

One can expect that the main features of the process, including the mode conversion with corresponding enhancement of the electric field, are not restricted to the WKB case. Equations (5) and (6) have stationary solutions in the opposite limiting case that describe the propagation of a slowed wave against the background of a strongly modulated density profile. There are also solutions with unlimited growth of the density modulation whose behaviour resembles a wave collapse /5/ (however, this "collapse" is due to a propagating rather than to a trapped wave).

We are now able to make two conlusions. First, the wave conversion in the early stage of the process means that the validity of the perturbation theory analysis may be much more restricted than it would seem at first glance. Second, the hypothesis can be suggested that the Trivelpiece-Gould mode with a sufficiently large amplitude does not exist in a plasma in its "normal" form but rather in the form of a slow short-wavelength plasma wave due to the conversion on the self-consistent density variations. This possibility is rather unfavourable for prospects of the low hybrid plasma heating.

The author is indebted to V.I.Fedorov and A.M. Rubenchick for a very useful discussion.

## REFERENCES

1. V.P.Silin. Параметрическое воздействие излучения большой мощности на плазму, М.,"Наука", I973г.
2. A.M.Rubenchick, I,Ribak, B.M.Sturman. ZhETF, 67, 1364, 1971.
3. A.D.Piliya, V.I.Fedorov. ZhETF, 57, 1198, 1969.
4. A.D.Piliya, V.I.Fedorov. ZhETF, 60, 299, 1971.
5. V.E.Saharov. ZhETF, 62, 1745, 1972.
6. W.Gekelsman, R.L.Stenzel. Phys.Rev.Lett., 35, 1708, 1976.

# SOME NONLINEAR EFFECTS IN TEARING MODE INSTABILITY

Predhiman Kaw

Plasma Physics Laboratory

Princeton University, Princeton, N.J. 08540

## ABSTRACT

Two nonlinear effects which may be of relevance in the development of tearing mode instability are discussed:

(a) effect of a given background turbulence of drift waves on the growth of tearing modes, and

(b) effect of a plasma flow towards the X points on nonlinear growth of tearing modes.

The possible relevance of the latter process to disruptive instability in tokamaks is also discussed.

## 1. INTRODUCTION

The tearing mode instability is widely believed to be associated with a number of interesting phenomena observed in nature; e.g., observed MHD oscillations of tokamak discharges in the laboratory,[1] energy release in solar flares,[2] field reconnection in the magnetosphere,[3] etc. The linear theory[4,5] of this instability was first worked out more than a decade and a half ago. However, the nonlinear theories are of a more recent origin. In this paper we consider two nonlinear effects which may have relevance to the development of tearing mode instability in a plasma. These are

(a) effect of given background turbulence of drift waves on the development of tearing modes (work done in collaboration with John Hsu), and

(b) effect of a plasma flow towards the X points on nonlinear growth of tearing modes and the possible relevance of this phenomenon to disruptive instability in tokamaks.

A good part of the discussion will be oriented toward tokamak application but the processes considered are quite general and should be applicable to other physical situations also.

The linear theory of tearing mode can best be understood in a sheet-current or slab model.[4] Imagine an infinite plane layer of current, of thickness a (along x axis), extending out of the plane of paper (the -z axis) and resulting in a reversal of $B_y$ at $x = 0$ . Let us investigate the stability of this layer to an incompressible perturbation, which is periodic in y . We can obviously carry out an ideal-MHD normal mode analysis of this problem. Demanding that the perturbations vanish infinitely far away from the layer (i.e., at $x = \pm\infty$) , we find that no solutions can be constructed which connect smoothly at $x = 0$ . Realizing that dissipative processes like resistivity $\eta$ play a crucial role in the neutral layer (near $x = 0$) and should therefore be incorporated in the basic equations, one can then find a smoothly connected normal mode solution. This solution shows that perturbations longer than the current layer thickness ($ka < 1$) are unstable with a growth rate proportional to $\eta^{1/3}$ . This is the tearing mode instability. Physically, its consequence is to break the original current layer, uniform in y , into parallel current filaments extending out of the plane of paper. The free energy source for the instability is the magnetic field associated with the current and one can show that energetically, it is favorable for the current to form maxima and minima along y . However, formation of such maxima and minima requires topological changes in the magnetic field configuration. This can come about only through reconnection of magnetic lines of force at certain points. Resistivity, or other non-ideal effects, thus play a crucial role in the tearing mode instability.

It should be emphasized that for the above discussion to apply, it is sufficient that only a component of $\tilde{B}$ vanish at $x = 0$ (the total $|B|$ need not vanish). As an example, in tokamak discharges, the quantity

$$\tilde{k} \cdot \tilde{B} = \frac{m}{r} B_\theta - \frac{n}{R} B_T = \frac{B_\theta}{r} \left[m - nq(r)\right] \tag{1}$$

changes sign at the mode rational surfaces, where the safety factor q(r) is equal to a ratio of two integers, m and n . These are the surfaces which can be "torn" apart by a tearing mode instability and can then result in magnetic island formation. This is again just a redistribution of current with formation of maxima and minima in the $\theta$ direction. Note that most of the calculations below will be done in the simple slab geometry. This is a reasonable approximation for the tokamak configuration also (identifying the z axis with the local magnetic field direction, y axis with the azimuthal coordinate $\theta$ , and the x axis with the radial coordinate) provided that the perturbations considered have a high m number and are localized near the mode rational surfaces.

## 2. EFFECT OF DRIFT WAVE TURBULENCE ON TEARING MODE INSTABILITY

It is well known that a tokamak plasma has two predominant sources of free energy. These are the free energy associated with the poloidal magnetic field which is responsible for MHD activity of the plasma and the free energy of expansion (associated with any confined plasma) which drives microinstabilities like drift waves. Since $\beta_\theta \simeq O(1)$ for tokamaks, these two sources of free energy are comparable in magnitude. MHD activity tends to involve long wavelengths and macroscopic motions of the discharge. Drift-wave activity, on the other hand, has a shorter wavelength and is believed to be responsible for anomalous transport coefficients like enhanced electron thermal conductivity, etc. The question that we ask here is: Can MHD oscillations (such as the tearing mode) be influenced by the source associated with expansion-free energy through some nonlinear coupling effects? Of special significance is the fact that drift waves are almost flute-like modes and therefore also tend to be localized in the neighborhood of mode-rational surfaces.

We shall now carry out a simple model calculation in the slab geometry to investigate the nonlinear coupling between drift waves and tearing modes. Consider a current layer equilibrium of the type discussed in the introduction. The basic equations describing it are

$$E_o \hat{e}_z = \eta J_o \hat{e}_z \; ; \quad \frac{d}{dx} B_{yo} = - \frac{4\pi}{c} J_o(x) \tag{2a,b}$$

$$\frac{d}{dx} p_o + \frac{J_o}{c} B_{yo} = \frac{J_{Do}}{c} B_o \tag{2c}$$

where $\underset{\sim}{B}_o = \hat{e}_z B_o$ is the toroidal magnetic field, $J_{Do} = \hat{e}_y J_{Do}$ is the diamagnetic drift current and the rest of the symbols have the usual meaning. Let us further imagine a saturated spectrum of turbulent, electrostatic drift waves superposed on this equilibrium. We assume the amplitude of drift wave turbulence to be small, and approximate it by a spectrum of randomly phased drift oscillations satisfying the linearized equations. We thus write

$$\underset{\sim}{E}_d = \sum_{k_o} \hat{e}_y E_{k_o} \exp\left[i(k_o y - \omega_o t)\right] + \text{c.c.} \tag{3a}$$

Assuming warm electrons and cold ions, the linearized velocities of the two fluids in drift wave fields may be written

$$v_{ed}^x \equiv \sum_{k_o} v_{ek_o}^x = \sum_{k_o} \left[ \frac{c}{B_o} E_{k_o} + \frac{cT_e}{eB_o} \frac{\partial}{\partial y} \left( n_{ek_o}/n_o \right) \right] \tag{3b}$$

$$v_{id}^x \equiv \sum_{k_o} v_{ik_o}^x = \sum_{k_o} \frac{c}{B_o} E_{k_o} \quad . \tag{3c}$$

The associated quasineutral density fluctuations are

$$n_{ed} \simeq n_{id} = \sum_{k_o} n_{k_o} = n_o \sum_{k_o} (e\phi_{k_o}/T_e) \tag{3d}$$

and the corresponding dispersion relation for the modes is

$$\omega_o \simeq k_o v_{Do} = k_o (cT_e/eB_o)(1/n_o)(dn_o/dx) \equiv \omega_{o*} \quad . \tag{3e}$$

For simplicity of calculation we shall use the local approximation for drift waves and assume $E_{k_o}$ to be independent of x . This is strictly not justified but should not produce a qualitative change in the results.

We now investigate the stability of this new equilibrium (i.e., current layer equilibrium plus drift oscillations) to tearing mode perturbations. Physically, the random spectrum of drift oscillations in the background acts as a spectrum of pump waves, which can cause parametric excitation or suppression of the tearing mode. The tearing mode perturbation considered by us is a drift-tearing perturbation,[6] which involves significant quasineutral density fluctuations $\delta n_t$ . It is easy to see that the nonlinear ion flux density $\Gamma_i^{NL}$

$$\Gamma_i^{NL} = \delta n_t \, v_{id}^x \, \hat{e}_x + n_{id} \, \delta v_t^y \, \hat{e}_y \tag{4}$$

(where the $\delta$ terms with subscript t refer to the tearing perturbation) has a finite divergence and through the continuity equation leads to the generation of sideband drift waves at the sum and difference frequencies. Noting that the $(v \cdot \nabla)v$ term gives a negligible contribution for this generation and combining the equations of continuity and motion for the ion and electron fluids we get

$$\left( \frac{\partial}{\partial t} + i\omega_{\pm *} \right) \delta v_{\pm} = \frac{i\omega_{\pm *}}{(dn_o/dx)} \left( -v_{ik_o}^x \frac{\partial}{\partial x} \delta n_t - n_{k_o} ik_y \delta v_{yt} \right) \hat{e}_x \tag{5a}$$

$$\delta n_{\pm} \simeq n_o e \delta\phi_{\pm}/T_e = i\left[ (dn_o/d_x)/\omega_{\pm *} \right] \delta v_{\pm} \tag{5b}$$

where $\omega_{\pm *} = (k_y \pm k_o)v_{Do}$ .

To describe the low-frequency drift-tearing perturbations, we use the single-fluid MHD equations, together with the Maxwell equations. The relevant equations are

$$\nabla_\perp^2 \delta B_{xt} + \frac{4\pi}{c} \frac{\partial}{\partial y} (\delta j_{\parallel t}) = 0 \tag{6a}$$

$$\frac{\partial}{\partial y} (\delta E_{\parallel t}) - \nabla_\parallel (\delta E_{yt}) = - \frac{1}{c} \frac{\partial}{\partial t} (\delta B_{xt}) \tag{6b}$$

$$\frac{\partial}{\partial t} (\delta n_t) + \delta v_{xt} \frac{dn_o}{dx} + \sum_{k_o} \left( v^x_{ik_o} \frac{d}{dx} \delta n_\pm \right) \simeq 0 \tag{6c}$$

$$\rho_o \frac{\partial}{\partial t} \nabla_\perp^2 \delta v_{xt} - \frac{(B_o \cdot \nabla)}{4\pi} \nabla_\perp^2 \delta B_{xt} = - \rho_o \sum_{k_o} \frac{\partial}{\partial y} \left( v^x_{ik_o} k_y^2 \delta v_\pm \right.$$

$$\left. + i \frac{v^x_{iko}}{k_\pm} \frac{\partial}{\partial x} \nabla_\perp^2 \delta v_\pm \right) \tag{6d}$$

$$\delta E_{\parallel t} + \frac{T_e}{n_o e} \nabla_\parallel \delta n_t - \eta \delta j_{\parallel t} + \frac{J_{Do}}{en_o c} \delta B_{xt} = \sum_{k_o} \frac{T_e}{e} \nabla_\parallel \left( \frac{n_{ko} \delta n_\pm}{n_o^2} \right) . \tag{6e}$$

The first three equations are self-explanatory. Equation (6d) is the x component of the curl of equation of motion. The right side of this equation contains the largest nonlinear term coming from the ponderomotive force ($v \cdot \nabla v$ terms) of drift waves. Equation (6e) is the parallel component of Ohm's law.

Proceeding further as in the usual analysis of drift-tearing instability and retaining only the largest nonlinear term from coupling to drift waves, one obtains the inner layer equations

$$\frac{d^2 n}{dX^2} - X^2 n + XB = \alpha \frac{d^4 n}{dX^4} \tag{7a}$$

$$\frac{d^2 B}{dX^2} - A(B - Xn) = 0 \tag{7b}$$

where

$$n = \delta n_t / n_o \ , \ B = i \frac{\delta B_{xt}}{B_o} \left( \frac{1}{n_o} \frac{dn_o}{dx} \right) \left( \frac{\tau_R}{\tau_A^2 \gamma} \right) \frac{1}{k'_\parallel x_R} \tag{7c,d}$$

$$X = x/x_R \ , \ A = 4\pi x_R^2(\gamma + i\omega_*)/c^2\eta \tag{7e,f}$$

$$x_R = \left[\gamma^2\tau_A/k_\parallel'^2\tau_R(\gamma + i\omega_*)\right]^{1/4} \ , \ \tau_A = a\sqrt{\rho}/B_o \ , \tag{7g,h}$$

$$\tau_R = a^2/c^2\eta \tag{7i}$$

and the nonlinear coefficient

$$\alpha = \sum_{k_o} \left(\frac{|v_{ik_o}|^2}{x_R^2}\right) \frac{\omega_{\pm*}\left[1 + (ik_\pm\gamma/k\omega_*)\right]}{i\gamma\left[\gamma - i\tilde{k}(\partial\omega_o/\partial\tilde{k}_o)\right]^2} \ . \tag{7j}$$

Note that the nonlinear term appears in the same manner as the introduction of ion viscosity with a complex ion viscous drag coefficient. Note also that the nonlinear term becomes important when $(cE_d/B_ox_R) \sim \omega_*$ , where $x_R$ is the typical resistive layer thickness. For a typical resistive layer thickness of order the ion Larmor radius $\rho_i$ , this gives $(cE_d/B_ov_{ith}) \sim \omega_*/\omega_{ic} << 1$ . Thus the nonlinear terms should become important for very small amplitudes of the background drift waves. Finally, the presence of a resonance in the denominator for the expression for $\alpha$ should make the drift turbulence even more effective. A detailed calculation is in progress now.

## 3. EFFECT OF PLASMA FLOW ON NONLINEAR TEARING MODE

Before turning to the second problem to be discussed in this paper, let us briefly recapitulate the current models of the nonlinear development of tearing mode instability. Following up on a suggestion due to Kadomtsev,[7] Rutherford[8] has quantitatively analyzed the following theoretical idea: Associated with the growth of tearing perturbation is a quasilinear modification of the basic current profile $\delta J_{zo} \sim \tilde{v}_y\tilde{B}_x/\eta$ , especially within the newly formed current filament. This current perturbation leads to a third order nonlinear force $\delta J_{zo}\tilde{B}_x/c$ which dominates over the linear inertial force $\rho\partial v_y/\partial t$ in the fluid equation of motion, even for small amplitudes of the growing magnetic perturbations. If one proceeds with the normal mode analysis with the nonlinear force replacing the inertial force, one finds a similarity solution

$$(B_x)^{1/2} \sim (kB_y')^{1/2} \ \eta \ \Delta' \ t$$

(where $B_y' \equiv dB_y/dx$ is related to the equilibrium current and $\Delta'$ denotes the usual jump in the logarithmic derivatives of $B_x$ at $x = 0$) ; i.e., the perturbation grows algebraically (rather than exponentially, as in linear theory) and that too with a very slow resistive growth rate. Extensive numerical calculations with the exact nonlinear equations have been carried out by Rosenbluth et al.,[9] and they have confirmed the Rutherford calculation.[8] Furthermore, the relatively slow growth of the $m = 2$ , 3 --- etc. oscillations observed in tokamak discharges can also be explained by the Rutherford theory.[8] The numerical calculations have also seen saturation of the tearing instability under certain conditions, presumably due to nonlinear effects not included in the Rutherford calculation.

One shortcoming of these theories is that they give no interpretation of the observed plasma disruption in tokamak discharges. Experimentally, a plasma disruption (i.e., plasma loss on a very rapid time scale, such as an order or more faster than the time-scale of linear growth of tearing modes) always seems to be preceded by large amplitude $m = 2$ MHD oscillations. It is in this context that we investigate the second problem viz., the effect of plasma flow towards the X points, on the nonlinear growth of tearing mode instability. We take our cue from certain models of magnetic field reconnection at X points, which have been developed in astrophysics[2,3] and a suggestion of Kulsrud[10] that this may be related to the nonlinear development of tearing mode instability.

In our discussion below, we wish to elaborate upon the relationship between the nonlinear tearing mode and the reconnection mechanisms of Sweet-Parker[2] and Petschek[2] and also present some arguments for the relevance of this discussion to disruptive instabilities in tokamaks. In astrophysics, the problem of release of energy stored in magnetic fields, by having two regions with oppositely directed fields impinge on each other, has attracted a lot of attention. Sweet-Parker[2] developed a magnetic annihilation model, which goes as follows.

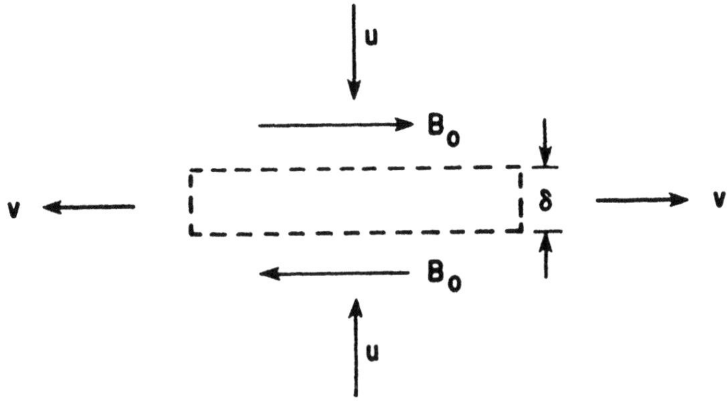


Fig. 1. Sweet-Parker mechanism of magnetic annihilation.

Consider a black box of length L and thickness $\delta$ , which basically is the annihilation region. Assuming an incompressible fluid, we have

$$uL = v\delta \ . \tag{8}$$

The fluid brings in the magnetic field into the box where it is annihilated by resistivity. The hot plasma then rushes out along the magnetic field with velocity v . Assuming that the pressure in the black box is p and outside it is $p_o$ , we can write pressure balance equations

$$p = p_o + B_o^2/8\pi \ ,$$

$$p = p_o + (1/2)\rho v^2 \ ,$$

$$v^2 = B_o^2/8\pi\rho = v_A^2 \ . \tag{9}$$

For a steady-state configuration, the magneitc field convection by the inward moving fluid must be balanced by the resistive diffusion giving

$$u = \eta/4\pi\delta \ . \tag{10}$$

Combining Eqs. (8), (9), and (10), we get

$$u = (\eta v_A/4\pi L)^{1/2} \ . \tag{11}$$

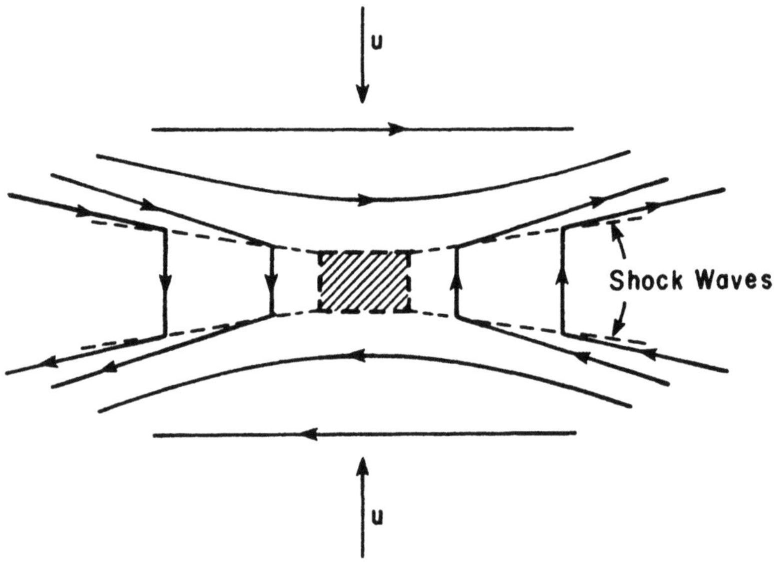


Fig. 2. Petschek mechanism of X point reconnection.

Thus resistivity determines the rate at which the magnetic field can be taken into the neutral regions and hence the rate of magnetic annihilation.

Historically, this concept was first used for the solar flare problem. It was found that the rate of energy release was too slow to account for observed solar flare time scales.[2] The same thing has since been found for the magnetospheric problem.[3] Thus the Sweet-Parker theory is in gross contradiction with observations.

Petschek[2] improved on the above model; he realized that the magnetic annihilation should be more properly treated as a field reconnection problem.

Petschek's model gives an elaborate account of what is in the black box. Two points are of interest.

1. There is an ideal MHD field reconnection arising through slow shock waves near the x points (dotted lines). Field reconnection also signifies annihilation because $\underset{\sim}{B}$ and plasma kinetic energy suffer changes at the shock front.

2. Any amount of fluid flow up to a certain maximum ($u_{max} \simeq 0.1\ v_A$) is permitted. Resistivity is only important in a very small region near the x point whose size adjusts itself to accommodate any flow (up to $u_{max}$) provided by the boundary conditions. Thus the magnetic field reconnection rate (and annihilation rate) is not limited by resistivity but is determined by the flow velocity u .

If we align several Petschek black boxes end to end to form a periodic structure, we can see a model for a growing magnetic island structure where the magnetic reconnection rate (and hence the growth of magnetic island) is determined not by resistivity but by the flow of matter towards the singular surface. This is the model we wish to pursue below, for disruptive instabilities in tokamaks.

Let us now give a simple description of the nonlinear tearing mode instability (see Rutherford[8]). Magnetic islands grow because of tearing and resistive reconnection of the lines of force near the singular surface. Note that only poloidal field perturbations are involved in the instability; i.e., $B_0 = B_{\theta 0}$ . The toroidal field primarily plays the role of making the transverse motion of the MHD fluid incompressible. Rutherford's nonlinear tearing mode state,[8] corresponding to an island whose width grows algebraically in time can be understood in terms of a Sweet-Parker like model as follows:

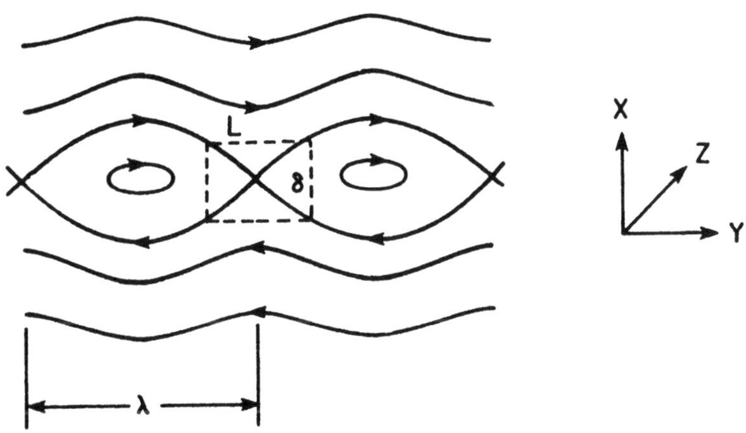


Fig. 3. Magnetic island growth with a diffusion region near X point.

Let there be a resistivity dominated region of length L and width $\delta$ near the x point. Matter flows in through length L with velocity u and then travels along the lines of force to fill up the island; this causes the island to inflate like a balloon. Since the fluid motion is incompressible in this plane, we may write

$$uL \sim \lambda \, dW/dt \tag{12}$$

where $\lambda$ is the length and W the width of the island. U is the flow velocity at width $\delta$

$$u \sim E/B_y' \delta \tag{13}$$

where it is assumed that outside $\delta$ ideal MHD dominates. We may also write the geometrical relation

$$L/\delta \sim \lambda/W \tag{14}$$

and the Maxwell equation

$$E/\lambda \sim \partial B_x/\partial t \quad . \tag{15}$$

Combining Eqs. (12) to (15), we get

$$W \sim (B_x \lambda / B_y')^{1/2} \quad . \tag{16}$$

Ampere's law gives

$$j_z \sim \nabla \times B \sim B_y/a \tag{17}$$

(a is the width of current layer). Within the resistive region

$$j_z \sim E/\eta \sim (\lambda/\eta)\partial B_x/\partial t \quad . \tag{18}$$

The fundamental assumption of Rutherford is that the entire island remains inside a resistivity dominated region so that

$$\delta \sim W \quad . \tag{19}$$

Combining (16) to (19), we now get

$$(\lambda/\eta)\partial B_x/\partial t \sim (B_y'/a)W$$

or

$$(B_x)^{1/2} \sim (B_y'/\lambda)^{1/2} \; \eta/a \quad t \quad . \tag{20}$$

Writing $B_x\lambda \sim \psi$ and $1/a \sim \Delta'$ , we get the Rutherford result

$$\sqrt{\psi} \sim \eta\Delta' (B_y')^{1/2}\ t\ . \tag{21}$$

It is instructive to understand the linear tearing mode on the basis of the above analysis. Equations (12) and (13) are valid but (14) is not, since the resistive layer is now much larger than the island thickness. Equations (12), (13), and (15) yield (with $L \sim \lambda$)

$$W \sim \lambda/\delta\ (B_x/B_y')\ . \tag{22}$$

Equations (17) and (18) may now be combined to give

$$(\lambda/\eta)\dot{B}_x \sim B_y/a \sim (B_y'/a)W$$

i.e.,

$$\dot{B}_x/B_x \sim \eta/a\delta\ . \tag{23}$$

To determine the resistive layer thickness $\delta$ we use the usual arguments (see Rutherford[8]) of balancing the torque due to linear $j \times B$ forces with that due to linear inertial forces giving

$$\delta \sim (\gamma\rho\eta)^{1/4}/(kB_y')^{1/2} \tag{24}$$

where $\gamma = \dot{B}_x/B_x$ is the growth rate and $k \sim \lambda^{-1}$ . Combining (23) with (24) gives the usual linear tearing growth.

The essential difference between the linear and nonlinear theories is worth noting in the above simple model. In the linear tearing mode, the resistive layer thickness $\delta \gg W$ , whereas in the Rutherford nonlinear theory we assume $\delta \approx W$ (we are calling $\delta \sim W$ , an assumption in the Rutherford calculation; actually, it is a consequence of his assumption that the torque due to nonlinear $\delta j \times B$ forces balances that due to the linear driving force). As pointed out below, the third possibility $\delta \ll W$ (as illustrated in Fig. 3) is likely to arise in a situation with plasma flow toward the neutral layer.

Rutherford's calculation is analogous to the Sweet-Parker calculation in that resistivity is playing a dominant role in the entire reconnection region. This may very well be perfectly valid for a natural resistive tearing mode. We now ask what happens to the tearing mode growth rate if matter is forced towards the X points, with a velocity u. In this case, Petchek's reconnection mechanism[2] should operate. Reconnection and hence island growth now occurs not only because of a little resistive region ($\delta \ll W$) near the x point but also because of reconnection across the shock waves in the ideal MHD region outside the diffusion region near the x point. The island growth is now determined by the flow velocity u (as long as $u < u_{max}$)

$$\lambda\ (dW/dt) \sim uL \tag{25}$$

where u is now treated as a given parameter set by external boundary conditions; L is the size of the shock region across which matter is flowing. Assuming $L \sim \lambda$ (as a crude guess) we get

$$W \sim ut$$

$$\text{or} \quad \sqrt{\psi} \sim (B_y')^{1/2} ut \quad . \tag{26}$$

So the island width grows at a rate independent of resistivity and dependent only on u . A detailed analytical and numerical investigation of this process is currently in progress.

We thus conclude that a plasma flow toward the neutral layer can increase the nonlinear growth rate of the tearing mode since the magnetic field reconnection rate is no longer limited by resistivity (in contrast to Rutherford's theory). Let us now speculate on the application of this concept to disruptive instability in tokamaks. Imagine a weakly growing or saturated $m = 2$ island on the $q = 2$ surface due to the natural resistive tearing instability. Suppose the plasma is suddenly forced outwards towards the x points and away from o points on $q = 2$ surface with velocity u . It will cause the $m = 2$ mode to grow very rapidly and result in disruption.

What could produce a plasma flow velocity u with the correct azimuthal variation near the $q = 2$ surface? We believe that an interaction between an internal kink mode ($m = 1$ , $n = 1$) and the poloidal variation of the toroidal field ($m = 1$ , $n = 0$) is a good candidate for producing this flow. Recent calculations by Edery et al.[11] have shown that small distortions of a circular cylindrical equilibrium (such as those produced by an $m = 2$ tearing mode) influences the internal kink mode in two significant ways: (i) it produces a strong destabilization of the mode, and (ii) the mode perturbations are no longer limited to within the $m = 1$ surface but extend over the whole discharge. Rough estimates show that velocity flows generated are of order

$$u \sim \delta^2 \varepsilon (v_A) \ ,$$

where $\varepsilon = r/R$ is the inverse aspect ratio, and $v_A$ is the poloidal Alfvén speed, $\delta$ is a measure of the elliptic distortion of the resonant $q = 1$ surface (due to the $m = 2$ mode). This then gives a typical time scale for the growth of m = 2 mode as

$$\delta^2 \varepsilon (v_A/a)$$

which can readily be of order the disruption time scale if $\delta$ is sufficiently large.

The ingredients of the proposed model are: the presence of a q = 2 surface where the natural resistive mode is present, a plasma flow towards the x points on this surface as a consequence of the modified internal m = 1 kink mode, and the proximity of the q = 2 surface to the plasma limiter so that a sudden growth of m = 2 island can lead to plasma disruption.

The following experimental facts are in favor of the model:

(a) m = 2 modes associated with external disruptions have a lot of high-frequency structure (on the x ray output). This might correspond to some fine sharp structures inside the m = 2 island such as would result from Petschek shocks.

(b) m = 2 modes seen without external disruption and with external disruption differ in one respect. The latter are more common at lower toroidal fields and higher electron densities. It is consistent with the model since the latter regime is more likely to have an m = 1 instability in the center.

(c) m = 2 and m = 1 modes are rarely seen together, presumably because plasma disrupts.

## 4. CONCLUSIONS

To summarize, we have investigated

(a) the effect of background drift wave turbulence on tearing mode instability and concluded that nonlinear coupling effects should be important when the ion excursion in the drift wave is comparable to the resistive layer thickness; i.e., $cE_d/B\omega_* x_R \sim 1$ .

(b) the effect of plasma flow towards the neutral layer on nonlinear growth of tearing mode and concluded that considerable enhancement in the tearing mode growth can be obtained.

## ACKNOWLEDGMENTS

I acknowledge the collaboration of John Hsu in the work presented in Sec. 2.

This work was supported by U. S. Energy Research and Development Administration Contract E(11-1)-3073, and by U. S. Air Force Office of Scientific Research Contract F44620-75-C-0037.


## REFERENCES


1. H. P. Furth, P. H. Rutherford, and H. Selberg, Phys. Fluids 16, 1054 (1973); S. V. Mirnov and I. V. Semenov, in Plasma Physics and Controlled Nuclear Fusion Research (International Atomic Energy Agency, Vienna, 1971), Vol. II, p. 401.

2. P. A. Sweet, in Electromagnetic Phenomena in Cosmical Physics, ed. by B. Lehnert, p. 123, Cambridge University Press, London, 1958; E. N. Parker, J. Geophys. Res. 62, 509 (1957); H. E. Petschek, AAS-NASA Symposium on the Physics of Solar Flares, NASA Spec. Publ. SP-50, 425 (1964).

3. V. M. Vasylianas, Revs. of Geophys. and Space Physics 13, 303 (1975); also private communication, 1976.

4. J. W. Dungey, Cosmic Electrodynamics (Cambridge University Press, New York, 1958), 98.

5. H. P. Furth, J. Killeen, and M. N. Rosenbluth, Phys. Fluids 6, 459 (1963).

6. P. H. Rutherford and H. P. Furth, Princeton PPL Report MATT-872 (1971).

7. B. B. Kadomtsev and O. P. Pogutse, in Reviews of Plasma Physics, ed. by M. A. Leontovich (Consultants Bureau, N.Y. 1970), 5 346.

8. P. H. Rutherford, Phys. Fluids, 16, 1903 (1973).

9. R. White, D. Monticello, M. N. Rosenbluth, H. Strauss, B. B. Kadomtsev, Fifth International Conference on Plasma Physics and Controlled Nuclear Fusion Research, A13-2, Tokyo (1974).

10. R. M. Kulsrud, Proc. of the International School of Physics Enrico Termi Course XXXIX (Academic Press, New York, 1967), 46.

11. D. Edery, G. Laval, R. Pellat, and J. L. Soule, Phys. Fluids 19, 260 (1976).

# NONLINEAR LANGMUIR WAVES IN THE SELF-CONSISTENT DENSITY PERTURBATION

K. Kato, K. Mima*, M. Watanabe and K. Nishikawa

Faculty of Science, Hiroshima University

Hiroshima, Japan

## §1. Introduction

Nonlinear propagation of Langmuir waves accompanied by an ion density perturbation has extensively been investigated as one of the simplest but physically interesting examples of nonlinear phenomena in a plasma. In this report, we present some of the recent eleborations on this problem carried out in our group. As in most of the works on this problem, we shall restrict ourselves to one-dimensional, long wavelength propagation ($k\lambda_D \ll 1$) in a medium of cold ions and hot electrons($(T_e \neq 0$ and $T_i = 0)$.

The paper consists of three parts. First, in §2, we consider a coherent localized Langmuir wave (Langmuir solitary wave) based on the fluid model for both the electrons and the ions. One can find many literatures concerning this problem[1-7], but to our view the relationships among various different descriptions of Langmuir solitary waves have not yet been fully understood. Here we classify these descriptions according to their range of validity in the space of various smallness parameters, such as those representing the slow space and time variations, the amplitudes of the Langmuir wave and the ion density perturbation, etc. We assess the relative magnitudes of these parameters in terms of the basic smallness parameter $\varepsilon = m/M$, the electron-to-ion mass ratio. In §3, we consider the case when there exist many Langmuir-wave eigenmodes localized in a coherent density cavity, still using the fluid model for both species of particles. Assuming that the frequency separation of these eigenmodes is sufficiently large as compared with the characteristic ion-acoustic frequency, we find a self-consistent set of solutions

* Bell Laboratories, Murray Hill, N.J., USA. Permanent Address: Institute of Laser Engineering, Osaka University, Osaka, Japan.

for these eigenmodes and the density cavity. Finally in §4, we investigate the effect of trapped ions on the nonlinear propagation of coherent Langmuir waves. A BGK-like solution[8] is obtained for the trapped-ion distribution for an arbitrary shape of the Langmuir wave profile, save for the condition that the trapped-ion distribution be positive definite. Two examples are considered specifically, namely, the case of a solitary wave of the form $E \sim$ sech Kx and that of a shock wave of the form $E \sim$ tanhKx, where E is the envelope of the Langmuir wave. In particular, we show that there exist two shock-like solutions, one corresponding to the usual ion-acoustic shock due to trapped ions[9] and the other to a shock produced by the ponderomotive force of the Langmuir wave confined in a half-space.

## §2. Classification of Coherent Solitary Waves in the Fluid Model[10]

We consider here a coherent Langmuir wave localized in a density cavity, using the fluid model for both the electrons and the ions. There exist several different descriptions for Langmuir solitary waves and each description has its own range of validity depending on the relative magnitudes of various smallness parameters. We show this range of validity by classifying these descriptions in terms of the basic smallness parameter, $\varepsilon = m/M$.

In the fluid model, the dominant nonlinear effect arising from a moderately strong Langmuir wave is a low-frequency density perturbation $\delta n$ created by the ponderomotive force of the high--frequency electric field. We denote this electric field as

$$E_1(\xi,t)\exp[i(kx-\omega t)] + c.c.,$$

where $\xi = x - Vt$, $V(=3k/\omega)$ being the group velocity of the Langmuir wave of wavenumber k and frequency $\omega$, and we choose $\lambda_D$, $\omega_{pe}^{-1}$ and $T/e\lambda_D$ as units of length, time and electric field, respectively, where $\lambda_D$ is the Debye length, $\omega_{pe}$ the electron plasma frequency and T the electron temperature. We assume that $E_1$ varies slowly in both space and time and that $k \ll 1$.

The basic equation which describes the Langmuir wave propagation is the Schrödinger-like equation,

$$i\omega\frac{\partial}{\partial t}E_1 + \frac{3}{2}\frac{\partial^2}{\partial\xi^2}E_1 + \Delta E_1 - \frac{\delta n}{2}E_1 = 0, \qquad (1)$$

where $\Delta(=[\omega^2-1-3k^3]/2)$ measures the frequency shift from the Bohm--Gross frequency. The density perturbation $\delta n$ obeys the ion--acoustic equation with the ponderomotive force acting as a source,

$$\left(\frac{\partial}{\partial t}-V\frac{\partial}{\partial \xi}\right)^2 \frac{\delta n}{\varepsilon} - \frac{\partial^2}{\partial \xi^2}\delta n = \frac{\partial^2}{\partial \xi^2}|E_1|^2 , \tag{2}$$

where we used the local charge neutrality. In this equation, the right-hand side represents the ponderomotive force, the last term on the left-hand side denotes the pressure gradient force and the term $(\partial/\partial t-V\partial/\partial\xi)^2\ \delta n/\varepsilon$ comes from the low-frequency electric field which drives the ion motion. Note that we have ignored the ion temperature.

We write various small quantities as

$$E_1 \sim \varepsilon^{\alpha},\quad \partial/\partial\xi \sim \varepsilon^{\beta},\quad \partial/\partial t \sim \varepsilon^{\gamma} \text{ and } k\sim\varepsilon^{\lambda}, \tag{3}$$

with $\alpha$, $\beta$, $\gamma$, $\lambda>0$. Langmuir solitary waves are characterized by the balance of the last three terms on the left-hand side of (1), so that we assume

$$\Delta\sim\varepsilon^{2\beta},\quad \delta n\sim\varepsilon^{2\beta} \text{ and } \gamma \gtrsim 2\beta . \tag{4}$$

Under this condition, we find four different regimes depending on the relative magnitudes of various terms in (2).

i) Inertial Regime: $\gamma=\alpha+1/2$, $\alpha<\beta$ and $\gamma<\lambda+\beta$.
In this case, both $(V\partial/\partial\xi)^2\delta n/\varepsilon$ and $\partial^2\delta n/\partial\xi^2$ are negligible and eq.(2) becomes

$$\frac{\partial^2\delta n}{\partial t^2} = \varepsilon\frac{\partial^2}{\partial\xi^2}|E_1|^2, \tag{5}$$

which describes a tendency of collapse, namely, $\partial^2\delta n/\partial t^2$ is negative in the region where $|E_1|^2$ assumes a maximum, whence the density dip in this region deepens with time.

ii) Compressional Regime: $\lambda=\alpha-\beta+1/2$, $\alpha<\beta$ and $\gamma\geq\lambda+\beta$.
In this case, $\partial^2\delta n/\partial\xi^2$ is negligible in eq.(2). Since $\lambda<1/2$, $V$ is supersonic, i.e. $V>\varepsilon^{1/2}$ (the ion-acoustic speed). In particular when $\gamma>\lambda+\beta$, eq.(2) can be solved as $|E_1|^2=V^2\delta n/\varepsilon$ which implies that the density has a hump in the region of high Langmuir wave intensity. Substitution of this relation into (1) yields a nonlinear Schrödinger equation for $E_1$,

$$i\omega \frac{\partial E_1}{\partial t} + \frac{3}{2}\frac{\partial^2}{\partial \xi^2} E_1 + \left(\Delta - \frac{\varepsilon^2}{2V^2}|E_1|^2\right) E_1 = 0, \qquad (6)$$

that admits a finite amplitude plane wave which is modulationally stable.[11] Local modulation of this plane wave yields an envelope hole.

iii) Static Regime: $\alpha=\beta$, $\lambda>1/2$ and $\gamma>\alpha+1/2$.

This is the case when the low-frequency electric field is negligible and the pressure gradient force balances the ponderomotive force, i.e. $|E_1|^2=-\delta n$. Substituting this into (1) yields

$$i\omega \frac{\partial}{\partial t} E_1 + \frac{3}{2}\frac{\partial^2}{\partial \xi^2} E_1 + \left(\Delta + \frac{|E_1|^2}{2}\right) E_1 = 0, \qquad (7)$$

whose plane wave solution is now modulationally unstable. Instead, we have a stable soliton solution for negative $\Delta$,

$$E_1 = 2(-\Delta)^{1/2} \operatorname{sech}\left[\left(\frac{-2\Delta}{3}\right)^{1/2} \xi\right]. \qquad (8)$$

iv) Acoustic Regime: $\lambda=1/2$, $\alpha\geq\beta$ and $\gamma\geq\beta+\lambda$

This is the case when all three terms in (2) become of comparable order. In particular when $\partial/\partial t$ is negligible, we have $\delta n=|E_1|^2/(V^2/\varepsilon-1)$ which derives a nonlinear Schrodinger equation of the form

$$i\omega \frac{\partial E_1}{\partial t} + \frac{3}{2}\frac{\partial^2}{\partial \xi^2} E_1 + \left(\Delta + \frac{|E_1|^2}{1-V^2/\varepsilon}\right) E_1 = 0. \qquad (9)$$

The plane wave solution is unstable when V is subsonic ($V^2<\varepsilon$), and we get a solitary wave solution for $\Delta<0$,

$$E_1 = 2\left[-\Delta\left(1-\frac{V^2}{\varepsilon}\right)\right]^{1/2} \operatorname{sech}\left[\left(-\frac{2\Delta}{3}\right)^{1/2} \xi\right]. \qquad (10)$$

In the particular case when $V\to\varepsilon^{1/2}$, $\delta n$ diverges and we have to invoke higher-order corrections to the low-frequency equation. Such a calculation was carried out[3,6] and the result is written as

$$\left(\frac{V^2}{\varepsilon}-1\right)\delta n - \frac{\partial^2}{\partial \xi^2}\delta n - \delta n^2 = |E_1|^2 + \text{const.} \qquad (11)$$

The coupled set of eqs.(1) and (11) have a new type of solitary wave solution for $\Delta<0$.

$$\delta n = 12\Delta \operatorname{sech}^2\left[\left(-\frac{2\Delta}{3}\right)^{1/2}\xi\right], \quad V/\varepsilon^{1/2} = 1 + 20\Delta/3$$
$$E_1 = (192)^{1/2}\Delta \operatorname{sech}\left[\left(-\frac{2\Delta}{3}\right)^{1/2}\xi\right]\tanh\left[\left(-\frac{2\Delta}{3}\right)^{1/2}\xi\right], \qquad (12)$$

which satisfies the parameter relation, $\lambda=1/2$, $\alpha=2\beta$ and $\gamma>\beta+1/2$.

These four different regimes are graphically represented in the parameter space of $\alpha$ and $\beta$ in Fig.1, where I, C, S and A, respectively, stand for the inertial, compressional, static and acoustic regimes. The line $\alpha=2\beta$ corresponds to the coupled soliton given by (12), while the line $\alpha=2\beta-1/2$ has two meanings: first the inertial regime is limited by the condition $\gamma=\alpha+1/2>2\beta$, and secondly the Langmuir soliton moves with its linear group velocity dragging the ion density perturbation when $\lambda=(\alpha-\beta)+1/2<\beta$, whereas in the case when $\lambda\geq\beta$ the Langmuir wave is trapped in the density hole and is unable to move with the linear group velocity.

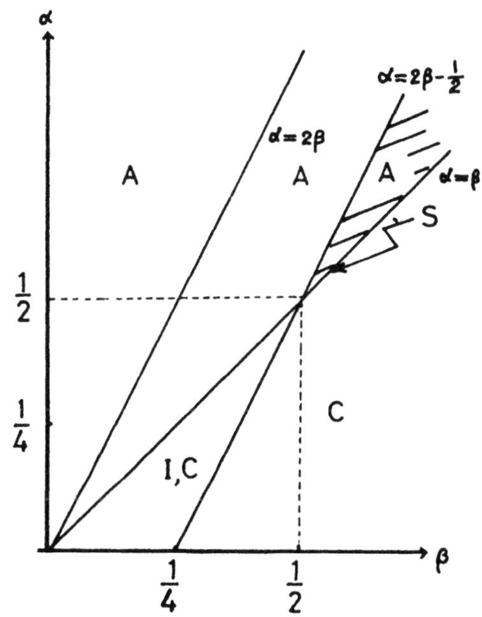


Fig.1. Graphical representation of different regimes of the Langmuir wave soliton; A: acoustic regime, S: static regime, I: inertial regime and C: com--pressional regime, with $E_1\sim\varepsilon^{\alpha}$, $\partial/\partial x\sim\varepsilon^{\beta}$, and $\varepsilon=m/M$. The meanings of the lines $\alpha=2\beta$ and $\alpha=2\beta-1/2$ are explained in the text.

We conclude this Section by pointing out that the inertial regime can include the case when $\alpha=0$, i.e. when $|E_1|^2/4\pi nT$ is of order unity. In this case, we have to include the contribution of both the second harmonies of the Langmuir wave and the local charge separation $\delta n_e-\delta n_i$. Taking both effects into account, we can show that eqs.(1) and (5) are unchanged by choosing $\delta n_i=\delta n$.[10)]

## §3. Self-Consistent Density Cavity with Many Bounded Langmuir Waves

We now consider the case when a spectrum of Langmuir waves are trapped in a local density cavity. For the case when both the frequency and the wavenumber spectra are broad as compared with the characteristic frequency and wavenumber of the density cavity, one can use the wave kinetic equation for the Langmuir waves. Such a situation was discussed in ref.12. Here we consider the situation when the frequency spectrum is broad, but the wavenumbers of both the Langmuir waves and the density cavity are comparable to each other. Physically, this corresponds to the situation when the density cavity admits a discrete set of bounded Langmuir eigenmodes of frequency separation being large as compared with the characteristic ion-acoustic frequency. The latter condition allows us to use the random-phase approximation (RPA) to calculate the ponderomotive force which creates the density cavity. This approximation is in turn admits us to consider a stationary density cavity in a moving frame, $\xi$=x-Vt, to which case we shall restrict ourselves.

We represent the Langmuir waves by the electric field

$$E(x,t)=\sum_k \left\{E_k(\xi)e^{i(kx-\omega_k t)} + c.c.\right\}, \tag{13}$$

where $\omega_k$ is the eigenfrequency which is to be determined from the eigenvalue equation for $E_k(\xi)$ and k is defined by the relation $k=V\omega_k/3$; note that we are using the same units as in §2. Since we are considering a stationary state, $E_k$ is a function of $\xi$ alone. The above choice of the wavenumber k yields the following equation for $E_k$ (see (1)):

$$\frac{3}{2}\frac{\partial^2}{\partial\xi^2}E_k + \left(\Delta_k-\frac{\delta n}{2}\right)E_k = 0, \tag{14}$$

where $\Delta_k=[\omega_k^2-(1+3k^2)]/2$. This is the eigenvalue equation for the eigenvalue $\Delta_k$. The density perturbation obeys the ion-acoustic equation (2) with $|E_1|^2$ being replaced by the average ponderomotive potential

$$\frac{1}{2}\langle|E(x,t)|^2\rangle = \sum_k |E_k(\xi)|^2 + \sum_k\sum_{k'} E_k(\xi)E_{k'}^*(\xi)\langle e^{i[(k-k')x-(\omega_k-\omega_{k'})t]}\rangle, \tag{15}$$

where the angular bracket depicts the time average. The RPA amounts to neglecting the last term, i.e. the interference term in (15). Using this approximation and restricting ourselves to a stationary localized solution, we obtain

$$\delta n = \left[\frac{\varepsilon}{V^2-\varepsilon}\right]\sum_k |E_k(\xi)|^2 . \tag{16}$$

For density cavity, $V^2<\varepsilon$. This equation can be regarded as the self-consistency relation which the solutions of the eigenvalue equation (14) have to satisfy.

The set of equations (14) and (16) can be solved in terms of elementary functions for the case when the density cavity takes the following form:

$$\delta n(\xi) = -3K^2 s(s+1)\,\mathrm{sech}^2 K\xi, \quad (s=1,2,3,\cdots), \tag{17}$$

where K is an arbitrary constant which represents the characteristic wavenumber of the density cavity. For s=1, the solution is reduced to (10) with $K=(-2\Delta/3)^{1/2}$. Substituting (17) into (14) and changing the variable from $\xi$ to $\zeta=\tanh K\xi$, we find

$$\left[\frac{d}{d\zeta}(1-\zeta^2)\frac{d}{d\zeta} + s(s+1) - \frac{\eta^2}{1-\zeta^2}\right]E_k(\zeta) = 0, \tag{18}$$

where $\eta^2=-2\Delta_k/3K^2$. The localized solutions of (10) are the s-th order associated Legendre functions, $P_s^n(\zeta)$ (n=0, 1, 2, ---, s-1), with the eigenvalues given by $\eta$=s-n. The corresponding eigenfrequencies are

$$\omega_k^2 = 1+\frac{V^2}{3} - 3K^2(s-n)^2, \qquad (V^2 \ll 1), \qquad (19)$$

In order to examine the self-consistency, we introduce the normalized associated Legendre function $\bar{P}_s^{s-n}(\zeta)$ by the relation

$$\int_{-\infty}^{\infty} d(K\xi)\left[\bar{P}_s^{s-n}(\zeta)\right]^2 = 1 \quad \text{or} \quad \bar{P}_s^{s-n}(\zeta) = \left[\frac{(s-n)\, n!}{(2s-n)!}\right]^{1/2} P_s^{s-n}(\zeta),$$

and expand the n-th eigenmode as

$$E_{k(n)} \equiv E_n = A_s^n\, \bar{P}_s^{s-n}(\zeta), \quad (n=0, 1, \cdots, s-1). \qquad (20)$$

Note that $A_s^n$ is considered the relative amplitude of the n-th eigenmode. Substituting (17) and (20) into (16), we find

$$3K^2 s(s+1)\left(1-\frac{V^2}{2}\right)(1-\zeta^2) = \sum_{n=0}^{s-1} |A_s^n|^2 \left[\bar{P}_s^{s-n}(\zeta)\right]^2. \qquad (21)$$

We expand the right-hand side in powers of $(1-\zeta^2)$ using the addition thorem for the associated Legendre functions as

$$\left[\bar{P}_s^{s-n}(\zeta)\right]^2 = (s-n)\sum_{m=s-n}^{s} (-1)^{s-n+m} \frac{(2m)!}{(m!)^2\, 2^{2m}} \frac{(s+m)!}{(s-m)!} \frac{(m+s-n)!}{(m-s+n)!} (1-\zeta^2)^m,$$

and equate the coefficient of each power of $(1-\zeta^2)$ in (21). After some calculation, we obtain

$$|A_s^n| = 12K^2(s-n)(1-\frac{V^2}{\varepsilon}). \qquad (22)$$

Substitution of (22) in (20) and then in (13) completes our solution.

As an example, we present the explicit expressions of the solutions for s=2. In this case, we have two eigenmodes which are given by

$$E_1(x,t) = [72(1-V^2/\varepsilon)]^{1/2} \, K \operatorname{sech}^2 K\xi \cdot \times \cos[(1-12K^2+\frac{V^2}{3})^{1/2}(t-\frac{Vx}{3})+\theta_1],$$

$$E_2(x,t) = [288(1-\frac{V^2}{\varepsilon})]^{1/2} \, K \operatorname{sech} K\xi \tanh K\xi \times \cos[(1-3K^2+\frac{V^2}{3})^{1/2}(t-\frac{Vx}{3})+\theta_2],$$

where $\theta_1$ and $\theta_2$ are arbitrary constants. One can see that the two eigenmodes oscillate with the beat frequency given by

$$(1-3K^2+\frac{V^2}{3})^{1/2} - (1-12K^2+\frac{V^2}{3})^{1/2} \doteq \frac{9K^2}{2}.$$

The RPA is justified when this frequency is much greater than the ion-acoustic frequency $K\varepsilon$, i.e. $9K/2 \gg \varepsilon$. Since the peak intensity of the Langmuir wave is given by $|E_1|^2 \sim 72K^2$, the condition for the validity of the RPA can alternatively be written as

$$|E_1|^2 \gg \frac{32m}{9M}. \qquad (23)$$

Similar conditions can be obtained for the cases s>2.

## §4. Trapped-Ion Effects on Coherent Langmuir Waves

In the collisionless fluid model which we used in the preceding two Sections, stationary localized solutions for the density perturbation $\delta n$ are always symmetric as functions of $\xi$, and the self-consistent Langmuir wave has an intensity which is determined by the width of the density perturbation. The situation is changed when trapped particles are taken into account. In ref. 13, we derived an envelope shock solution by making allowance for the electrons trapped by the electrostatic and ponderomotive potentials. Here we consider the effect of trapped ions, neglecting electron trapping, i.e. by treating the electrons to be isothermal.

Since ions can be trapped only by the electrostatic potential $\varphi(\xi)$, one can use the BGK solution for the trapped ion distribution function $F_t(w)$[8];

$$F_t(w) = \frac{1}{\pi} \int_{\phi_m}^{w} d\phi \, \frac{dg(\phi)}{d\phi} \, [\phi - w]^{-1/2} , \tag{24}$$

where $w(=Mv^2/2 + e\varphi(\xi))$ is the total ion energy, $\phi_m$ is the maximum value of the nondimensionalized potential, $e\varphi/T$, and $g(\phi)$ is given by

$$g(\phi) = -\frac{d^2}{d\xi^2}\phi + \delta n_e + 1 - \int_{\phi_m}^{\infty} dw \, F_{ut}(w) \Big/ \sqrt{w - \phi} , \tag{25}$$

$F_{ut}(w)$ being the untrapped ion distribution. We use the Boltzmann distribution for the electron density in the presence of the electrostatic and the ponderomotive potential, i.e.

$$\delta n_e = \exp\{\phi - |E_1|^2\} - 1 \doteq \phi - |E_1|^2 , \tag{26}$$

where the Langmuir-wave field $E_1$ obeys the equation (cf.(1))

$$\frac{3}{2}\frac{\partial^2}{\partial \xi^2} E_1 + \left(\Delta - \frac{\delta n_e}{2}\right) E_1 = 0 . \tag{27}$$

From these set of equations, one can immediately infer that a self-consistent trapped ion distribution can be found for arbitrary shape of the potential, save for the condition that $F_t(w)$ be nonnegative.

We illustrate such solutions by two examples; that is, the case of a solitary wave of the form $E_1=E_0\text{sech}K\xi$ and that of a shock wave of the form $E_1=E_0(1-\tanh K\xi)$.

Case i) $E_1=E_0\text{sech}K\xi$.

The self-consistent electron density profile $\delta n_e$ and the frequency shift $\Delta$ are determined from (27) as

$$\delta n_e = -6K^2 \text{sech}^2 K\xi\,, \quad \Delta = -\frac{3K^2}{2}. \tag{28}$$

Using (26), we then find to the lowest order

$$\delta n_e = \frac{6K^2\phi}{6K^2 - |E_0|^2}. \tag{29}$$

Neglecting $d^2\phi/d\xi^2$ and choosing $\phi_m=\phi(\pm\infty)=0$, we find from (25)

$$g(\phi) = 1 + \frac{\phi\cdot 6K^2}{6K^2-|E_0|^2} - \int_0^\infty dw\, F_{ut}(w)\Big/\sqrt{w-\phi^2}. \tag{30}$$

Since the last integral is rather insensitive to the exact shape of the distribution of untrapped ions, we assume for simplicity that they are cold, i.e.

$$F_{ut}(w) = \frac{n_{ut}}{\sqrt{2}}\,\delta\left[\frac{V}{\sqrt{\varepsilon}} - \sqrt{2w}\right], \tag{31}$$

where $n_{ut}$ is determined to be unity from the condition that $F_t(w)=0$ at $\phi=\phi_m=0$, or $g(0)=0$. Differentiating (30) with respect to $\phi$ and using (31), we obtain

$$\frac{dg(\phi)}{d\phi} = \frac{6K^2}{6K^2-|E_0|^2} - \frac{V}{\sqrt{\varepsilon}}\,\frac{1}{[V/\sqrt{\varepsilon}-2\phi]^{3/2}}, \tag{32}$$

from which the trapped ion distribution is calculated as

$$F_t(w) = \frac{2}{\pi}\sqrt{-w}\left\{\frac{\varepsilon}{V^2} + \frac{6K^2}{|E_0|^2 - 6K^2}\right\}, \tag{33}$$

where we have ignored the terms of the order $w^{3/2}$. Since $F_t(w)$ has to be nonnegative, the coefficient of $(-w)^{1/2}$ must be positive, so that we have

$$|E_0|^2 \leq 6\left(1 - \frac{V^2}{\varepsilon}\right)K^2. \tag{34}$$

The equality corresponds to the case of no trapped ions and hence to the solution (10). We therefore conclude that the allowance for trapped ions decreases the width of the solitary wave for given values of the amplitude and the propagation velocity. Conversely, if K and V are given, the amplitude of the Langmuir wave is decreased by the trapped ions.

Case ii) $E_1 = E_0(1 - \tanh K\xi)$.

In this case, eq.(27) determines $\delta n_e$ and $\Delta$ as

$$\begin{aligned} \delta n_e &= 6K^2\left[\tanh^2 K\xi + \tanh K\xi - 1\right] \\ \Delta &= -12K^2, \end{aligned} \tag{35}$$

with $\delta n_e(\pm\infty) = \pm 6K^2$. Using (26), we then find

$$\phi = \left[|E_0|^2 + 6K^2\right]\tilde{E}_1^2 - 18K^2\tilde{E}_1 + 6K^2, \tag{36}$$

where $\tilde{E}_1 \equiv E/E_0 = 1 - \tanh K\xi$. As a function of $\phi$, there are two solutions for $E_1$, corresponding to whether the potential $\phi$ has a maximum or a minimum at $\xi \to \infty$ where the Langmuir-wave amplitude vanishes. We shall refer to these two cases as case I (when $\phi(\infty) = \phi_m$) and case II (when $\phi(-\infty) = \phi_m$). From (36), we find that $\phi_m = 6K^2$ and $|E_0|^2 \leq 3K^2$ in case I and $\phi_m = 4|E_0|^2 - 6K^2$ and $|E_0|^2 > 3K^2$ in case II. Using (31) and assuming that $K^2 \sim |E_0|^2 \sim |\phi| << V^2/\varepsilon$ and $(\phi_m - \phi) << K^2$, we obtain

$$g(\phi) = \begin{cases} 1+\phi - n_{ut}\left[1-(\phi_m-\phi)\varepsilon/V^2\right], & \text{(case I)}, \\ 1+\phi - 4|E_0|^2 + 2|E_0|^2(\phi_m-\phi)/(3K^2+2|E_0|^2) & \\ \quad - n_{ut}\left[1-(\phi_m-\phi)\varepsilon/V^2\right], & \text{(case II)}. \end{cases} \tag{37}$$

Condition that $g(\phi_m)=0$ then yields $n_{ut}=1\pm 6K^2$ with plus and minus signs for cases I and II, respectively. Substituting (37) into (24), we find for the trapped ion distribution the following expressions,

$$F_t(w) = \begin{cases} \frac{2}{\pi}\left[\frac{1+6K^2}{V^2/\varepsilon} - 1\right]\sqrt{\phi_m - w} & \text{(case I)}, \\ \frac{2}{\pi}\left[\frac{1-6K^2}{V^2/\varepsilon} - \frac{3K^2}{3K^2+2|E_0|^2}\right]\sqrt{\phi_m - w} & \text{(case II)}. \end{cases} \tag{38}$$

The condition that $F_t(w)\geq 0$ yields $|V^2|\leq\varepsilon$ for case I and $|V^2|\leq 1+2|E_0|^2/3K^2$ for case II.

The above results can be explained as follows. When $|E_0|^2=0$, our solution for case I is reduced to the ion-acoustic shock wave obtained by Moiseev and Sagdeev.[9] When we switch on the Langmuir wave, the qualitative feature of the shock wave is unchanged as far as $|E_0|^2\leq 3K^2$. That is, ions are trapped in the potential trough in the upstream region ($\xi<0$) and raises the effective ion temperature there as compared with the downstream ($\xi>0$), so that the plasma density is depressed in the upstream by the ion kinetic pressure. On the other hand, when $|E_0|^2>3K^2$(case II), the ponderomotive force becomes stronger than the kinetic pressure force and hence the density is depressed in the downstream where the Langmuir wave intensity is large. Since the ponderomotive force acts only on the electrons, the potential in the downstream becomes higher than in the upstream and hence traps the ions in the upstream.

## References


1) V.E.Zakharov, Sov.Phys.-JETP 35 (1972) 908.
2) G.J.Morales and Y.C.Lee, Phys.Rev.Letters 33 (1974) 1016, 1534.
3) K.Nishikawa, H.Hojo, K.Mima and H.Ikezi, Phys.Rev.Letters 33 (1974) 148.
4) Kh.O.Abdulloev, I.L.Bogoljubskii and V.G.Makhankov, Phys.Letters 48A (1974) 161; Nuclear Fusion 15 (1975) 21.
5) L.M.Degtyarev, V.G.Makhankov and L.I.Rudakov, Sov.Phys.-JETP 40 (1975) 264.
6) V.G.Makhankov, Phys.Letters 50A (1974) 42; JINR, E5-8389, Dubna (1974).
7) V.I.Karpman, Physica Scripta 11 (1975) 263.
8) Ira B.Bernstein, J.M.Greene and M.D.Kruskal, Phys.Rev. 108 (1957) 546.
9) S.S.Moiseev and R.Z.Sagdeev, J.Nuclear Energy Py.C.t (1963) 43.
10) M.Watanabe and K.Nishikawa, to be published in J.Phys.Soc.Japan
11) See, for instance, A.Hasegawa, Plasma Instabilities and Nonlinear Effects (Springer-Verlag, Berlin, Heidelberg, New York, 1975) §4.4.C.
12) P.K.Kaw, K.Mima and K.Nishikawa, Phys.Rev.Letters 34 (1975) 803.
13) K.Mima, K.Nishikawa and H.Ikezi, Phys.Rev.Letters 35 (1975) 726.

# PARAMETRIC DECAY PROCESSES AND HEATING IN A BOUNDED MAGNETIZED PLASMA

G. van Oost, V.P. Bhatnagar, A.M. Messiaen,
A.S. Paithanker* and P.E. Vandenplas

Paper presented by P.E. Vandenplas

Laboratoire de Physique des Plasmas - Laboratorium voor Plasmafysica
Association"Euratom-Etat belge"- Associatie"Euratom-Belgische Staat"
Ecole Royale Militaire-1040 Brussels-Koninklijke Militaire School

Abstract : Parametric interactions are studied on the present plasma set-up characterized by the presence of many normal modes, previously investigated in a small-signal study. Two types of resonant parametric instabilities have been simultaneously observed above a certain threshold of the pump power at frequency $f_p$ ($f_{LH} < f_p << f_{pe}$). In both processes the electron decay product is a Trivelpiece-Gould (TG) mode. In the first type the ion decay is an ion-acoustic (IA) wave ($f < f_{ci}$) and electron and ion heating are observed. In the second type the ion decay mode is an ion-cyclotron (IC) wave ($f_{ci} < f < f_{pi}$) the spectral width of which is much larger than that of the IA spectrum and it is only accompanied by electron heating. All decay modes have been identified after careful comparison with the results of the study of the linear wave resonances of the system. Enhanced diffusion is furthermore observed.

## 1. INTRODUCTION

During the last few years plasma heating via the excitation of parametric instabilities has been of considerable theoretical and experimental interest. A detailed linear investigation of the

---

* On leave of absence from the "Bhabha Atomic Research Centre", Bombay, India.

bounded plasma system investigated in this paper has been performed (Bhatnagar et al., 1975a and 1976). The dispersion diagram of this warm-electron, cold-ion, magnetized plasma waveguide (plasma radius = a, metal tube radius = d) is shown in Fig. 1 for the azimuthal mode n = o. $V_S$, $V_e$ and $V_A$ represent plasma sound speed, electron thermal velocity and Alfvén velocity respectively. Warm plasma effects appear above the $V_e$ line. There are four sets of modes viz. T-G modes, surface waves, IC modes and IA modes that exist in the frequency domain shown, i.e. from $\omega < k_{\parallel} V_S$ to $\omega_{LH} < \omega < \omega_{pe}$, the so-called magneto-ion domain. The existence of these plasma modes and their dispersion characteristics have been experimentally verified by linear excitation.

In the present work nonlinear effects are investigated. This plasma column lends itself to parametric interactions due to the presence of all the above mentioned normal modes. Parametric instabilities above the lower-hybrid frequency and associated heating of the particles are of considerable importance from the point of view of radio-frequency heating of fusion plasmas. By applying an RF "pump" wave to the plasma the different modes can be parametrically excited. The decay modes are simultaneously excited if the pump amplitude exceeds a threshold, determined by the damping rates of the decay products. The pump power absorbed in the plasma at threshold is of the order of 500 mW. Above a first threshold sharp sidebands and low-frequency signals (kHz range) appear in the spectrum. Above a second threshold (which can be lower or higher than the first one depending on plasma parameters), a rather broad lower sideband and a broad low-frequency decay mode (MHz range) appear as shown in Fig. 2. This type of parametric excitation occurs by virtue of interaction between the motions of the electrons and those of the ions. Because of the high density of modes the processes are not strongly subject to resonance detuning. The decay products are identified by means of the dispersion diagram obtained from the linear investigation. Preliminary nonlinear results have been published elsewhere (Bhatnagar et al., 1975b).

A description of the experimental setup is given in Section 2. The experimental results of the nonlinear investigation and their theoretical interpretation are described in Section 3. Conclusions are given in Section 4.

## 2. EXPERIMENTAL ARRANGEMENT

A schematic diagram of the experimental setup is shown in Fig. 3. The plasma column is located in a 10 cm dia and 90 cm long cylindrical stainless steel vacuum chamber. A turbomolecular pump system at one end of the machine evacuates the chamber to a base pressure of $8 \times 10^{-7}$ Torr. An RF-helix discharge system (Aubert et

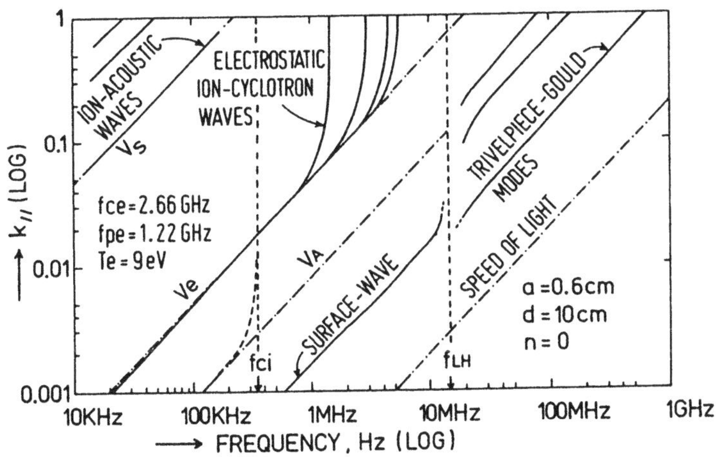


Fig. 1. Theoretically computed dispersion diagram(f vs. $k_{\|}$) of the bounded plasma system including the ion-acoustic waves ($f \lesssim f_{ci}$), surface wave, electrostatic ion-cyclotron waves ($f_{ci} < f < f_{LH}$) and Trivelpiece-Gould mode($f_{LH} < f < f_{pe}$) Warm plasma effects appear above the $V_e$ line. Plasma parameters are $f_{ce} = 2.66$ GHz, $f_{pe} = 1.22$ GHz, a = 0.6 cm, d = 5 cm, Azimuthal mode n = o, Te = 9 eV. Note : $k_{\|}$ in $cm^{-1}$.

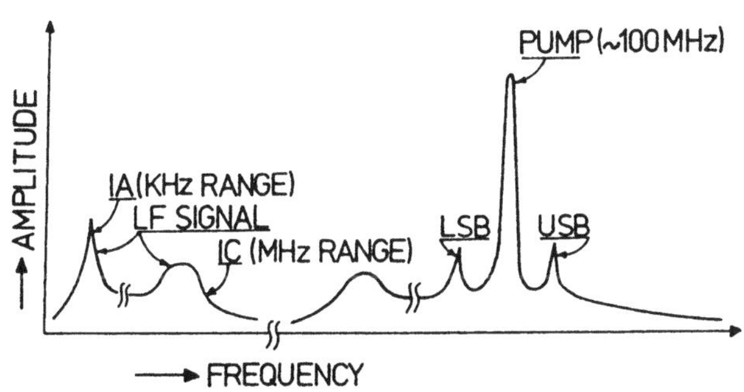


Fig. 2. Schematic spectrum of the parametric decay processes.

al., 1971a and b) is located at the other end where helium gas is continuously introduced by means of a leak valve. A microwave generator (2.45 GHz) delivering an output power of 20-100 W CW feeds the helix system. The plasma created in the helix diffuses along a steady magnetic field (900-2200 G) produced by a solenoid, and is terminated by an end plate that is either grounded or floating.

Typical plasma parameters are : $f_{pe} < f_{ce}$ ; density on the axis, $5 \times 10^9 - 8 \times 10^{10} cm^{-3}$ ; half-width of the radial density distribution, 1 - 1.2 cm ; helium gas pressure, $1 - 5 \times 10^{-3}$ Torr ; electron temperature, 4 - 15 eV ; ion temperature, 0.3 - 0.8 eV. Plasma density is measured with a Langmuir probe. Electron temperature is determined spectroscopically by the helium line-intensity ratio technique (Latimer et al., 1970), the Langmuir probe method and by means of the ion-acoustic phase velocity and the axial drift velocity. Ion temperature is obtained from Langmuir probe data (Sato, 1972) which are corroborated spectroscopically by Doppler-broadening of the 4686 Å-line (Sato et al., 1973).

A split-cylindrical slow-wave structure with several sections along the plasma column is provided for excitation of waves in the plasma. In the present experiment only one section has been fed (axisymmetric excitation). The well balanced axially and radially movable double probes are used to detect waves in the plasma. The detected signal, after suitable processing and amplification is either fed to a spectrum analyzer which provides the amplitude, or to a very selective superheterodyne system to filter out pump or sideband signal and afterwards to a double balanced mixer for phase detection. It should be underlined that in this experiment the same $k_{\parallel}$ (propagation constant parallel to the external steady magnetic field) is obtained by both methods. Since a vacuum electromagnetic field at the applied RF frequency with a wavelength large compared with the length of the machine is always present and is modulated by the waves we are interested in, and since reflected waves are only present near the ends, the distance between two maxima or minima in the curve of the amplitude versus axial position z gives $\lambda_{\parallel}$ .

The measurement of the RF power absorbed by the plasma is explained in Fig. 4.

A probe for measuring azimuthal wave numbers n is not available but(through stimulation) one can indirectly deduce that the n of the pump (axisymmetric excitation) and of the decay modes are all zero (conservation of angular momentum). Deducing n from the theoretical dispersion curves is not possible since the curves for n = o ; ± 1 ; etc... are quite near each other and no distinction

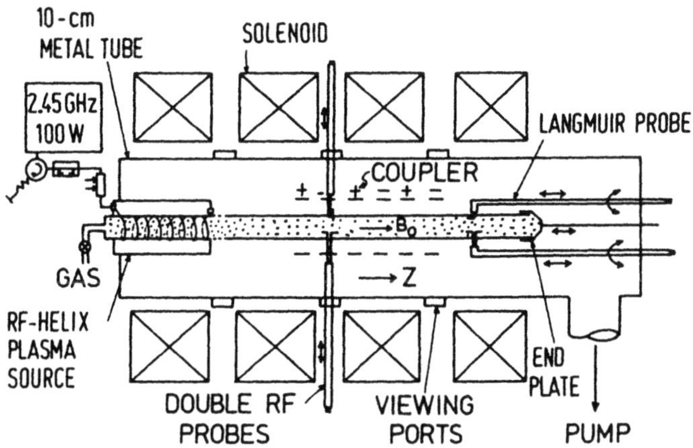


Fig. 3. Schematic diagram of the experimental apparatus.

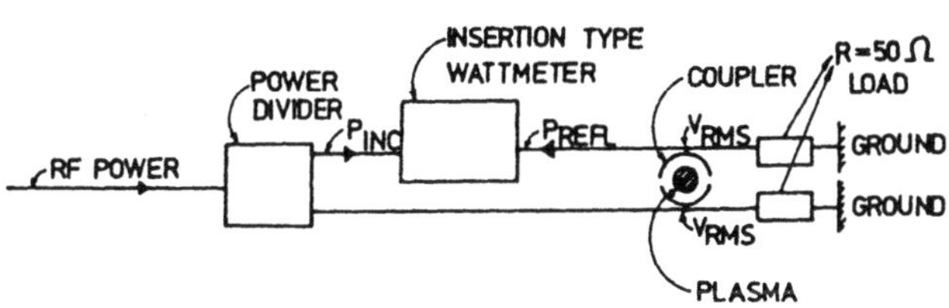


Fig. 4. Measurement of $P_{abs}$, the RF power absorbed by the plasma

$$\frac{1}{2} P_{abs} = P_{inc} - P_{refl} - \frac{V_{rms}^2}{R}$$

can be made within the errors of the particle density measurement. However, investigation of the decay products by comparison with modes directly excited via the couplers is not always possible. In direct excitation the important point is the radial field distribution. With our localized structure it is not possible to excite one single radial mode and the lowest order mode is always dominant. In parametric excitation the decay products are determined by the selection rules and higher order radial modes can be excited if they fulfil these rules. The electrostatic ion-cyclotron mode is difficult to excite directly but easy to excite parametrically.

## 3. EXPERIMENTAL RESULTS

The following subscripts are introduced : p, pump ; LSB : lower sideband ; USB : upper sideband ; LF : low frequency. The results will present an in-depth study of the observed parametric effects, by which a pump field (Electromagnetic or T-G mode) pumps the parametric decay instability into a sideband electron mode (T-G) and a LF ion mode (ion-acoustic mode, $f < f_{ci}$ ; or electrostatic ion-cyclotron mode, $f_{ci} < f < f_{pi}$). The excited waves can have different radial wavenumbers. These parametric processes are resonant, i.e. $\bar{k}_p = \bar{k}_{SB} + \bar{k}_{LF}$ ; $\omega_p = \omega_{SB} + \omega_{LF}$ ; $D(\omega_p, k_p) = D(\omega_{SB}, k_{SB}) = D(\omega_{LF}, k_{LF}) = o$ , each of the excited modes satisfies a linear dispersion relation. For a plasma with cylindrical symmetry the vector selection rule $\bar{k}_p = k_{SB} + \bar{k}_{LF}$ (valid for plane waves) is replaced by two scalar relations , one for the axial propagation constants $k_{\parallel}$ , and one for the azimuthal wave numbers n (Barett). The setup together with the plasma column constitutes a plasma system possessing resonances ; the resonance $k_{\perp}$'s are thus determined by the boundary conditions of the system. There is no selection rule for the $k_{\perp}$'s.

### 3.1. The Ion Decay Product is an IA Mode

The dispersion relation of the IA modes involved is (Messiaen and Vandenplas, 1973),

$$k_{\perp}^2 = -\,\omega_{ci}^2 \left[ \frac{1}{V_S^2} - \frac{k_{\parallel}^2}{\omega^2} \right] \quad , \qquad \omega^2 \ll \omega_{ci}^2 \ll \omega_{pi}^2 \ , \ V_S^2 = \frac{kTe}{m_i}$$

These waves propagate almost along the magnetic fiel $\bar{B}_o$ : the coupling mechanism to the IA branch is electron drift along $E_{p\parallel}$ (Chu et al., 1974). $V_S$ is the ion-acoustic speed. There exists an ambipolar drift velocity $v_d$ of the order of $10^5$ cm/s from the plasma source towards the end plate. This drift velocity is not taken into account in the above dispersion relation.

A typical decay spectrum is shown in Fig. 5. This resonant parametric decay process has been investigated for a broad range of plasma conditions (see Section 2) and pump frequencies $f_p$ ($f_{LH} < f_p << f_{pe} < f_{ce}$). Fig. 6 shows the amplitude of the pump and the ion decay wave. The shape of the latter one will be explained later. Comparison of pump and LF amplitude is not straightforward because of the frequency-dependent capacitive coupling of the probe. The increase in ion temperature is given in the same figure since it is only due to the IA and not to the IC instability; for discussion see 3.1.b.

a) Selection rules. Depending on the pump frequency, two types of pump mode have been observed : electromagnetic (EM) mode for $f_p >> f_{LH}$ and Trivelpiece-Gould (T-G) mode for $f_p$ closer to - but not below - $f_{LH}$.

(i) EM pump($f_p >> f_{LH}$). In the neighbourhood of the coupler there exists a vacuum field $E_p(\omega_p, k_{\parallel p} \approx o)$ where $k_{\parallel p}$ is the real part of the axial wavenumber. This field decreases rapidly since its axial wavenumber has an imaginary part $\gamma$ such that $\gamma^2 >> k^2_{\parallel p} = \omega^2/c^2$. The small axial evanescence length (typically of the order of a few cm) is clearly due to the fact that we have a localized EM excitation with vacuum wavelength far greater than the radius of the waveguide. This field is responsible for the parametric excitation of a LSB mode (T-G) and a LF mode (IA) which satisfy the selection rules for frequencies and real parts of axial wavenumbers.

$$\omega_p = \omega_{LSB} + \omega_{LF}$$

$$k_{\parallel p} \approx o = |\bar{k}_{\parallel LSB}| - |\bar{k}_{\parallel LF}|$$ within the errors of the measurements (accuracy about 10 %)

The LSB mode is characterized by n = o and propagates both upstream and downstream with its $k_\parallel$ not significantly changed by the axial ambipolar drift velocity of the plasma. The IA waves on the other hand are heavily damped upstream, their phase velocity with respect to the laboratory is given by : $v_{ph\ up} - v_{ph\ down} = 2v_d$ (difference of the order of 30 to 40 %) and their $k_\parallel$ is far larger than downstream. This means that for a given collision frequency $\nu$ the damping is much stronger than in the downstream case.
The IA wave vector is therefore essentially directed downstream

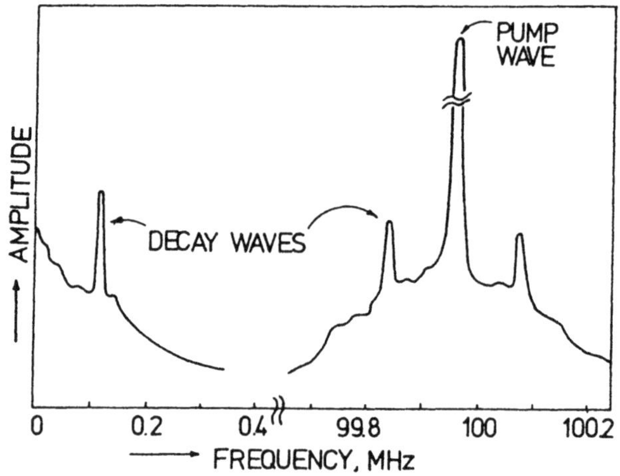


Fig. 5. Spectrum of the ion-acoustic instability. Parameters $B_o$ = 940 G, $N_o$ = 2.5 x $10^{10}cm^{-3}$, Teo = 5.1 eV, Tio = 0.35 eV, $P_{MW}$ = 60 W, He Gas = 2 x $10^{-3}$Torr.

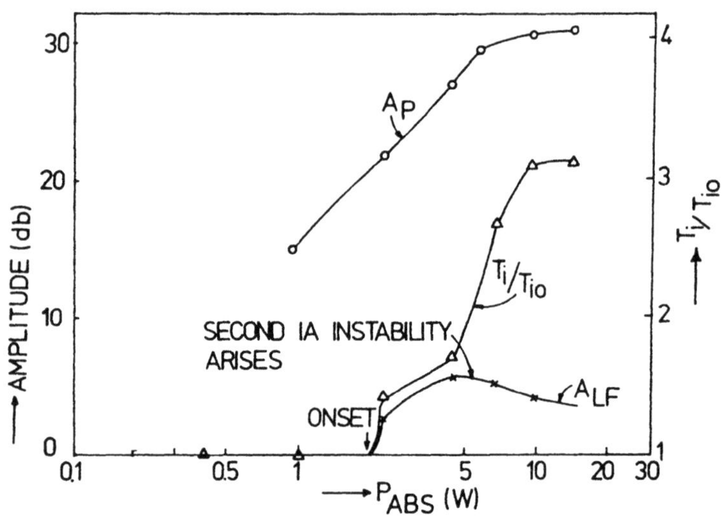


Fig. 6. Amplitude of the pump and the IA decay wave. The ion temperature increase as a function of the power absorbed by the plasma. Same parameters as in Fig. 5.

and we have the ensuing vector diagram, which has been experimentally verified.

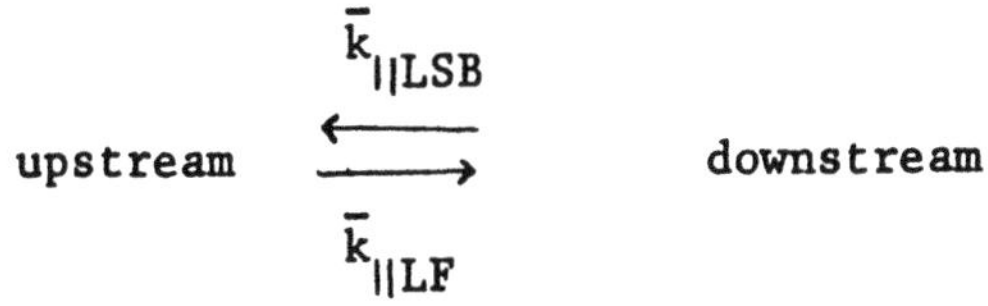


since $|\bar{k}_{\|p}| \simeq o = |\bar{k}_{\|LSB}| - |\bar{k}_{\|LF}|$ in the EM case.

Along the column there also exists a second component of the pump field, namely the field $E'_p$ $(\omega_p, k'_{\|p})$ of the T-G mode described in (ii) below with $k'_{\|p}$ determined by the relevant dispersion equation. By mode-mode coupling between the latter mode and the previously excited IA mode an USB arises with

$$\omega_{USB} = \omega_p + \omega_{LF}$$

$|\bar{k}_{\|USB}| = |\bar{k}_{\|LF}| \pm |\bar{k}'_{\|p}|$ with $\bar{k}_{\|LF}$ always directed downstream

This coupling effect causes continuous energy transfer among the modes involved. The USB is growing at the expense of the IA mode (energy transfer from LF to HF waves), is in general sharper than the LSB and does not show saturation even for pump powers far above threshold. Its amplitude is proportional to the product of the amplitudes of pump and IA mode and since $A_p >> A_{LF}$, it will grow as long as the pump does not saturate. Two vector diagrams were experimentally found. The large $|\bar{k}_{\|USB}|$ of the first diagram is possible because the T-G mode is excited on a higher radial mode (see Fig. 1) than that of the T-G pump mode to be examined below in (ii) .

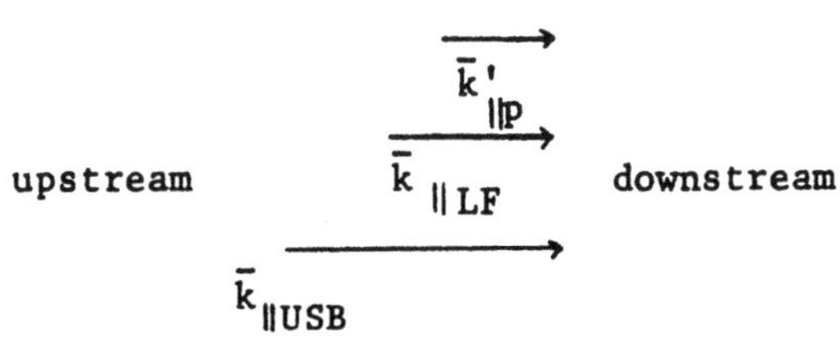


; with $|\bar{k}_{\|USB}| = |\bar{k}_{\|LF}| + |\bar{k}'_{\|p}|$

or

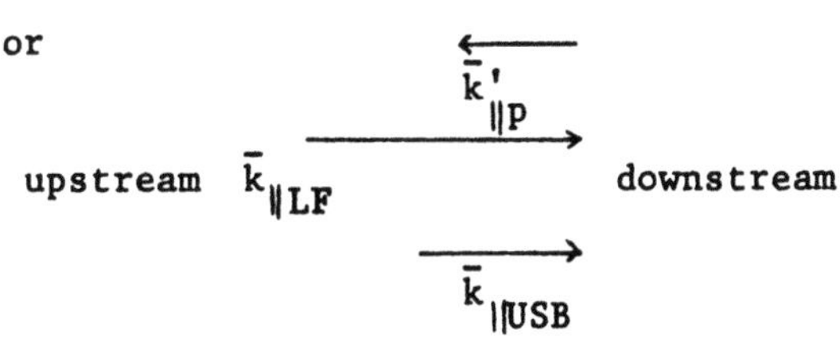

with $|\bar{k}_{\parallel USB}| = |\bar{k}_{\parallel LF}| - |\bar{k}_{\parallel p}|$

Experiments show that either one or the other is observed depending on the frequency of the pump and on plasma parameters.

(ii) T-G pump. For lower pump frequencies it has been observed that the mode responsible for parametric excitation is no longer the one with $k_{\parallel p} \simeq 0$, but the T-G mode at $\omega_p$ with a $k_{\parallel}$ imposed by the dispersion equation. This T-G mode ($\omega_p$ ; $k_{\parallel p}$) propagates and disintegrates along the column into another T-G mode (LSB) and an IA mode (LF). The selection rules are observed to be fulfilled as follows :

$$\omega_p = \omega_{LSB} + \omega_{LF}$$

$$|\bar{k}_{\parallel p}| = |\bar{k}_{\parallel LF}| - |\bar{k}_{\parallel LSB}| \text{ again within experimental error}$$

with $|\bar{k}_{\parallel LSB}| \simeq |\bar{k}_{\parallel p}|$

$$|\bar{k}_{\parallel LF}| \simeq 2|\bar{k}_{\parallel p}|$$

Pump and lower sideband are T-G modes and belong to the same dispersion curve. The high-frequency decay wave is similar in nature to the pump wave but has a different frequency and a slightly different axial propagation constant. The vector diagram is :

$\longrightarrow$ $\bar{k}_{\parallel p}$

upstream $\longleftarrow$ $\bar{k}_{\parallel LSB}$ downstream

$\longrightarrow$ $\bar{k}_{\parallel LF}$

All the azimuthal wavenumbers involved are zero (see Section 3.2.a). As far as the upper sideband is concerned, the wave coupling phenomenon is analogous to the one described at the end of (i)

(iii) Discussion. To understand the transition from the case described under (i) to the one described under(ii) the dispersion diagram (Fig. 1) has to be studied in more detail. When the pump frequency comes closer to the lower-hybrid frequency, $k_{\parallel}$ of the T-G mode can no longer match $k_{\parallel}$ of the IA decay mode and the selection rule $|\bar{k}_{\parallel LF}| = |\bar{k}_{\parallel LSB}|$ can no longer be fulfilled. In the decay process described under(ii) on the contrary, $|\bar{k}_{\parallel LF}| \simeq 2|\bar{k}_{\parallel LSB}|$ .

It can be mentioned incidentally that for T-G excitation the decay waves originate at about the same position where the pump wave begins to propagate (there is an important initial decay of the pump signal near the excited coupler before the wave propagation starts). This indicates that probe effects can be clearly excluded from being responsible for the decay process (Porkolab and Chang, 1969).

b) Heating. When the RF pump power is increased, a broadening of the decay spectrum and a shift of the IA frequency towards higher values is obtained in correspondence with the increase in electron temperature. Sometimes splitting of the IA frequency into the original one and a higher one occurs, e.g. the second slope of the rise in ion temperature shown in Fig. 6 corresponds to such a splitting, the amplitude of the original decay mode decreases (Fig. 6). For several conditions harmonics of the original IA frequency arise and are explained by direct nonlinearity of the IA mode (Mix et al., 1972). These IA harmonics couple furthermore to the pump and give rise to new sidebands. Thus the generation of higher order sidebands is due to the interaction of the LF signal back on the pump.
One can state reasonably generally that waves do not continue to exist if the wave potential energy exceeds the mean thermal energy of the electrons (Franklin, 1974). Saturation occurs because of channeling of energy into the enhanced fluctuations around $\omega_{LF}$ and $\omega_{SB}$. In general, saturation of an instability occurs when the energy input into the instability balances the energy loss to the particles.

(i) Electron heating. This has to be treated together with the occurrence of the IC instability, since the threshold of the latter is sometimes higher and sometimes lower than the IA threshold and since distinction between electron heating due to the two types of instability is not possible. Both the RF pump and the sidebands have $k_{\parallel}$'s of the same order of magnitude. Since the RF pump amplitude is of course by far the largest this means that it will mostly be the RF pump power which will heat the electrons. Moreover since $v_{phSB} \gg V_e$ there will be little or no Landau damping and not much enhanced heating can be expected.
The increase in electron temperature has been measured spectroscopically (Latimer et al., 1970). Local measurements by means of a Langmuir probe showed uniform electron heating throughout the plasma column.
For this plasma $\omega_{pe} < \omega_{ce}$, hence body waves are mainly excited and since those (T-G) waves are responsible for electron heating, heating of the entire column is expected (Porkolab et al., 1973), the more so as the plasma diameter is quite small. Furthermore the Langmuir probe data show a Maxwellian distribution before and after heating, thus the bulk of the electrons is heated.
Fig. 7 shows the amplitude of the pump and of the two ion decay waves, together with the electron heating as a function of the pump

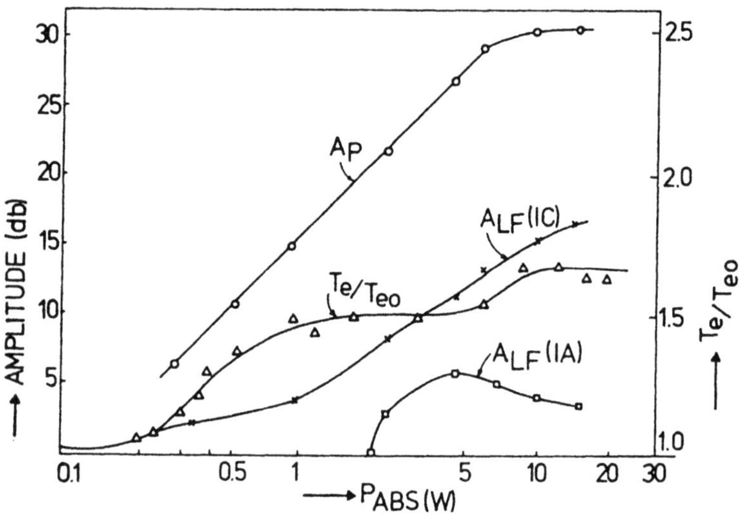


Fig. 7. Amplitude of the pump and the two ion decay waves. The electron temperature increase as a function of the power absorbed by the plasma. Same parameters as in Fig. 5.

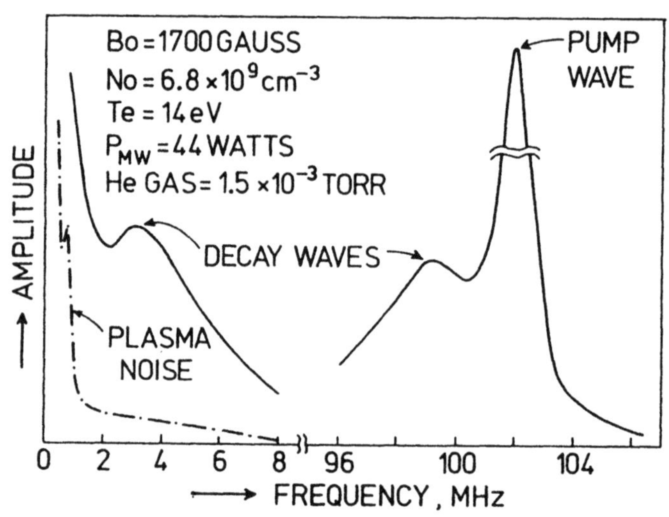


Fig. 8. Spectrum of the ion-cyclotron instability. Parameters $B_o$ = 1700 G, $N_o$ = 6.8 x $10^9 cm^{-3}$, Te ≐ 14 eV, $P_{MW}$ = 44 W, He Gas = 1.5 x $10^{-3}$Torr.

power absorbed by the plasma. The increase in electron temperature Te/Teo ranges from 1.5 to 2.

(ii) Ion heating. The ion heating is observed to be uniform throughout the column and has been determined by Langmuir probe technique (Sato, 1972) and by Doppler-broadening (Sato et al., 1973), with comparable results. A typical result is shown in Fig. 6, Ti/Tio ranges from 1.5 to 3. The measurement of Ti by Doppler-broadening technique should reflect the bulk ion temperature and should not be sensitive to the tail of the velocity distribution (Okabayashi et al., 1973).
The ion heating is ascribed to ion Landau damping of the IA decay waves. Typical values of the IA phase velocity and of the average ion thermal velocity are $10^6$cm/s and 3 x $10^5$cm/s respectively. The observed initial ion temperature without RF applied (0.3 to 0.8 eV) is high with respect to the value one expects in such a device (Ti < 0.1 eV).

A similar Tio has been found recently in hydrogen gas (Brown, 1976) and the presence of a tail of high energy ions is explained by Frank-Condon excitation of the hydrogen molecule and consequent ionization of energetic neutrals by the hotter electron component, and partial cooling via charge exchange on the background gas. The ion temperature can be controlled by variation of the background gas pressure, in agreement with our measurements. Besides this explanation there is also the fact that nonlinear effects occur in our plasma source which could also be responsible for the high ion temperature.

Anyhow this rather high ion temperature is, of course, favourable for further heating via ion Landau damping of the IA waves. On the other hand, a low electron temperature also favours ion heating since the IA waves then have lower phase velocity matching more adequately the ion thermal velocity.

### 3.2. The Ion Decay Product is an IC Mode.

The dispersion relation of these modes is given by (Messiaen and Vandenplas, 1973)

$$k_\perp^2 = \frac{\omega^2 - \omega_{ci}^2}{V_S^2} \quad , \qquad k_\parallel^2 V_S^2 << \omega_{ci}^2 << \omega^2 << \omega_{pi}^2$$

These modes propagate nearly perpendicular to the magnetic field ; the coupling mechanism to the IC branch is the $\bar{E}_{p\perp} \times \bar{B}_o$ drift (Chu et al., 1974).

This parametric process has also been investigated for many plasma conditions and pump frequencies $f_p (f_{LH} < f_p << f_{pe} < f_{ce})$. For $f_p \lesssim 3f_{LH}$, roughly, the decay process does no longer occur which is in agreement with Corti et al., 1975 and with theory (Porkolab, 1974). Fig. 8 shows the amplitude of the pump and the decay waves in a typical spectrum. Here the decay spectrum is broader (width of 2 to 4 MHz) than in the IA case.

a) Selection rules. The pump field at $\omega_p$ responsible for the decay process has been found to be always that of the T-G mode with a $k_{\|}$ imposed by the dispersion equation, thus $k_{\|p}$ is always different from zero (no vacuum EM pump wave). An upper sideband does in general not occur. The spectrum around the pump frequency shows the presence of a lower sideband which is the image of the ion decay spectrum. The selection rules for frequencies and wavenumbers are satisfied as follows :

$$\omega_p = \omega_{LSB} + \omega_{LF}$$

$$|\bar{k}_{\|p}| = |\bar{k}_{\|LSB}| + |\bar{k}_{\|LF}|$$

The lower sideband mode and the pump mode belong to the same T-G dispersion curve, the values of $k_{\|p}$ and $k_{\|LSB}$ are close to each other. In general $k_{\|LF}$ is difficult to measure because of the parallel wavelength that is comparable to or larger than the length of the machine. Nevertheless for conditions of low eletron temperature $\lambda_{\|LF}$ becomes small enough to perform a measurement (see Fig. 1). However the difference between $k_{\|p}$ and $k_{\|LSB}$ is in the range of the theoretical $k_{\|}$-values of the IC modes, whatever the plasma conditions are. Both directions of this now small $k_{\|LF}$ are possible. The following vector diagrams can be considered.

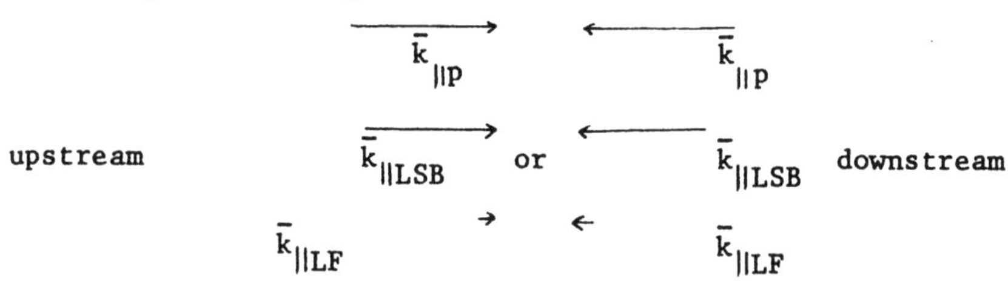


The broadness of the decay spectra can be explained as follows :

$$|\bar{k}_{\|p}| = |\bar{k}_{\|LSB}| + |\bar{k}_{\|LF}| \pm \Delta|\bar{k}_{\|}| \qquad \Delta|\bar{k}_{\|}|: \text{resonance defect.}$$

This selection rule is easily satisfied in a $k_{\|}$-range because $|\bar{k}_{\|P}|$ and $|\bar{k}_{\|LSB}|$ are close to each other and because $|\bar{k}_{\|LF}|$ and $\Delta|\bar{k}_{\|}| << |\bar{k}_{\|LSB}|$ in a region where the density of IC modes is high. The broad decay spectrum is therefore not surprising. This reasoning is no longer valid when $f_p$ approaches $f_{LH}$ since then $|\bar{k}_{\|LSB}| \simeq |\bar{k}_{\|LF}|$. Experimentally the IC decay instability disappears when $f_p \lesssim 3f_{LH}$.

Notice that the foregoing conditions leading to broad IC decay are not fulfilled for the IA parametric decay process. Indeed, $|\bar{k}_{\|P}| \simeq o$ for an EM pump and although $|\bar{k}_{\|LSB}| \simeq |\bar{k}_{\|P}|$ for a T-G pump, $|\bar{k}_{\|LF}|$ is in both cases of the same order of magnitude as $|\bar{k}_{\|LSB}|$. The density of IA modes is also much lower than in the IC case. This explains the basically different width of the IC decay spectrum as compared to the IA one.

An experimental difficulty occurs in the IC measurement. Because of the interference between signals at different frequency in the broad decay spectrum,it is difficult to measure a $k_{\|}$ corresponding to a single frequency. In order to obtain good selectivity, a stimulation method has been used (Edgley et al., 1975). A single frequency $f_{LF}$ out of the broad ion decay spectrum is applied to a coupler different from the one to which the pump frequency is applied. In this manner a signal superimposed on the broad sideband spectrum is observed for the frequency corresponding to $f_p - f_{LF}$ as long as $f_{LF}$ belongs to the ion decay spectrum. The interaction through stimulation is only observed if the pump field amplitude is above threshold. Below threshold there is no interaction, however high the amplitude of the stimulating LF signal is ; the RF pump signal changes the plasma conditions and the dispersion diagrams in such a manner that it is only above threshold that the selection rules are satisfied.

The azimuthal selection rule is fulfilled since all n = o. It is precisely the stimulation method which is used (with n = o excitation) in order to show that the LF component of the decay spectrum is indeed characterized by n = o. Since both the pump and LF have an n = o azimuthal number, the selection rule entails that the LSB (T-G) mode has the same n = o value.

b) Heating. No ion heating due to the IC parametric instability has been observed. The electrons are heated by the T-G modes (mainly by the pump of course). But they can besides that also be heated by the ion decay waves (remember that this is not the case in the

IA decay process). Linear Landau damping of the IC waves can occur and also nonlinear Landau damping through nonlinear interaction of two coherent IC waves ($\omega_1$, $k_{\parallel 1}$) and ($\omega_2$, $k_{\parallel 2}$) with electrons of velocity $V_e = \frac{\omega_1 - \omega_2}{k_{\parallel 1} - k_{\parallel 2}}$ ; typical value of this quotient being of the same order of magnitude as the mean $V_e$.

### 3.3. Diffusion

In a device as this one the loss of electrons and ions from the plasma comes through ambipolar diffusion to the walls.

When the pump signal is applied to the coupler it is found that there is an increase in the axial density gradient (from the source towards the end plate). A possible explanation for this effect is the following. The electric field of the T-G mode is mainly perpendicular to $\bar{B}_o$ causing increased $T_{e\perp}$ which results in higher radial diffusion. Since more particles are lost radially, by continuity, the axial density gradient increases.

Moreover, the observed enhancement of the noise fluctuations and the broadening of the decay spectra at higher pump powers may lead to enhanced particle cross-field diffusion.

## 4. CONCLUSIONS

Work is now in progress in order to check and extend the obtained experimental results by different methods. The measured values of the ion temperature will be checked by means of a gridded electrostatic energy analyzer. Pulsed measurements will be performed soon in order to determine growth and heating rates.

In conclusion, two parametric instabilities have been simultaneously observed in this linear machine. The first instability results in a T-G electron decay mode and in an IA ion decay mode and is accompanied by an increase in electron and ion temperature ; the pump mode is a vacuum EM wave or a T-G mode depending on the pump frequency. The upper sideband is due to further coupling between the excited IA mode and the T-G mode at the pump frequency. The second parametric instability shows a much broader spectrum and results in a T-G electron decay mode and in an IC ion decay mode and is only accompanied by an increase in the electron temperature. Both parametric decay processes are resonant. Nonresonant ion modes (called quasimodes in Chang and Prokolab, 1974) have not been observed in this experiment. Note that these so-called quasimodes are in fact the result of the scattering of the pump wave off the ions (Compton effect of the pump wave on the ions).

Acknowledgment. We thank Dr P. Leprince of the Laboratoire de Physique des Plasmas, Université de Paris-Sud for interesting discussions and suggesting the explanation for the upper sideband.

## REFERENCES

A bibliography on parametric phenomena can be found in :
PORKOLAB M. (1975) Proceedings of the Twelfth International Conference on Phenomena in Ionized Gases, Eindhoven, The Netherlands, Part 2, p. 86, Department of Physics, Eindhoven University of Technology.

AUBERT A.E., MESSIAEN A.M. and VANDENPLAS P.E. (1971a) Appl. Phys. Lett. 18, 63.

AUBERT A.E., MESSIAEN A.M. and VANDENPLAS P.E. (1971b) Proceedings of the 4th Conference on Plasma Physics and Controlled Nuclear Fusion, Madison, 1971, vol. III p.513, IAEA Vienna.

BARETT P.J. Thesis, Culham Laboratory, Abingdon, Berkshire (England)

BHATNAGAR V.P., VAN OOST G., MESSIAEN A.M., VANDENPLAS P.E. (1975a) Proceedings of the Second International Congress on Waves and Instabilities in Plasmas, Innsbruck, Austria, p. H4, Institut für Theoretische Physik, Universität Innsbruck.

BHATNAGAR V.P., VAN OOST G., MESSIAEN A.M., VANDENPLAS P.E., PAITHANKAR A.S. and VAN HOVE W. (1975b) Proceedings of the VIIth European Conference on Controlled Fusion and Plasma Physics, Lausanne, Vol. I p. 149, Centre de Recherches en Physique des Plasmas, Ecole Polytechnique Fédérale de Lausanne, Switzerland.

BHATNAGAR V.P., VAN OOST G., MESSIAEN A.M., VANDENPLAS P.E. (1976) Plasma Phys. 18, 525.

BROWN I.G. (1976) Plasma Phys. 18, 205.

CHANG R.P.H., PORKOLAB M. (1974) Phys. Rev. Lett. 32, 1227.

CHU T.K., LEE Y.C., PORKOLAB M. (1974) Symposium on Plasma Heating in Toroidal Devices, Varenna, Italy, Lectures and Seminars, Editrice Compositori, Bologna, 1974.

CORTI S., FONTANESI M., GROSSO G., SINDONI E. (1975) Proceedings of the Twelfth International Conference on Phenomena in Ionized Gases, Eindhoven, The Netherlands, Part 1 p. 319, Department of Physics, Eindhoven University of Technology.

EDGLEY P.D., FRANKLIN R.N., HAMBERGER S.M., MOTLEY R.W. (1975) Phys. Rev. Lett. 34, 1269.

FRANKLIN R.N. (1974) Lectures from the Plasma Physics Summer Schools, Culham Laboratory, Abingdon, England.

LATIMER I.D., MILLS J.I. and DAY R.I. (1970) J. Quant. Spectroscop. Radiat. Transfer, 10, 629.

MESSIAEN A.M. and VANDENPLAS P.E. (1973) Plasma Phys. 15, 505.
MIX L.P., LITZENBERGER L.N. and BEKEFI G. (1972) Physics Fuids 15, 2020.
OKABAYASHI M., CHEN K., PORKOLAB M. (1973) Phys. Rev. Lett. 31, 1113
PORKOLAB M. and CHANG R.P.H. (1969) Phys. Rev. Lett. 22, 826
PORKOLAB M., ARUNASALAM V., LUHMANN, Jr., N.C. (1973) Agard Conference Proceedings No.138, 8-1, on Non-linear Effects in Electromagnetic Wave Propagation.
PORKOLAB M. (1974) Symposium on Plasma Heating in Toroidal Devices, Varenna, Italy, Lectures and Seminars, Editrice Compositori, Bologna, 1974.
SATO M. (1972) Physics Fluids 15, 2427.
SATO T., DE KOCK L.C.J.M., WINKEL Th. G.A. (1973) Plasma Phys. 15, 921.

# FLUCTUATIONS AND NONLINEAR WAVE PROCESSES IN A PLASMA

A. G. Sitenko

Institute for Theoretical Physics
Academy of Sciences of the Ukrainian SSR
252130, Kiev, USSR

Electromagnetic fluctuations in a nonequilibrium plasma are investigated taking into account nonlinear wave interactions. The spectral distribution of stationary field fluctuations is determined and the possibility of induced fluctuation enhancement by external fields is shown. The scattering and transformation of waves and radiation in nonequilibrium plasmas are studied on the basis of a generalized kinetic equation allowing for interaction between waves and fluctuation fields in plasmas.

## 1. NONLINEAR EQUATION FOR THE FLUCTUATION FIELD

It is well known that the spectral distribution of electric field fluctuations in a plasma is characterized by a wide maximum in the low-frequency domain, the result of random motion of charged particles, and a set of narrow maxima at higher frequencies. These sharp maxima appear to be due to collective fluctuations, i.e. stochastic plasma eigenoscillations[1]. The collective fluctuation level is determined by the temperature for a plasma in thermodynamic equilibrium. However, this level may grow to very high values, and even tends to infinity in linear theory[2] for a nonequilibrium plasma, especially when the plasma state is close to a kinetic instability. Unlimited growth of the fluctuation oscillation amplitude shows that the linear approximation is invalid under such conditions. That is why it is necessary to take nonlinear effects into account when fluctuation oscillations near a plasma kinetic stability boundary are under consideration. In particular, nonlinear effects govern the investigation of the stationary fluctuation spectrum in a turbulent plasma[3,4].

When one considers fluctuations in a nonequilibrium plasma, it is very convenient to use a nonlinear equation for the electric field that may be derived directly from Maxwell's equations and the equation for microscopic density, which describes the motion of individual charged particles. Separating out the fluctuation part of the microscopic density and solving the above-mentioned set of equations, it is easy to derive the following nonlinear equation for the fluctuation electric field in plasmas:

$$\Lambda_{ij}(k)\,E_j(k) + \sum_{1+2=k} \varkappa^{(2)}_{ijk}(1,2)\,E_j(1)\,E_k(2) + \sum_{1+2+3=k} \varkappa^{(3)}_{ijkl}(1,2,3)\,E_j(1)\,E_k(2)\,E_l(3) + \ldots$$

$$+ \sum_{k'} \delta\epsilon_{ij}(k,k')\,E_j(k') + \sum \delta\varkappa^{(2)}_{ijk}(1,2;k')\,E_j(1)\,E_k(k') + \ldots =$$

$$= -\frac{4\pi i}{\omega}\, j^{\circ}_i(k) \;, \tag{1}$$

where

$$\Lambda_{ij}(k) \equiv \epsilon_{ij}(k) - \left(\frac{k_i k_j}{k^2} - \delta_{ij}\right)\eta^2 \quad \left(\eta^2 = \frac{k^2 c^2}{\omega^2}\right).$$

$\epsilon_{ij}(k)$ is the dielectric permittivity tensor, $\varkappa^{(2)}_{ijk}(1,2)$ and $\varkappa^{(3)}_{ijkl}(1,2,3)$ are nonlinear tensor susceptibilities (k stands for $\omega$ and $\mathbf{k}$, i.e. $k \equiv \omega, \mathbf{k}$; correspondingly $1 \equiv \omega_1, \mathbf{k}_1$; summation in (1) is taken over indexes appearing twice).

Explicit expressions for the dielectric permittivity tensor and nonlinear tensor susceptibilities are as follows:

$$\epsilon_{ij}(k) = \delta_{ij} + \sum \frac{4\pi e^2}{m}\frac{1}{\omega}\int dv\, \frac{v_i}{\omega - kv + io}\, \beta_{jj'}(k)\frac{\partial}{\partial v_{j'}} f_0(v) \;, \tag{2}$$

$$\varkappa^{(2)}_{ijk}(1,2) = \sum \frac{(-i)}{2!}\, \mathcal{P}\, \frac{4\pi e^2}{m}\,\frac{e}{m}\,\frac{1}{\omega_1+\omega_2}\int dv\, \frac{v_i}{\omega_1+\omega_2-(k_1+k_2)v+io}\, \beta_{jj'}(1)\,\cdot$$

$$\cdot \frac{\partial}{\partial v_{j'}}\,\frac{1}{\omega_2 - k_2 v + io}\,\beta_{kk'}(2)\frac{\partial}{\partial v_{k'}} f_0(v) \;, \tag{3}$$

$$\chi^{(3)}_{ijkl}(1,2,3) = \sum \frac{(-i)^2}{3!} \mathcal{P} \frac{4\pi e^2}{m}\left(\frac{e}{m}\right)^2 \frac{1}{\omega_1+\omega_2+\omega_3} \int dv \, \frac{v_i}{\omega_1+\omega_2+\omega_3-(k_1+k_2+k_3)v+i0} \cdot$$

$$\cdot \beta_{jj'}(1) \frac{\partial}{\partial v_{j'}} \frac{1}{\omega_2+\omega_3-(k_2+k_3)v+i0} \beta_{kk'}(2) \frac{\partial}{\partial v_{k'}} \frac{1}{\omega_3-k_3 v+i0} \beta_{ll'}(3) \frac{\partial}{\partial v_{l'}} f_0(v) \,, \quad (4)$$

where

$$\beta_{ij}(k) \equiv \left(1 - \frac{kv}{\omega}\right)\delta_{ij} + \frac{v_i k_j}{\omega} \,.$$

The tensor nonlinear susceptibilities (3) and (4) are symmetric relative to permutations of the arguments and the corresponding indexes.*

We shall take into account in (1) only the first-, second-, and third-order terms in the fluctuation field. Equation (1) links the fluctuation field intensity $E_i(k)$ with the fluctuation current density $j^\circ_i(k)$ arising from the stochastic motion of plasma particles:

$$j^\circ_i(k) = \sum e \int dv \, v_i \, \delta f^\circ_k(v) \,. \quad (5)$$

Random particle motion also causes inhomogeneity of the plasma dielectric properties, because the dielectric permittivity and nonlinear electric susceptibilities depend on the plasma distribution function. Due to this

---

*Nonlinear plasma susceptibilities were first introduced in References 5-7.

effect the two last terms on the lefthand side of (1) appear. The quantities $\delta\varepsilon_{ij}$ and $\delta\chi^{(2)}_{ijk}$ represent variations of the dielectric permittivity and second-order electric susceptibility due to distribution function random fluctuations $\delta f^{\circ}_{k}(v)$.

Using the multiple time scale perturbation analysis, one can derive from (1) a hierarchy of equations which describes the time evolution of different nonlinear processes in plasmas.

## 2. SPECTRAL CORRELATION FUNCTIONS

As follows from its definition, the field fluctuation mean value is equal to zero. The fluctuation field in the linear approximation is described by the pair correlation function $\langle E_i E_j \rangle_k$. To describe field fluctuations in a plasma with a nonlinear wave interaction one must know also the hgiher order correlation functions $\langle E_i E_j E_k \rangle_{k,k'}$ and $\langle E_i E_j E_k E_l \rangle_{k,k',k''}$.

To find a sequence of higher order correlation functions one can use equation (1), where all terms containing fluctuation variations of the dielectric permittivity and nonlinear susceptibility must be transferred to the right-hand side. By multiplying the left and right sides by themselves and then averaging, it is not difficult to derive a set of nonlinear integral equations which determine unambiguously a sequence of correlation functions for field fluctuations in plasmas. The averaging of nonuniform parts of these equations may be performed, due to the correlation weakening principle, using the noncorrelated Liouville distribution. That is why fluctuation sources are expressed directly in terms of the distribution function and current density fluctuation correlation functions for a system of noninteracting particles.

The equation for the field fluctuation pair correlation function $\langle E_i E_j \rangle_k$ derived by such a method is as follows:

$$\Lambda_{ij}(k)\langle E_j E_{j'}\rangle_k - 2\Lambda^{*-1}_{j'i'}(k)\sum_1 \varkappa^{(2)}_{ijk}(1,k-1)\,\varkappa^{(2)*}_{i'j''k'}(1,k-1)\langle E_j E_{j''}\rangle_1 \langle E_k E_{k'}\rangle_{k-1} -$$

$$-\sum_1 a_{ijkk'}(k,1)\langle E_k E_{k'}\rangle_1 \langle E_j E_{j'}\rangle_k =$$

$$= \Lambda^{*-1}_{j'i'}(k)\left\{\sum_1 b_{ii'kk'}(k,1)\langle E_k E_{k'}\rangle_1 + q'_{ii'}(k)\right\}, \qquad (6)$$

where the following notations are used:

$$a_{ijkk'}(k,1) \equiv 4\varkappa^{(2)}_{ij'k}(k-1,1)\,\Lambda^{-1}_{j'i'}(k-1)\,\varkappa^{(2)}_{i'jk'}(k,-1) +$$

$$+ 2\varkappa^{(2)}_{ij'k}(k,0)\,\Lambda^{-1}_{j'i'}(0)\,\varkappa^{(2)}_{i'jk'}(1,-1) - 3\varkappa^{(3)}_{ijkk'}(k,1,-1),$$

$$b_{ii'kk'}(k,1) \equiv \langle \delta\epsilon_{ik}(k,1)\,\delta\epsilon_{i'k'}(k,1)\rangle^{\circ}_{k-1} -$$

$$-\frac{8\pi i}{\omega-\omega_1}\left\{\chi^{(2)}_{ijk}(k-1,1)\Lambda^{-1}_{jl}(k-1)\langle j_l\,\delta\epsilon_{i'k'}(k,1)\rangle^{\circ}_{k-1} - \chi^{(2)}_{i'j'k'}(k-1,1)\Lambda^{*-1}_{j'l}(k-1)\langle\delta\epsilon_{ik}(k,1)\,j_l\rangle^{\circ}_{k-1}\right\},$$

$$q'_{ii'}(k) = \frac{16\pi^2}{\omega^2}\left\{\langle j_i j_{i'}\rangle^{\circ}_{k} + \Lambda^{*-1}_{j i'}(k)\sum_{1} a^{*}_{i'jkk'}(k,1)\langle E_k E_{k'}\rangle^{*}_{1}\langle j_i j_{i'}\rangle^{\circ *}_{k}\right\} + \ldots$$

Supposing that (i) fluctuation fields are small and (ii) the field fluctuation spectral distributions are stationary, one can solve Eq. (6) by means of successive approximations.

## 3. STATIONARY FLUCTUATION SPECTRAL DISTRIBUTIONS AND INDUCED ENHANCEMENT OF FLUCTUATIONS BY AN EXTERNAL FIELD

In the linear approximation the spectral distribution for the fluctuation electric field is determined by the following expression:

$$\langle E_i E_j\rangle_k = \frac{16\pi^2}{\omega^2}\Lambda^{-1}_{ik}(k)\,\Lambda^{*-1}_{jl}(k)\langle j_k j_l\rangle^{\circ}_{k}\,, \qquad (7)$$

i.e. spectral distributions for noninteracting fluctuation electric fields are entirely determined by a pair correlation function for the current density fluctuations in a system of noninteracting particles $\langle j_i j_j \rangle_k^\circ$.

In the plasma transparency domain near eigenfrequencies, (7) takes the form:

$$\langle E_i E_j \rangle_{k\omega} = \pi I_k \{ \delta(\omega - \omega_k) + \delta(\omega + \omega_k) \} e_i e_j^* , \qquad (8)$$

where $e_i$ is a polarization vector for oscillations with frequency $\omega_k$ and

$$I_k = 16\pi^2 \frac{\alpha_k}{\omega_k^2} \frac{\lambda_{ij}^*(\omega_k, k) \langle j_i j_j \rangle_{k\omega_k}^\circ}{| \operatorname{Im} \Lambda(\omega_k, k) |} , \qquad \alpha_k \equiv \frac{\operatorname{Sp} \lambda(\omega_k, k)}{\frac{\partial \operatorname{Re} \Lambda(\omega_k, k)}{\partial \omega_k}} .$$

The quantities $\lambda_{ij}(\omega,k)$ and $\Lambda(\omega,k)$ are defined by the relation

$$\Lambda_{ij} \lambda_{jk} = \Lambda \delta_{ik} .$$

When pair and ternary interactions between the fluctuation fields is taken into account, it appears that the spectral distribution $\langle E_i E_j \rangle_k$ depends not only on the pair correlation function $\langle j_i j_j \rangle_k^\circ$, but also on the ternary and quaternary correlation functions for the current density fluctuations and distribution functions in a system of noninteracting particles. We note that the correction to the spectral distribution (7), caused by a ternary field interaction, is characterized not only by the maxima at eigenfrequencies $\omega_k$, but also by the additional maxima at combination frequencies corresponding to sums and differences of the plasma eigenfrequencies.

When a plasma is in thermal equilibrium, the corrections in the fluctuation spectrum, caused by a fluctuation field nonlinear interaction, are small. But the nonlinear interaction is definitely present in a non-equilibrium plasma, especially when the plasma state is close to a kinetic instability.

It is worth mentioning that the nonlinear wave interaction in plasmas leads to enhancement of the fluctuation level when some eigenoscillations are excited by external sources. We denote the spectral distribution of eigenoscillations excited by external sources by $\langle E_i E_j \rangle_k^{\circ}$. When damping is neglected, $\langle E_i E_j \rangle_k^{\circ}$ has the form of a delta-like set of maxima corresponding to excited oscillation frequencies. Using (6) and taking into account the nonlinear interaction between the fluctuation field and the excited oscillation field in the lowest order (allowing for ternary interaction if necessary), we have for the fluctuation field spectral distribution $\langle \widetilde{E_i E_j} \rangle_k$:

$$\langle \widetilde{E_i E_j} \rangle_k = \left\{ 1 + 2 \operatorname{Re} \Lambda^{-1}_{j'i'}(k) \sum_1 a_{i'j'kk'}(k,1) \langle E_k E_{k'} \rangle_1^{\circ} \right\} \langle E_i E_j \rangle_k , \tag{9}$$

where $\langle E_i E_j \rangle_k$ is the fluctuation field spectral distribution in the absence of oscillations excited by external sources. When the intensity of the excited oscillations

$\langle E_k E_{k'} \rangle^{\circ}_{k}$ is high enough, the fluctuation oscillation intensity (9) can significantly exceed the thermal fluctuation intensity. We notice that potential fluctuation induced enhancement is possible as a result both of potential and nonpotential oscillation excitement. In particular, induced enhancement of potential fluctuations will occur when electromagnetic radiation is pumped into plasmas.

## 4. NONLINEAR FREQUENCY SHIFT AND FLUCTUATION LEVEL SATURATION IN NONEQUILIBRIUM PLASMAS

One of the most important manifestations of the nonlinear wave interaction is a nonlinear frequency shift. Taking this effect into account is important for the investigation of fluctuation oscillations near eigenfrequencies in a nonequilibrium plasma, when fluctuation oscillation intensities are high.

One can consider plasma eigenoscillations using equation (6) with a vanishing righthand side. After contracting this equation, we set the real part of the resulting expression equal to zero and thus obtain a relation which determines the eigenfrequencies with nonlinear corrections:

$$\mathrm{Re}\left\{ \Lambda_{ij}(k) \langle E_j E_i \rangle_k - \right.$$

$$- 2 \Lambda^{*-1}_{ii'}(k) \sum_{1} \chi^{(2)}_{ijk}(1,k-1)\, \chi^{(2)*}_{i'j'k'}(1,k-1) \langle E_j E_{j'} \rangle_1 \langle E_k E_{k'} \rangle_{k-1} -$$

$$- \sum_{1} a_{ijkk'}(k,1) \langle E_k E_{k'} \rangle_1 \langle E_j E_i \rangle_k \Big\} = 0 . \qquad (10)$$

We denote eigenoscillation frequencies taking the nonlinear interaction into account by $\tilde{\omega}_k$. It is obvious that

$$\tilde{\omega}_k = \omega_k + \Delta \omega_k ,$$

where $\Delta\omega_k$ is a contribution to the eigenfrequency due to the nonlinear wave interaction. Supposing that corrections are small compared with the eigenfrequencies, one can easily derive an approximate expression for $\Delta\omega_k$:

$$\Delta \omega_k = \alpha_k \sum_{\omega', k'} e_i^* e_j \, a_{ijkk'}(\omega_k, k; \omega', k') \langle E_k E_{k'} \rangle_{k'\omega'} , \qquad (11)$$

The nonlinear shift (11) corresponds to the resonant fourwave interaction. The shift value is determined by the wave intensity in plasmas. In the interaction of certain types of waves, it may be shown that the main contribution to (11) is given by values of k' close to k. So one can put approximately

$$\Delta \omega_k \simeq \beta_k I_k , \qquad (12)$$

where $\beta_k$ is a coefficient determined by dispersion of the waves under consideration, and $I_k$ is the wave intensity. For example, in the case of longitudinal Langmuir waves we obtain

$$\beta_k = \left(\frac{\partial \operatorname{Re}\epsilon(\omega_k,k)}{\partial\omega_k}\right)^{-1} \operatorname{Re}\left\{2\,\frac{\chi^{(2)}(2\omega_k,2k;-\omega_k,-k)\,\chi^{(2)}(\omega_k,k;-\omega_k,-k)}{\epsilon(2\omega_k,2k)} + \right.$$

$$\left. + 4\,\frac{\chi^{(2)}(\omega_k,k;0,0)\,\chi^{(2)}(\omega_k,k;-\omega_k,-k)}{\epsilon(0,0)} - 3\chi^{(3)}(\omega_k,k;\omega_k,k;-\omega_k,-k)\right\} \quad (13)$$

In the hydrodynamic limit, corresponding to the temperature T = 0, the frequency shift for Langmuir waves is equal to zero[8]. Taking thermal particle motion into account one obtains from (13)*

$$\beta_k = \frac{3}{4\pi^2}\,\frac{e^2 k^7}{m^3\Omega^5}\,T \; .$$

The wave intensity growth causes an increase in the shift value, so taking the frequency shift into account is essential when one studies critical fluctuations in nonequilibrium plasmas with intensities growing infinitely in the linear approximation[10]. The field fluctuation spectral distribution allowing for the nonlinear interaction near plasma eigenfrequencies is determined by the following expression:

$$\langle E_i E_j \rangle_{k\omega} = \pi I_k \{\delta(\omega-\omega_k) + \delta(\omega+\omega_k)\} e_i e_j^* , \quad (14)$$

---

*The nonlinear correction to the Langmuir eigenfrequency, taking electron thermal motion into account, was found in Reference 9.

where

$$I_k = 16\pi^2 \frac{\alpha_k}{\omega_k^2} \frac{\lambda^*_{ij}(\tilde{\omega}_k, k)\langle j_i j_j \rangle^0_{k\tilde{\omega}_k}}{|\mathrm{Im}\,\Lambda(\tilde{\omega}_k, k)|} . \tag{15}$$

As follows from (14), the nonlinear wave interaction causes a shift of the maxima in the electric field fluctuation spectral distribution. The eigenfrequency with the nonlinear correction $\tilde{\omega}_k$ is a function of $I_k$, so relation (15) may be considered as an equation for the fluctuation intensity $I_k$ in the plasma instability domain. The existence of a solution of (15) means that the wave interaction results in the establishment of a stationary turbulent state in a nonequilibrium plasma (unstable in the linear approximation). The solution of (15) then determines a stationary level of the turbulent electromagnetic fluctuations. The field fluctuation intensity level in the instability domain can considerably exceed the thermal one.

## 5. TIME DEPENDENCE OF THE FIELD SPECTRAL CORRELATION FUNCTION

The fluctuation field spectral distribution in an unstable plasma can vary with time both due to linear damping (or enhancement), caused by resonant interaction with particles, and to nonlinear oscillation interactions. If the fluctuation oscillation intensity reaches a high enough level, one has to take into account particle diffusion arising from particle scattering by fluctuation oscillations and leading to a quasilinear relaxation of the particle distribution.

In order to find the time dependence of the fluctuation field spectral distribution one can apply a multiple time scale expansion.* Here a hierarchy of times is introduced, each of which increases successively by the factor $\alpha^{-1}$, where $\alpha$ is a dimensionless interaction constant. Removal of time secularities on a fast time scale leads to an equation which determines the time development of the system on a slower time scale. Applied to the nonlinear equation for the spectral correlation function (6), such a procedure reduces to the formal substituion

$$\Lambda_{ij}(k)\langle E_j E_i\rangle_k \rightarrow \left\{\Lambda_{ij}(k)e_i^* e_j + i\frac{\partial}{\partial\omega}\left(\frac{\partial \mathrm{Re}(\epsilon_{ij}(k)e_i^* e_j)}{\partial t}\right) + \frac{i}{2}\frac{\partial \mathrm{Re}(\Lambda_{ij}(k)e_i^* e_j)}{\partial\omega}\frac{\partial}{\partial t}\right\}\langle E^2\rangle_k . \quad (16)$$

(The spectral distribution $\langle E_i E_j\rangle_k$ near eigenfrequencies may be presented in the form

$$\langle E_i E_j\rangle_k = \langle E^2\rangle_k e_i e_j^* \quad ).$$

Equating the imaginary part of the resulting relation to zero we obtain an equation which determines the field

---

*See e.g. Reference 11.

fluctuation spectral distribution time dependence caused by the fluctuation field nonlinear interaction and by linear dissipation:

$$\frac{1}{2}\frac{\partial \mathrm{Re}\left(\Lambda_{ij}(k)e_i^* e_j\right)}{\partial\omega}\frac{\partial}{\partial t}\langle E^2\rangle_k = -\left\{\mathrm{Im}\left(\epsilon_{ij}(k)e_i^* e_j\right)+\frac{\partial}{\partial\omega}\frac{\partial \mathrm{Re}\left(\epsilon_{ij}(k)e_i^* e_j\right)}{\partial t}\right\}\langle E^2\rangle_k +$$

$$+\mathrm{Im}\sum_1 a_{ijkk'}(k,1)\langle E_k E_{k'}\rangle_1\langle E_j E_i\rangle_k +$$

$$+\mathrm{Im}\,\Lambda^{*-1}_{ii'}(k)\left\{\sum_1\left(2\varkappa^{(2)}_{ijk}(1,k-1)\varkappa^{(2)*}_{i'j'k'}(1,k-1)\langle E_k E_{k'}\rangle_{k-1}+b_{ii'jj'}(k,1)\right)\langle E_j E_{j'}\rangle_1 + q'_{ii'}(k)\right\}. \quad (17)$$

One should consider (17) together with the kinetic equations for particles which govern the distribution function time dependence. Restricting ourselves to the consideration of a stationary distribution of the particles we shall, however, be interested in the time evolution of the field spectral distribution. In the stationary case, equation (17) reduces to (6). Neglecting the fluctuation field sources, i.e. setting $b_{ii'jj'}(k,1)$ and $q_{ii'}(k)$ equal to zero, one obtains from (17) an equation discussed in Reference 6.

The general solution of (17) with neglect of the nonlinear interaction may be presented in the form

$$\langle E_i E_j\rangle_k = \langle E_i E_j\rangle^{\circ}_k + \pi\sum I_k(t)\left\{\delta(\omega-\omega_k)+\delta(\omega+\omega_k)\right\} e_i e_j^*, \quad (18)$$

where $\langle E_i E_j\rangle^{\circ}_k$ is a fluctuation field stationary distribution, determined by the plasma state, and

$$I_k(t) = I_k(0)\, e^{-2\gamma_k t}. \quad (19)$$

$I_k(0)$ is the spectral density, taken at the initial moment of time, and $\gamma_k$ is an eigenoscillation damping coefficient. The summation in (18) should be performed over different eigenoscillations. The exponential time dependence of the eigenoscillation spectral density $I_k(t)$ in (18) results from neglect of the nonlinear interaction.

The fluctuation field spectral distribution $\langle E_i E_j \rangle_k$ in a stable plasma, characterized by stationary particle distributions, tends to a stationary level $\langle E_i E_j \rangle^{\circ}_k$ as $t \to \infty$. The stationary spectrum $\langle E_i E_j \rangle^{\circ}_k$ contains besides noncoherent fluctuations also eigenfluctuation oscillations. The eigenfluctuation oscillation intensity in equilibrium plasmas is determined by the temperature. The stationary fluctuation level in a nonequilibrium plasma is determined by the corresponding solution of (18) and can considerably exceed the thermal one. If the distribution functions are such that $\gamma_k<0$, then the intensity $I_k(t)$ in (18) can reach high values. However neglect of the nonlinear terms in (17) is then unjustified and the $I_k(t)$ time dependence is essentially determined by the nonlinear wave interaction.

## 6. GENERALIZED KINETIC EQUATION FOR WAVES

When the particle distribution functions are fixed and there exists a stationary fluctuation level $\langle E_i E_j \rangle^{\circ}_k$ then it is not difficult to derive from (17) a kinetic equation for waves. Substituting the electric field spectral distribution (18) into (17), one obtains the following equation for the intensity $I_k$ of certain types of oscillations:

$$\frac{1}{2}\frac{\partial I_k}{\partial t} = -\gamma_k I_k +$$

$$+\alpha_k \mathrm{Jm}\sum_{\omega'}\sum_{k'} a_{ii'jj'}(\omega_k,k;\omega',k')e_i^* e_{i'}\left(\pi\delta(\omega'-\omega_{k'})e_j' e_{j'}'^* I_{k'} + \langle E_j E_{j'}\rangle^{\circ}_{k'\omega'}\right)I_k +$$

$$+\pi\alpha_k^2\sum_{k'}\left|\chi^{(2)}_{ijk}(\omega_{k'},k';\omega_{k''},k'')e_i^* e_j' e_k''\right|^2 \delta(\omega_k-\omega_{k'}-\omega_{k''})I_{k'}I_{k''}+$$

$$+\frac{1}{2}\alpha_k^2\sum_{k'}\left\{b_{ii'jj'}(\omega_k,k;\omega_{k'},k')+4\chi^{(2)}_{ijk}(\omega_{k'},k';\omega_{k''},k'')\chi^{(2)*}_{i'j'k'}(\omega_{k'},k';\omega_{k''},k'')\langle E_k E_{k'}\rangle^{\circ}_{k''\omega_{k''}}\right\}\cdot$$

$$\cdot e_i^* e_{i'} e_j' e_{j'}'^* I_{k'}\ , \tag{20}$$

where $k'' = k - k'$ and $\alpha_k = \mathrm{Sp}\lambda(\omega_k,k)\Big/\frac{\partial \mathrm{Re}\Lambda(\omega_k,k)}{\partial\omega_k}$

The wave energy density $W_k$ is linked with $I_k$ by the relation

$$W_k = \frac{1}{16\pi}\frac{\omega_k}{\alpha_k} I_k\ . \tag{21}$$

Equation (20) describes the electromagnetic wave dynamics in a plasma taking into account the wave interaction both with waves and with fluctuation oscillations in the plasma. With neglect of the fluctuation oscillation intensity $\langle E_i E_j\rangle^{\circ}_{k\omega}$ and of the value $b_{ii'jj'}(\omega_k,k;\omega_{k'},k')$, corresponding to the particle distribution fluctuations, (20) reduces to a well-known kinetic equation for waves[5,12,13]. In contrast to the usual kinetic equation, taking into account the wave intensity variations resulting from nonlinear dissipation, decay interaction, and induced wave scattering on plasma particles, the generalized kinetic equation governs also the interaction of waves with fluctuation fields and particle distribution function fluctuations. These effects are described by a part of the second term and by the last term on the righthand side of (20).

The second term on the righthand side of Eq. (20) describes nonlinear wave damping. According to (20), nonlinear interaction between waves and the fluctuation field causes additional nonlinear Landau damping. The last term on the righthand side of (20) describes wave scattering and transformation by the fluctuations of the fields and the particle distributions. In a nonequilibrium plasma with a high fluctuation level such processes can play an essential role.

## 7. SCATTERING AND RADIATION OF WAVES

When wave scattering in a plasma is considered, a current, which induces scattered waves, is usually constructed and used to determine the intensity and distribution of scattered waves[1]. The generalized kinetic equation (20) makes it possible to investigate these processes directly without introducing an inducing current. If the intensities of the incident waves are not too high, one can neglect induced scattering and transformation in (20) and take account only of spontaneous processes, which are described by the last term on the right-hand side of (20). As a result one can derive the following general expression for the scattering or transformation process:

$$d\Sigma = \frac{k^2}{8\pi^3} \frac{\alpha_k \alpha_{k_0}}{\frac{\partial \omega}{\partial k} \frac{\partial \omega_0}{\partial k_0}} \frac{\omega}{\omega_0} \Big\{ \ell_{ii'jj'}(\omega, k; \omega_0, k_0) +$$

$$+ 4\chi^{(2)}_{ijk}(\omega_0, k_0; \omega-\omega_0, k-k_0)\chi^{(2)*}_{i'j'k'}(\omega_0, k_0; \omega-\omega_0, k-k_0)\langle E_j E_{j'}\rangle^{\circ}_{k-k_0, \omega-\omega_0} \Big\} e^{*}_{i} e_{i'} e^{\circ}_{j} e^{\circ *}_{j'} \, d\omega \, do, \tag{22}$$

where $\omega_0, k_0$ and $\omega, k$ are the frequencies and wave vectors for the incident and scattered waves, respectively.

Considering some specific scattering or transformation process one must use explicit expressions for the nonlinear tensor susceptibility $\chi^{(2)}_{ijk}$ and the tensor element $\ell_{ii'jj'}$. When the wave phase velocities are large compared with electron thermal velocities, it is necessary to use a series expansion of the relation between these two velocities. Cross-sections are expressed in terms of the density and electron velocity fluctuation spectral distributions in plasmas. Expression (22) determines wave scattering and transformation cross-sections in the general case and is valid for the description of the above-mentioned processes both with and without external fields.

The generalized kinetic equation (20) makes it possible to describe not only wave scattering and transformation by density and particle velocity fluctuations, but also scattering of charged particles by electric field fluctuations in plasmas. Such charged particle scattering is accompanied by radiation. It is nothing but plasma bremsstrahlung, for electric field fluctuations are directly connected with the microscopic motion of individual charged particles. Besides bremsstrahlung resulting from particle scattering by electric field fluctuations, additional radiation can also be present in a nonequilibrium plasma, caused by transformation of fluctuation longitudinal waves into transverse electromagnetic waves. Such spontaneous emission is possible only in a nonequilibrium plasma, because all radiation in equilibrium plasmas is of the Rayleigh type. Spontaneous emission intensity reaches anomalous values when the nonequilibrium plasma state is close to instability.

The author expresses his sincere gratitude to O. D. Kocherga for her help in preparation of this paper.

REFERENCES

1. A. G. Sitenko. Electromagnetic Fluctuations in Plasma. Academic Press, New York, 1967.
2. S. Ichimaru, D. Pines, N. Rostoker. Phys. Rev. Letters 8, 231, 1962.

3. A. G. Sitenko. Plasma Physics (USSR) 1, 45, 1975.

4. A. G. Sitenko, O. D. Kocherga. Preprint ITP-76-10P, Kiev, 1976 (in Russian).

5. B. B. Kadomtsev. Plasma Turbulence. Academic Press, New York, 1965.

6. L. M. Gorbunov, V. V. Pustovalov, V. P. Silin. JETP 47, 1437, 1964 (in Russian).

7. A. A. Galeev, V. I. Karpman, R. Z. Sagdeev, Nucl. Fusion 5, 20, 1965.

8. A. I. Akheizer, G. Ja. Liubarsky. Reports of the Academy of Sciences of the USSR, 80, 193, 1951; R. V. Polovin. JETP 31, 354, 1956 (in Russian).

9. H. Wilhelmsson. Phys. Fluids 4, 335, 1961.

10. A. G. Sitenko. Physica Scripta 7, 193, 1973.

11. R. C. Davidson. Methods in Nonlinear Plasma Theory. Academic Press, New York, 1972

12. R. Z. Sagdeev, A. A. Galeev. Nonlinear Plasma Theory. Benjamin, New York, 1969.

13. V. N. Tsytovich. Turbulent Plasma Theory. Atomizdat, Moscow, 1971 (in Russian).

# A KINETIC FOUNDATION OF THE THEORY OF PARAMETRIC INSTABILITIES IN A PLASMA

R. Balescu

Faculté des Sciences, C.P. 231. Université Libre de Bruxelles. Association Euratom-Etat Belge, 1050 Bruxelles

I. Paiva-Veretennicoff

Fakulteit van de Wetenschappen, Vrije Universiteit Brussel. Pleinlaan 2, 1050 Brussel

## 1. INTRODUCTION

This work is a first step in a long range programme on the foundation of the theory of parametric instabilities on a firm statistical mechanical basis. The present study is restricted to the parametric instabilities resulting from the coupling of plasma-dynamical modes through an external, time-dependent electromagnetic field. The plasmadynamical modes (P.D-modes) are defined, in absence of the field, to be the least damped collective excitations of the system , in the limit of small wave-vectors. In presence of the field, the coupling of these modes potentially leads to the instabilities with the lowest thresholds.

Section 2 of this paper is devoted to the description of the fourteen plasmadynamical modes on a microscopic level. The starting point is a complete set of kinetic equations including the Vlassov and collision operators, linearized around thermal equilibrium, in absence of the external field. The fourteen lowest eigenvalues of this set can be shown to be in one-to-one correspondence with those of the macroscopic linearized plasmadynamical equations. This correspondence is only achieved if the complete set of conservation equations, -including the energy conservation- is used and if all the dissipative processes (friction, viscosity, heat conduction and -transfer) are properly taken into account. The P.D.-modes can be classified into wave-like and zero-frequency modes. The latter, which are rarely discussed in the literature, are candidates in coupling processes leading to possible parametric instabilities. (see section 4).

Section 3 deals with the derivation of the P.D-mode coupling equations. These are easily obtained from the full kinetic equations, using the assumption that the plasma is in a "P.D-state", when the field is switched on. Two different sets of coupling equations arise according to the order of magnitude of the overall acceleration produced by the field. Both are illustrated in the simplest possible cases.

In section 4 the couplings of longitudinal plasma waves and (transverse zero frequency) shear modes will be shown to giverise to a new type of instability : the "zero-mode" instability.

Section 5 is devoted to the coupling of longitudinal plasma waves to ion-acoustic waves. The role of the inclusion of the energy conservation equation is discussed in some detail. It is shown that quantitative differences appear as compared to the traditional treatments (1).
The present paper is a short review of our methods and results. The details of the calculations will be published elsewhere (2).

## 2. THE PLASMADYNAMICAL MODES

We consider an infinite, fully ionized plasma, consisting of electrons (mass : $m_e$, charge : -e) and singly charged ions (mass : $m_i$, charge : +e), characterized by their positions $\underline{q}$ and velocities $\underline{v}$. In order to specify completely the statistical state of the plasma, the transverse part of the self-consistent field must also be treated as an independent entity (3). The latter is described by a set of oscillators. For each value of the wavevector $\underline{\kappa}$, they are characterized by two action variables $(\eta_1, \eta_2)$ and two angle variables $(\xi_1, \xi_2)$ corresponding to the two transverse polarisations. It is assumed that the state of the system is completely determined by the knowledge of the distribution functions of the electrons, $f^e(\underline{q},\underline{v};t)$, of the ions, $f^i(\underline{q}\underline{v};t)$ and of the field oscillators, $f^F(\xi_1, \xi_2, \eta_1, \eta_2;\underline{\kappa};t)$. The evolution of the distribution functions in presence of the external field is governed by the following set of coupled kinetic equations :

$$\partial_t f^\alpha = E_\alpha(t) f^\alpha + \Phi_\alpha f^\alpha + \sum_{\beta=e,i} V_{\alpha\beta} f^\alpha f^\beta$$

$$+ V_{\alpha F} f^\alpha f^F + \sum_{\beta=e,i} K^E_{\alpha\beta}(t) f^\alpha f^\beta + K_{\alpha F} f^\alpha f^F \qquad (2.1)$$

$$\partial_t f^F = \Phi_F f^F + \sum_{\alpha=e,i} V_{F\alpha} f^F f^\alpha + \sum_{\alpha=e,i} K_{F\alpha} f^F f^\alpha \qquad (2.2)$$

The operator $E_\alpha(t)$ is the <u>external field term</u> :

$$E_\alpha(t) f^\alpha = -\frac{e_\alpha}{m_\alpha} \underline{E}^{ext}(t) \cdot \frac{\partial}{\partial \underline{v}} f^\alpha(\underline{q},\underline{v};t) \qquad (2.3)$$

where $\underline{E}^{ext}(t)$ is an external, time-dependent, spatially homogeneous electric field.

The symbols $\Phi_\alpha$ and $\Phi_F$ denote the <u>free-flow operator</u>, describing the motion of the free particles and of free field oscillators :

$$\Phi_\alpha f^\alpha = -\underline{v} \cdot \frac{\partial}{\partial \underline{q}} f^\alpha(\underline{q},\underline{v};t) \qquad (2.4)$$

$$\Phi_F f^F = -\frac{c\kappa}{2\pi} \left(\frac{\partial}{\partial \xi_1} + \frac{\partial}{\partial \xi_2}\right) f^F(\xi_1, \xi_2, \eta_1, \eta_2;\underline{\kappa};t) \qquad (2.5)$$

where c is the speed of light in vacuo.
The <u>Vlassov operator</u> is defined by the set of integro-differential operators $V_{\alpha\beta}$, $V_{\alpha F}$, $V_{F\alpha}$ given in ref. (2). For brevity we shall only recall here the particle-particle operator

$$V_{\alpha\beta} f^\alpha f^\beta = \frac{e_\alpha e_\beta}{m_\alpha} \frac{\partial}{\partial \underline{q}} \cdot \frac{\partial}{\partial \underline{v}} \int d^3\underline{q}_2 \, d^3\underline{v}_2 \, |\underline{q}-\underline{q}_2|^{-1}$$

$$f^\alpha(\underline{q},\underline{v};t) f^\beta(\underline{q}_2,\underline{v}_2;t) \qquad (2.6)$$

The symbol $K^{E}_{\alpha\beta}(t)$ denotes the particle-particle collision operator, whereas $K_{\alpha F}$ and $K_{F\alpha}$ describe the "collisions" involving particles and field oscillators. The latter will not be discussed here. Interparticle collisions will be treated in the Landau approximation of weak coupling. In the presence of an intense external field the perturbation of the particle trajectories during the collision process must be included, with the result :

$$K^{E}_{\alpha\beta}(t)f^{\alpha}f^{\beta}=2e^{2}_{\alpha}e^{2}_{\beta}\int_{0}^{\infty}d\tau\int d^{3}\underline{v}_{2}\int d^{3}\underline{l}\; l^{-4}$$

$$\underline{l}.\left(\frac{1}{m_{\alpha}}\frac{\partial}{\partial\underline{v}}-\frac{1}{m_{\beta}}\frac{\partial}{\partial\underline{v}_{2}}\right)\exp\left[i\underline{l}.(\underline{v}-\underline{v}_{2})\tau-i\underline{l}.\underline{\gamma}_{\alpha\beta}(t,t-\tau)\right]$$

$$\underline{l}.\left(\frac{1}{m_{\alpha}}\frac{\partial}{\partial\underline{v}}-\frac{1}{m_{\beta}}\frac{\partial}{\partial\underline{v}_{2}}\right)f^{\alpha}(\underline{q},\underline{v};t)f^{\beta}(\underline{q},\underline{v}_{2};t) \qquad (2.7)$$

where

$$\underline{\gamma}_{\alpha\beta}(t,t-\tau)=\left(\frac{e_{\alpha}}{m_{\alpha}}-\frac{e_{\beta}}{m_{\beta}}\right)\int_{t-\tau}^{t}d\sigma\int_{\sigma}^{t}d\theta\;\underline{E}^{ext}(\theta) \qquad (2.8)$$

In order to define the plasmadynamical (P.D.) modes, we need to study the evolution of the plasma in absence of the external field. For a system near thermal equilibrium, the distribution functions, localised in physical space, can be written as follows :

$$f^{\alpha}(\underline{q},\underline{v};\underline{x},t)=f^{\alpha}_{eq}(v)\{1+\chi^{\alpha}(\underline{q},\underline{v};x,t)\}$$

$$f^{F}(\xi_{1},\xi_{2},\eta_{1},\eta_{2};\underline{K};t)=f^{F}_{eq}(\eta_{1},\eta_{2};\underline{K})\{1+\chi^{F}(\xi_{1},\xi_{2},\eta_{1},\eta_{2};\underline{K},t)\}$$

$$(2.9)$$

The small deviations, $\chi^{\alpha}$ and $\chi^{F}$, can be grouped in a vector $\underline{\chi}(t)$, which evolves according to a matrix (linearized) kinetic equation :

$$\partial_{t}\underline{\chi}=(\underline{\underline{\Phi}}+\underline{\underline{V}}^{L}+\underline{\underline{K}}^{L})\underline{\chi} \qquad (2.10)$$

The explicit forms of these operators are given in ref. (2).

The P.D-modes are the eigenstates of the kinetic matrix -in the Fourier representation- which have the longest life time. By definition, these are the only ones surviving in the plasmadynamical (P.D.) regime.

As first shown by Resibois (4) in the case of neutral fluids, they arise from the extension of the collisional invariants of the homogeneous collision operator, into the inhomogeneous region. The P.D. eigenvalues are in one-to-one correspondence with the eigenvalues of the macroscopic linearized P.D. equations, describing the evolution of the particle number densities $n^{\alpha}$, momentum densities $g_{r}^{\alpha}$, energy densities $w^{\alpha}$, and the transverse, self consistent electric and magnetic fields.

The explicit expressions of the P.D-modes, $\overline{\Psi}_{n,k}$ (n=1,...14), are obtained by a perturbation method valid for small wave vectors of the excitations and small ion-electron collision frequencies : they appear as linear combinations of the collisional invariants $\underline{\Psi}_n^o$ (2). The P.D-modes are classified as waves or zero-frequency modes, according to the imaginary part of their eigenvalue $\Omega_n$.

The wave-like P.D-modes are well-known :
two longitudinal plasma waves, $\Omega_{1,2} = \mp i\omega_p - \lambda_p$,
four transverse plasma waves, $\Omega_{7,8} = \Omega_{11,12} = \mp i\omega_T - \lambda_T$
two ion-acoustic waves, $\Omega_{3,4} = \mp i\,\omega_s - \lambda_s$.

The zero-frequency modes describe pure decay processes having well known analogues in hydrodynamics :
one heat conduction mode $\Omega_5 = -\lambda_5$
one heat transfer mode $\Omega_6 = -\lambda_6$
two friction modes $\Omega_9 = \Omega_{13} = -\lambda_F$
two shear modes $\Omega_{10} = \Omega_{14} = -\lambda_V$

The damping rates and frequencies appearing in these expressions have been evaluated in detail (2).

## 3. THE MODE-COUPLING EQUATIONS

The derivation of the mode-coupling equations, is based on the fundamental assumption that, even in presence of the external field, the plasma is in a P.D. state. This means that the distribution functions are fully determined by the fourteen independent amplitudes of the microscopic P.D-modes, instead of an infinite number of degrees of freedom. In other words, the strongly damped non-P.D-modes cannot be excited by the external field, at least for not too high intensities.

Two different pictures arise according to the importance of the velocity of the particles in the external field, $\underline{\beta}^{\alpha}(t)$ :

$$\underline{\beta}^{\alpha}(t)=\frac{e_{\alpha}}{m_{\alpha}}\int_{0}^{t} d\tau\ \underline{E}^{ext}(\tau) \tag{3.1}$$

If $\underline{\beta}^{\alpha}(t)$ can be considered as a first order perturbation, the distribution functions $f^{\alpha}$ are represented as follows :

$$f^{\alpha}(\underline{q},\underline{v};\underline{x},t)=f^{\alpha}_{eq}(v)+f^{\alpha}_{eq}(v)\sum_{n=1}^{14}\int d^{3}k\ e^{i\underline{k}\cdot\underline{x}}\ .$$

$$\cdot\ a_{n\underline{k}}(t)\ e^{-i\omega_{n\underline{k}}t}\ \Psi^{\alpha}_{n\underline{k}}(\underline{q},\underline{v}) \tag{3.2}$$

If, on the other hand, $\underline{\beta}^{\alpha}(t)$ is a zeroth order quantity, the distribution functions are represented in a different way :

$$f^{\alpha}(\underline{q},\underline{v};\underline{x},t)=\exp\left[-\underline{\beta}^{\alpha}\cdot\frac{\partial}{\partial\underline{v}}\right]f^{\alpha}_{eq}(v)$$

$$+\exp\left[-\underline{\beta}^{\alpha}\cdot\frac{\partial}{\partial\underline{v}}\right]f^{\alpha}_{eq}(v)\sum_{n=1}^{14}\int d^{3}\underline{k}\,e^{i\underline{k}\underline{x}}\,a_{n\underline{k}}(t)e^{-i\omega_{n\underline{k}}t}$$

$$\Psi^{\alpha}_{n\underline{k}}(\underline{q},\underline{v}) \tag{3.3}$$

In (3.2) and (3.3) $\Psi^{\alpha}_{n\underline{k}}(\underline{q},\underline{v})$ represent the α-th component of the n-th P.D. eigenvector (similar formulas hold for the transverse oscillator distribution function).

The P.D. amplitudes $a_{n\underline{k}}(t)$ (n=1,...14) satisfy a set of equations which are easily obtained by substitution of (3.2) into the full kinetic equations (2.1) and (2.2) and use of the orthogonality properties of the P.D. eigenvectors (2) :

$$\dot{a}_{n\underline{k}}+\lambda_{n\underline{k}}a_{n\underline{k}}=\sum_{m=1}^{14}\ e^{i(\omega_{n\underline{k}}-\omega_{m\underline{k}})t}\ .$$

$$\cdot\left[\Phi_{n\underline{k},m\underline{k}}+E_{n\underline{k},m\underline{k}}(t)+\Lambda^{E}_{n\underline{k},m\underline{k}}(t)\right]a_{m\underline{k}}$$

$$+\sum_{m,s=1}^{14}\sum\int d^{3}\underline{k}'e^{i(\omega_{n\underline{k}}-\omega_{m\underline{k}'}-\omega_{s\underline{k}-\underline{k}'})t}$$

$$\cdot\left[\Delta+\Gamma^{E}(t)\right]_{n\underline{k},m\underline{k}',s\underline{k}-\underline{k}'}a_{m\underline{k}'}a_{s\underline{k}-\underline{k}'}+$$

$$+\delta(\underline{k})e^{i\omega_{n\underline{k}}t}\ \Sigma^{E}_{n\underline{k}}(t) \tag{3.4}$$

Here $\lambda_{n\underline{k}}$ are the linear damping rates of the modes $\Phi_{n\underline{k},m\underline{k}}$ originates from the flow-term, $E_{n\underline{k},m\underline{k}}(t)$ describes the direct action of the field, $\Delta_{n\underline{k},m\underline{k}',s\underline{k}-\underline{k}'}$ represents the non-linear Vlassov and collisional couplings, and all the other coefficients take into account the collisional effects modified by the external field (2).

The amplitudes $\bar{a}_{n\underline{k}}(t)$ in the second picture satisfy equations of precisely the same form as (3.4). However the coupling coefficients have quite different values. For instance, in the representation (3.2), all contributions of the flow-term $\Phi_{n\underline{k},m\underline{k}}$ vanish. On the contrary, in the representation (3.3) the matrixelements $E_{n\underline{k},m\underline{k}}$ vanish. The values of some of the coupling coefficients will be given in the following sections.

Eqs. (3.4) are of the general type of mode-coupling equations as derived e.g. by Wilhelmsson (5), Stenflo (6) or Lashmore-Davies (7). They have a rather peculiar appearance, being in some respect intermediate between a microcopic and a macroscopic description. As they deal with amplitudes depending only on time, not on the phase space variables, they look like those which could be derived from the macroscopic two-fluid equations. Yet, several features could not possibly be derived from a fluid theory. It should also be stressed that the whole microscopic information within the P.D. approximation, is contained in eqs. (3.4). In particular, the deviations of the distribution function from the reference-state, can be determined in detail from a knowledge of the P.D-amplitudes, through eqs. (3.2) or (3.3).

In the present work, we shall illustrate some of the many processes contained in the mode-coupling equations. To this effect, we shall neglect all the non linear couplings among the modes, and also all the collisional effects. We therefore only retain the first two terms in the r.h.s. of eq. (3.4).
We also assume that the external field has a particular time-dependence of the form :

$$\underline{E}^{ext}(t) = 2\underline{E}\cos\omega_0 t \tag{3.5}$$

where $\underline{E}$ is a real constant vector. Such a function simulates an electromagnetic wave in the dipole approximation.

It is well known that the most effective coupling is obtained for interactions of normal modes, the frequencies of which satisfy a quasi-resonance condition. In the first two terms of the r.h.s. of eqs. (3.4) combined with (3.5), this condition amounts to

$$\omega_{n\underline{k}} - \omega_{m\underline{k}} \pm \omega_o \simeq 0 \qquad (3.6)$$

If only these interactions are retained, eq. (3.4) reduces to the simple form :

$$\dot{a}_n + \lambda_n a_n = \sum_m^{(R)} E_{n,m} a_m \qquad (3.7)$$

Similarly, the equations for the amplitudes $\bar{a}_n$ in the second representation (3.3) take the form :

$$\dot{\bar{a}}_n + \lambda_n \bar{a}_n = \sum_m^{(R)} \phi_{n,m} \bar{a}_m \qquad (3.8)$$

The wave-vector being the same in all coefficients appearing in these equations, the subscript $\underline{k}$ will henceforth be dropped.
In the next sections, these equations will be illustrated in two particular cases.

## 4. THE ZERO-MODE INSTABILITY

The "traditional" parametric instabilities involve mutual couplings of wave-like P.D. modes. There is however no reason to exclude the zero-frequency modes from the coupling game. They contribute to eqs. (3.7) on the same footing as the oscillating ones, provided the coupling coefficients are strong enough.
Obviously, the resonance condition (3.6) can be easily satisfied for interactions involving one zero-frequency mode.

For frequencies $\omega_o$ of the external field in the infrared or visible domain, the resonance condition can be satisfied in usual experimental situations by a coupling of a longitudinal (n=1,2) or a transverse (n=7,8, 11,12) wave to any of the six zero-frequency modes (m=5,6,9,10,13,14). It was shown in ref. 2 that an instability can arise from the interaction of the two longitudinal plasma waves (n=1,2) with the shear modes (m=10,14). Restricting ourselves to this process, eqs. (3.7) reduce to :

$$\omega_e^{-1}\dot{a}_{\substack{1\\2}}+\bar{\lambda}_p a_{\substack{1\\2}}=-\frac{1}{3}(2\mu)^{1/2}\kappa(\varepsilon_1 a_{10}+\varepsilon_2 a_{14})$$

$$\omega_e^{-1}\dot{a}_{\substack{10\\14}} + \bar{\lambda}_V\, a_{\substack{10\\14}} = -(\mu/2)^{1/2}\kappa\varepsilon_{\substack{1\\2}}\ (a_1+a_2) \tag{4.1}$$

The coefficients appearing in the r.h.s. of these equations result from the explicit evaluation of the matrix elements $E_{n,m}$ (2). To dominant order in $K=kc_e/\omega_e$, where $\omega_e$ is the electron-plasma frequency, and $c_e$ is the electron thermal speed, the damping rates $\bar{\lambda}_p$ of the longitudinal plasma waves are (2) :

$$\bar{\lambda}_p = \frac{1}{2} \qquad (nm_e\omega_e)^{-1}\zeta \tag{4.2}$$

and those of the shear modes $\bar{\lambda}_V$ :

$$\bar{\lambda}_V = \mu\kappa^2(\eta_e+\eta_i)(\omega_e/nm_e c_e^2) \tag{4.3}$$

Here $\zeta$ is the electron-ion friction coefficient and $\eta_e$ ($\eta_i$) is the electron (ion) viscosity. $\mu$ is the mass ratio $m_e/m_i$.

The external field components $E_j$ ($j=1,2,3$) enter through the dimensionless parameters :

$$\varepsilon_j = \frac{e\,E_j}{\omega_e m_e c_e} \qquad j = 1,2,3 \tag{4.4}$$

The direction $x_3$ is chosen parallel to the wave-vector $\underline{k}$ of the normal modes.

The eigenvalue problem associated with eq. (4.1) leads to the following dispersion equation :

$$(\Lambda+\bar{\lambda}_p)\ (\Lambda+\bar{\lambda}_V)-\frac{2}{3}\mu\kappa^2(\varepsilon_1^2 + \varepsilon_2^2) = 0 \tag{4.5}$$

Obviously, there exists an unstable root to this equation, provided that the transverse external field intensity $\varepsilon_1^2 + \varepsilon_2^2$ exceeds the threshold value

$$\varepsilon_o^2 = \frac{3}{2}\ \frac{\bar{\lambda}_V\bar{\lambda}_p}{\mu\kappa^2} \tag{4.6}$$

This effect will be called the <u>zero-mode instability</u>.

The threshold value $\varepsilon_o^2$ can be expressed explicitly in terms of the transport coefficients, by using eqs. (4.2) and (4.3) :

$$\varepsilon_o^2 = \frac{3}{4}\,\zeta(\eta_e + \eta_i)\,(n\,m_e\,c_e)^{-2} \qquad (4.7)$$

It is independent of $\kappa^2$.

The growth rate of the zero-mode instability near the threshold is given by :

$$\Lambda = \frac{4}{3}\,\frac{\mu}{\zeta}\,(n\,m_e\,\omega_e)\kappa^2\,(\varepsilon_1^2 + \varepsilon_2^2 - \varepsilon_o^2) \qquad (4.8)$$

It is independent of the viscosity and is proportional to $\kappa^2$ : long waves grow slower.

Numerical estimates of the threshold intensity can be obtained by using Braginsky's expressions for the transport coefficients (2). It is found that these thresholds are rather high, but well within the reach of the present-day lasers.

The zero-mode instability might be considered as a typical "thermal effect". The usual macroscopic two-fluid model, involving only the conservation equations for number density and momentum density, cannot account for this effect. The addition of the energy equation is quite essential : it is the Joule dissipation term that provides the coupling mechanism for the onset of this new instability.

## 5. THERMAL EFFECTS ON NISHIKAWA'S INSTABILITIES

Eq. (3.8) is illustrated on the example of the resonant couplings of longitudinal plasmawaves (n=1,2) to the ion-acoustic waves (m=3,4). (1) (7)
A mismatch $\delta$ is introduced by :

$$\omega_e\delta = \omega_o - \omega_e \qquad (5.1)$$

After the explicit evaluation of the coupling coefficients and a redefinition of the P.D. mode amplitudes, according to ref. 7 :

$$A_{1,2} = e^{\pm i\delta\omega_e t}\,\bar{a}_{1,2}$$
$$A_{3,4} = e^{\mp i\omega_s t}\,\bar{a}_{3,4} \qquad (5.2)$$

Eqs. (3.8) reduce to :

$$\omega_e^{-1}\,\dot{A}_{\frac{1}{2}} + (\bar{\lambda}_p \mp i\delta)A_{\frac{1}{2}} = \mp\frac{\varepsilon_3\kappa}{2}(A_3+A_4)$$

$$\omega_e^{-1}\,\dot{A}_{\frac{3}{4}} + (\bar{\lambda}_s \pm \frac{i\omega_s}{\omega_e})A_{\frac{3}{4}} = \pm\,\frac{3}{20}\,\varepsilon_3\,\frac{\omega_s}{\kappa\omega_e}(A_1+A_2) \qquad (5.3)$$

Most of the quantities appearing in this set of equations have been defined in section 4. The frequency and damping rate of the ion-sound modes are given by (2) :

$$\omega_s = \sqrt{\frac{10}{3}}\sqrt{\mu}\,\kappa\,\omega_e \qquad (5.4)$$

and

$$\bar{\lambda}_s = \frac{4}{3}\mu\,(\eta_e+\eta_i)(\,\omega_e/nm_e c_e^2)\kappa^2$$

$$+\frac{2}{15}(\lambda_e+\lambda_i)\;\;(\omega_e/nk_B c_e^2)k^2 \qquad (5.5)$$

where $\lambda_e$ ($\lambda_i$) is the electron (ion) heat conductivity. The dispersion equation obtained from (5.3) is, to dominant order in $\kappa$ :

$$\left[(\Lambda+\bar{\lambda}_p)^2+\delta^2\right]\left[(\Lambda+\bar{\lambda}s)^2+(\frac{\omega_s}{\omega_e})^2\right]+\frac{1}{2}\,\varepsilon_3^{\,2}(\frac{\omega_s}{\omega_e})^2\;\delta = 0 \quad (5.6)$$

This equation agrees exactly with the one first obtained by Nishikawa (1), from a macroscopic two-fluid model, excluding the energy equation. An elegant way of studying the stability properties associated with eq. (5.6), is the application of the extended Routh-Hurwitz theorem (9). It provides directly the number of unstable roots of any algebraic equation.
The results of this analysis are well-known. For $\delta<0$ there is an aperiodic instability whenever $\varepsilon_3^2$ exceeds a threshold value $\varepsilon_{IN}^2$ whose minimum value is

$$\varepsilon_{IN}^2 = 4\;\bar{\lambda}_p \qquad (5.7)$$

and the corresponding growth rate, near the threshold is :

$$\gamma_{IN} = \frac{1}{4}\;(\varepsilon_3^2 - \varepsilon_{IN}^2)\;\omega_e \qquad (5.8)$$

On the other hand, for $\delta>0$, there is a pair of imaginary unstable roots, corresponding to the parametric decay instability. Its minimum threshold is :

$$\varepsilon^2_{II\ N} = \frac{32\sqrt{3}}{9} \frac{\overline{\lambda}^2_p \overline{\lambda}_s}{\omega_s^2} \omega_e^2 \tag{5.9}$$

and the growthrate, near this threshold :

$$\gamma_{II\ N} = \frac{9}{32\sqrt{3}} \frac{\omega_s^2}{\overline{\lambda}^2_p \omega_e^2} (\varepsilon_3^2 - \varepsilon^2_{II\ N}) \tag{5.10}$$

One might be tempted to interpret these results as an absence of non-trivial thermal effects in this coupling process. However, interesting features appear, when the connection between the amplitudes $\overline{a}_{n\underline{k}}$ and the macroscopic variables is more closely analysed. As pointed out in section 2, the P.D. modes $\overline{\Psi}_{n\underline{k}}$ are linear combinations of the collisional invariants $\underline{\Psi}^o_n$, in Fourier picture.
This means that the distribution functions f $(\underline{q},\underline{v};\underline{x},t)$ can also be represented, instead of (3.3) by the equation :

$$f^\alpha(\underline{q},\underline{v};\underline{x},t) = \exp\left[-\underline{\beta}^\alpha \cdot \frac{\partial}{\partial \underline{v}}\right] f^\alpha_{eq}(v)$$

$$+\exp\left[-\underline{\beta}^\alpha \cdot \frac{\partial}{\partial \underline{v}}\right] f^\alpha_{eq}(v) \sum_{n=1}^{14} c_n(\underline{x}+\underline{q};t)\overline{\Psi}_n^{o\alpha}(\underline{v}) \tag{5.11}$$

where (2) $$\Psi^{o\alpha}_{n,\underline{k}}(\underline{v},\underline{q}) = e^{i\underline{k}\underline{q}}\ \overline{\Psi}^{o\alpha}_n(\underline{v}) \tag{5.12}$$

The coefficients $c_n(\underline{x},t)$ can be shown to be very simply related to the macroscopic plasmadynamical variables, as defined in section 2.
For instance :

$$\delta n^e \equiv n^e - n = c_1(\underline{x},t)n$$

$$\delta g_r^e \equiv g_r^e - m_e n \beta_r^e(t) = nc_e m_e \; c_{r+1}(\underline{x},t) + nm_e \beta_r^e(t) c_1(\underline{x},t)$$

$$\delta W^e \equiv W^e - \frac{1}{2} m_e n \beta^{e2}(t) - \frac{3}{2} nk_B T$$

$$= \frac{3}{2} m_e n c_e^2 c_5(\underline{x},t) + m_e n c_e \beta_r^e(t) c_{r+1}(\underline{x},t)$$

$$+ (\frac{1}{2} m\beta^{e2}(t) + \frac{3}{2} k_B T) n \; c_1(\underline{x},t) \qquad (5.13)$$

$n$ and $T$ are, respectively, the equilibrium number density and temperature.

In absence of the field, or when $\underline{\beta}^\alpha(t)$ are small quantities, (of the order of $\delta \underline{g}^\alpha / nm_\alpha$), the coefficients $c_{r+1}$ and $c_5$ are exactly equal resp. to the momentum density fluctuations $\delta g_r^e$ and the properly reduced <u>total energy fluctuation</u> :

$$\delta w^e = \delta W^e - \frac{3}{2} k_B T \; \delta n^e \qquad (5.14)$$

But, when $\underline{\beta}^\alpha(t)$ is a finite quantity, this correspondence fails. The $c_n(\underline{x},t)$ appear as the "internal" macroscopic variables per species. For instance, $c_5(\underline{x},t)$ is now associated with the <u>internal energy</u>.

The mode-coupling equations (3.8) are clearly equivalent to those obtained from a macroscopic treatment in terms of number densities $\delta n^\alpha$, the momentum densities $\delta g_r^\alpha$, and the <u>temperatures</u> $\delta T^\alpha$.(✱)
The coupling coefficients $\Phi_{n,m}$ in (3.8) thus exclude all couplings between the P.D. modes arising from the acceleration of the particles in the field. The latter would be obtained from a macroscopic treatment in terms of the densities $\delta n^\alpha, \delta g_r^\alpha$ and <u>total energy density</u> $\delta w^\alpha$. This is also clear from (5.13) as the link between the two sets of variables involves the external field, through the time dependent velocities $\underline{\beta}^\alpha(t)$.
The P.D. mode analysis, starting from the new set of variables, $(\delta n^\alpha, \delta g_r^\alpha, \delta w^\alpha)$ yield the following set of coupling equations, which replace eqs. (5.3) :

(✱) This result was first obtained from a macroscopic treatment by C.N. Lashmore-Davies (private communication).

$$\omega_e^{-1}\dot{B}_{\frac{1}{2}}+(\bar{\lambda}_p \mp i\delta)B_{\frac{1}{2}} = \mp\frac{\varepsilon_3 \kappa}{2}(B_3+B_4) \qquad (5.15)$$

$$\omega_e^{-1}\dot{B}_{\frac{3}{4}}+(\bar{\lambda}_s \pm \frac{i\omega_s}{\omega_e})B_{\frac{3}{4}} = \frac{\varepsilon_3\delta}{10\kappa}(-B_1+B_2) \pm \frac{1}{20}\frac{\varepsilon_3\omega_s}{\kappa\omega_e}(B_1+B_2) \qquad (5.16)$$

The corresponding dispersion equation is quite different from (5.6)

$$\left[(\Lambda+\bar{\lambda}_p)^2+\delta^2\right]\left[(\Lambda+\bar{\lambda}_s)^2+\frac{\omega_s^2}{\omega_e^2}\right] - (\Lambda+\bar{\lambda}_p)(\Lambda+\bar{\lambda}_s)\frac{\varepsilon_3^2\delta}{5}$$

$$+ \frac{1}{10}\varepsilon_3^2\,\delta\,\frac{\omega_s^2}{\omega_e^2} = 0 \qquad (5.17)$$

The discrepancy between eqs. (5.6) and (5.17) is mainly due to the first term on the r.h.s. of (5.16). This term can be traced back to the Joule dissipation term of the energy balance equation, a term which is absent in the temperature balance equation. It is interesting to note that the same term gives rise to the zero-mode instability, as pointed out in sec. 4.
The stability analysis associated with eq. (5.17) yields quantitative differences with respect to Nishikawa's results.
For $\delta<0$, there is again an aperiodic instability, whenever $\varepsilon_3^2$ exceeds a threshold $\varepsilon_I^2$, whose minimum value is

$$\varepsilon_I^2 = 20\,\bar{\lambda}_p \qquad (5.18)$$

The corresponding growth rate, near the threshold is :

$$\gamma_I = \frac{3}{40}\frac{\omega_s^2}{\omega_e^2\bar{\lambda}_p^2}(\varepsilon_3^2-\varepsilon_I^2)\,\omega_e \qquad (5.19)$$

For $\delta>0$, we find again the parametric decay instability, with a new minimum threshold :

$$\varepsilon_{II}^2 = 20\,\bar{\lambda}_s \qquad (5.20)$$

The corresponding growth rate, near the threshold is :

$$\gamma_{II} = \frac{1}{20}(\varepsilon_3^2 - \varepsilon_{II}^2) \qquad (5.21)$$

The results of this section show how striking are the effects of the inclusion of the energy equation -rather than the temperature equation- in the description of a plasma under the action of a strong external electric field.

a) Aperiodic Instability ($\delta<0$)

In both models the thresholds are independent of $\kappa$ and are numerically of the same order. But the growth rate is independent of $\kappa$ in the temperature (or Nishikawa) model, while it is of order $\kappa^2$ in the energy model.

b) Parametric Decay Instability ($\delta>0$)

The threshold which is independent of $\kappa$ in the temperature model, is proportional to $\kappa^2$ in the energy model. Moreover, the growth rate is independent of $\kappa$ as opposed to the $\kappa^2$ dependence in the temperature model. Thus already for low values of the field, long wavelength waves are excited, their growth rate being the same as the one of the shorter wavelength modes, which appear for higher values of the external field intensity.

## 6. CONCLUSIONS

In this work we aimed at a kinetic foundation of the mode-coupling equations appearing in the theory of parametric instabilities. The whole theory rests on the assumption of the existence of a P.D. regime, where collisions play an important role. In this regime we demonstrated the importance of the thermal effects, determined by the total energy balance equation. On one hand, a new type of instability appears, on the other, the characteristics of the classical instabilities can be seriously modified.

The theory should be refined in various ways. First, the collision operator should include non-Markovian contributions (10). Second, the transport coefficients might become frequency-dependent. Finally, an extension of the theory towards Vlassov dominated regimes should be envisaged.

Acknowledgments

We ackowledge the very fruitful discussions with Dr. C.N. Lashmore-Davies.

References

(1) Nishikawa, K., J. Phys. Soc. Japan, 24, 916, 1152 (1968).

(2) Balescu, R. and Paiva-Veretennicoff, I., three papers in press, J. Plasma Phys., 1976.

(3) Balescu, R., Paiva-Veretennicoff, I. and Brenig, L., Physica, 81A, 1, 17, (1975).

(4) Resibois, P., J. Statist. Phys. 2, 21 (1970).

(5) Wilhelmsson, H., J. Plasma Phys., 3, 215 (1969).

(6) Stenflo, L., Plasma Phys., 12, 509 (1970).

(7) Lashmore-Davies, C., Plasma Phys., 17, 281 (1975).

(8) Braginsky, S.I., in Reviews of Plasma Physics, Consultants Bureau, N.Y., 1965.

(9) Marden, M., The geometry of the zeroes of a polynomial in a complex variable, Math. Surveys n° III, Amer. Math. Soc., New York 1949.

(10) Baus, M., Physica 79A, 377 (1975).

# SOME REMARKS ON COHERENT NONLINEAR COUPLING OF WAVES IN PLASMAS

Hans Wilhelmsson

Institute for Electromagnetic Field Theory
and EURATOM-FUSION Research
Chalmers University of Technology
S-402 20 Göteborg, Sweden

ASSOCIATION EURATOM-CEA
Departement de Physique du Plasma et
de la Fusion Contrôlée
Service IGn - Centre d'Etudes Nucléaires
85X
38041 Grenoble Cedex, France

## INTRODUCTION

During the last decades active and increasingly important research has been carried out in the field of nonlinear plasma physics, like in the field of nonlinear optics.

To interpret new experiments carried out at high intensities of radiation, the results of which have recently become available, many nonlinear processes have been analysed, and methods developed, which may be regarded as unifying theoretical approaches to describe various nonlinear processes.

A detailed description of the nonlinear evolution of a wave in its interaction with other waves should account for the changes in the amplitude as well as in the phase of the wave, generally in space and time. For many situations of practical interest, such as those where the plasma exhibits turbulent features, the description of the plasma becomes so complicated that such detailed phase-analysis is generally not possible, and certain simplifying assumptions are needed. The random-phase method, which has been used extensively to describe such situations, introduces an ensemble averaging over the phases, as suggested by the physical situation. On the other hand experiments on nonlinear interaction of waves in plasmas indicate that in many cases coherent features are preserved, at least to some extent, in the nonlinear interaction. Detailed studies of the basic system of equations describing coherent nonlinear evolution of waves are therefore of primary interest to modern plasma physics and its applications, for example to astrophysics, space science or fusion-plasma research.

The purpose of the present contribution is to discuss various possibilities of analysing the basic set of coupled equations and to point out inherent difficulties associated with the description of the nonlinear interaction between different types of waves, and for general coupling situations, including deviations from perfect matching of frequencies and wave-numbers. We apply our considerations to specific examples in the field of stimulated excitation of waves, namely to parametric excitation of hybrid resonances in hot magnetized multi-ion component plasma and to laser-plasma interaction.

## BASIC SYSTEM OF EQUATIONS AND APPROACHES OF SOLUTION

The basic equations for the complex amplitudes $a_i$ of three coupled waves, propagating in the x-direction, are

$$\frac{\partial a_0}{\partial t} + \nu_0 a_0 + v_0 \frac{\partial a_0}{\partial x} = c_{12}^* a_1 a_2 \exp(-i\Delta\Phi) , \quad (1)$$

$$\frac{\partial a_1}{\partial t} + \nu_1 a_1 + v_1 \frac{\partial a_1}{\partial x} = c_{02} a_0 a_2^* \exp(i\Delta\Phi) , \quad (2)$$

$$\frac{\partial a_2}{\partial t} + \nu_2 a_2 + v_2 \frac{\partial a_2}{\partial x} = c_{01} a_0 a_1^* \exp(i\Delta\Phi) , \quad (3)$$

where

$$\nu_i = \mathrm{Im}(\omega_i)$$

is the linear damping or growth of a wave with group velocity

$$v_i = \frac{\partial \omega_i}{\partial k_i} ,$$

and

$$\Delta\Phi = (\Delta\omega)t - \int^x \Delta k(x')dx' ,$$

with

$$\Delta\omega = \mathrm{Re}(\omega_0) - \mathrm{Re}(\omega_1) - \mathrm{Re}(\omega_2) ,$$

$$\Delta k = k_0 - k_1 - k_2 ,$$

and where $c_{ij}$ are the coupling coefficients.

Here $\Delta\omega$ represents a frequency mismatch, which we regard as fixed, determined by the frequencies $\omega_i$, whereas $\Delta k$ represents a wave-number mismatch, which in an inhomogeneous medium (including also long scale-length random inhomogeneities) depends on x.

A complete study of this system of waves, considering space and time variations of the amplitudes is connected with serious difficulties, in particular with regard to those introduced by differences in the parameters describing damping and group-velocity of the waves. Such differences do indeed exist in physical systems. Even the technique of inverse scattering does not allow for a solution of the full system. It is therefore customary to consider simplified model equations, where

i) One of the waves is sufficiently strong as compared to the others, so that its amplitude can be considered as constant and one can use a parametric approach. Useful information about the thresholds and growth-rates for various processes can be obtained from such an approach, which has been used frequently to describe stimulated scattering and decay processes, e.g. in laser-plasma interaction. Considering a case where, for example, two of the modes are electromagnetic waves with frequencies $\omega_0$ (the pump wave) and $\omega_2$, whereas the third wave, corresponds to a collective excitation in the medium with a characteristic frequency $\omega_L$, one obtains as is well known (Nishikawa 1968) an equation of the form

$$(\omega-\omega_L)^2 - [i(\Gamma_1+\Gamma_2) + \Delta\omega](\omega-\omega_L) + i\Delta\omega\Gamma_1 - \Gamma_1\Gamma_2 + \frac{G}{4\omega_L\omega_2} = 0 , \tag{4}$$

where the frequency mismatch

$$\Delta\omega = \omega_0 - \omega_L - \omega_2$$

and where $\Gamma_1$ and $\Gamma_2$ refer to linear damping rates of the parametrically excited waves. The quantity G is proportional to the power of the pump wave (see example below).

Considering first the case, where $\Gamma_1 = \Gamma_2 = 0$, we obtain by solving Eq. (4) for the root corresponding to growth (time factor exp($i\omega t$), and assuming $G/(\omega_L\omega_2) > (\Delta\omega)^2$) the expression

$$\omega_1 = \omega_L + \frac{\Delta\omega}{2} - \frac{i}{2}\left[G/(\omega_L\omega_2) - (\Delta\omega)^2\right]^{1/2}, \tag{5}$$

which clearly exhibits the important and well known fact that the threshold value for the pump power to surmount the effect of mismatch is determined by

$$G = \omega_L\omega_2(\Delta\omega)^2 \tag{6}$$

whereas there is no growth for

$$|\Delta\omega| > 2\gamma_0 \tag{7}$$

where

$$\gamma_0 = \frac{1}{2}\left(G/(\omega_L\omega_2)\right)^{1/2} \tag{8}$$

is the growth-rate in the absence of mismatch (the maximum growth-rate), i.e. for $\Delta\omega = 0$.

Taking into account the damping rates $\Gamma_1$ and $\Gamma_2$ in Eq. (4) the condition for threshold becomes (Nishikawa 1968),

$$G = 4\omega_L\omega_2\Gamma_1\Gamma_2\left[1 + (\Delta\omega)^2/(\Gamma_1+\Gamma_2)^2\right], \tag{9}$$

and the maximum growth-rate

$$\gamma_0 = \frac{1}{2}\left\{\left[G/(\omega_L\omega_2) + (\Gamma_1-\Gamma_2)^2\right]^{1/2} - (\Gamma_1+\Gamma_2)\right\} \tag{10}$$

We have reviewed here, some of the basic formulas because we need them in the following, when we are going to use them for some specific applications.

ii) The mismatch or damping of one of the waves is so large that the corresponding term in one of the equations dominates the linear part of the equation, which becomes algebraic and allows for direct solution for one of the unknown amplitudes in terms of the others. By substitution of this amplitude into the remaining equations, one is left with two nonlinear coupled equations, more suited for detailed discussion.

Introducing the following expressions for the normalized intensities

$$I_0 = s_0 \mathrm{Re}\left(\frac{c_{01}^* c_{02}}{\nu_2 - i(\Delta\omega - v_2\Delta k)}\right)\frac{2}{v_1 - v_0}\,|a_0|^2 \,, \tag{11a}$$

$$I_1 = s_1 \mathrm{Re}\left(\frac{c_{12}^* c_{01}}{\nu_2 + i(\Delta\omega - v_2\Delta k)}\right)\frac{2}{v_0 - v_1}\,|a_1|^2 \,, \tag{11b}$$

where $s_0$ and $s_1$ account for the signs of the Re( ) factors in (11a,11b), so that $I_0$ and $I_1$ are positive quantities, and furthermore introducing the variables

$$\xi = x - v_0 t \,,$$

$$\eta = x - v_1 t \,,$$

and

$$\alpha_0 = \frac{2\nu_0}{v_0 - v_1} \,,$$

$$\alpha_1 = \frac{2\nu_1}{v_1 - v_0} \,,$$

we obtain the following coupled equations

$$\frac{\partial}{\partial\eta} I_0 + \alpha_0 I_0 = s_1 I_0 I_1 \,, \tag{12a}$$

$$\frac{\partial}{\partial\xi} I_1 + \alpha_1 I_1 = s_0 I_0 I_1 \,. \tag{12b}$$

The system (12a,12b) is by far not trivial as a result of the presence of $\alpha_0$ and $\alpha_1$ and in view of the two independent variables $\xi$ and $\eta$. Let us here only note that for $s_0 = s_1 = 1$, and considering a stationary case, the eqs.(12a,12b) exhibit solutions of the explosive type for which $I_0 \sim I_1 \sim (\xi_0 - \xi)^{-1}$. This dependence is characteristic for an incoherent interaction, and corresponds to the results of a random phase approach, for which the corresponding three coupled equations are of the form

$$\frac{\partial N_0}{\partial t} + \alpha_0 N_0 = N_1 N_2 + N_0 N_2 + N_0 N_1 \,,$$

where $N_i$ are the quasi-particle densities, and where signs refer to a nonlinearly (explosively) unstable case. A solution of Eqs. (1-3) for $\Delta\Phi = 0$ and considering the phase-dynamics of the complex amplitudes would correspond to $I_0 \sim I_1 \sim (\xi_0-\xi)^{-1/2}$. Apparently the result mirrors the destruction of coherent interaction accompanying the assumption of large $\Delta\Phi$. For $I_0$ and $I_1$ corresponding to electromagnetic waves, coupled by plasmons such an explosive state should be possible in a medium, containing molecules with inverted populations.

iii) Only the dependence on one independent variable (x or t) is considered when studying the coupled system of the three (slowly) varying amplitudes, and mismatches are neglected. Even so, the system can be given analytic solutions in terms of elliptic functions only when all the coefficients of dissipation $\nu_j$ are equal. The nonlinear wave interaction may lead to either nonlinearly stable or unstable solutions depending on the properties of the interacting waves (in the case of real coupling constants nonlinear (explosive) instabilities may occur for interaction between negative - and positive - energy waves).

It has been demonstrated recently (Nakach and Wilhelmsson 1976) that, when two of the coefficients of dissipation are equal, the set of three coupled equations can be reduced to a single equivalent equation, which in the nonlinearly unstable case, where one wave is undamped, asymptotically takes the form of an equation defining the third Painlevé transcendent. The general solution of even the stationary form of the system, and for perfect matching, i.e. $\Delta\Phi = 0$, therefore belongs to a higher class of functions which reduces to the Jacobian elliptic type only when all coefficients of dissipation are equal.

iv) Use of computer is necessary to study the full system of equations (1-3), which may then also be supplemented by higher order terms (cf. Oraevskii et al 1973, Weiland and Wilhelmsson 1973, 1976). As a transition between coherent and incoherent dynamics of the system coherent interaction between partial subsystems of three wave-packets have been studied (Weiland et al 1975, Weiland and Wilhelmsson 1976). The stationary form of the Eqs. (5a,5b) have also been studied by means of computer, including effects of higher order terms (White et al 1972, Anderson and Bondeson 1976).

After this survey on the possibilities of studying wave-wave coupling from a coherent interaction point of view let us briefly discuss some specific examples which are related to experiments in magnetized plasmas and in laser-plasma interaction.

## PARAMETRIC EXCITATION OF A LOWER-HYBRID RESONANCE IN A MULTI-ION COMPONENT PLASMA

As an example of parametric interaction let us consider the coupling between a strong ordinary electromagnetic wave with a lower-hybrid wave (Stix 1965) and an other ordinary electromagnetic wave (Nguyen The Hung 1975), for the specific case where all waves propagate perpendicularly to the magnetic field. We pay particular attention to the presence of multi-ion components in the plasma, corresponding to experimental situations in present fusion-plasma devices (Wilhelmsson 1976).

Denoting by $C_k$ the fractional concentration and by $Z_k$ the charge of the impurity ions of the kind denoted by suffix k, and assuming that in the presence of the impurity ions, we have charge conservation the ion and electron number densities of a hydrogen plasma are related by

$$N_{oi} = N_{oe}(1 + \Sigma\, C_k Z_k)^{-1} \tag{13}$$

The characteristic frequencies of a fusion plasma fulfil the following condition

$$\omega_{Hi}^2 << \omega_{pi}^2 \approx \omega_{LH}^2 << \omega_{He}^2 \simeq \omega_{pe}^2 , \tag{14}$$

where $\omega_{LH}$ denotes the lower hybrid frequency, and the other notations refer to Larmor and plasma frequencies of the protons and electrons.

From the hot plasma dielectric tensor (See for example Ichimaru 1973) and considering the multi-ion species (Wilhelmsson 1976) we obtain the following relation for the lower-hybrid frequency

$$\omega_{LH}^2 = \frac{\omega_{He}^2 + k^2 u_e^2}{\omega_{pe}^2 + k^2 u_e^2 + \omega_{He}^2} \cdot \frac{m_e}{m_i} \cdot \omega_{pe}^2 (1 + \Sigma\, C_k Z_k)^{-1} \cdot$$

$$\cdot \left[ 1 + \frac{k^2 u_i^2}{(\omega_{LH}^o)^2} + \Sigma\, C_k Z_k^2 \frac{m_i}{m_k} \left( 1 + \frac{k^2 u_k^2}{(\omega_{LH}^o)^2} \right) \right] , \tag{15}$$

where we introduced the notation $\omega_{LH}^o$ for the lower-hybrid frequency in the absence of impurity and temperature corrections, i.e.

$$(\omega_{LH}^{o})^2 = \frac{\omega_{pi}^2 \cdot \omega_{He}^2}{\omega_{He}^2 + \omega_{pe}^2}$$

From the quantity $G/(\omega_{LH}\omega_2)$ which enter the threshold and growth-rate equations (9,10) we obtain ($\omega_L = \omega_{LH}$),

$$\frac{G}{\omega_{LH}\omega_2} = k^2\left(\frac{e}{m_e\omega_0}\right)^2 \cdot E_0^2 \cdot \frac{\omega_{LH}\omega_{pe}^4}{\omega_2(\omega_{pe}^2 + k^2u_e^2 + \omega_{He}^2)(\omega_{He}^2 + k^2u_e^2)} \tag{16}$$

where $E_0$ is the electric field amplitude of the pump wave, and where other notations are standard. Introducing collision frequencies $\nu_e$ and $\nu_i$ we have $2\Gamma_1 = \nu_i$ and $2\Gamma_2 = \nu_e(\omega_{pe}^2/\omega^2)$. For realistic impurity concentrations of hot fusion plasmas it turns out that the value of the lower-hybrid frequency may be decreased by about 20% as a result of the presence of impurities. As a consequence the threshold of the process here considered will increase by about 20% whereas the maximum growth-rate will decrease by about 10%. It should be added that different impurities may be located in different parts of the plasma, adding their contributions to $\Delta k(x)$ and thereby to the mismatches in the nonlinear couplings.

For vanishing static magnetic field the results apply to the stimulated excitation of ion-acoustic waves, i.e. to stimulated Brillouin scattering in the presence of multi-ion specia.

## EFFECT OF FREQUENCY BROADENING IN LASER-PLASMA INTERACTION

For the nonlinear coupling processes in laser-plasma interaction it is important to consider spectral broadening effects (Valeo and Oberman, 1973, Anderson and Wilhelmsson 1976).

We notice that the threshold condition (9) of parametric instability can be written

$$(\Delta\omega)^2 + (\Gamma_1+\Gamma_2)^2 = \frac{(\Gamma_1+\Gamma_2)^2}{\Gamma_1\Gamma_2}\gamma_0^2 \tag{9a}$$

with $\gamma_0$ defined by relation (12). When "blurring" effects on the left hand side become larger than the expression on right hand side the coupling process is no longer efficient.

Let us represent the influence of the coupling process on the frequency width of the backscattered wave, due to stimulated scattering, by a normalized Gaussian function $K(|\omega-\omega'|)$ of width $\sigma_{NL}$. We furthermore assume that the spectra of incident and reflected radiation can be represented by Gaussian functions of widths $D_I$ and $D_R$. The frequency broadening process can then be described as follows in terms of the powers of the reflected and incident radiation

$$P_R(\omega) = \int K(|\omega-\omega'|)P_I(\omega')d\omega' \tag{17}$$

(cf. Wilhelmsson 1963)

and

$$D_R^2 = D_I^2 + \sigma^2_{NL} \tag{18}$$

where we take

$$\sigma_{NL}^2 = (\Delta\omega_{cr})^2 + (\Gamma_1+\Gamma_2)^2 = \frac{(\Gamma_1+\Gamma_2)^2}{\Gamma_1\Gamma_2}\gamma_0^2 , \tag{19}$$

with $\Delta\omega_{cr}$ denoting the critical frequency deviation for marginal fulfilment of relation (17). We then have

$$\left(\frac{D_R}{D_I}\right)^2 = 1 + \frac{\sigma_{NL}^2}{D_I^2} , \tag{20}$$

or if in relation (19) we indentify $\Gamma_1$, the damping rate of the reflected wave, with the width of the reflected spectrum and furthermore assume $\Gamma_2 > \Gamma_1$ we obtain (Anderson and Wilhelmsson 1976)

$$\frac{D_R}{D_I} = \left(1 + \alpha\frac{W_L}{D_R/D_I}\right)^{1/2} , \tag{21}$$

where $W_L$ is the energy of the incident laser pulse and $\alpha$ a proportionality constant.

In figure 1 we depict the relative bandwidth $D_R/D_I$ as a function of $\alpha W_L$. One of the points ($W_L$ = 1,4 J, $D_R/D_I \approx$ 2,5) has been determined from available experimental data (Baldis, Pepin and Johnston 1975), and as could be seen in the figure the theoretical curve then follows closely the other experimental points.

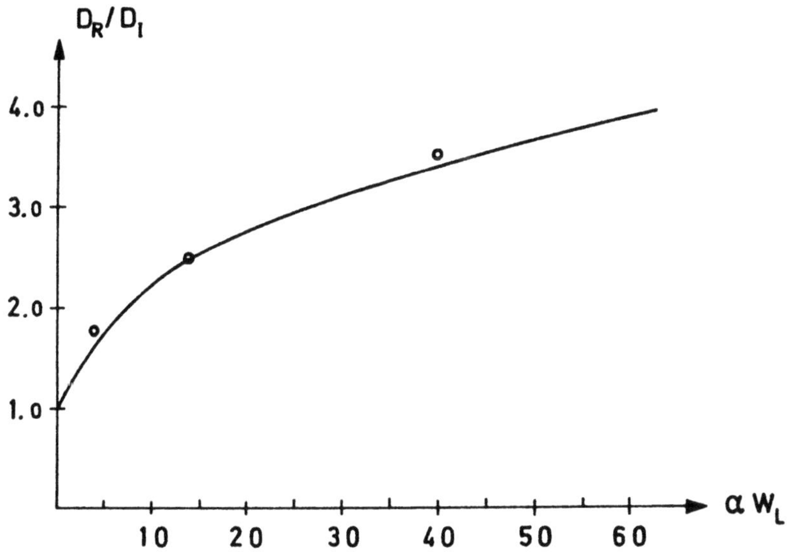


Fig. 1. The theoretical curve (Anderson and Wilhelmsson 1976) corresponding to relation (21), normalized by and compared with the experimental results on laser reflection of Baldis, Pepin and Johnston 1975.

It should furthermore be mentioned that recent measurements on finite bandwidth effects on the parametric decay instability in the radiofrequency and HF-domain (Oberschain, Luhmann, and Greiling 1976) give results which for wide-bandwidth pumps ($\Delta\omega >> \gamma$) are consistent with theoretical predictions.

## REFERENCES

1. Aliev, A.M., V.P. Silin, and C. Watson, Parametric Resonance in a Plasma in a Magnetic Field, Sov. Phys. JETP, 23, 626, 1966.

2. Akhiezer, A.I., I.A. Akhiezer, R.V. Polovin, A.G. Sitenko, and K.N. Stepanov, Plasma Electrodynamics 2. Non-Linear Theory and Fluctuations, Pergamon, Oxford - New York, 1975.

3. Anderson, D., and H. Wilhelmsson, Effect of Mismatch on Stimulated Scattering in Laser-Plasma Interactions, Nucl. Fusion, 15, 387, 1975.

4. Anderson, D. and H. Wilhelmsson, Nonlinear Spectral Broadening of Backscattered Light from Laser-Plasma Interactions, Phys. Lett. 56A, 37, 1976.

5. Anderson, D., and A. Bondeson, See paper presented at this symposium.

6. Anderson, D., and H. Wilhelmsson, Coherent Nonlinear Backscattering by Laser-Plasma Interactions, Physica Scripta (Sweden) 11, 341, 1975.

7. Baldis, H.A., H. Pepin, and T.W. Johnston, Reflectivity Measurements from a Nanosecond $CO_2$-Laser Produced Plasma, Optics Comm., 15, 311, 1975.

8. Bezzerides, B., and D.F. DuBois, Coupling Saturation in the Nonlinear Theory of Parametric Decay Instabilities, Phys. Rev. Lett., 36, 729, 1976.

9. Bloembergen, N., Nonlinear Optics, Benjamin, Inc., New York, 1965.

10. Bornatici, M., Parametric Backscattering and Absorptive Instabilities in Homogeneous Unmagnetized Plasmas, J. Plasma Phys., 14, 105, 1975.

11. Chu, Flora Y.F., and A.C. Scott, Inverse Scattering Transform for Wave-Wave Scattering, Phys. Rev. A, 12, 2060, 1975.

12. Chu, Flora Y.F., Bäcklund, Transformation for the Wave-Wave Scattering Equations, Phys. Rev. A, 12, 2065, 1975.

13. Cohen, B.I., Space-Time Interaction of Opposed Transverse Waves in a Plasma, Phys. Fluids, 17, 496, 1974.

14. Coppi, B., M.N. Rosenbluth, and R.N. Sudan, Nonlinear Interactions of Positive and Negative Energy Modes in Rarefied Plasmas (1), Ann Phys. (N.Y.), 55, 207, 1969.

15. Davidson, R.C., Methods in Nonlinear Plasma Theory, Academic, New York - London, 1972.

16. Dikasov, V.M., L.I. Rudakov, and D.D. Ryutov, Interaction of Negative-Energy Waves in a Weakly Turbulent Plasma, Sov. Phys. JETP, 21, 605, 1965.

17. DuBois, D.F. and M.V. Goldman, Radiation-Induced Instability of Electron Plasma Oscillations, Phys. Rev. Lett., 14, 544, 1965.

18. DuBois, D.F., Nonlinear Parametric Excitation of Plasma Fluctuations, Statistical Physics of Charged Particle Systems (Ed. by R. Kubo and T. Kihara), Syokabo and Benjamin, Inc., 87, 1969. See also paper presented at this symposium.

19. DuBois, D.F. and M.V. Goldman, Nonlinear Saturation of Parametric Instability: Basic Theory and Application to the Ionosphere, Phys. Fluids, 15, 919, 1972

20. Engelmann, F., and H. Wilhelmsson, Phase Effects in the Nonlinear Interaction of "Negative"-Energy Waves, Z. Naturforsch., 24a, 206, 1969.

21. Forslund, D.W., J.M. Kindel, and E.L. Lindman, Nonlinear Behaviour of Stimulated Brillouin and Raman Scattering, Phys. Rev. Lett., 30, 739, 1973.

22. Fuchs, V., and G. Beaudry, Effect of Damping on Nonlinear Three-Wave Interaction, J. Math. Phys., 16, 616, 1975.

23. Fuchs, V., and G. Beaudry, Stability of Nonlinear Parametric-Decay Interactions in Finite Homogeneous Plasma, J. Math. Phys., 17, 208, 1976.

24. Fuchs, V., The Influence of Linear Damping on Nonlinearly Coupled Positive and Negative Energy Waves, J. Math. Phys., 16, 1388, 1975.

25. Galeev, A.A., and R.Z. Sagdeev, Parametric Phenomena in a Plasma Nucl. Fusion, 13, 603, 1973.

26. Guzdar, P.N., Effect of Langmuir Turbulence on Stimulated Brillouin Backscattering, Phys. Rev. Lett., 35, 1635, 1975.

27. Hasegawa, A., Plasma Instabilities and Nonlinear Effects, Springer, Berlin, 1975.

28. Ichimaru, S., Basic Principles of Plasma Physics. A Statistical Approach, Benjamin, Inc., London - Tokyo, 1973.

29. Kadomtzev, B.B., A.B. Mikhailovskii, and A.V. Timofeev, Negative Energy Waves in Dispersive Media, Sov. Phys. JETP, 20, 1517, 1965.

30, Karpman, V.N., Non-Linear Waves in Dispersive Media, Pergamon, Oxford - New York, 1975.

31. Kaufman, A.N., See papers presented at this symposium.

32. Kaw, P.K., Parametric Excitation of Electrostatic Waves in a Magnetized Plasma, Adv. Plasma Phys., 6, 179, 1976.

33. Kaw, P.K., Parametric Excitation of Electromagnetic Waves in Magnetized Plasmas, Adv. Plasma Phys., 6, 207, 1976

34. Kruer, W.L., K.G. Estabrook, and K.H. Sinz, Instability-Generated Laser Reflection in Plasmas, Nucl. Fusion, 13, 952, 1973.

35. Kruer, W.L., Saturation and Nonlinear Effects of Parametric Instability, Adv. Plasma Phys., 6, 237, 1976.

36. Larsson, J., and L. Stenflo, Three-Wave Interaction in Magnetized Plasmas, Beitr. Plasmaphys., 13, 169, 1973.

37. Larsson, J., and L. Stenflo, Threshold Fields of Parametric Instabilities, Beitr. Plasmaphys., 14, 7, 1974; See also paper presented at this symposium.

38. Laval, G., R. Pellat, and D. Pesme, Absolute Parametric Excitation by an Imperfect Pump or by Turbulence in an Inhomogeneous Plasma, Phys. Rev. Lett., 36, 192, 1976.

39. Liu, C.S., M.N. Rosenbluth, and R.B. White, Parametric Scattering Instabilities in Inhomogeneous Plasmas, Phys. Rev. Lett., 31, 697, 1973.

40. Liu, C.S., and R.E. Aamodt, Explosive Instability of Drift-Cone Modes in Mirror Machines, Phys. Rev. Lett., 36, 95, 1976.

41. Liu, C.S., Parametric Instabilities in an Inhomogeneous Unmagnetized Plasma, Adv. Plasma Phys., 6, 121, 1976.

42. Maier, M., W. Kaiser, and J.A. Giordmaine, Backward Stimulated Raman Scattering, Phys. Rev., 177, 580, 1969.

43. Nakach, R., and H. Wilhelmsson, Solution of the Equations for Nonlinear Interaction, Phys. Rev. A, 14, 451, 1976.

44. Nguyen The Hung, Parametric Instabilities of Ordinary and Hybrid Waves, Plasma Physics, 17, 633, 1975.

45. Nishikawa, K., Parametric Excitation of Coupled Waves. I. General Formulation, J. Phys. Soc. Japan, 24, 916, 1968; II. Parametric Plasmon-Photon Interaction, J. Phys. Soc. Japan, 24, 1152, 1968.

46. Nishikawa, K., and C.S. Liu, General Formalism of Parametric Excitation, Adv. Plasma Phys., 6, 3, 1976.

47. Oberschain, S.P., N.C. Luhmann, Jr and P.T. Greiling, Effects of Finite-Bandwidth Driver Pumps on the Parametric-Decay Instabilities, Phys. Rev. Lett. 36, 1309, 1976.

48. Oraevsky, V.N., V.P. Pavlenko, H. Wilhelmsson, and E.Ya. Kogan, Stabilization of Explosive Instabilities by Nonlinear Frequency Shifts, Phys. Rev. Lett., 30, 49, 1973

49. Pesme, D., G. Laval, and R. Pellat, Parametric Instabilities in Bounded Plasmas, Phys. Rev. Lett. 31, 203, 1973

50. Porkolab, M., High Frequency Parametric Wave Phenomena and Plasma Heating: A Review, Physica (The Netherlands) 82C, 1976.

51. Rosenbluth, M.N., B. Coppi, and R.N. Sudan, Nonlinear Interaction of Positive and Negative Energy Modes in Rarefied Plasmas (II), Ann. Phys. (N.Y.), 55, 248, 1969.

52. Stenflo, L., Non-Linear Interaction between Three Ordinary Electromagnetic Waves, J. Plasma Physics, 5, 413, 1971; See alos paper presented at this symposium.

53. Stix, T.H., Radiation and Absorption via Mode Conversion in an Inhomogeneous Collision-Free Plasma, Phys. Rev. Lett., 15, 878, 1965.

54. Stern, R.A., and N. Tzoar, Parametric Coupling between Electron-Plasma and Ion-Acoustic Oscillations, Phys. Rev. Lett., 17, 903, 1966.

55. Tang, C.L., Saturation and Spectral Characteristics of the Stokes Emission in the Stimulated Brillouin Porcess, J. Appl. Phys. 37, 2945, 1966.

56. Thomson, J.J., Finite-Bandwidth Effects on the Parametric Instability in an Inhomogeneous Plasma, Nucl. Fusion, 15, 237, 1975.

57. Tsytovich, V.N., Nonlinear Effects in Plasma, Plenum, New York-London, 1970.

58. Tsytovich, V.N., and H. Wilhelmsson, Nonlinear Enhancement of Radiation in Optically Amplifying Media, Physica Scripta (Sweden), 7, 251, 1973.

59. Tsytovich, V.N., Solitons, Cavitons and Strong Langmuir Plasma Turbulence, Comments on Plasma Phys. Controlled Fusion, 2, 127, 1976; See also paper presented at this symposium.

60. Valeo, E.J., and C.R. Oberman, Model of Parametric Excitation by an Imperfect Pump, Phys. Rev. Lett., 30, 1035, 1973.

61. Weiland, J. and H. Wilhelmsson, Repetitive Explosive Instabilities, Physica Scripta (Sweden), 7, 222, 1973.

62. Weiland, J. and H. Wilhelmsson, A.N. Kaufman, and V.N. Tsytovich, Phase-broadening and Phase-focusing Effects in Nonlinear Partially Coherent Interaction, Physica Scripta (Sweden), 11, 275, 1975.

63. Weiland, J., and H. Wilhelmsson, Coherent Nonlinear Interaction of Waves in Plasmas, Pergamon, Oxford, 1976.

64. White, R.B., Y.C. Lee, and K. Nishikawa, Nonlinear Mode Coupling and Relaxation Oscillations, Phys. Rev. Lett., 29, 1315, 1972.

65. Wilhelmsson, H., A Nonlinear Coupling Mechanism for Generation of Very Low Frequency Waves, Astronomical Notes, Section of Astronomy, University of Gothenburg, (Sweden), 8, 3, 1963.

66. Wilhelmsson, H., Accounting for Smoothing Effects in 21 cm Observations, Arkiv för Astronomi, 3, 187, 1963.

67. Wilhelmsson, H., Nonlinear Coupling of Waves in a Magnetized Plasma with Particle Drift Motions, J. Plasma Physics, 3, 215, 1969.

68. Wilhelmsson, H., L. Stenflo, and F. Engelmann, Explosive Instabilities in the Well-Defined Phase Description, J. Math. Phys., 11, 1738, 1970.

69. Wilhelmsson, H., and K. Östberg, Phase-locking of Coupled Modes in Nonlinearly Unstable Plasmas, Physica Scripta (Sweden), 1, 267, 1970.

70. Wilhelmsson, H., On the Explosive Instabilities in the Presence of Linear Damping or Growth, Physica Scripta (Sweden), 2, 113, 1970.

71. Wilhelmsson, H., Evolution of Explosively Unstable Systems, Phys. Rev. A, 6, 1973, 1972.

72. Wilhelmsson, H., Wave-Wave Interaction in Plasmas, Physica (The Netherlands), 82C, 52, 1976.

73. Wilhelmsson, H., The Lower Hybrid Resonance - Its Dependence on Impurities and Ion-Larmor Radii in Hot Tokamak Plasmas, Physica Scripta (Sweden), 14, 60, 1976.

74. Zakharov, V.E. and S.V. Manakov, Resonant Interaction of Wave Packets in Nonlinear Media, Sov. Phys. JETP lett., 18, 243, 1973.

# EFFECT OF FINITE BEAT-WAVE RELAXATION TIME AND GROWTH-RATE MODULATION ON THE NONLINEAR MODE COUPLING BETWEEN STABLE AND UNSTABLE PLASMA WAVES

Dan Anderson and Anders Bondeson

Institute for Electromagnetic Field Theory
and EURATOM-FUSION Research
Chalmers University of Technology
S-402 20 Göteborg, Sweden

## ABSTRACT

Starting from a situation where an unstable plasma wave is saturated by means of nonlinear Landau damping, an investigation is made of the complicating effects of (i) finite beat wave relaxation time and (ii) an external modulation of the growth rate of the unstable wave. It is shown for case (i) that the finite damping time of the beat mode causes damping or growth of the relaxation oscillations associated with the saturation behaviour, depending on whether the damping rate is larger or smaller than the beat frequency deviation from exact resonance. For case (ii) the parametric instability resulting from a proper choice of modulation frequency is studied in detail and the corresponding growth-rate, threshold, and final saturation level are determined.

## INTRODUCTION

An important saturating mechanism for plasma wave instabilities is mode coupling by means of nonlinear Landau damping [1,2], which transfers energy from unstable to stable waves using Landau damping of the corresponding beat mode. This interaction may be viewed as a nonlinear three wave process involving two primary waves (the unstable and stable plasma waves) and a secondary wave (the associated nonresonant beat wave). The differential equation for the secondary wave is usually approximated by an algebraic relation giving the beat wave amplitude in terms of the amplitudes of the primary waves. This approximation when used in the equations for the primary waves yields the nonlinear mode coupling equations.

In the present study we first investigate the consequences of allowing for a finite damping time of the beat mode, an effect which is shown to imply additional terms in the algebraic approximation. It is then demonstrated that these correction terms give rise to damping or growth of the relaxation oscillations associated with the saturation behaviour, depending on whether the damping rate of the beat mode is larger or smaller than the beat frequency deviation from exact resonance.

In a second modification of the mode coupling equations we consider the case when the growth-rate of the unstable plasma wave is modulated externally, a situation which arises e.g. if the growth-rate is provided by an amplitude modulated pump wave. The parametric instability, which results for properly tuned modulation frequency, gives rise to enhanced excitation of the plasma waves. This problem was previously mentioned in [1], but will here be given a more thorough analytical treatment, which includes the stabilization of the instability by means of nonlinear frequency shifts. The results of this analysis include the growth-rate, threshold, and saturation level of the parametric instability.

Finally, in both studied cases, the results are supported by numerical analysis.

## THE NONLINEAR MODE COUPLING EQUATIONS

An often used equation modelling the mode coupling of unstable and stable plasma waves by means of nonlinear Landau damping is, [2]

$$\frac{dI_0}{dt} = \gamma_0 I_0 - \alpha I_0 I_1$$

$$\frac{dI_1}{dt} = -\gamma_1 I_1 + \alpha I_0 I_1 \tag{1}$$

where $\gamma_0$ and $\gamma_1$ denote the growth- and damping-rates of the unstable and stable waves (intensities $I_0$ and $I_1$) and $\alpha$ is a nonlinear coupling coefficient. The system (1) has been studied e.g. in [1] where it was shown to lead to relaxation oscillations, which damp towards a saturating value if terms corresponding to spontaneous emission are included.

## EFFECT OF FINITE BEAT WAVE DAMPING

We shall study a modification of eq. (1), which results from accounting for the finite damping time of the beat wave. For that purpose we consider a nonlinear three wave interaction process involving two primary waves (amplitudes $a_0$ and $a_1$) and their corresponding beat wave (amplitude $a_2$), viz.

$$\left(\frac{d}{dt} - \gamma_0/2\right) a_0 = -\beta a_1 a_2$$

$$\left(\frac{d}{dt} + \gamma_1/2\right) a_1 = \beta a_0 a_2^*$$

$$\left(\frac{d}{dt} - i\Delta\omega\right) a_2 = \beta a_0 a_1^* \qquad (2)$$

Here $\Delta\omega = \Omega + i\Gamma_2$, where $\Omega$ is the beat frequency deviation from exact resonance and $\Gamma_2$ the damping rate of the beat mode. Writing eq. (2c) as

$$a_2 = \frac{i\beta}{\Delta\omega}\left(1 + \frac{i}{\Delta\omega}\frac{d}{dt}\right)^{-1} a_0 a_1^* \qquad (3)$$

we directly obtain an approximation for $a_2$ which extends the conventional algebraic approximation

$$a_2 \simeq \frac{i\beta}{\Delta\omega} a_0 a_1^* + \frac{\beta}{(\Delta\omega)^2}\frac{d}{dt}(a_0 a_1^*) \qquad (4)$$

where the last term can be simplified by means of eq. (2a-b) to read

$$\frac{d}{dt}(a_0 a_1^*) \simeq \frac{1}{2}(\gamma_0 - \gamma_1) a_0 a_1^* + \frac{i\beta^2}{\Delta\omega} a_0 a_1^* (|a_0|^2 - |a_1|^2) \qquad (5)$$

By means of eq.s (2), (4), and (5) we obtain the following modification of eq. (1):

$$\frac{dI_0}{dt} = \gamma_0 I_0 - \alpha I_0 I_1 [1 + \varepsilon_1 + \varepsilon_2 (I_0 - I_1)]$$

$$\frac{dI_1}{dt} = -\gamma_1 I_1 + \alpha I_0 I_1 [1 + \varepsilon_1 + \varepsilon_2(I_0 - I_1)] \tag{6}$$

where

$$\varepsilon_1 = \frac{\gamma_0-\gamma_1}{2\Gamma_2} \; \frac{\Omega^2-\Gamma_2^2}{\Omega^2+\Gamma_2^2} \quad ; \quad \varepsilon_2 = \frac{\alpha}{2\Gamma_2} \; \frac{3\Omega^2-\Gamma_2^2}{\Omega^2+\Gamma_2^2}$$

We conclude that the series expansion of eq. (3) is valid if $2\Gamma_2 \gg \gamma_0-\gamma_1$, i.e. the damping rate of the beat mode must be sufficiently large.

In order to see the consequences of the correction terms for the relaxation oscillations we linearize eq. (6) around the steady-state solutions $I_0^s \simeq \gamma_1[1-\varepsilon_1-\varepsilon_2(\gamma_1-\gamma_0)/\alpha]/\alpha$ and $I_1^s \simeq \gamma_0[1-\varepsilon_1-\varepsilon_2(\gamma_1-\gamma_0)/\alpha]/\alpha$. We then find damped oscillations with frequency $\omega_0 = (\gamma_0\gamma_1)^{1/2}[1+\varepsilon_2(\gamma_1-\gamma_0)/(2\alpha)]$ and damping rate $\gamma \simeq \varepsilon_2\omega_0^2/\alpha$ indicating that when $\varepsilon_2 > 0$ the system will damp towards the steady state solutions and for $\varepsilon_2 < 0$ the small amplitude disturbances become unstable, cf. fig. 1.

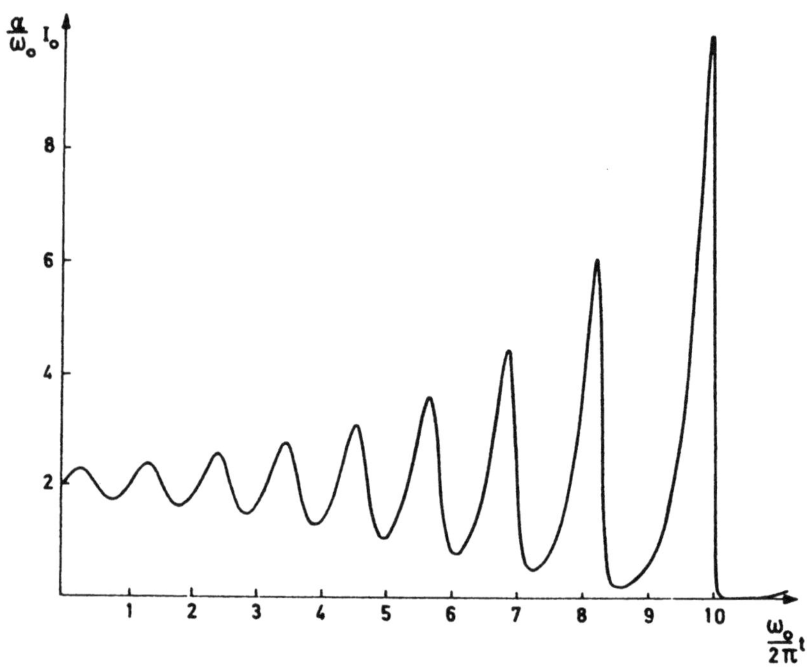


Fig. 1. $I_0(t)$ in the case $\gamma_0/2\Gamma_2 = .025$, $\gamma_1/2\Gamma_2 = .1$ and $\Omega = 0$.

This stability criterion is also valid for large amplitude oscillations. When $\varepsilon_2 = 0$ one has the following constant of motion, C, for the system (6) $C = \alpha(1+\varepsilon_1)(I_0+I_1) - \gamma_1 \ln I_0 - \gamma_0 \ln I_1$ implying that the solutions are periodic, cf. [2]. When $\varepsilon_2 \neq 0$ we introduce the quasi constant $\overline{C}(t)$ given by

$$\overline{C}(t) = \alpha(1+\varepsilon_1)(I_0+I_1) - \gamma_1 \ln I_0 - \gamma_0 \ln_1 + $$

$$+ \frac{\gamma_1-\gamma_0}{2(\gamma_1+\gamma_0)} \alpha\varepsilon_2(I_0 + I_1)^2 \tag{7}$$

One easily verifies that

$$\frac{d\overline{C}}{dt} = - \frac{2\alpha\varepsilon_2}{\gamma_0+\gamma_1} (\gamma_0 I_0 - \gamma_1 I_1)^2$$

indicating that the maximum oscillation amplitudes are monotonically increasing if $\varepsilon_2 < 0$ and decreasing if $\varepsilon_2 > 0$.

This difference in the behaviour of the system when $\varepsilon_2 \lesseqgtr 0$ is readily interpreted in physical terms, since the sign of $\varepsilon_2$ depends on whether $\Gamma_2$, which represents the width of the beat wave spectrum, is smaller or larger than $\Omega$, i.e. the deviation of the beat frequency from the resonance frequency. In the former case the beat wave spectrum is well separated from the resonance domain, leading to characteristic off-resonant behaviour. However, in the latter case the large overlap between the beat wave spectrum and the resonance domain results in an essentially resonant behaviour.

## EFFECT OF EXTERNAL GROWTH-RATE MODULATION

We will now investigate the consequences of an external modulation of the growth-rate for the unstable wave, i.e. we write

$$\gamma_0 \to \gamma_0(1 + z\cos\Omega t) \tag{8}$$

where z denotes the degree of modulation and $\Omega$ is the modulation frequency ($\Omega \sim 2\omega_0$).

Eq. (1) is rewritten by means of the substitution $I_j = I_j^s + i_j$, which together with (8) yields

$$\frac{di_0}{dt} = -\gamma_1 i_1 + \gamma_0 z i_0 \cos\Omega t + \frac{z\omega_0^2}{\alpha}\cos\Omega t - \alpha i_0 i_1$$

$$\frac{di_1}{dt} = \gamma_0 i_0 + \alpha i_0 i_1 \qquad (9)$$

This equation is conveniently analyzed by means of the multiple time scale technique. We introduce an ordering parameter $\varepsilon$ in such a way as to make possible a balance between parametric growth and nonlinear stabilizing effects. This is accomplished by choosing $i_0$, $i_1 \sim \varepsilon$ and $z \sim \varepsilon^2$.

Expanding

$$i_j = i_{j0}(t_0,t_1,t_2) + \varepsilon i_{j_1}(t_0,t_1,t_2) + \varepsilon^2 i_{j_2}(t_0,t_1,t_2)$$

where the slow time scales $t_j$ are related to the original time t according to $dt_j/dt = \varepsilon^j$ one obtains

$$i_{00} = A(t_2)\exp(i\omega_0 t_0) + \text{c.c.}$$

$$i_{10} = -\frac{i\omega_0}{\gamma_1}A(t_2)\exp(i\omega_0 t_0) + \text{c.c.}$$

$$i_{01} = -\frac{i\alpha A^2}{3\gamma_1\omega_0}\exp(2i\omega_0 t_0) - \frac{iz\omega_0^2\Omega}{2\alpha(\Omega^2-\omega_0^2)}\exp(i\Omega t_0) + \text{c.c.}$$

$$i_{11} = -\frac{\alpha A^2}{3\gamma_1^2}(2\gamma_1+i\omega_0)\exp(2i\omega_0 t_0) - \frac{2\gamma_0}{6\alpha}\exp(i\Omega t_0) + \text{c.c.}$$

In order to avoid secular terms the amplitude $A(t_2)$ or more conveniently $a(t_2) = A(t_2)\exp[-i(\Omega/2-\omega_0)t_2 + i\phi]$ must satisfy the equation

$$\left(\frac{\partial}{\partial t_2} + i\delta + i\kappa|a|^2\right)a = \gamma a^* \qquad (10)$$

where

$$\delta = \Omega/2 - \omega_0 \quad ; \quad 2\phi = \arg(2\omega_0 - i\gamma_1)$$

$$\kappa = \frac{\alpha^2}{6\omega_0\gamma_1}(\gamma_0+\gamma_1) \quad ; \quad \gamma = \frac{\omega_0 z}{12}\left(1 + 4\frac{\gamma_0}{\gamma_1}\right)^{1/2}$$

We note that eq. (10) predicts parametric amplification with growth-rate $(\gamma^2-\delta^2)^{1/2}$ for small amplitudes and that eventually the parametric growth ceases when the nonlinear frequency shift $\kappa|a|^2$ becomes sufficiently large.

## POTENTIAL FUNCTION DESCRIPTION

The nonlinear dynamics of eq. (10) can be further elucidated using a potential function description. For that purpose we substitute $a = u \exp i\phi$, ($u$ and $\phi$ real). After separating the real and imaginary parts of eq. (10) one finds a constant of motion

$$C = \gamma u^2 \sin 2\phi + \delta u^2 + \frac{\kappa}{2} u^4 \tag{11}$$

which enables eq. (10) to be expressed in potential form as

$$\left(\frac{\partial u}{\partial t}\right)^2 + \pi(u) = 0 \tag{12}$$

where the potential function $\pi(u)$ is given by

$$\pi(u) = u^{-2}[C - (\gamma+\delta)^2 - \frac{\kappa}{2} u^4][C + (\gamma-\delta)u^2 - \frac{\kappa}{2} u^4] \tag{13}$$

We shall concentrate on the case $C = 0$ implied e.g. by $u(0) = 0$. $\pi(u)$ then simplifies to

$$\pi(u) = -u^2[\gamma^2 - (\delta + \frac{\kappa}{2} u^2)^2] \tag{14}$$

indicating instability for $u = 0$ if $\gamma > |\delta|$. The maximal oscillation amplitude in this case is given by $u_+ = [2(\gamma-\delta)/\kappa]^{1/2}$ and the dynamic behaviour of the amplitude is a bounded oscillation in the potential valley between $u = 0$ and $u = u_+$.

The case of nonzero initial amplitude gives rise to several different possibilities depending on the value of C. An instructive situation occurs if C = 0 and $\delta < -\gamma$, which implies the possibility of solutions oscillating in a new potential valley between the zeroes $u_\pm$ of $\pi(u)$

$$u_\pm = [\frac{2}{\kappa}(\pm\gamma-\delta)]^{1/2} \tag{15}$$

In this situation, although the oscillator is stable for u(0) = 0, an initial nonzero amplitude may contribute a nonlinear frequency shift, which places the effective frequency within the resonance domain with subsequent parametric growth.

## INFLUENCE OF SPONTANEOUS EMISSION

In the presence of spontaneous emission terms (R and S) eq. (1) becomes

$$\frac{dI_0}{dt} = \gamma_0 I_0 - \alpha I_0 I_1 + R$$

$$\frac{dI_1}{dt} = -\gamma_1 I_1 + \alpha I_0 I_1 + S \tag{16}$$

It has previously been shown, [1], that this modification makes the relaxation oscillations approach the corresponding steady state values, with a damping rate $\gamma_s = \alpha(R/\gamma_1+S/\gamma_0)/2$ and a frequency shift $\delta_s = \alpha(R-S)/2\omega_0$. Including the spontaneous emission terms in eq. (9), one obtains the following secularity equation

$$\left[\frac{\partial}{\partial t_2} + i(\delta-\delta_s) + \gamma_s + i\kappa|a|^2\right] a = \gamma a^* \tag{17}$$

which implies a saturation level $a_s$ given by

$$|a_s|^2 = \frac{1}{\kappa}\left[(\gamma^2-\gamma_s^2)^{1/2} - (\delta-\delta_s)\right] \tag{18}$$

Summarizing, we have obtained all relevant quantities, which characterize the parametric instability, viz. threshold, initial growthrate and final saturation level, as explicit functions of the physical parameters of the system. A numerical study of the parametric system has also been made which gives very good agreement with the

the predictions of the theoretical analysis, cf fig. 2

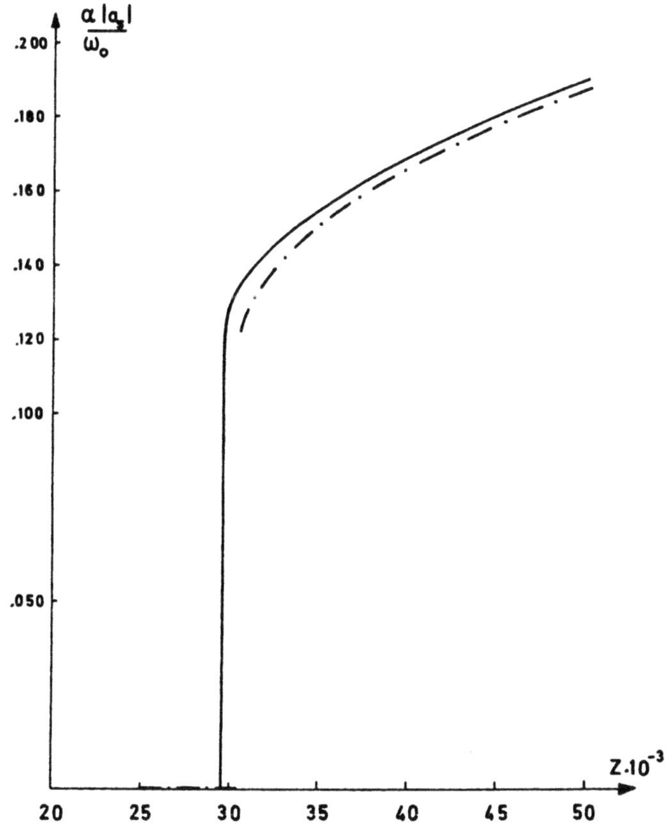


Fig. 2. Comparison between the theoretical (——) and numerical (— • —) results for the saturation amplitude ($\alpha/\omega_0|a_s|$) as a function of the degree of modulation (z) for the case $\gamma_1/\gamma_0 = 1$, $\alpha/\omega_0^2\ R = 0{,}01$, $\alpha/\omega_0^2\ S = 0{,}001$ and $\Omega/\omega_0 = 2$.

REFERENCES


1. White, R.B., Y.C. Lee, K. Nishikawa, Phys. Rev. Lett. 29, 1315 (1972).

2. Hasegawa, A., Plasma Instabilities and Nonlinear Effects, pp 183-188, Springer Verlag, Berlin 1975.

# SELF-SIMILAR SOLUTIONS OF NONLINEAR EVOLUTION EQUATIONS OF PHYSICAL SIGNIFICANCE

R. Nakach

ASSOCIATION EURATOM-CEA
Département de Physique du Plasma et de la Fusion Controlée
Service IGn - Centre d'Etudes Nucléaires
85X - 38041 GRENOBLE CEDEX, France

## INTRODUCTION

One of the most remarkable recent advances in mathematical physics is a method of solution for certain classes of nonlinear partial differential equations which arise naturally in many scientific areas, and particularly in nonlinear plasma physics.

This method which can, in fact, be reduced to an inverse scattering problem was invented in 1967 by Gardner, Greene, Kruskal and Miura [1], for the solution of the initial-value problem of the Korteweg - de Vries (KdV) equation. It was generalized in 1968 by Lax [2] who was able to show that the same technique could be applied to a large class of nonlinear differential equations.

Consider a nonlinear evolution equation of the form

$$u_t = K(u) \qquad (1)$$

where u is the unknown vector function and K denotes a nonlinear operator acting on a suitable space of functions.

If it is possible to find two linear operators $\hat{L}$ and $\hat{A}$, containing linearly the function u(x,t), and for which the equation

$$L_t = i[\hat{L}, \hat{A}] = i(\hat{L}\hat{A} - \hat{A}\hat{L}) \qquad (2)$$

is satisfied when u evolves according to Eq. (1), then it turns out that the Eq. (1) can in certain cases by solved exactly by the inverse scattering method.

Let us consider the eigen-value problem for the $\hat{L}$ operator

$$\hat{L}\psi = \lambda\psi \tag{3}$$

As can easily be demonstrated the eigen-values $\lambda$ are independent of time and are accordingly constants of motion of Eq. (1).

Furthermore, the eigen-functions $\psi$ evolve in time as described by the equation

$$\psi_t = -i\hat{A}\psi + a(t)\psi \tag{4}$$

where a(t) is an arbitrary function of time which accounts for the fact that the normalization of the eigen-functions may vary arbitrarily in time.

Assuming the initial function u(x,0) given, the solution u(x,t) may in principle be determined in three steps, each including only linear calculations, as follows:

i) The solution of the direct scattering problem consists in determining the scattering parameters (reflexion and transmission coefficients associated with the operator $\hat{L}$) for the eigen-functions $\psi$ at $x \to \pm\infty$ and for t = 0, for the particular initial function u(x,0), the time playing the rôle of a parameter.

ii) the temporal evolution of these scattering data can be determined by means of Eq. (4) and the asymptotic form of the operator $\hat{A}$ when $x \to \pm\infty$.

iii) Finally the inverse scattering problem consists in reconstructing the potential u(x,t) from the knowledge of the scattering parameters for which the time dependence has been settled.

It should be noticed that when the two operators $\hat{L}$ and $\hat{A}$ are self-adjoint the Eq. (2) means that there exists a unitary equivalent relation between the various values of the operator $\hat{L}$ as a function of time.

Certain fundamental equations which describe nonlinear and dispersive effects in plasmas can be expressed in the form of the operator equation (2), for example the well-known KdV equation for which the $\hat{L}$ associated operator is the one-dimensional Schrödinger operator.

Besides, Zakharov and Shabat [3], [4], have established a formal theory of self-modulation and self-focusing phenomena for plane waves in a one-dimensional and nonlinear medium, by reducing the so-called nonlinear Schrödinger equation

$$iu_t + u_{xx} + k|u|^2 u = 0 \tag{5}$$

to an inverse scattering problem, and by giving the $\hat{L}$ and $\hat{A}$ associated operators needed for the solution of the complete inverse scattering problem. In this case the $\hat{L}$ operator is again a differential second-order operator (the Dirac matrix operator).

An other example has been given by Zakharov [5] who succeeded in interpreting remarkably the problem of the anomalously slow stochastization of one-dimensional chains of nonlinear coupled oscillators which is generally refered to as the Fermi-Pasta-Ulam problem [6].

For one of the most important chains, he has considered the continuous analogue equation that is the following nonlinear string equation

$$u_{tt} = u_{xx} + (u^2)_{xx} + \frac{1}{4} u_{xxxx} \tag{6}$$

Zakharov [5] has proved that this last Eq. (6) could still be solved by the inverse scattering technique, and he has given the form of the operators $\hat{L}$ and $\hat{A}$, but in spite of this not been successful in obtaining a full explicit solution to the problem. This is due to the fact that the $\hat{L}$ operator is in this case of third differential order.

Zakharov and Manakov [7] have also investigated, by the same method, the problem of the nonlinear resonant interaction of three wave-packets. In this case, the $\hat{L}$ operator is still of third differential order.

Finally, a method for solving the "sine-Gordon" equation

$$\frac{\partial^2 u}{\partial x \partial t} = \sin u , \tag{7}$$

which arises in many fields of nonlinear physics, has been established in an analogous way by Ablowitz et al. [8].

In this case, the $\hat{L}$ operator is a matrix differential second-order operator and the full solution of the problem could be obtained.

It is a well-known fact that the direct scattering problem could be given an explicit solution only when the $\hat{L}$ operator is of second order (for example the one-dimensional Schrödinger operator, or the matrix Dirac operator), but the inverse scattering theory for a third-order operator is a difficult and as yet unsolved mathematical problem.

The aim of the present paper is to show that the nonlinear wave-equations of mathematical physics which can be solved by the inverse scattering technique as an initial-value problem, possess also other interesting properties namely the following:

In addition to the well-known stationary solutions in the form of solitary waves, these equations admit non-stationary self-similar solutions and it seems furthermore that, in the case where the $\hat{L}$ operator associated with these equations, is of second order and only then, the solutions of the ordinary self-similar differential equations can be expressed in terms of the various Painlevé transcendents[9], which means that the general solution of these equations is free from movable essential singularities.

It thus seems that the difficulties associated with the solution of a scattering problem, when the operator is of more than second order, are connected with the existence of singularities of more complex nature than those of the Painlevé transcendents.

In order to elucidate this question, we here investigate some examples of nonlinear evolution equations relevant to the inverse scattering method and particularly interesting from the point of view of nonlinear physics.

i) Korteweg-de Vries Equation and Modified Korteweg-de Vries Equation

The non-stationary self-similar solutions of these equations are connected with the second Painlevé transcendent.

ii) "Nonlinear Schrödinger" Equation

The self-similar solution of this equation is connected with the forth Painlevé transcendent.

iii) "Sine-Gordon" and "Sine-hyperbolic-Gordon" Equations

The self-similar solutions can be expressed in terms of the third Painlevé transcendent.

iv. Nonlinear String Equations

For the Eq. (6) which has been investigated by Zakharov in the Fermi - Pasta - Ulam problem, the L operator is of third differential order. It will be shown that the self-similar solution exists but cannot be expressed in terms of the Painlevé transcendents except asymptotically (the solution is then given in terms of the first Painlevé transcendent).

v. Nonlinear System of Three-Wave Packets

For the problem of the resonant interaction of three-wave packets in a nonlinear and dissipationless medium, the $\hat{L}$ operator is a matrix third differential order operator.

In this case it does not seem possible to express the self-similar solutions in terms of the Painlevé transcendents. On the other hand, if we suppose that two of the three group velocities are equal, then the L operator turns out to be of second-order and the self-similar solution can be reduced to the "sine-Gordon" equation.

## I. KORTEWEG-DE VRIES AND MODIFIED KORTEWEG-DE VRIES EQUATIONS

As regards the theory of the inverse scattering the KdV and mKdV equations were the first investigated. Among the equations of the form:

$$u_t + u^p u_x + u_{xxx} = 0 \qquad (p = 1,2....) \tag{8}$$

only the KdV and mKdV equations, corresponding respectively to p = 1 and p = 2, possess an infinite set of "polynomial laws" of conservation and can be solved by the inverse scattering method.

As shown by Miura [10] a remarkable explicit nonlinear transformation relating the solutions of the two equations can be given. In fact, if v satisfies Eq. (8) with p = 2 (mKdV equation) and u the Eq. (8) with p = 1 (KdV equation), we have then the following relation between the two solutions:

$$u \equiv v^2 \pm (-\sigma)^{1/2}\, v_x \tag{9}$$

As we will see, we will find again the same result for the self-similar solutions.

Notice here that in the case of the KdV equation, the nonlinear term is given by $uu_x$ and the $\hat{L}$ associated operator is the one-dimensional linear Schrödinger operator [1], [2]

$$L = \frac{\partial^2}{\partial x^2} + \frac{1}{6} u(x,t)$$

On the other hand, for the mKdV equation the nonlinear term is given by $u^2 u_x$ and corresponds physically to the propagation of conic but not plane perturbations. In this case the $\hat{L}$ operator is the following [12], [13] matrix second order operator:

$$\hat{L} = \begin{pmatrix} 0 & -\frac{\partial}{\partial x} + iu(x,t) \\ \frac{\partial}{\partial x} + iu(x,t) & 0 \end{pmatrix}$$

and is not self-adjoint.

It has been shown [14] that the non-stationary self-similar solution of the KdV equation written in the form:

$$u_t + uu_x - \frac{1}{3} u_{xxx} = 0 \qquad (10)$$

are given by the following expressions:

$$u = (t_0-t)^{-2/3} F(X) \quad ,$$

$$X = (x-x_0)(t_0-t)^{-1/3} \quad ,$$

and

$$F = 2(Z' + Z^2) \quad , \qquad (11)$$

where the function Z(X) is any solution of the following equation which defines the second Painlevé transcendent [9]

$$Z'' = 2Z^3 + ZX + \mu \qquad (\mu = \text{constant}) \qquad (12)$$

and the prime sign denotes the derivative of the function F with respect to the self-similar variable X, a notation that will be adopted in the following.

Let us consider now the self-similar solution of the mKdV equation which can be conveniently written in the form:

$$u_t + 2u^2 u_x - \frac{1}{3} u_{xxx} = 0 \tag{13}$$

This equation obeys, like all the nonlinear evolution equations that we will study, a similarity principle which can be found by dimensionality considerations. However, a systematic way for obtaining the self-similar equations associated with the nonlinear evolution equation is given by the following substitution:

$$u = (t_0 - t)^{\alpha} F(X) \quad , \qquad X = (x - x_0)(t_0 - t)^{\beta} \quad , \tag{14}$$

$\alpha$ and $\beta$ being coefficients to be determined.

In the case of Eq. (13) we obtain:

$$\alpha = -\frac{1}{3} \quad , \quad \beta = -\frac{1}{3}$$

and the following ordinary differential equation for the function F(X):

$$F''' - 6F^2F' - F - XF' = 0 \tag{15}$$

The solution of this last equation (15) is immediately written

$$F = Z(X) \tag{16}$$

where Z(X) is any solution of Eq. (12) in which $\mu$ is a constant of integration. As can easily be seen, there exists an analogous relation to the expression (9) between the expressions (11) and (16) for the self-similar solutions.

## II. NONLINEAR SCHRÖDINGER EQUATION

A great deal of literature has been devoted to the nonlinear interaction of monochromatic plane waves in a dispersive medium [15], [16], [17]. A simplified form of the so-called "parabolic" equation which arises in nonlinear optics, is the "nonlinear Schrödinger equation". In a one-dimensional system this equation describes the self-modulation and self-focusing phenomena and can be written in the standard form:

$$iu_t + u_{xx} + k|u|^2u = 0 \tag{5}$$

In this equation k is the coefficient associated with the nonlinear effects and the positive values of k correspond to the instability of waves and the negative values of k to propagation in a stable medium.

Zakharov and Shabat [3], [4], have shown that Eq. (5) could be written in the form of Eq. (2). The operators $\hat{L}$ and $\hat{A}$ are then given by the following expressions:

$$\hat{L} = \begin{pmatrix} i(1+p)\dfrac{\partial}{\partial x} & u \\ u & i(1-p)\dfrac{\partial}{\partial x} \end{pmatrix} \qquad \hat{A} = \begin{pmatrix} -p\dfrac{\partial^2}{\partial x^2} + \dfrac{|u|^2}{1+p} & iu_x \\ -iu_x & -p\dfrac{\partial^2}{\partial x^2} - \dfrac{|u|^2}{1-p} \end{pmatrix}$$

in which $k = 2/(1-p^2)$ (p = real).

Consider the self-similar solution of Eq. (5). By applying the substitution (14) we find the values $\alpha = -\frac{1}{2}$, $\beta = -\frac{1}{2}$ and the following ordinary equation for the function F(X):

$$\frac{i}{2}(F + XF') + F'' + k|F|^2F = 0 \tag{17}$$

By putting

$$F = \rho\, e^{i\Theta} \tag{18}$$

where ρ and Θ are real functions of X, Eq. (17) is equivalent to the following system:

$$\rho'' - \rho\Theta'^2 - \frac{X}{2}\rho\Theta' + k\rho^3 = 0$$

$$\rho\Theta'' + 2\rho'\Theta' + \frac{\rho}{2} + \frac{X\rho'}{2} = 0 \tag{19}$$

By making the change of variable

$$\Theta' + \frac{X}{2} = \phi \tag{20}$$

The system (19) becomes:

$$\rho'' - \rho\phi^2 + k\rho^3 + \frac{\rho X\phi}{2} = 0$$

$$\rho\phi' + 2\rho'\phi - \frac{X\rho'}{2} = 0 \tag{21}$$

This last system (21) has the following exact first integral

$$\rho'^2 + \rho^2\phi^2 + \frac{k\rho^4}{2} = \lambda_0 \tag{22}$$

where $\lambda_0$ is a constant of integration. If we now suppose that this last constant is zero, corresponding for example to $\rho = \rho' = 0$ at some point (let us say at infinity), it is then possible to establish a unique differential equation for the variable φ. By taking the derivative of the second equation of system (21) and eliminating the variables ρ, ρ', ρ", using Eqs. (21) and (22), we obtain

$$-\frac{k}{2}\rho^2 = \phi^2 + \frac{\phi'^2}{(2\phi - \frac{X}{2})^2} \tag{23}$$

and the following nonlinear differential equation for the variable φ:

$$(2\phi-\frac{X}{2})\,\phi''-\phi'^2-\frac{\phi'}{2}+\phi(2\phi-\frac{X}{2})^2(3\phi-\frac{X}{2})=0 \quad . \tag{24}$$

By making the change of variable:

$$\phi=\psi+\frac{X}{4} \tag{25}$$

Eq. (24) becomes:

$$\psi''=\frac{\psi'^2}{2\psi}-\frac{1}{32\psi}-2\psi(3\psi^2+X\psi+X^2/16) \quad . \tag{26}$$

One can verify that this last equation (26) satisfies a special case of the following equation which defines the fourth Painlevé transcendent [9]:

$$y''=\frac{y'^2}{2y}-\frac{a^2}{2y}+2(\xi^2-b)y+4\xi y^2+\frac{3}{2}y^3, \tag{27}$$

where the prime sign in Eq. (27) denotes the derivative of the y function with respect to the variable $\xi$, and where

$$b=0,\ a=2,\ y=4(i)^{-1/2}\psi,\ \xi=\frac{(i)^{-1/2}}{2}X.$$

## III. THE SINE-GORDON EQUATION

The sine-Gordon equation arises in many fields of mathematical physics.

Written conveniently in the form:

$$u_{xt}=\sin u$$

this equation has been given a solution due to Ablowitz et al. [8] by the inverse scattering method in a similar manner as have Zakharov and Shabat given for the nonlinear Schrödinger equation.

In this case, the $\hat{L}$ operator is still of second-order, and the explicit expressions for the $\hat{L}$ and $\hat{A}$ operators of the equation (2), associated with the sine-Gordon equation are given by Kato [12], in the following form:

$$\hat{L} = i \begin{pmatrix} 2\frac{\partial}{\partial x} & u_x \\ u_x & -2\frac{\partial}{\partial x} \end{pmatrix} \text{ and } \hat{A} = \frac{1}{2} \begin{pmatrix} \cos u & \sin u \\ \sin u & -\cos u \end{pmatrix} \hat{L}^{-1} .$$

The fact that the equation (7) is invariant under the transformation $\bar{x} = ax$, $\bar{t} = at$ implies the existence of the similarity variable $X = (x - x_0)(t - t_0)$.

By putting

$$u = F(X)$$

$$X = (x - x_0)(t - t_0) \tag{28}$$

the Eq. (7) is transformed into the following ordinary differential equation:

$$XF'' + F' = \sin F \tag{29}$$

By making now the change of variable $e^{iF} = Z$, the Eq. (29) becomes:

$$Z'' = \frac{Z'^2}{Z} - \frac{Z'}{X} + \frac{Z^2 - 1}{2X} \tag{30}$$

This last equation (30) represents, as can easily been seen, a special case of the following equation which defines the third Painlevé transcendent [9]

$$Z'' = \frac{Z'^2}{Z} - \frac{Z'}{X} + \frac{a}{Z} + \frac{b}{X} + c\,\frac{Z^2}{X} + d\,Z^3 .$$

It is interesting to notice that Eq. (29) also describes the nonlinear interaction of three linearly damped monochromatic waves. In fact, it has been shown [19] that, in the case where two of the three linear damping coefficients are equal, the solution for the evolution of the envelope is given in terms of a unique function $\psi(t)$ which, in the stable case, satisfies the following equation:

$$\frac{\partial^2\psi}{\partial t^2} + \nu_0 \frac{\partial\psi}{\partial t} - \Omega^2 \exp(-2\nu t)\ \sin\psi = \frac{16\Gamma^2}{\Omega^4} \exp(-2\nu_0 t)\ \frac{\cos\psi}{\sin^3\psi} \quad (31)$$

and in the unstable case, the following equation:

$$\frac{\partial^2\psi}{\partial t^2} + \nu_0 \frac{\partial\psi}{\partial t} - \Omega^2 \exp(-2\nu t)\ \mathrm{sh}\psi = \frac{16\Gamma^2}{\Omega^4} \exp(-2\nu_0 t)\ \frac{\mathrm{ch}\psi}{\mathrm{sh}^3\psi} \quad (32)$$

In these equations, $\nu$ is the identical damping coefficient for the two waves, $\nu_0$ is the coefficient of the third wave, $\Omega$ and $\Gamma$ are constants of motion which can be determined by the initial conditions.

If we now suppose that in Eq. (31) $\Gamma = 0$ and $\nu_0 = 0$, and if we make the change of variable $\exp(-2\nu t) = X$, we obtain the self-similar equation of the sine-Gordon equation:

$$X\psi'' + \psi' = \frac{\Omega^2}{4\nu^2}\ \sin\psi$$

where the prime sign denotes the derivative of $\psi$ with respect to X.

The solution of this last equation is thus connected with the third Painlevé transcendent.

## IV. NONLINEAR STRING EQUATION

In the Fermi - Pasta - Ulam [6] problem concerning the stochastization of nonlinear one-dimensional coupled chains of oscillators where a quadratic coupling term is assumed, Zakharov [5] has investigated the continuous analogue equation of a particular chain for which the most clearly pronounced "anomalous" behaviour was demonstrated. The nonlinear string equation obtained is conveniently written

$$u_{tt} = u_{xx} + (u^2)_{xx} + \frac{1}{4}\, u_{xxxx} \quad (6)$$

The equation (6) can be rewritten in the form of a system

$$\begin{aligned} u_t &= \phi_{xx} \\ \phi_t &= u + u^2 + \frac{1}{4}\, u_{xx} \end{aligned} \quad (33)$$

Zakharov [5] has given the expression for the two linear differential operators L and A corresponding to the system (33):

$$\hat{L} = i\frac{\partial^3}{\partial x^3} + i\left(u\frac{\partial}{\partial x} + \frac{\partial}{\partial x}u\right) + i\frac{\partial}{\partial x} - \left(\frac{4}{3}\right)^{1/2}\phi_x ,$$

$$\hat{A} = \left(\frac{3}{4}\right)^{1/2}\frac{\partial^2}{\partial x^2} + \left(\frac{4}{3}\right)^{1/2} u .$$

Consider the self-similar solutions of Eq. (6). It is convenient to make the simple change of variable

$$u + \frac{1}{2} = v ,$$

in order to obtain the following equation which admits self-similar solutions:

$$v_{tt} = 2v_x^2 + 2vv_{xx} + \frac{1}{4}v_{xxxx} \qquad (34)$$

By applying the substitution (14), we find in the case of Eq. (34), the coefficients $\alpha = -1$, $\beta = -\frac{1}{2}$.

Thus, we look for solutions of the form:

$$v = (t_0 - t)^{-1}F(X)$$
$$X = (x - x_0)(t_0 - t)^{-1/2} \qquad (35)$$

The Eq. (34) then becomes the following nonlinear ordinary equation for the function F(X):

$$\frac{1}{4}F'''' + 2F'^2 + 2FF'' - \frac{1}{4}X^2F'' - \frac{7}{4}XF' - 2F = 0 \qquad (36)$$

This last equation (36) admits in fact two exact first integrals. The first one can easily be found by multiplying Eq. (36) by X, and one obtains after integration

$$XF''' - F'' + 8XFF' - 4F^2 - X^3F' - 4X^2F = \lambda_0$$

where $\lambda_0$ is a constant.

In order to find the second integral, it is convenient to make now the change of variable:

$$F = G(X) + \frac{X^2}{2} \quad . \tag{38}$$

The Eq. (37) becomes:

$$XG''' - G'' + 8XGG' - 4G^2 + 3X^3G' = \lambda_0 + 1 \quad . \tag{39}$$

By putting

$$U = G'' + 4G^2 + \lambda_0 + 1 \quad , \tag{40}$$

the Eq. (39) can be written:

$$XU' - U + 3X^3G' = 0 \quad . \tag{41}$$

By multiplying now Eq. (41) by $2U/X^3$ one obtains after integration:

$$\frac{U^2}{X^2} + 3G'^2 + 8G^3 + 6(\lambda_0 + 1)G = \lambda_1 \quad , \tag{42}$$

where $\lambda_1$ is a second constant of integration.

By putting finally

$$U = X(H - 3XG) \quad , \tag{43}$$

where H is a new function of X, one obtains the following differential equation for the function H(X)

$$H''^2 + \frac{8}{9}H'^3 + 3(H - XH')^2 + 6(\lambda_0 + 1)H' - \lambda_1 = 0 \quad , \tag{44}$$

and the solution is given by

$$F = G + \frac{X^2}{2} = \frac{X^2}{2} + H'(X) \quad , \tag{45}$$

where H(X) is a solution of Eq. (44).

In spite of the fact that the first term of Eq. (44) is the square of a second derivative, it can be shown that Eq. (44) belongs to the class of nonlinear second order ordinary differential equations of polynomial class but does not belong to the subclass of the equations for which the solution is free from movable singularities, and thus the solution of Eq. (44) cannot be expressed in terms of the Painlevé transcendents.

On the other hand, it is easy to show that the asymptotic solution for $t \to t_0$, that is to say $X \to 0$, is given by an equation which defines the first Painlevé transcendent.

If in Eq. (41) one drops the term $3X^3G'$, one obtains $U = \lambda_2 X$ where $\lambda_2$ is a constant of integration and the Eq. (40) reads:

$$G'' + 4G^2 + \lambda_0 + 1 = \lambda_2 X \quad , \tag{46}$$

which is nearly the canonical equation for the first Painlevé transcendent.

## V. RESONANT INTERACTION OF WAVE PACKETS IN NONLINEAR MEDIA

We consider here the self-similar solutions of the problem of resonant interaction of three wave packets with frequencies $\omega_1$, $\omega_2$, $\omega_3$ and wave-vectors $\vec{k}_1$, $\vec{k}_2$, $\vec{k}_3$. As is well-known, the two types of interaction are possible, decay and explosive.

In decay interaction, the frequencies and wave vectors of the packets satisfy the relations [7]

$$\omega_1 = \omega_2 + \omega_3 \quad , \quad \vec{k}_1 = \vec{k}_2 + \vec{k}_3 \quad ,$$

and the complex envelopes $u_1$, $u_2$, $u_3$ of the packets in a one-dimensional dissipationless medium satisfy the system of equations

$$\begin{aligned} u_{1t} + v_1 u_{1x} &= u_2 u_3 \quad , \\ u_{2t} + v_2 u_{2x} &= u_1 u_3^* \quad , \\ u_{3t} + v_3 u_{3x} &= u_1 u_2^* \quad . \end{aligned} \tag{47}$$

In the nonlinearly "explosive" interaction, the frequencies and wave vectors of the packets fulfil the relations

$$\omega_1 + \omega_2 + \omega_3 = 0 \quad , \quad \vec{k}_1 + \vec{k}_2 + \vec{k}_3 = 0 \quad ,$$

while the complex envelopes satisfy the system

$$\begin{aligned} u_{1t} + v_1 u_{1x} &= u_2^* u_3^* \quad , \\ u_{2t} + v_2 u_{2x} &= u_1^* u_3^* \quad , \\ u_{3t} + v_3 u_{3x} &= u_1^* u_2^* \quad . \end{aligned} \tag{48}$$

Zakharov and Manakov [7] have shown that the inverse scattering method is also applicable to the systems (47) and (48), however, the $\hat{L}$ operator is in these cases of third differential order and they did not give the solution of the initial value problem.

Consider the self-similar solutions of the system (48) for example("explosive" case), we find the substitution:

$$\begin{aligned} u_i &= \frac{1}{t_0 - t} U_i(X) \quad , \\ X &= \frac{x - x_0}{t_0 - t} \quad . \end{aligned} \tag{49}$$

The system (48) becomes the following ordinary system of differential equations for the functions $U_i(X)$:

$$\begin{aligned} U_1 + (V_1 + X)\, U_1' &= U_2^* U_3^* \quad , \\ U_2 + (V_2 + X)\, U_2' &= U_1^* U_3^* \quad , \\ U_3 + (V_3 + X)\, U_3' &= U_1^* U_2^* \quad . \end{aligned} \tag{50}$$

In the case where all the group velocities $V_i$ are different, the system (50) does not admit any first integral and its analytical solution seem exceedingly difficult. Let us investigate the case where two group velocities are equal ($V_2 = V_3 = V$). In this case, the system (48) has a first integral and the $\hat{L}$ operator can be reduced to second order. The same is true for the self-similar system (50), and we obtain the integral:

$$(V + X)^2(U_2U_2^* - U_3U_3^*) = \lambda^2 \quad , \tag{51}$$

where $\lambda$ is a constant.

If besides we suppose, that the envelopes are real quantities, the system (50) becomes:

$$\begin{aligned} U_1 + (V_1 + X)\, U_1' &= U_2U_3 \quad , \\ U_2 + (V + X)\, U_2' &= U_1U_3 \quad , \\ U_3 + (V + X)\, U_3' &= U_1U_2 \quad , \end{aligned} \tag{52}$$

and the integral (51) reduces to

$$(V + X)^2(U_2^2 - U_3^2) = \lambda^2 \quad . \tag{53}$$

By making the following change of variable

$$\begin{aligned} U_2 &= \frac{\lambda\, \mathrm{ch}\, \Theta/2}{V+X} \quad , \\ U_3 &= \frac{\lambda\, \mathrm{sh}\, \Theta/2}{V+X} \quad , \end{aligned} \tag{54}$$

one obtains:

$$U_1 = \left(\frac{V + X}{2}\right) \frac{d\Theta}{dX} \quad , \tag{55}$$

and the differential equation for the function $\Theta(X)$:

$$(V_1 + X)(V + X)\frac{d^2\Theta}{dX^2} + (2X + V + V_1)\frac{d\Theta}{dX} = \frac{\lambda^2 \mathrm{sh}\Theta}{(V+X)^2} \quad . \tag{56}$$

By putting

$$\xi = \frac{V_1+X}{V+X} \tag{57}$$

the Eq. (56) reads:

$$\xi \frac{d^2\Theta}{d\xi^2} + \frac{d\Theta}{d\xi} = \frac{\lambda^2}{(V-V_1)^2} \,\mathrm{sh}\,\Theta \quad . \tag{58}$$

As can easily be seen, Eq. (58) is nothing but the self-similar sine-hyperbolic-Gordon equation. As in the case for the sine-Gordon equation,its solution can be expressed in terms of the third Painlevé transcendent.

This last example shows once more that the self-similar solutions of nonlinear evolution equations, which are relevant to the inverse scattering theory, and for which the associated $\hat{L}$ operator is of second order, can be expressed in terms of the various Painlevé transcendents.

REFERENCES

1. Gardner, C.S., J.M. Greene, M.D. Kruskal, R.M. Miura, Phys. Rev. Letters, 19, 1095, (1967).

2. Lax, P.D., Comm. on Pure and Appl. Math., 21, 467, (1968).

3. Zakharov, V.E., A.B. Shabat, Zh. Eksp. Teor. Fiz., 61, 118 (1971), (Sov. Phys. JETP, 34, 62, (1972)).

4. Zakharov, V.E., A.B. Shabat, Zh. Eksp. Teor. Fiz., 64, 1627, (1973), (Sov. Phys. JETP, 37, 823, (1973)).

5. Zakharov, V.E., Zh. Eksp. Teor. Fiz., 65, 219 (1973) (Sov. Phys. JETP, 38, 108 (1974)).

6. Fermi, E., J. Pasta, S. Ulam, Los Alamos Scientific Report LA-1940, (1955).

7. Zakharov, V.E., S.U. Manakov, Zh. Eksp. Teor. Fiz. Pis'ma Red., 18, 413, (1973)(JETP Lett., 18, 243, (1973)).

8. Ablowitz, M.J., D.J. Kaup, A.C. Newell, H. Segur, Phys. Rev. Letters, 30, 1262, (1973).

9. Ince, E.L., Ordinary Differential Equations, p. 345, (Dover 1956).

10. Miura, R.M., Journal of Math. Physics, 9, 1202, (1968).

11. Ablowitz, M.J., D.J. Kaup, A.C. Newell, H. Segur, Phys. Rev. Letters, 31, 125, (1973).

12. Kato, Y., Suppl. of the Progress of Theor. Physics, 55, 247, (1974).

13. Kodama, Y., Progress of Theor. Physics., 54, 669, (1975).

14. Nakach, R.D., Report EUR-CEA-FC 803, (1975).

15. Kadomtsev, B.B., V.I. Karpman, Sov. Phys. USPEKHI, 14, 40, (1971).

16. Bespalov, V.I., A.G. Litvak, V.I. Talanov, (Second All-union Symp. on Nonlinear Optics), 1966, collection of papers, NAUKA, 1968.

17. Karpman, V.I., E.M. Krushkal, Zh. Eksp. Teor. Fiz., 55, 530, (1968), (Sov. Phys. JETP, 28, 277, (1969)).

18. Lamb, Jr., G.L., Review of Modern Physics, 43, Part I, 99, (1971).

19. Nakach, R.D., H. Wilhelmsson, Phys. Review, Serie A, (To be published July 1976).

20. Rosenbluth, M.N., B. Coppi, R. Sudan, Proc. 3rd Intern. Conf. Plasma Physics and Contr. Nuclear Fusion Research, Novosibirsk, (1968).

# DIFFUSION DUE TO A SINGLE WAVE IN A MAGNETIZED PLASMA

Gary R. Smith and Allan N. Kaufman

Department of Physics and Lawrence Berkeley Laboratory

University of California, Berkeley, California 94720

The nature of charged-particle motion in the presence of a spectrum of waves usually depends on the width of the spectrum. In a narrow spectrum (modeled as a single wave), particles may be trapped in the potential wells of the wave and thereby have a limited acceleration. In a broad spectrum, resonant particles diffuse in velocity space, and thereby undergo a more extensive (stochastic) acceleration. In contrast to these well-known results we find[1] that a single wave in a magnetized plasma may cause particle diffusion.

In a magnetized plasma, a wave propagating at an oblique angle $\theta = \tan^{-1}(k_\perp/k_z)$ to a uniform magnetostatic field $B_o\hat{z}$ has a <u>set</u> of resonant parallel velocities

$$V_\ell = (\omega + \ell\Omega)/k_z \ , \ \ell = 0, \pm 1, \pm 2, \cdots ,$$

where the gyrofrequency $\Omega = eB_o/mc$ . Near each $V_\ell$ there is a trapping layer $(V_\ell \pm w_\ell)$ , with a half-width

$$w_\ell = 2 \mid e\Phi_o \, J_\ell(k_\perp \rho)/m \mid^{\frac{1}{2}} ,$$

for an electrostatic wave of amplitude $\Phi_o$ . For small $\Phi_o$ the oscillations of a particle's parallel velocity are limited to the trapping width $2w_\ell$ . For $\Phi_o$ large enough that the trapping layers overlap, i.e., roughly when

$$2w_\ell \gtrsim |v_{\ell+1} - v_\ell| = \Omega/k_z \, , \qquad (1)$$

a particle can move from one resonance region to another, changing its parallel velocity much more than $2w_\ell$ . In the latter case we call the particle motion stochastic.

We use Hamiltonian methods to study the motion of a particle in the field $B_o\hat{z}$ and a single electrostatic wave. We work in the wave frame, which moves with velocity $(\omega/k_z)\hat{z}$ with respect to the plasma. (We consider only waves with $\omega \ll k_z c$ .) The Hamiltonian is then

$$H = (\underline{p} + m\Omega y\hat{x})^2/2m + e\Phi_o \sin(k_z z + k_\perp y) \, .$$

Instead of the Cartesian variables $(x,y,p_x,p_y)$ we use a more convenient set of canonical variables. The gyrophase, $\phi = \tan^{-1}(v_x/v_y)$ , is conjugate to the angular momentum $p_\phi = \frac{1}{2} m\Omega\rho^2 = mv_\perp^2/2\Omega = (mc/e)\mu$ , where $\rho$ is the gyroradius and $\mu$ is the magnetic moment. The y-component of the guiding center, $Y = y + \rho \sin\phi$ , is conjugate to $m\Omega X = m\Omega(x - \rho\cos\phi)$ . In terms of these variables

$$H = p_z^2/2m + p_\phi\Omega + e\Phi_o \sin(k_z z + k_\perp Y - k_\perp\rho\sin\phi) \, .$$

We redefine the origin of $z$ by performing a canonical transformation to the new variables

$$z' = z + k_\perp Y/k_z \, , \; p_z' = p_z$$

$$Y' = Y \, , \; X' = X - k_\perp p_z/k_z m\Omega \, .$$

Since $Y'$ and $X'$ do not appear in the transformed Hamiltonian, they are each constants of the motion. Dropping the primes on $p_z'$ and $z'$ , we write

$$H = p_z^2/2m + p_\phi\Omega + e\Phi_o \sin(k_z z - k_\perp\rho\sin\phi) \, . \qquad (2)$$

If the wave amplitude is small, perturbation methods can be used to study the particle motion. There exists an additional constant of the motion $I$ ; Taylor and Laing[2] have found the

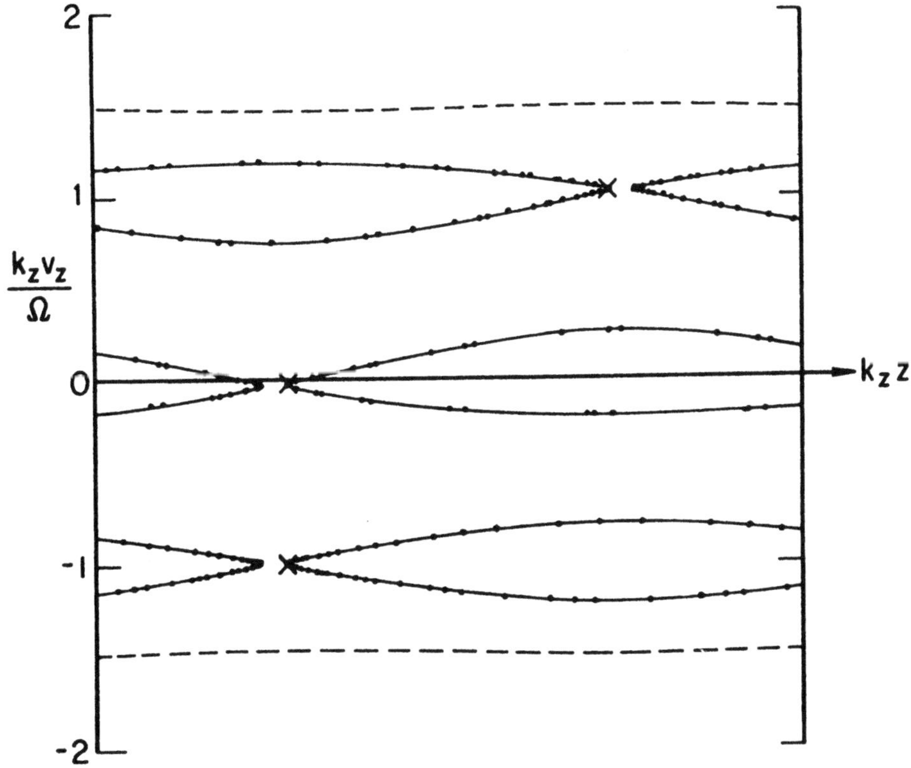


Fig. 1. Particle trajectories, calculated from Hamiltonian (2), represented by dots at the values of $z$ and $v_z$ whenever $\Phi = (2N + 1)\pi$, $N = 1,2,\cdots$. The three initial conditions, indicated by crosses, were chosen near the separatrices. The parameters are the same for all three trajectories: $k_\perp \rho_E \equiv k_\perp (2E/m)^{\frac{1}{2}}/\Omega = 1.48$, $(\tilde{\omega}/\Omega)^2 \equiv k_z^{\,2} |e\Phi_o|/m\Omega^2 = 0.025$, $\theta = 45^\circ$.

expression

$$I = \cos\left(\frac{\pi k_z v_z}{\Omega}\right) - \pi\left(\frac{k_z^{\,2} e\Phi_o}{m\Omega^2}\right)\sin\left(\frac{\pi k_z v_z}{\Omega}\right)\sum_{\ell} J_\ell \frac{\sin(k_z z - \ell\Phi)}{(k_z v_z/\Omega) - \ell} ,$$

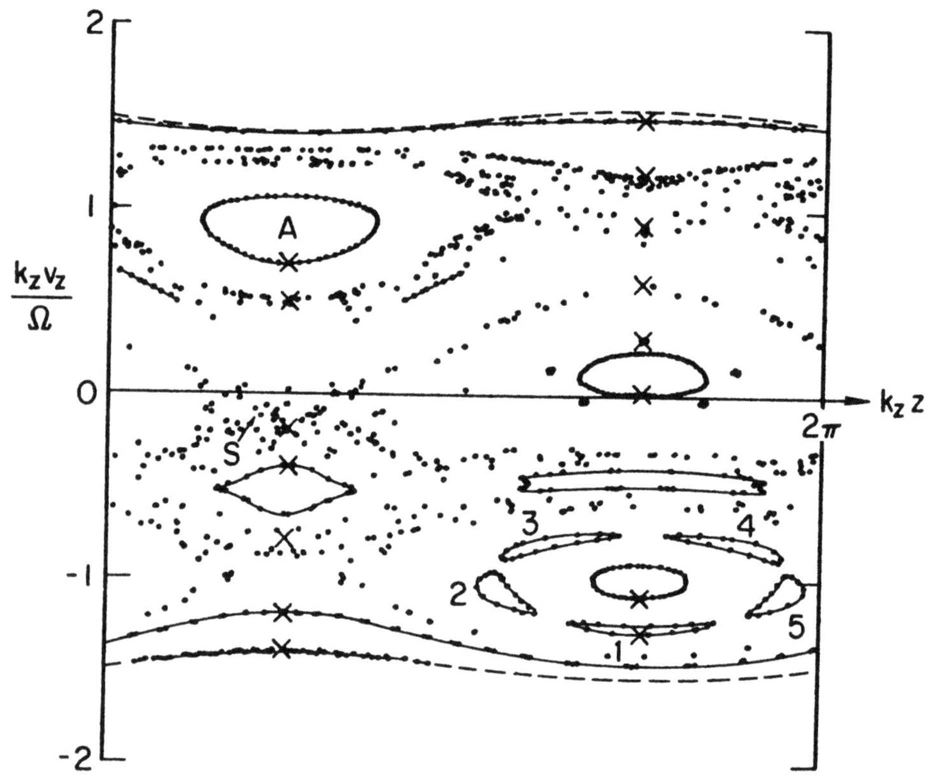


Fig. 2. Similar to Fig. 1 but for a larger wave amplitude, $(\tilde{\omega}/\Omega)^2 = 0.1$ . The plane is divided into adiabatic regions (like A) and stochastic regions (like S) . A chain of five islands is indicated by the numbers 1-5.

accurate to first order in $\Phi_0$ . The existence of the constant I is verified by the nature of the particle trajectories shown in Fig. 1. The dots representing a trajectory lie on a smooth curve only if a constant I exists.

We study the particle motion by numerically integrating the equations of motion derivable from (2) . A transition to stochastic motion occurs when the wave amplitude satisfies the threshold condition (1) . Near the threshold, the motion has the interesting character illustrated in Fig. 2. This type of

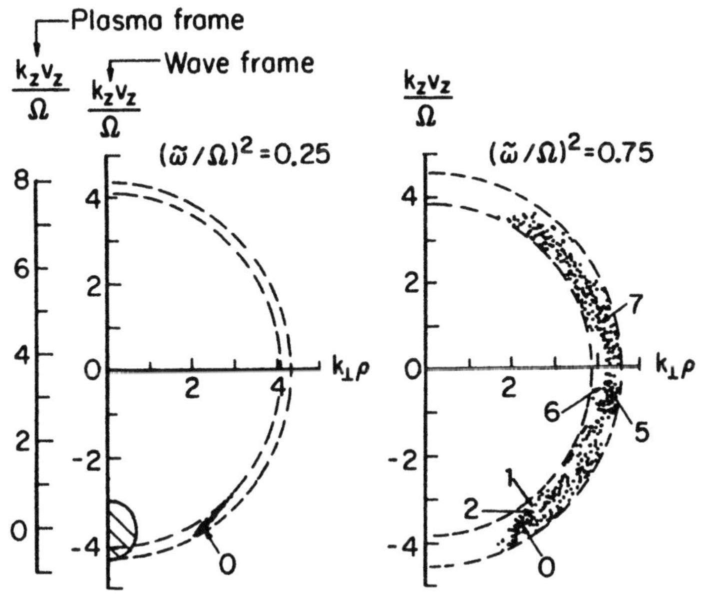


Fig. 3. Particle trajectories represented by $v_z$ and $\rho = v_\perp/\Omega$ whenever $2N\pi = \Phi - \pi \approx \Omega t$. The numbers 0,1,2,5,6,7 are the values of N for a trajectory initiated with a particular $z(t = 0)$. The wave amplitude is given by the parameter $(\tilde{\omega}/\Omega)^2$ ; $\theta = 45°$ and $k_z v/\Omega = 4.24$, where v is the initial speed in the wave frame.

motion has been observed for several other problems.[3-7]

Fig. 3 shows the particle motion in velocity space. In the small wave amplitude case $[(\tilde{\omega}/\Omega)^2 = 0.25]$, particles with initial conditions at the point marked 0 are accelerated from the initial parallel and perpendicular velocities by only small amounts. In the large wave amplitude case $[(\tilde{\omega}/\Omega)^2 = 0.75]$, the particle acceleration is substantial. The particle remains

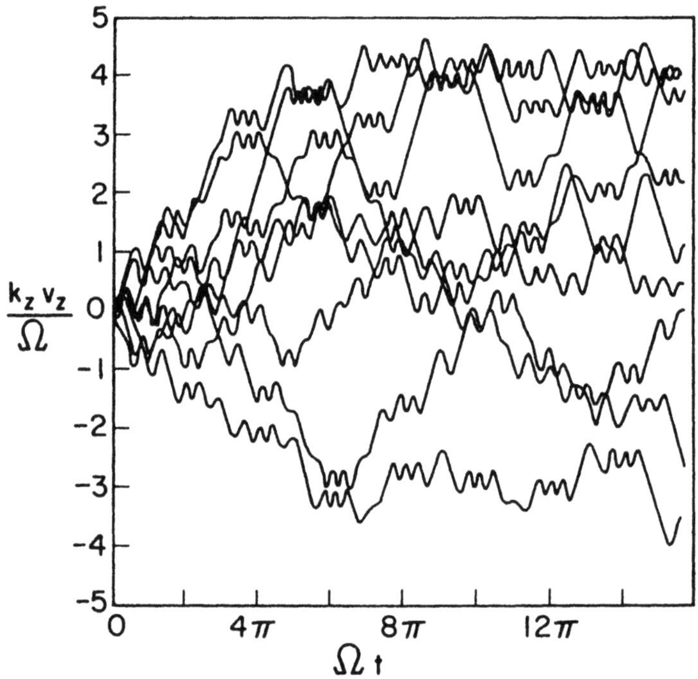


Fig. 4. Particle trajectories represented by $v_z$ vs. time. $(\tilde{\omega}/\Omega)^2 = 0.75$, $\theta = 45^\circ$, $k_z v/\Omega = 5$ and $v_z(t = 0) = 0$.

near a curve of constant speed in the wave frame (the dashed semicircles in Fig. 3), because the Hamiltonian (2) is a constant of the motion. In the plasma frame the particle's parallel and perpendicular energies can increase substantially.

The stochastic particle motion has the nature of a diffusion process. We establish this fact by numerically calculating a set of 50 to 200 trajectories. The initial velocity components ($v_z$ and $v_\perp$) are the same for the whole set, but the initial phases are distributed throughout $0 \leq k_z z$, $\Phi < 2\pi$. A subset

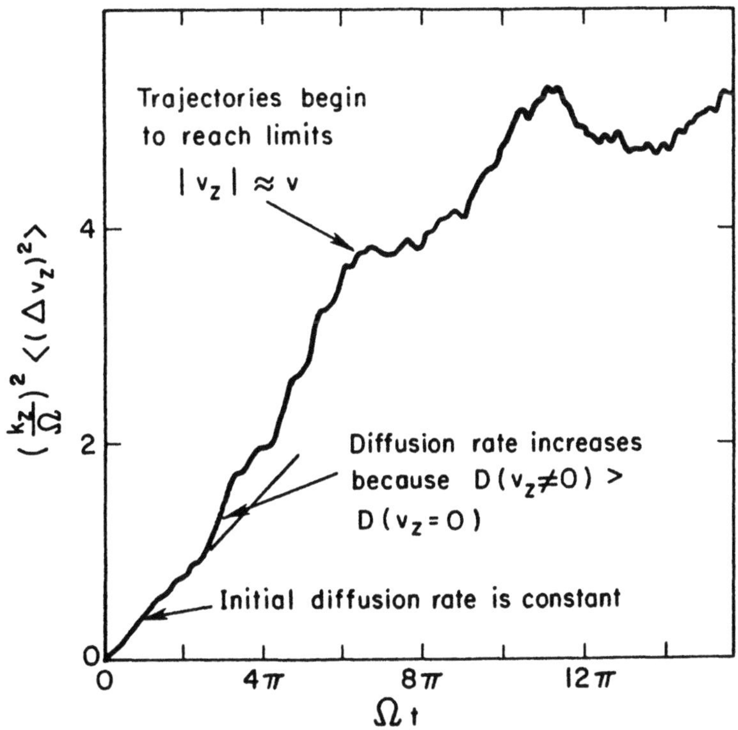


Fig. 5. Mean square deviation in parallel velocity vs. time for the same parameters as in Fig. 4.

of the trajectories is shown in Fig. 4. From the set of trajectories we calculate the mean square deviation in parallel velocity, $\langle[\Delta v_z(t)]^2\rangle$, and find that it increases linearly with time, as shown in Fig. 5.

To explain analytically the observed diffusion rates we use methods developed for the general theory of stochastic instability of nonlinear oscillations.[8] Using (2) and a Bessel function identity, we write the potential due to the wave as

$$\Phi(z,\phi,\rho) = \Phi_0 \sum_{\ell} J_{\ell}(k_{\perp}\rho) \sin(k_z z - \ell\phi) .$$

We replace $J_{\ell}$ by an average, $\bar{J}_{\ell}$, over $\ell$, with resonant values of $\ell$ (those near $k_z v_z/\Omega$) weighted most heavily. We then use another identity to write

$$\Phi(t) \approx \Phi_0 \bar{J}_{\ell} \sin[k_z z(t)] T \sum_{n} \delta(t - nT + \phi_0/\Omega) , \qquad (3)$$

where $T \equiv 2\pi/\Omega$ is the gyroperiod. The parallel velocity undergoes a sudden jump once each gyroperiod in this approximation. (Some trajectories in Fig. 4 show rapid changes in $v_z$ each gyroperiod.) The jump in $v_z$ is

$$\Delta v_z = -\frac{e}{m}\int_0^T dt \, \frac{\partial\Phi}{\partial z} . \qquad (4)$$

The diffusivity D is defined by

$$D \equiv \langle [\Delta v_z(t)]^2 \rangle / 2t , \qquad (5)$$

where the average is over the initial phases $z$ and $\phi$. Combining (3) - (5) and choosing the typical value 0.2 for $\bar{J}_{\ell}$, we find

$$D \sim (0.1)(k_z e\Phi_0/m)^2/\Omega .$$

The particle motion has a diffusive nature only when (1) is satisfied, which implies roughly

$$k_{\perp}\rho \gtrsim 1$$

$$|e\Phi_0| \gtrsim m(\Omega/k_z)^2 .$$

A more sophisticated theory is necessary to explain the dependence of the diffusivity D on $v_z$ and $\rho$. A rigorous theory for problems involving stochastic instability does not yet exist.[8] As a model to describe our observations we take

$$D \approx \frac{1}{2}(k_z e\Phi_0/m)^2 \sum_{\ell} J_{\ell}^{\,2}(k_{\perp}\rho) R_{\ell}(k_z v_z - \ell\Omega) . \qquad (6)$$

A quasilinear theory would yield a resonance function

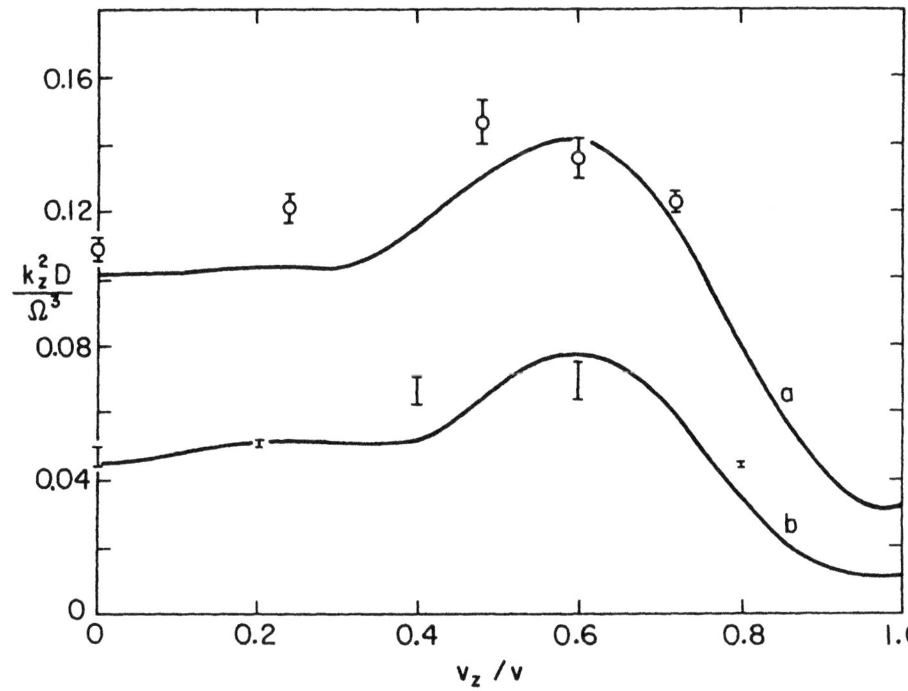


Fig. 6. Agreement between model (6) for the diffusivity D and the numerically observed diffusion, for several values of $v_z(t = 0)$ . $\theta = 45^o$ .

(a) $k_z v/\Omega = 5$ , $(\tilde{\omega}/\Omega)^2 = 1.00$ .

(b) $k_z v/\Omega = 6$ , $(\tilde{\omega}/\Omega)^2 = 0.75$ .

The measurement of the observed points and the error bars is somewhat subjective.

$$R_\ell(\omega) = \int_0^\infty d\tau \cos \omega\tau = \pi \delta(\omega) .$$

Instead we allow the resonance to be broadened:

$$R_\ell(\omega) = \int_0^\infty d\tau \cos \omega\tau \exp(-\nu_\ell \tau)$$
$$= \nu_\ell(\omega^2 + \nu_\ell^2)^{-1} .$$

The amount of broadening is determined by the trapping half-width:

$$\nu_\ell = k_z w_\ell \ .$$

(Conventional resonance broadening theories[9] do not adequately explain our observations.) Shown in Fig. 6 is the agreement between this model and the numerically observed diffusivity.

A specific wave which would lead to heating of an ion distribution is an ion-acoustic wave with frequency $\omega \approx kc_s > \Omega_i$ and oblique propagation angle $\theta$ (e.g., $45^\circ$). The cross-hatched region in Fig. 3 shows the extent of the thermal ions for $\omega = 3.6\ \Omega_i$ and $T_e = 16\ T_i$ . Ions in the tail of an initially Maxwellian distribution would be stochastically accelerated by a wave with $\delta n/n \gtrsim 0.1$ . A Langmuir wave with analogous properties would lead to heating of the tail of the electron distribution.

This work was supported by the U. S. Energy Research and Development Administration.

## References

1. G. R. Smith and A. N. Kaufman, Phys. Rev. Lett. 34, 1613 (1975).

2. J. B. Taylor and E. W. Laing, Phys. Rev. Lett. 35, 1306 (1975).

3. M. Hénon and C. Heiles, Astron. J. 69, 73 (1964). Particle in an anisotropic two-dimensional potential well.

4. D. A. Dunnett, E. W. Laing and J. B. Taylor, J. Math. Phys. 9, 1819 (1968). Particle in a spatially modulated magnetic field.

5. F. Jaeger, A. J. Lichtenberg and M. A. Lieberman, Plasma Phys. 14, 1073 (1972). Electron cyclotron resonance heating.

6. R. E. Aamodt, Phys. Rev. Lett. 27, 135 (1971); M. N. Rosenbluth, Phys. Rev. Lett. 29, 408 (1972); A. V. Timofeev, Nucl. Fusion 14, 165 (1974). Superadiabaticity in mirror machines.

7. J. M. Finn, Nucl. Fusion 15, 845 (1975). Magnetic field lines in tokamaks in the presence of resistive tearing modes.

8. G. M. Zaslavskii and B. V. Chirikov, Usp. Fiz. Nauk 105, 3 (1971) [Sov. Phys. Usp. 14, 549 (1972)].

9. T. H. Dupree, Phys. Fluids 9, 1773 (1966); J. Weinstock, Phys. Fluids 11, 1977 (1968).

# THE FREE ELECTRON LASER*

T. Kwan, J. M. Dawson, and A. T. Lin

Department of Physics, University of California

Los Angeles, California 90024

## ABSTRACT

We have investigated the possibility of a new type of laser by computer simulation using a 1-2/2 dimensional fully relativistic electromagnetic particle code. By passing a relativistic electron beam over a rippled static magnetic field, high frequency electromagnetic radiation is generated. If the ripple wavelength is $\lambda_o$, the lasing wavelength is roughly $\lambda_o/2\gamma^2$. Thus such a laser is continuously tunable by varying $\gamma$. We also find that as much as 35% of the beam energy can be converted into radiation, of which as much as nearly 90% can be in the most rapidly growing mode. A theory of the coupling between the negative energy plasma wave and electromagnetic radiation by means of the rippled magnetic field is presented. We have obtained good agreements between the simulation and the theory. The saturation mechanism is found to be the trapping of the beam by the unstable plasma wave. We are also able to estimate theoretically the amount of energy that can be converted into radiation from the electron beam.

## I. INTRODUCTION

Sources of tunable coherent electromagnetic radiation are highly desirable in the application of many areas, particularly in plasma heating and diagnosis as well as in laser-pellet fusion. In recent years, extensive efforts have been devoted to finding new mechanisms for the production of electromagnetic radiation by relativistic electron beam interacting with a spatially modulated static magnetic field. Recent experiments of L. R. Elias and his collaborators at Stanford[1,2] demonstrated that a relativistic electron beam passing over a rippled magnetic field can amplify light waves of 10.6 μm. Also, generation of large amounts of microwave power has been detected in similar experiments at NRL.[3-6] There has been some theoretical work on the microwave generation by an intense relativistic electron beam in a rippled magnetic field inside a waveguide. W. M. Manheimer and E. Ott investigated the stability of a waveguide which encloses a relativistic electron beam in a rippled magnetic field.[7] P. Sprangle theoretically analyzed the Ubitron which is a microwave generator based upon fast-wave excitation of a magnetically modulated relativistic electron beam.[8] P. Sprangle, et al., studied the scattering of an electromagnetic wave by a magnetized relativistic electron beam and showed that large frequency conversion could occur for the forward-scattered electromagnetic wave.[9] The interaction of a nonrelativistic electron beam with a rippled magnetic field was investigated by Lin and Dawson using computer simulation.[10] They found that 30% of the beam energy could be converted into radiation which had frequency near the plasma frequency.

In this work, we investigate the possibility of this new type of laser by computer simulation using a 1-2/2 dimensional (1 r coordinate, 3 velocity coordinates) fully relativistic electromagnetic particle code and by theoretical analysis. We passed a relativistic electron beam over a helical static magnetic field and we observed the generation of electromagnetic radiation with the predicted frequency (as large as 14 $\omega_{pe}$ in one of the simulations). The physical situation is shown schematically in Fig. 1 where $\lambda_o$ is the wavelength of the helical magnetic field. From kinematic

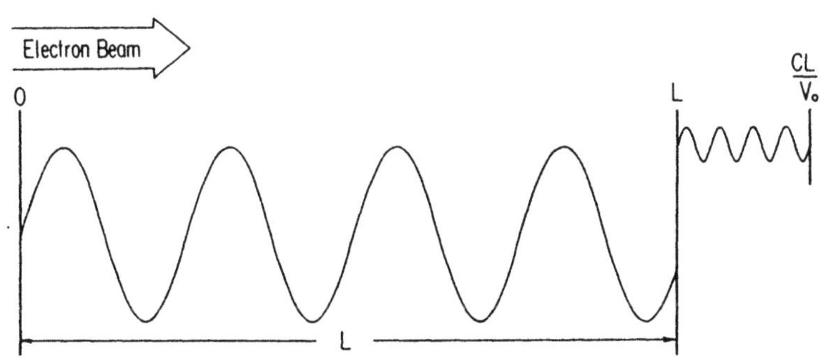


Figure 1. Schematic drawing of the interaction of a relativistic electron beam with a rippled magnetic field. The lasing wavelength is $\lambda = \lambda_o(c/v_o - 1) \simeq \lambda_o/2\gamma^2$, where $\lambda_o$ is the wavelength of the rippled magnetic field.

argument, the lasing wavelength is $\lambda = \lambda_o(c/v_o - 1) \simeq \lambda_o/2\gamma^2$. In our computer experiments, we have found as much as 35% of the beam energy converted into radiation, of which as much as 90% was in the most rapidly growing mode.

In section II, we formulate the theory of the coupling between the negative energy plasma wave and the electromagnetic wave through the rippled magnetic field. Numerical solution of the dispersion relation is presented and growth rates are calculated for various parameters. In section III, results from computer experiments are given and compared to the theory in details. Nonlinear saturation of the instability is discussed in section IV. An estimate of the saturation level is given and found to predict the observed value accurately.

## II. THEORY

In the following analysis, we often write equations in momentum space rather than velocity space for the sake of convenience. The momentum of a relativistic particle is given in terms of its velocity by

$$\underline{p} = m_o\gamma\underline{v} = \frac{m_o\underline{v}}{(1 - v^2/c^2)^{1/2}} \tag{1}$$

and $\underline{v}$ is given in terms of $\underline{p}$ by

$$\underline{v} = \frac{\underline{p}}{m_o(1+p^2/m_o^2c^2)^{1/2}} = \frac{\underline{p}}{m_o\gamma} \tag{2}$$

where

$$\gamma = 1/(1 - v^2/c^2)^{1/2} = (1 + p^2/m_o^2c^2)^{1/2} \quad .$$

We take the externally imposed static magnetic field to be

$$\underline{B}_{o\perp} = B_o(\hat{\underline{e}}_y \cos k_ox + \hat{\underline{e}}_z \sin k_ox) \tag{3}$$

where $B_o$ is a constant which measures the strength of the magnetic field and $k_o$ is the wave number. This is a model for an externally imposed magnetic field which is not curl free. The intense relativistic electron beam is traversing in the x-direction as shown in Fig. 1. In the beam frame, the static helical magnetic field looks like an intense electromagnetic pump wave and under proper circumstances, parametric instability can take place. The instability will give rise to simultaneous growth of the negative energy longitudinal plasma waves and an electromagnetic wave. Treating the ions as an immobile neutralizing background, we will derive the dispersion relation. Then the dispersion relation is solved approximately and we obtain an analytic form of the growth rate.

The electric and magnetic field quantities can be derived from scalar and vector potentials.

$$\underline{E} = -\underline{\nabla}\phi - \frac{1}{c}\frac{\partial}{\partial t}\underline{A} \tag{4}$$

$$\underline{B} = \underline{\nabla} \times \underline{A} \quad . \tag{5}$$

The Coulomb gauge is chosen so that $\underline{\nabla} \cdot \underline{A} = 0$. From Ampere's law, we obtain the wave equation describing the transverse component of the vector potential.

$$\frac{\partial^2}{\partial x^2}\underline{A_\perp} - \frac{1}{c^2}\frac{\partial^2 A_\perp}{\partial^2 t} = -\frac{4\pi}{c}\underline{J_\perp} \tag{6}$$

We linearize writing $A_\perp = \underline{A}_{o\perp} + \underline{A_\perp}$, where $A_{o\perp}$ and $A_\perp$ are the zero order and the first order vector potential respectively. Since the transverse coordinates (y,z) are ignorable coordinates in the problem, the transverse canonical momentum of a particle is a constant of motion. Thus we have

$$\underline{p_\perp} = \frac{e}{c}\underline{A_\perp} \tag{7}$$

where $p_\perp$ is the first order perturbation of the transverse kinetic momentum. The current density is given by

$$\underline{J_\perp} = -en\underline{v_\perp} \quad . \tag{8}$$

Substituting Eq. (2) into (8) and linearizing, we obtain

$$\underline{J_\perp} = -\frac{en_o}{m_o\gamma_o}\underline{p_\perp} - \frac{en}{m_o\gamma_o}\underline{p}_{o\perp} \tag{9}$$

where $\underline{J}_{o\perp}$ is the first order transverse current density, $n_o$ and n are the zeroth order and the first order electron density, respectively, and

$$\gamma_o = \frac{1}{[1-(v_o^2/c^2)]^{1/2}} = [1 + (p_{ox}^2/m_o^2c^2)]^{1/2} \quad .$$

In the linearization, the relativistic factor γ is approximated by

$$\gamma = \gamma_o(1 + \gamma_o^2\underline{v}_o\cdot\underline{v}/c^2)$$

where $\underline{v}$ is the perturbation in the electron velocity. The linearized transverse current density given by Eq. (9) is an approximation only, but it is sufficiently accurate for the analysis. It can be shown that the contributions from the other linearized terms to the current density is negligibly small. Hence, the linearized wave equation is

$$\left(\frac{\partial^2}{\partial x^2} - \frac{1}{c^2}\frac{\partial^2}{\partial t^2} - \frac{\omega_{pe}^2}{\gamma_o c^2}\right) A_\perp = \frac{4\pi e p_{o\perp}}{\gamma_o m_o c} n \qquad (10)$$

where $\omega_{pe}$ is defined to be $4\pi n_o e^2/m_o$.

The relativistic equation of motion is

$$\frac{d}{dt}\underline{p} = -e\left[\underline{E} + \frac{\underline{p} \times \underline{B}}{\gamma m_o c}\right] - \underline{\nabla} \cdot \underline{\underline{\Pi}} \qquad (11)$$

where $\underline{\underline{\Pi}}$ is the pressure tensor.

The zeroth order equation of motion is

$$\frac{d}{dt}\underline{p}_o = -e\frac{\underline{p}_o \times \underline{B}_o}{\gamma_o m_o c} \qquad (12)$$

and the linearized equation of motion is

$$\frac{d}{dt}\underline{p} = -e\left[\underline{E} + \frac{\underline{p}_o \times \underline{B} + \underline{p} \times \underline{B}_o}{\gamma_o m_o c} - \frac{(\underline{p}_o \times \underline{B}_o) p_{ox} p_x}{\gamma_o^3 m_o^3 c^3}\right]$$

$$- \frac{3T}{n_o}\frac{\partial n}{\partial x}\hat{\underline{e}}_x \qquad (13)$$

where $n_o$ and $T$ are the density and temperature of the electron beam in the lab frame. The steady state solution of Eq. (12) can be obtained immediately upon integration,

$$\underline{p}_o = p_{ox}\hat{\underline{e}}_x - \frac{m_o \omega_{ce}}{k_o}[\hat{\underline{e}}_y \cos k_o x + \hat{\underline{e}}_z \sin k_o x] \qquad (14)$$

where $\omega_{ce} = eB_o/m_o c$. The x-component of Eq. (13) describes the longitudinal motion of the electron fluid under the action of the static rippled magnetic field and the perturbed electric and magnetic fields; the transverse component only reviews the fact that the transverse canonical momentum is a constant of motion. Since $\underline{B} = \underline{\nabla}\times\underline{A}$ by Eqs. (3) and (14), we obtain

$$\left(\frac{\partial}{\partial t} + v_o \frac{\partial}{\partial x}\right) p_x = -eE_x + \frac{e\omega_{ce}}{\gamma_o kc}\left(\cos k_o x \frac{\partial}{\partial x} A_y + \sin k_o x \frac{\partial}{\partial x} A_z\right)$$

$$- \frac{e\omega_{ce}}{\gamma_o c}\left(A_y \sin k_o x - A_z \cos k_o x\right) - \frac{3T}{n_o}\frac{\partial n}{\partial x} . \tag{15}$$

Together with the linearized equation of continuity,

$$\frac{\partial}{\partial t} n + v_o \frac{\partial}{\partial x} n + \frac{n_o}{\gamma_o^3 m_o} \frac{\partial}{\partial x} p_x = 0 \tag{16}$$

Poisson equation

$$\frac{\partial}{\partial x} E_x = - 4\pi en \tag{17}$$

and the linearized wave equation, Eq. (10), we have a closed set of equations. We may employ the following dimensionless variables:

$$x \rightarrow x/\lambda_D, \qquad v \rightarrow v/\omega_{pe}\lambda_D, \qquad t \rightarrow \omega_{pe} t$$

$$n \rightarrow n/n_o, \qquad E \rightarrow eE/m_o\omega_{pe}c, \qquad A \rightarrow eA/m_o c^2$$

$$p \rightarrow p/m_o c, \qquad \omega \rightarrow \omega/\omega_{pe}$$

where $\omega_{pe}^2 = 4\pi n_o e^2/m_o$ and $\lambda_D{}^2 = T/4\pi n_o e^2$, and rewrite the set of equations as

$$\left(\frac{\partial^2}{\partial x^2} - \frac{1}{c^2}\frac{\partial^2}{\partial t^2} - \frac{1}{\gamma_o c^2}\right) A_\perp = \frac{p_{o\perp}}{\gamma_o c^2} n \tag{18}$$

$$\left(\frac{\partial}{\partial t} + v_o \frac{\partial}{\partial x}\right) p_x = -E_x + \frac{\omega_{ce}}{\gamma_o k_o}\left(\cos k_o x \frac{\partial}{\partial x} A_y + \sin k_o x \frac{\partial}{\partial x} A_z\right)$$

$$- \frac{\omega_{ce}}{\gamma_o}\left(A_y \sin k_o x - A_z \cos k_o x\right) - \frac{3}{c}\frac{\partial}{\partial x} n \tag{19}$$

$$\left(\frac{\partial}{\partial t} + v_o \frac{\partial}{\partial x}\right) n + \frac{c}{\gamma_o^3} \frac{\partial}{\partial x} p_x = 0 \tag{20}$$

$$\frac{\partial}{\partial x} E_x = -\frac{n}{c} . \tag{21}$$

We now assume the solution has the form

$$\underline{A_\perp} = \sum_k \underline{A_{\perp k}} e^{i(kx-\omega t)} \tag{22a}$$

$$n = \sum_k n_k e^{i(kx-\omega t)} \tag{22b}$$

$$p_x = \sum_k p_{xk} e^{i(kx-\omega t)} \tag{22c}$$

$$E_x = \sum_k E_{xk} e^{i(kx-\omega t)} \tag{22d}$$

After substituting into Eq. (18), we obtain

$$A_{yk} = -\frac{\omega_{ce}}{2\gamma_o k_o c[\omega^2 - k^2c^2 - (1/\gamma_o)]} (n_{k-k_o} + n_{k+k_o}) \tag{23a}$$

$$A_{zk} = \frac{i\omega_{ce}}{2\gamma_o k_o c[\omega^2 - k^2c^2 - (1/\gamma_o)]} (n_{k-k_o} - n_{k+k_o}) . \tag{23b}$$

From the equation of continuity (20) and Poisson equation (21), we get

$$p_{x,k+k_o} = i \gamma_o^3 \left(\omega - (k+k_o)v_o\right) E_{x,k+k_o} . \tag{24}$$

Substituting Eq. (22) into the equation of motion (19) and combining Eq. (23) and Eq. (24), we obtain the dispersion relation

$$\left[\left(\omega - (k+k_o)v_o\right)^2 - \frac{1}{\gamma_o^3}\left(1 + 3(k+k_o)^2\right)\right]\left[\omega^2 - k^2c^2 - \frac{1}{\gamma_o}\right] = \frac{\omega_{ce}^2 (k+k_o)^2}{2\gamma_o^5 k_o^2} . \tag{25}$$

The dispersion relation shows that the plasma oscillations are coupled to the electromagnetic waves through the rippled magnetic field. The dispersion diagram showing the uncoupled plasma oscillations and the electromagnetic waves is drawn in Fig. 2.

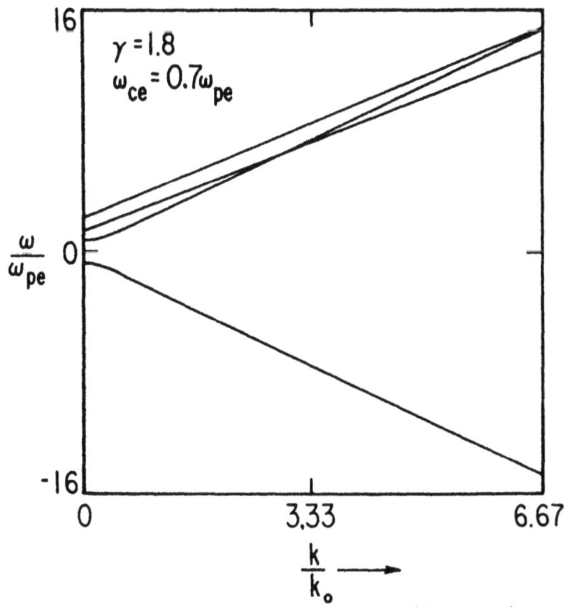


Figure 2. The uncoupled electromagnetic dispersion relation, $\omega^2 = \omega_{pe}^2/\gamma + k^2c^2$, and the electrostatic dispersion relation $\left(\omega - (k+k_o)v_o\right)^2 = \omega_{pe}^2\left(1 + (k+k_o)^2\lambda_D{}^2\right)/\gamma^3$.

Since the slow wave of plasma oscillation is a negative energy wave, we expect that its interaction with the electromagnetic wave will lead to instability. The unstable regions are the intercepts of the slow wave and the electromagnetic wave as shown. The dispersion relation is a fourth order equation in $\omega$. It is sufficiently accurate to obtain the solution by expanding the dispersion relation about $(\bar{\omega}_o, \bar{k}_o)$ provided the coupling term

$$Q = \frac{\omega_{ce}^2 (k+k_o)^2}{2\gamma_o{}^5 \, k_o{}^2}$$

is much less than unity. $(\bar{\omega}_o, \bar{k}_o)$ are the frequency and wave number at which the dispersion curves for the slow space-charge beam mode and the electromagnetic wave are simultaneously satisfied. The approximate solution is

$$\omega = T + \frac{-\mu \pm i(Q/TL - \mu^2)^{1/2}}{2} \qquad (26)$$

where

$$T = \left(\frac{1}{\gamma_o} + c^2k^2\right)^{1/2}$$

$$L = \frac{1}{\gamma_o^{3/2}} \left(1 + 3(k+k_o)^2\right)^{1/2}$$

$$D = (k+k_o)v_o$$

and

$$\mu = T + L - D \quad .$$

Equation (26) shows that for a given mismatch in frequencies, instability will occur if the strength of the rippled magnetic field is greater than a critical value. The threshold value of the magnetic field is given by

$$\omega_{ce} = \left(\frac{k_o}{k+k_o}\right) \mu \, \gamma_o^{5/2} \sqrt{2TL} \qquad (27)$$

and the growth rate is

$$\omega_i = \frac{1}{2}\left(\frac{Q}{TL} - \mu^2\right)^{1/2} \quad . \qquad (28)$$

One may also interpret Eq. (28) as giving the width of the unstable spectrum. For $\gamma_o >> 1$, $\omega_i \simeq \omega_{ce}/2\sqrt{k_o c}\gamma^{3/4}$.

A numerical study of the analytic result is carried out in parameter space ($\gamma_o$, $\omega_{ce}$). At first we study the dependence of the growth rate on the relativistic factor, $\gamma$. In Fig. 3, the growth rates are plotted against wavenumbers for a range of values of $\gamma$ from 1.6 to 2.2. The upper and lower set of curves correspond to cases with $\omega_{ce} = 0.7$ and $\omega_{ce} = 0.5$ respectively. The growth rate decreases monotonically as a function of $\gamma$ while the width of the unstable spectrum narrows infinitesimally. Next, we illustrate the dependence of the growth rate on the strength of the rippled magnetic field in Fig. 4. The growth rate and the width of the unstable spectrum are nearly linearly dependent on the magnitude of the rippled magnetic field.

In order to have high efficiency in energy conversion from the beam to the electromagnetic radiation, it is necessary that a large fraction of the energy loss from the beam during the instability ends up in the electromagnetic waves rather than the electrostatic

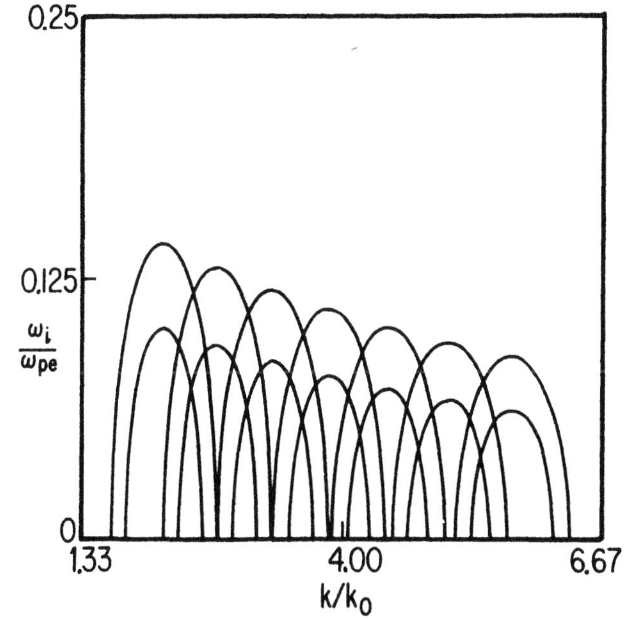


Figure 3. Theoretical growth rates versus wavelength numbers for $\gamma$ = 1.6 to 2.2. The upper and the lower sets of curves correspond to cases with $\omega_{ce}$ = 0.7 and $\omega_{ce}$ = 0.5.

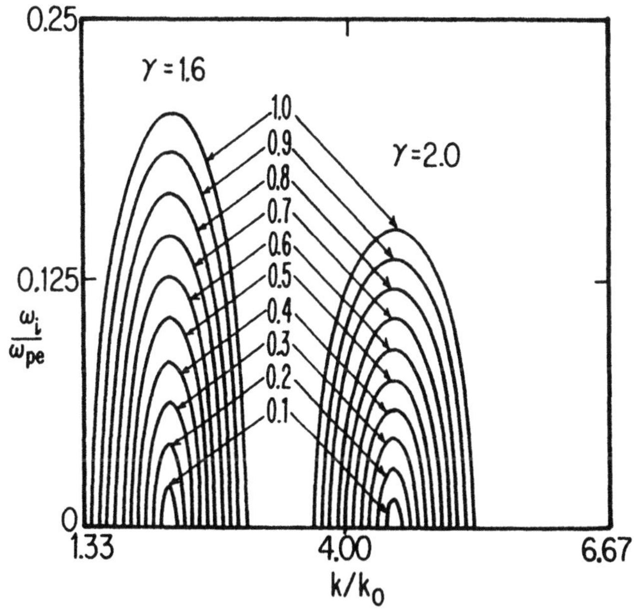


Figure 4. Theoretical growth rates for various strength of the magnetic field from $\omega_{ce}$ = 0.1 to 1.0 with $\gamma$ = 1.6 and $\gamma$ = 2.0.

waves. From Eqs. (4), (21), and (23), the amplitude of the electromagnetic wave can be written in terms of the amplitude of the electrostatic wave,

$$E_{y,k} = \frac{i\omega_{ce}(k+k_o)}{2\gamma_o k_o \omega \varepsilon_{em}(k,\omega)} E_{x,k+k_o}$$

and

$$E_{z,k} = E_{y,k}$$

where

$$\varepsilon_{em}(k,\omega) = \frac{k^2c^2}{\omega^2} + \frac{1}{\gamma_o \omega^2} - 1 \quad .$$

Thus,

$$W_{em} = \frac{\omega_{ce}^2 (k+k_o)^2}{\gamma_o^2\ k_o^2\ \omega^2\ |\varepsilon_{em}(k,\omega)|^2}$$

where $W_{em} = \frac{1}{4\pi}\,[E_{y,k}^2 + E_{z,k}^2]$ and $W_{es} = \frac{E_{x,k+k_o}^2}{8\pi}$ are the energies of the electromagnetic and the electrostatic waves. Since the unstable electromagnetic waves satisfy the electromagnetic dispersion relation approximately, we can write

$$\varepsilon_{em}(k,\omega) = \varepsilon_{em}(\bar{k},\bar{\omega}) + i\omega_i \left.\frac{\partial \varepsilon_{em}(k,\omega)}{\partial \omega}\right|_{\bar{k},\bar{\omega}}$$

where $\varepsilon_{em}(\bar{k},\bar{\omega}) = 0$ and $\omega_i$ is the growth rate. Thus

$$\varepsilon_{em}(k,\omega) = i\omega_i \left.\frac{\partial \varepsilon_{em}(k,\omega)}{\partial \omega}\right|_{\bar{k},\bar{\omega}} = -\frac{2i\omega_i}{\bar{\omega}} \quad .$$

Hence the energy of the unstable electromagnetic wave is

$$W_{em} = \frac{\omega_{ce}^2 (k+k_o)^2}{4\gamma_o^2\ k_o^2\ \omega_i^2} W_{es} \quad .$$

The ratio of the electromagnetic and the electrostatic wave energy is

$$\frac{W_{em}}{W_{es}} = \frac{\omega_{ce}^2 (k+k_o)^2}{4\gamma_o^2 \, k_o^2 \, \omega_i^2}$$

which is much larger than unity for all the cases we consider. For large $\gamma$,

$$\frac{W_{em}}{W_{es}} \simeq 4 \, k_o c \gamma^{7/2} \quad .$$

Therefore, almost all the energy loss of the electron beam is converted into radiation. This has been observed in our computer simulations.

## III. RESULTS OF COMPUTER SIMULATIONS

Using the 1-2/2 dimensional fully relativistic electromagnetic particle code, we ran many cases with different parameters to investigate the dependence of the instability and its saturation level. Since the code is written in momentum space, we initiate the momenta of particles such that they have the same momentum spread in x, y, and z while they have a mean x momentum corresponding to a mean value of $\gamma$. In all the cases, ions are treated as a fixed background; the system size is chosen to 256 grid length and 2560 particles are followed in time. Velocities are normalized to $\omega_{pe}\delta$, where $\delta$ is the grid length. The velocity of light is chosen to be 15. The static rippled field was such that $k_o c = 2.2\ \omega_{pe}$.

Figure 5 shows the phase space (x,p) of particles at the beginning. The particles have a drift velocity corresponding to $\gamma = 2.0$. In Fig. 6, the corresponding electromagnetic and electrostatic spectra are shown at $\omega_{pe} t = 40$. By this time the instability has already started and the unstable spectrum has grown to considerable amplitude. The wavenumber matching condition, $k_p = k_{em} + k_o$, is satisfied between the unstable modes in both the electromagnetic and the electrostatic spectrum. The electromagnetic waves are coupled to the plasma oscillations through the rippled magnetic field. The time evolution of the electromagnetic and the electrostatic energy, as well as the longitudinal current, are plotted in Fig. 7. At the time of saturation, the current was decreased by 36% while roughly 30% of the beam energy has been converted into radiation; of this, half is in the most unstable electromagnetic spectrum. In the computer experiments, we find that as the initial unstable spectrum grows to high amplitude, longer wavelength modes become unstable. The latter process is believed to be caused by nonlinear mode coupling at the initial unstable waves. The phase space diagram at $\omega_{pe} t = 160$ is shown in Fig. 8; this reveals that

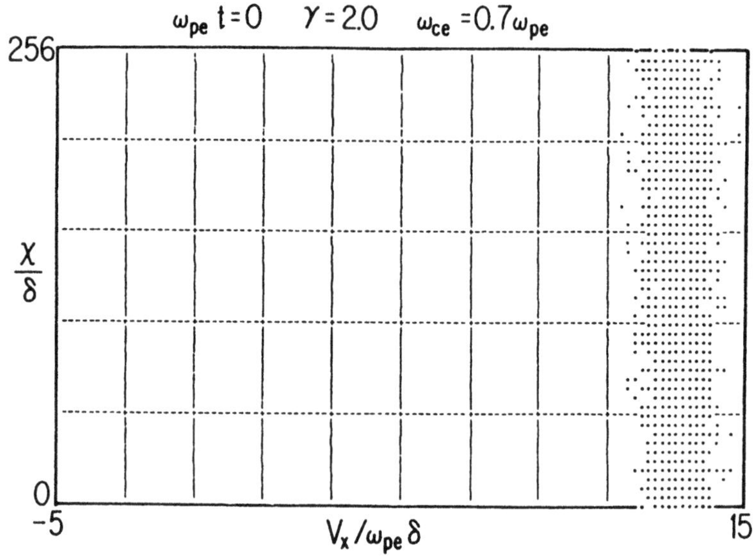


Figure 5. Electron phase space at $\omega_{pe}t = 0$ for the case with $\gamma = 2.0$ and $\omega_{ce} = 0.7\ \omega_{pe}$.

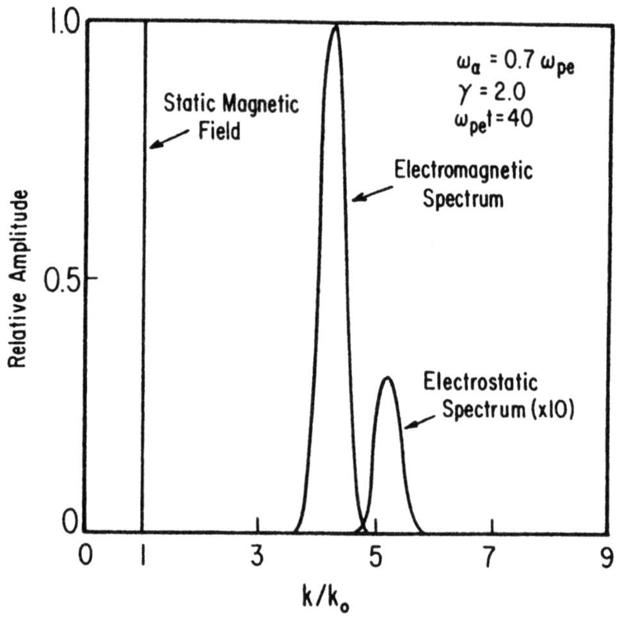


Figure 6. The rippled magnetic field, the unstable electromagnetic and electrostatic spectrum satisfy the wavenumber matching condition.

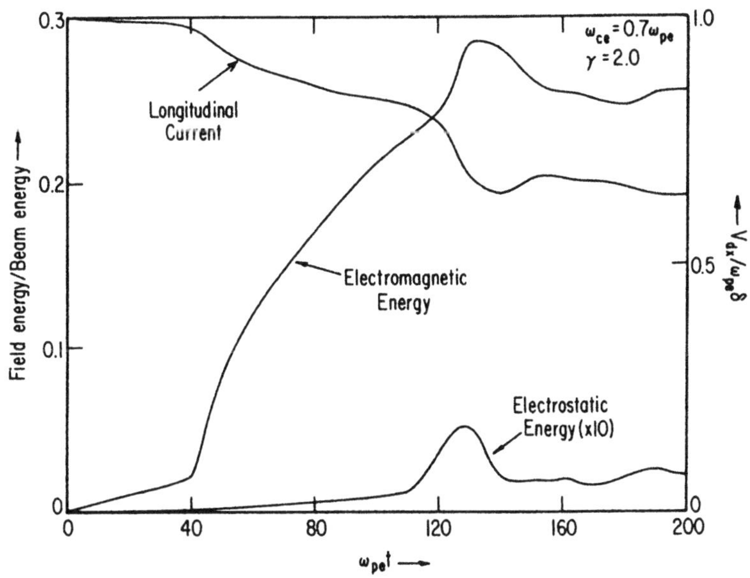


Figure 7. Time evolution of the longitudinal current, the electromagnetic and the electrostatic energy for the case with $\gamma = 2.0$, $\omega_{ce} = 0.7$.

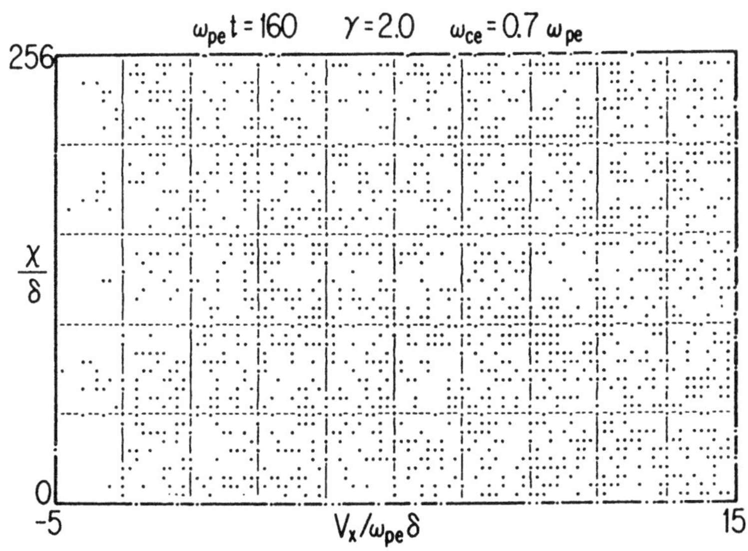


Figure 8. Electron phase space at $\omega_{pe}t = 160$ for the case with $\gamma = 2.0$ and $\omega_{ce} = 0.7\ \omega_{pe}$.

the electron beam has been considerably slowed down. The difference in beam energy goes into electromagnetic radiation. Figure 9 shows the time evolution of the longitudinal current, the electromagnetic and the electrostatic energy for a case with $\gamma = 1.6$. We see that the instability saturates much earlier than the case with $\gamma = 2.0$. This is due to the fact that the growth rate decreases as $\gamma^{-3/4}$ for increasing $\gamma$. The maximum growth rates are plotted versus $\gamma$ for three different strengths of the rippled magnetic field in Fig. 10. The solid curves are the maximum growth rates as predicted by the theory for the three cases. For all the cases we ran, the experimental data agree quite well with the theory. For the case with $\gamma = 1.8$ and $\omega_{ce} = 0.7\ \omega_{pe}$ the growth rates for various mode number obtained from the simulation are compared to the theory in Fig. 11. The simulation results show a slight shift to the long wavelength modes with respect to the curve obtained by solving Eq. (28). The maximum growth rate is closely in value to that predicted by theory. The discrepancy may be due to the fact that Landau damping is not included in the theory and the short wavelength modes are more heavily damped. Finite size particle effects have been taken into account in the theory by appropriately including the form factor $e^{-k^2a^2}$ in the dispersion relation. In any case, the difference is quite small and we have not tried very hard to explain it.

We summarize all the simulation results for the different parameters investigated in Table I. Since the energy in the electrostatic waves is negligibly small by comparison to the electromagnetic energy, the decrease in the beam energy is essentially converted into radiation. However, at the time of saturation, a fair amount of energy is deposited in the long wavelength electromagnetic waves which have grown to substantial level, presumably due to some nonlinear mode coupling process.

The percentage of beam energy deposited in the initial unstable spectrum is given as a function of $\gamma$ in Fig. 12 for three different strengths of the rippled magnetic field. We find that the efficiency of energy conversion decreases for increasing $\gamma$. Besides being a monotonic decreasing function of $\gamma$, it is also sensitive to the wavelength of the rippled magnetic field. We ran two cases with $\gamma = 1.8$ and $\gamma = 2.0$ for a ripple such that $k_o c \simeq 1.5\ \omega_{pe}$ rather than $2.2\ \omega_{pe}$ as in the previously described results. We find that the energy of the electromagnetic spectrum concentrates in the most unstable mode. For the case of $\gamma = 1.8$, the most rapidly growing mode (mode 9) has nearly 90% of the total electromagnetic energy. Therefore, the process could become very efficiency in amplifying light wave with a particular frequency by choosing the appropriate wavelength of the rippled magnetic field. Since the lasing wavelength is roughly given by $\lambda_o/2\gamma^2$, where $\lambda_o$ is the wavelength of the

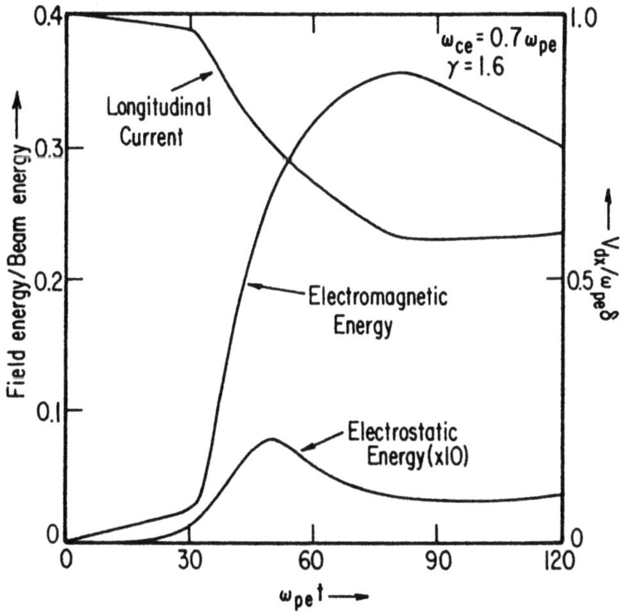


Figure 9. Time evaluation of the longitudinal current, the electromagnetic and the electrostatic energy for the case with $\gamma = 1.6$ and $\omega_{ce} = 0.7\ \omega_{pe}$.

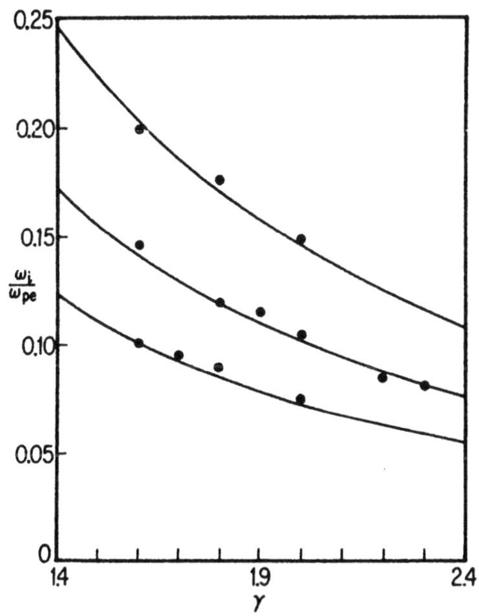


Figure 10. Theoretical maximum growth rate versus $\gamma$ and the maximum growth rates obtained from simulations.

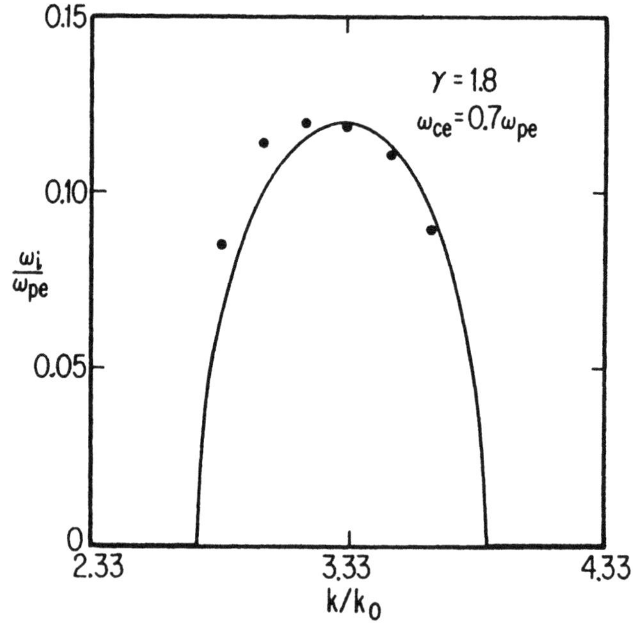


Figure 11. Theoretical growth rate versus wavenumber and the growth ratio obtained from simulation for the case with $\gamma = 1.8$ and $\omega_{ce} = 0.7\ \omega_{pe}$.

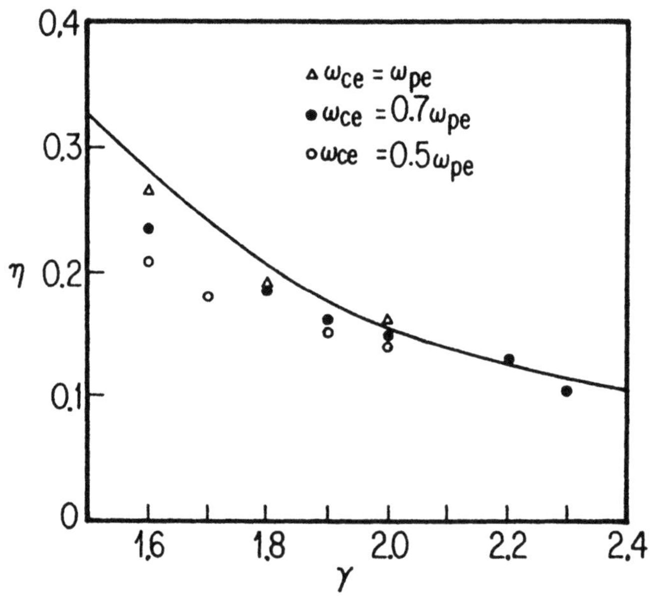


Figure 12. Theoretical efficiency of energy transfer from the relativistic electron beam to electromagnetic radiation versus $\gamma$ and the results obtained from simulations.

TABLE I

| $\gamma$ | Maximum growth rate ($\omega_{pe}^{-1}$) | | | Initial Electromagnetic Unstable Spectrum | | | Saturation Time ($\omega_{pe}t$) | | | Decrease in Beam Energy (%) | | |
|---|---|---|---|---|---|---|---|---|---|---|---|---|
| | $\omega_c$=0.5 | $\omega_c$=0.7 | $\omega_c$=1.0 | $\omega_c$=0.5 | $\omega_c$=0.7 | $\omega_c$=1.0 | $\omega_c$=0.5 | $\omega_c$=0.7 | $\omega_c$=1.0 | $\omega_c$=0.5 | $\omega_c$=0.7 | $\omega_c$=1.0 |
| 1.6 | 0.101 | 0.146 | 0.200 | 2.0-2.5 | 1.5-2.5 | 1.2-2.3 | 100 | 84 | 45 | 32.5 | 35.5 | 31.0 |
| 1.7 | 0.095 | | | 2.5-3.2 | | | 108 | | | 28.8 | | |
| 1.8 | 0.090 | 0.120 | 0.176 | 3.2-3.7 | 2.7-3.7 | 2.0-3.7 | 148 | 96 | 60 | 24.5 | 29.4 | 31.8 |
| 1.9 | | 0.116 | | | 3.2-4.3 | | | 112 | | | 28.1 | |
| 2.0 | 0.075 | 0.105 | 0.149 | 4.2-4.8 | 4.0-4.7 | 3.0-4.7 | 240 | 135 | 78 | 25.3 | 28.9 | 31.1 |
| 2.2 | | 0.085 | | | 4.8-5.8 | | | 216 | | | 30.6 | |
| 2.3 | | 0.083 | | | 5.8-7.0 | | | 240 | | | 31.0 | |

ripple, the frequency of the radiation is reduced when we use a rippled magnetic field with longer wavelength. In the case with $\gamma = 2.3$, radiation with frequencies up to 14 $\omega_{pe}$ has been observed in the simulation. Generation of even higher frequency electromagnetic wave is possible by increasing $\gamma$; however, simulating such high frequency wave becomes difficult because of the requirement that we use a sufficiently small time step with $\omega_{max}\Delta t < 0.5$.

## IV. NONLINEAR SATURATION

Of considerable importance is the saturation level, as this determines the potential efficiency of radiation production. The simulations indicate that saturation is due to electron trapping in the electrostatic wave which is generated. We now examine this question.

The dispersion relation given by Eq. (25) indicates that the instability arises from the coupling between the electromagnetic and electrostatic plasma waves through the rippled magnetic field. The electrostatic waves involved in the instability satisfy the dispersion relation

$$\left(\omega - (k+k_o)v_o\right)^2 = \frac{1}{\gamma_o^3}\left(1 + 3(k+k_o)^2\right) \quad .$$

The unstable spectrum is quite localized in Fourier space. Furthermore, all these modes have roughly the same velocity, $v_o$. Therefore, they maintain a coherent wave form in the plasma. Thus, we can approximate the spectrum as a single wave. As the wave grows and the beam loses energy during the instability, it will eventually reach such an amplitude that it can trap most of the electrons in the beam. After the trapping process sets in, the instability eventually saturates. At this time the electron beam is on the average slowed down to the phase velocity of the beam mode. Then the change in energy of the electron beam is given by

$$\Delta W = m_o c^2 (\gamma_o - \gamma_{ph})$$

where $\gamma_o$ applies to the initial beam velocity, while $\gamma_{ph}$ is the $\gamma$ corresponding to the phase velocity of the wave. Assuming all the energy loss from the electron beam converted into electromagnetic radiation, the efficiency, $\eta$, is then

$$\eta = \frac{|\Delta W|}{(\gamma_o - 1)m_o c^2} = \frac{(\gamma_o - \gamma_{ph})}{(\gamma_o - 1)} \quad .$$

Figure 12 illustrates its dependence on the relativistic $\gamma$. For large $\gamma$,

$$\eta \simeq \frac{1}{2k_o c\gamma^{3/2}} \quad .$$

A comparison of the computer simulations with the estimated values is also shown in Fig. 12. The experimental data show the same characteristic dependence on $\gamma$. For large $\gamma$, the efficiency of energy conversion seems to be independent of the strength of the rippled magnetic field.

## V. DISCUSSIONS AND CONCLUSION

We have presented the fluid theory and the computer simulation on the interaction of a relativistic electron beam with a static rippled magnetic field. The fluid analysis, of course, is not able to predict the effect of Landau damping on the growth of the unstable plasma waves. If the unstable plasma waves are such that $k\lambda_D \sim 1$, the waves are heavily damped and the effect of Landau damping becomes important in the coupling between the plasma wave and the electromagnetic wave. For $k\lambda_D \gtrsim 1$, Landau damping may be so dominant that the mode coupling process would fail to give rise to an instability. We have investigated this effect by computer simulation with increased thermal velocity of the electron beam. The instability seems to be weakened and the growth rates of the unstable modes are smaller in value than those obtained previously. However, nearly 30% of the beam energy can still be converted into electromagnetic radiation.

Let's take an Astron beam as an example to illustrate the feasibility of the idea of operating the free electron laser described above. The density of an Astron beam is $3 \times 10^{12}$ cm$^{-3}$ and $\gamma = 10$. The energy spread of the beam is $\Delta\gamma/\gamma = 10^{-3}$. The velocity spread, $\Delta v$, is roughly $c\Delta\gamma/\gamma^3$ and the Debye length, $\lambda_D$, is given by $\gamma^{3/2}\Delta v/\omega_{pe} \simeq 1$ $\mu$m. To amplify radiation of 100 $\mu$m, $k\lambda_D = 2\pi(\lambda_D/\lambda) << 1$. Hence, the unstable plasma wave is essentially undamped and the effect of Landau damping can be ignored. For a static ripped magnetic field of strength 5 KG, we have $\omega_{ce} \simeq \omega_{pe}$. With $\gamma = 10$, the growth rate is $\omega_i \simeq 0.05\ \omega_{pe}$. Amplifying the signal by ten e-folding, the interaction time is $\tau = 10/\omega_i \simeq 2 \times 10^{-9}$ sec, and the length of the interaction region is $L \simeq c\tau = 60$ cm. These parameters can be easily realized in present day laboratory.

It is difficult to compare our work with the experiment at Stanford in which 10.6 $\mu$m radiation is amplified because the experiment is in a different parameter regime from our calculation. The density of the electron beam used in the experiment is too low ($n \simeq 10^8$ $cm^{-3}$, $\gamma = 50$) to have any effect due to collective plasma oscillations. Since the time during which the radiation is amplified is much less than the period of the collective plasma oscillation, the process is mainly magnetic bremsstrahlung due to the interaction of single particle with the rippled magnetic field. The experiment also has a guiding magnetic field. The existence of the guiding field in our theory would make the cyclotron beam mode available, in addition to the plasma beam mode, to couple with the electromagnetic wave. It is of interest to find out which mode gives a higher efficiency in the production of electromagnetic radiation. We are investigating this problem and the results will appear in the next paper.

ACKNOWLEDGMENT

One of the authors (T.K.) wishes to thank Dr. J. F. Drake for useful discussions.

REFERENCES

* Supported in part by the National Science Foundation, Contract # PHY 74-15233A01.

1. Science 187, No. 4181, March 21, 1975, Research News.
2. Luis R. Elias, et al., Phys. Rev. Lett. 36, 717 (1976).
3. M. Friedman and M. Herndon, Phys. Rev. Lett. 28, 210 (1971).
4. M. Friedman and M. Herndon, Phys. Rev. Lett. 29, 55 (1972).
5. M. Friedman and M. Herndon, Appl. Phys. Lett. 22, 658 (1973).
6. P. Sprangle and V. L. Granatstein, Appl. Phys. Lett. 25, 377 (1974).
7. W. M. Manheimer and E. Ott, Phys. of Fluids 17, 463 (1974).
8. P. Sprangle, J. Plasma Phys. 11, 299 (1974).
9. P. Strangle, V. L. Granatstein, and L. Baker, Phys. Rev. A 12, 1697 (1974).
10. A. T. Lin, J. M. Dawson, and H. Okuda, Phys. of Fluids 17, 1995 (1974).

## PARTICIPANTS

| | |
|---|---|
| Anderson, D | Institute for Electromagnetic Field Theory, Chalmers University of Technology, Fack, S-402 20 Göteborg, Sweden |
| Balescu, R | Université Libre de Bruxelles, Faculté des Sciences, Service de Chimie Physique 2, Avenue F.-D. Roosevelt 50, 1050 Bruxelles, Belgium |
| Bobin, J L | Commissariat a l'énergie atomique, Boite Postale 27, 94 - Villeneuve St-Georges, France |
| Bondeson, A | Institute for Electromagnetic Field Theory, Chalmers University of Technology, Fack, S-402 20 Göteborg, Sweden |
| Brueckner, K A | University of California, San Diego, Department of Physics, La Jolla, Cal. 92093, U.S.A. |
| Chen, F F | University of California, Los Angeles School of Engineering and Applied Science Los Angeles, Cal. 90024, U.S.A. |
| Cotsaftis, M | CEN Assoc Euratom CEA, Fusion Dept Phys. Plasma & Fusion Controlée, 92260 Fontenay aux Roses, France |
| Davidson, R C | Department of Physics and Astronomy, University of Maryland, College Park, Maryland 20742, U.S.A. |
| DuBois, D F | University of California, Los Alamos Scientific Laboratory, P.O. Box 1663, Los Alamos, New Mexico 87544, U.S.A. |
| Dysthe, K | Nordlysobservatoriet, Tromsø, Norway |
| Engelmann, F | FOM-Instituut voor Plasmafysica, Overeindseweg 2 - Postbus 7, Nieuwegein, The Netherlands |
| ter Haar, D | Department of Theoretical Physics, 12 Parks Road, Oxford OX1-3PQ, England |
| Ichimaru, S | Department of Physics, University of Tokyo 3-1 Hongo 7-Chome, Bunkyo-Ku, Tokyo, (Postal Code 113), Japan |
| Johnson, R R | KMS Fusion, Inc., 3941 Research Park Drive, P.O. Box 1567, Ann Arbor, Michigan 48106 313/769-8500, U.S.A. |

| | |
|---|---|
| Kaw, P K | Plasma Physics Laboratory, James Forrestal Campus, P.O. Box 451, Princeton, N.J. 08540, U.S.A. |
| Kaufman, A N | Lawrence Berkeley Laboratory, University of California, Berkeley, Cal. 94720, U.S.A. |
| Kovrizhnykh, L M | P.N. Lebedev Physical Institute, USSR Academy of Sciences, Moscow, USSR |
| Kruer, W L | Lawrence Livermore Laboratory, University of California, P.O. Box 808, Livermore, Cal. 94550, U.S.A. |
| Larsson, J | Department of Plasma Physics, Umeå University, S-901 87 Umeå, Sweden |
| Liu, C S | Department of Physics and Astronomy, University of Maryland, College Park, Maryland 20742, U.S.A. |
| Montes, C | Observatoire de Nice, Le Mont-Gros, 06300 Nice, France |
| Nakach, R | C.E.N.G. - Fusion, 85X, 38041 Grenoble, France |
| Nishikawa, K | Department of Physics, Hiroshima University, Higashisenda - machi, Hiroshima City, Japan. |
| Ohlsson, D | Division of Plasma Physics and Fusion Research, The Royal Institute of Technology, S-100 44 Stockholm 70, Sweden |
| Ryutov, D D | Nuclear Physics Institute, Siberian Branch of USSR, Academy of Sciences, Novosibirsk, USSR |
| Sigel, R | Max-Planck-Institute für Plasma Physik, 8046 Garching bei München, Germany |
| Stenflo, L | Department of Plasma Physics, Umeå University, S-901 87 Umeå, Sweden |
| Tsytovich, V N | P.N. Lebedev Physical Institute, USSR Academy of Sciences, Moscow, USSR |
| Vandenplas, P | Laboratoire de Physique des Plasmas, Association Euratom-Etat Belge, Ecole Royale Militaire 1040 Bruxelles, Belgium |
| Paiva-Veretennicoff, I | Faculteit der Wetenschappen, Vrije Universiteit Brussel, B-1050 Brussels, Belgium |

Weiland, J — Institute for Electromagnetic Field Theory,
Chalmers University of Technology,
Fack, S-402 20 Göteborg, Sweden

Wilhelmsson, H — Institute for Electromagnetic Field Theory,
Chalmers University of Technology,
Fack, S-402 20 Göteborg, Sweden

MIX
Papier aus verantwortungsvollen Quellen
Paper from responsible sources
FSC® C105338

If you have any concerns about our products,
you can contact us on
**ProductSafety@springernature.com**

In case Publisher is established outside the EU,
the EU authorized representative is:
**Springer Nature Customer Service Center GmbH**
**Europaplatz 3, 69115 Heidelberg, Germany**

Printed by Libri Plureos GmbH
in Hamburg, Germany